Ageing of composites

Related titles:

Fatigue in composites: science and technology of the fatigue response of fibre-reinforced plastics
(ISBN 978-1-85573-608-5)
Fibre composites, like metals, exhibit a form of degradation in service which may be described as 'fatigue'. The mechanisms by which this deterioration occurs in composites are quite different from, and much more complicated than, those that are responsible for fatigue phenomena in metals, but the problems facing the designer are similar. The challenge for the engineer is to specify materials and use them in such a way as to avoid failures within the design life of a component or structure. This major handbook is an authoritative survey of current knowledge of fatigue behaviour of composites.

Multi-scale modelling of composite material systems: the art of predictive damage modelling
(ISBN 978-1-85573-936-9)
Predictive modelling provides the opportunity both to understand better how composites behave in different conditions and to develop materials with enhanced performance for particular industrial applications. This important book focuses on the fundamental understanding of composite materials at the microscopic scale, from designing microstructural features to the predictive equations of the functional behaviour of the structure for a specific end-application. Chapters discuss stress- and temperature-related behavioural phenomena based on knowledge of the physics of microstructure and microstructural change over time.

Delamination behaviour of composites
(ISBN 978-1-84569-244-5)
Delamination is a phenomenon that is of critical importance to the composite industry. It involves a breakdown in the bond between the reinforcement and the matrix material of the composite. With growing use of composites in aerospace and other sectors, understanding delamination is essential for preventing catastrophic failures. Part I focuses on delamination as a mode of failure. Part II covers testing of delamination resistance, while Part III analyses detection and characterisation. Further parts cover analysis of delamination behaviour from tests, modelling delamination, and prevention and mitigation of delamination.

Details of these and other Woodhead Publishing materials books, as well as materials books from Maney Publishing, can be obtained by:

- visiting our website at www.woodheadpublishing.com
- contacting Customer Services (e-mail: sales@woodhead-publishing.com; fax: +44 (0) 1223 893694; tel: +44 (0) 1223 891358 ext. 130; address: Woodhead Publishing Ltd, Abington Hall, Granta Park, Great Abington, Cambridge CB21 6AH, England)

If you would like to receive information on forthcoming titles, please send your address details to: Francis Dodds (address, tel. and fax as above; e-mail: francisd@woodhead-publishing.com). Please confirm which subject areas you are interested in.

Maney currently publishes 16 peer-reviewed materials science and engineering journals. For further information visit www.maney.co.uk/journals

Ageing of composites

Edited by
Rod Martin

Woodhead Publishing and Maney Publishing
on behalf of
The Institute of Materials, Minerals & Mining

CRC Press
Boca Raton Boston New York Washington, DC

WOODHEAD PUBLISHING LIMITED
Cambridge England

Woodhead Publishing Limited and Maney Publishing Limited on behalf of The Institute of Materials, Minerals & Mining

Woodhead Publishing Limited, Abington Hall, Granta Park, Great Abington
Cambridge CB21 6AH, England
www.woodheadpublishing.com

Published in North America by CRC Press LLC, 6000 Broken Sound Parkway, NW, Suite 300, Boca Raton, FL 33487, USA

First published 2008, Woodhead Publishing Limited and CRC Press LLC

British Library Cataloguing in Publication Data
A catalogue record for this book is available from the British Library.

Library of Congress Cataloging in Publication Data
A catalog record for this book is available from the Library of Congress.

Woodhead Publishing ISBN 978-1-84569-352-7 (book)
Woodhead Publishing ISBN 978-1-84569-493-7 (e-book)
CRC Press ISBN 978-1-4200-8776-5
CRC Press order number: WP8776

The publishers' policy is to use permanent paper from mills that operate a sustainable forestry policy, and which has been manufactured from pulp which is processed using acid-free and elementary chlorine-free practices. Furthermore, the publishers ensure that the text paper and cover board used have met acceptable environmental accreditation standards.

Typeset by SNP Best-set Typesetter Ltd., Hong Kong
Printed by TJ International Limited, Padstow, Cornwall, England

Contents

Contributor contact details

(* = main contact)

Editor

Dr Rod Martin
Materials Engineering Research Laboratory Ltd
Wilbury Way
Hitchin
Hertfordshire
SG4 0TW
UK
E-mail: rmartin@merl-ltd.co.uk

Chapter 2

Dr Kevin Plucknett
Materials Engineering Program
Department of Process Engineering and Applied Science
Dalhousie University
1360 Barrington Street
Halifax
Nova Scotia
B3J 1Z1
Canada
E-mail: kevin.plucknett@dal.ca

Chapter 3

Dr Heidi Cuypers*
Mechanics of Materials and Constructions
Vrije Universiteit Brussel
Pleinlaan 2
B-1050 Brussels
Belgium
E-mail: heidi.cuypers@vub.ac.be

Jeanette Orlowsky
Institut für Bauforschung der
RWTH Aachen
Schinkelstraße 3
52062 Aachen
Germany

Chapter 4

Dr Antoine Chateauminois
Laboratoire de Physico-Chimie des Polymères et des Milieux Dispersés
Ecole Supérieure de Physique et Chimie Industrielles (ESPCI)
10 rue Vauquelin
75231 Paris Cedex 05
France
E-mail: antoine.chateauminois@espci.fr

Chapter 5

Dr Thomas K. Tsotsis
The Boeing Company
5301 Bolsa Avenue
M/C H021-F120
Huntington Beach
California 92647
USA
E-mail: thomas.k.tsotsis@boeing.com

Chapter 6

Dr Roger W Jones* and Dr John McClelland
Ames Laboratory and Center for Nondestructive Evaluation
Iowa State University
Ames
Iowa 50011
USA
E-mail: jonesrw@ameslab.gov; mcclelland@ameslab.gov

Chapter 7

Huiwen Hu
Associate Professor
Department of Vehicle Engineering
National Pingtung University of Science and Technology
1 Hesuh-Fu Road
Neipu
Pingtung 91201
Taiwan
E-mail: huiwen@mail.npust.edu.tw

Chapter 8

Professor Stefanos M. Skolianos
Aristotle University of Thessaloniki
School of Engineering
Department of Mechanical Engineering
Thessaloniki 541 24
Greece
E-mail: sko@eng.auth.gr; skol@auth.gr

Chapter 9

Giuseppe Mensitieri
Italian Interuniversity Consortium on Materials Science and Technology (INSTM) – Reference Center (CR) for Transformation Technology of Polymeric and Composite Materials
Research Unit: Department of Materials and Production Engineering
University of Naples Federico II
Piazzale Tecchio 80
Naples 80125
Italy
E-mail: mensitie@unina.it

Michele Iannone
Alenia Aeronautica s.p.a.
Laboratorio Tecnologie, Materiali, Processi e CND
viale dell'Aeronautica
Pomigliano D'Arco (Na)
80038 Italy
E-mail: miannone@aeronautica.alenia.it

Chapter 10

Professor Kwang Bok Shin
Division of Mechanical Engineering
HANBAT National University
San 16-1
Dukmyung-Dong
Yuseong-Gu
Daejon
305-719
Korea
E-mail: shin955@hanbat.ac.kr

Chapter 11

Captain Krzysztof Dragan
Polish Air Force Institute of Technology
Bldg Z-31
Ksboleslawa 6
Warsaw
Poland 01-494
E-mail: krzysztof.dragan@itwl.pl

Chapter 12

Peter Davies and Dominique Choqueuse
Materials and Structures Group
IFREMER Brest Centre
29280 Plouzané
France
E-mail: peter.davies@ifremer.fr; dominique.choqueuse@ifremer.fr

Chapter 13

Dr Saverio Affatato
Laboratorio di Tecnologia Medica,
Istituti Ortopedici Rizzoli,
Via di Barbiano, 1/10
40136 Bologna
Italy
E-mail: affatato@tecno.ior.it

Chapter 14

Dr Simon Frost
ESR Technology Ltd
16 North Central 127
Milton Park
Abingdon
Oxfordshire
OX14 4SA
UK
E-mail: simon.frost@esrtechnology.com

Chapter 15

Dr Sue Halliwell
Tapton Park Innovation Centre Brimington Road
NetComposites Ltd
Chesterfield
S41 0TZ
UK
E-mail: sue.halliwell@netcomposites.com

Chapter 16

Stanislaw M. Gubanski
Department of Materials and Manufacturing Technology
Chalmers University of Technology
412 96 Gothenburg
Sweden
E-mail: stanislaw.gubanski@chalmers.se

Chapter 17

Dr Rod Martin
Materials Engineering Research Laboratory Ltd
Wilbury Way
Hitchin
Hertfordshire
SG4 0TW
UK
E-mail: rmartin@merl-ltd.co.uk

Chapter 18

Dominique Choqueuse and Peter Davies
Materials and Structures Group
IFREMER Brest Centre
29280 Plouzané
France
E-mail: dominique.choqueuse@ifremer.fr; peter.davies@ifremer.fr

This book is dedicated to Dr Tom Gates who passed away between the drafting of his chapter and the final publication. Tom was a personal friend of mine, we shared an office at NASA Langley Research Center and worked together for several years on the ageing of composite materials for the next generation of supersonic commercial aircraft. Tom was one of the first to submit his draft chapter for the publication and this was typical of his professionalism.

Rod Martin

Introduction

R. Martin, Materials Engineering Research Laboratory Ltd, UK

Composite materials offer many advantages over conventional structural materials. This includes their high strength and stiffness to weight ratios, their resistance to chemical attack and their tailorability. Much of this publication covers composite materials that are fibre reinforced polymers, but also included are the higher end composite materials in which the matrix is metallic or the fibres are those such as silicon carbide.

The use of these materials has seen considerable growth in many industry sectors in the latter part of the twentieth century and this growth has continued into the twenty-first century. The interest in composite materials is driven both by performance factors and environmental factors. The material's higher strength and stiffness has catapulted the use of these materials into the civilian aerospace market. These same properties along with energy absorption have made vast improvements to performances in many sports, including Grand Prix racing, skiing, golf and tennis. Composite materials are perhaps the only material of choice in certain highly corrosive environments such as those experienced in the petrochemical industry. These materials are also now being selected for environmental reasons because their low specific weight leads to fuel savings in the transport industry, allows for the design of large wind-turbine blades and can be used in construction projects for longer product lives.

However, despite this growth in the demand for and use of composite materials, their long-term properties when exposed to a combination of in-service loads and environments are still not well characterised. The effect of exposure to heat, moisture, solvents, acids, ozone, hydrocarbons, loads, etc., and more importantly a combination of these parameters, may degrade the material's stiffness and strength leading to cracks and ultimately the material failing to meet its purpose. The lack of long-term data or of an accelerated ageing methodology that will predict the effect such degradation might have on the residual properties and future life are two of the major issues still hindering the wider use of composites and leads to over design.

The generic term 'ageing' can range from the more benign physical ageing effects – such as swelling from moisture absorption – that are largely reversible, to the more serious chemical ageing, which is irreversible. Additionally, ageing from mechanical loading, such as creep, needs to be considered in isolation (or in addition) to that associated with the environment. Environmental ageing of composite materials occurs from the surface or edge inwards and requires time to penetrate into the material's centre. This is analogous to fluid diffusion and can be anisotropic, and the rate can be dependent on temperature and load. This makes predicting ageing around detailed geometries such as stress concentrations, non-trivial. Representing the true service history for long-term structural life prediction is a vital step to validate any short-term, coupon-based methodology. The coupon tests must reflect the effect of ageing on the polymer (i.e. matrix-dominated properties) because the fibres may mask any property loss in the resin. However, fibres can sometimes degrade quicker than the polymer in certain environments and, additionally, the fibre–matrix interface can be attacked.

The above description of the complexity of ageing of composite materials was the main reason for the publication of this book. The aim was to gather as much knowledge from a materials perspective and an end-use perspective in one place so that the different aspects of ageing can be understood. While this publication does not claim to be an all-inclusive encyclopaedia on the subject of ageing of composite materials, it brings many aspects of the subject together in one volume with international contributions.

The scope of this publication is to cover the aspects of ageing of composite materials from a fundamental level for different materials systems and from an industrial view point covering a wide variety of different industry sectors.

Part I of the publication addresses the fundamental aspects. Dr Gates of NASA Langley Research Center addresses polymeric-based composites and brings together the time–temperature dependency of physical, chemical and mechanical ageing. The chapter addresses early viscoelasticity work by Professor Richard Schapery in the 1970s right up to the ageing of composite materials in modern supersonic aircraft concepts.

Dr Plucknett focuses on the ageing of ceramic reinforced composites and particularly fibre reinforced glass ceramics. The degradation of these materials needs to be investigated at the micro-mechanics level. The chapter describes how knowledge of interfacial behaviour is essential to understand how this material class performs at the very high temperatures in which they operate.

The ageing of glass fibre reinforced concrete is the subject of Chapter 3 headed by Dr Cuypers. This material has an unusual type of ageing in that the fibres are aged by the matrix material itself. This chapter focuses on chemical attack of glass fibres in the absence of mechanical load (stress corrosion cracking is covered in Chapter 4) and discusses the modelling and experimental methods for characterising and predicting fibre degradation.

While 'corrosion' is a word that should be avoided in discussion of the ageing of composite materials, the term 'stress corrosion cracking' is well established. This is the topic of Chapter 4 by Dr Chateauminois who describes this phenomena in glass reinforced polymeric composites in static and cyclic loading conditions, in acidic and alkaline environments.

Chapter 5 written by Dr Tsotsis addresses thermo-oxidative ageing of polymeric composites. The general focus of this work is composite materials operating in air, at high temperatures for long periods of time. The ageing characteristics of a range of materials and some of the methods used to characterise this form of ageing are presented.

It is important to understand the mechanisms of ageing in polymeric composites and to do this, detailed investigation of the degradation at the polymer level is required. The use of Fourier transform infrared photoacoustic spectroscopy is the topic of the chapter headed by Dr Jones.

The modelling and understanding of physical ageing is the topic of the chapter written by Professor Hu. This work thoroughly explains the phenomenon of physical ageing, particularly creep and relaxation; it starts from the well-regarded work by Struik and supplements the work described in Chapter 1 of this publication.

The ageing of silicon carbide (SiC) composites is discussed in Chapter 8 by Professor Skolianos. These materials are used in very high temperature, load-bearing applications in the transport and propulsion industry. This chapter describes the change of properties with time in SiC reinforced composites and also describes other degradation phenomena such as wear and corrosion. Much of this needs to be investigated at the micro-structural level to understand the effects of grain size, porosity and matrix composition.

Part I of this publication concludes with a chapter authored by Professor Mensitieri. This is an in-depth review of the many aspects of the modelling and ageing of composite materials. The chapter overlaps, complements and adds to information in other chapters reinforcing this overall topic and the fundamental understanding of the physical, chemical and mechanical ageing of composite materials.

The publication then switches to transport applications in Part II. The specific topic of aerospace as a transport mechanism is not covered because it is inherently discussed in several of the above chapters where the fundamental work was done for the aerospace industry. Part II opens with a chapter on composite materials in the rail industry written by Professor Shin. This industry is seeing a growth in the use of composite materials primarily to reduce rolling stock weight. The main environmental ageing parameters are moisture, ultraviolet light and temperature, and this chapter discusses the evaluation of materials at the coupon and structural level in order to understand the effect of environment on the structure.

Captain Dragan then describes the issues of ageing of composites in the rotorcraft industry; much of the focus is on the effect of moisture degradation in the blades and the need to address the more mechanical degradation modes of impact and fatigue.

Chapter 12, led by Dr Davies, investigates the ageing of composite materials in marine vessels from leisure craft to underwater vessels. Clearly the main environmental degradation is that of sea water and this chapter gives a thorough review on the subject and discusses methods to characterise degradation – such as property changes, osmosis and blistering.

The publication then moves to non-transport applications in Part III. The first chapter, authored by Dr Affatato describes the background and use of polyethylene composites in medical devices. Two of the main ageing issues are oxidation and the generation of wear debris that can lead to osteolysis. The chapter describes some of the methods used to characterise these ageing mechanisms in an accelerated fashion.

The oil and gas industry is very demanding on materials in terms of the hostile environments worked in and the hostile conditions and fluids that are involved. Dr Frost's chapter describes ageing in this industry with the definition that ageing represents a reduction in performance of a component from chemical species ingress, elevated operating temperature and length of time of load application. He describes a model to predict ageing and damage from matrix cracking within composite components.

Dr Halliwell describes the ageing of composite materials in the construction industry. This chapter begins with a definition of the materials and the performance required. Many of the issues of ageing of composites in the construction industry are related to weathering, but ageing from the use of chemicals, liquids and temperature are also addressed.

A rare insight into the use of composite materials as high-voltage insulators is given in the chapter authored by Professor Gubanski. Here the ageing concerns are related to exposures to corona and dry-band discharges as well as biological growth. Modern composite insulators comprise a glass fibre reinforced resin-bonded core (rod or pipe) onto which two metal end-fittings are attached. Ageing can occur in the insulator core and at the interfaces.

Dr Martin then addresses the ageing of composites in the very hostile environments of the chemical processing industry. These materials are used quite extensively in piping and storage vessels for concentrated acids, amines, etc., often for decades. The ageing mechanisms begin as diffusion and quickly develop into severe chemical degradation including material loss. This chapter describes some of the methods used to characterise this form of ageing to give some information on long-term use.

Dr Choqueuse then concludes this part and the book with some case studies of the ageing of composites in marine environments including composite tubes and deep sea applications such as power turbines, and the use of foams as part of the composite.

In summary, this is both a detailed and an eclectic gathering of information on the subject of ageing of composites from the fundamental science level to the practicalities of industrial pragmatism.

Part I

Ageing of composites – processes and modelling

1 The physical and chemical ageing of polymeric composites

T. GATES, formerly NASA Langley Research Center, USA

1.1 Introduction

In aerospace vehicles, the durability of a material is ultimately an issue that directly relates to the operational cost of the vehicle. By taking into account durability during the lifetime of the vehicle one can minimize these operational costs. Cost issues aside, studying and understanding the processes related to durability in high-performance aerospace materials are critical to the safe design, construction, and operation of the vehicle. Unfortunately, despite close attention to details and using the best design methods, the long-term exposure of advanced polymer matrix composite (PMC) materials to the use-environment will eventually result in irreversible change(s) in the original properties of the material and effectively limit operating life.

This process of change in properties over time in PMCs is loosely referred to as 'ageing'. Ageing may be broadly categorized by three primary mechanisms: chemical, physical and mechanical. The interaction (if any) between these three areas is highly dependent on two variables: material characteristics, and ageing environment. Material ageing may translate to structural changes in mission-critical components which for an aerospace vehicle can have a potentially catastrophic effect on both the vehicle and its payload (Bristow, 2001; Nuss, 2001; Lawford, 2002). The three ageing mechanisms may be additive or subtractive depending on material type, the environment, and the mechanical loads. It is the objective of this chapter to present the descriptions, test techniques, and analysis methods necessary to understand these ageing mechanisms, in particular physical ageing and chemical ageing.

In order to address the primary issue, reduced vehicle lifetime costs, verified test and analysis methods are needed to quantify ageing, provide guidance for materials development, and screen materials, or accurately assess ageing tendencies of new and candidate materials. The issue of ageing in PMCs can be construed in different ways, therefore it is important that a common definition of some important terms be established before going into further detail. Three terms are of particular importance: environmental degradation factor, critical degradation mechanism, and accelerated ageing.

Environmental degradation factor is the general term for specific use-environment conditions. Heat, moisture, mechanical load, etc. are all environmental degradation factors. Critical degradation mechanism refers to the fact that polymer materials are susceptible to attack by a specific set of environmental degradation factors which include influences due to chemical, physical, and mechanical processes. The critical degradation mechanism is the mechanism that occurs due to this attack and results in a significant change in one or more bulk physical property of the material system. It is assumed that the critical degradation mechanism occurs when the environmental degradation factors are inside the boundaries of the use-environment. For example, critical degradation due to moisture would be assumed to occur only in environments where the relative humidity is typical of operating conditions. Accelerated ageing is defined as the process or processes required to accelerate a specific critical degradation mechanism or mechanisms relative to a baseline ageing condition; thereby resulting in the material reaching the same aged end-state as a real-time aged material, but in less time.

In general, material testing is a costly process that often involves many materials-related disciplines and a wide variety of laboratory equipment. It is recognized that, while long-term, real-time testing is required to assess the durability of materials fully, accelerated ageing may reduce the expense and time involved by significantly narrowing or screening the field of acceptable candidate materials that would go into long-term qualification tests. In addition to materials screening, accelerated testing may help determine residual service life of existing structures and suggest directions for product improvements. This type of information may then lead to changes in the standard practice for materials selection and provide quantitative rationale for manufacturers and fabricators to follow new and improved specific procedures. Typically, a largely empirical approach such as presented in Sargent (2005) or Murray *et al.* (2003) is used for accelerated ageing studies.

The empirical methods for accelerated testing may address the concerns for specific applications and environments, but the need for predicting performance in broader service conditions will require the development of empirical techniques coupled with analytical methods. Overlying this process is the development of laboratory-guided ageing methods that define critical environmental degradation factors and their interactions. It is a goal, therefore, that all of the testing that is undertaken should provide insight into how a material behaves and establish input for the development of analysis methods to predict material performance under various conditions of load, temperature, and environment. Validation of ageing methods takes place through a comparison of mechanical properties, damage mechanisms, and physical parameters (e.g. weight loss, changes in glass transition or fracture toughness) determined from accelerated testing with those from real-time ageing tests.

Because so much of the ageing process in PMCs is dependent on temperature, a clear understanding of temperature-related behavior is important. Most polymers have distinct material phases defined by temperature. In particular, they have a second-order transition temperature below the melting temperature called the glass transition temperature, T_g. This temperature marks the division between rubbery and glassy behavior for the material and it is a measure of the ease of torsion of the backbone of the polymer chain. At the glass transition temperature, discontinuities exist in the values of heat capacity and thermal coefficient of expansion. Correspondingly, there is a change in slope of the specific volume versus temperature: at temperatures above T_g, Brownian motion of the molecules is rapid such that an increase/decrease of temperature causes an increase/decrease of volume in the time scale of the temperature change. At lower temperatures, however, the slow molecular motion is such that a change in temperature is not immediately reflected by a corresponding change in volume of the material. Experimentally, T_g is often measured using dynamic mechanical analysis (DMA) and characterized as the temperature corresponding to the peak in the tan δ value where tan δ is defined as the ratio between the loss and storage modulus (Kampf, 1986). These measurements are sensitive to heating rate, sample preparation, and loading mode.

Chemical ageing is related to irreversible changes in the polymer chain/network through mechanisms such as cross-linking or chain scission. Chemical degradation mechanisms include thermo-oxidative, thermal, and hydrolytic ageing. At typical PMC operating temperatures, cross-linking and oxidation are the dominant chemical ageing mechanisms. Thermo-oxidative degradation becomes increasingly important as the exposure temperature and time increase. Frequently, chemical ageing results in an increase in cross-linking density that can lead to changes in material densification and increases the T_g, which in turn will influence mechanical properties such as strength and stiffness.

Physical ageing will occur when a polymer is rapidly cooled below its T_g and the material evolves toward thermodynamic equilibrium. This evolution is characterized by changes in the free volume, enthalpy, and entropy of the polymer and will produce measurable changes in the mechanical properties (Struik, 1978; McKenna, 1994; Hutchinson, 1995. As an example, the volume evolution as a function of temperature is shown in Fig. 1.1.

Referring to Fig. 1.1, the slope of the volume–temperature curve is dependent upon the rate of cooling of the material. A polymer held isothermally

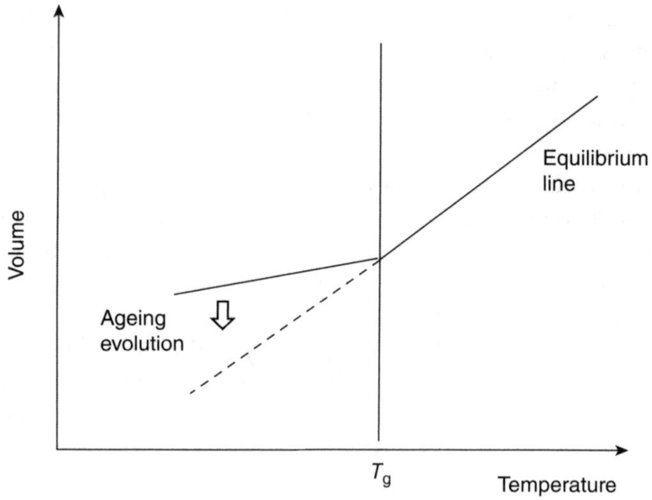

1.1 As an example, the volume evolution as a function of temperature is illustrated.

below T_g will experience a slow continuous decrease in volume associated with ageing of the material. This response occurs as the material evolves towards the desired equilibrium volume state, an evolution that occurs instantaneously for $T > T_g$. Physical ageing is responsible for changes over time of modulus, strength, and ductility for polymers in the glassy range. Since most polymer composite structures are used in the glassy range of the polymeric matrix, physical ageing has an important impact on long-term durability of composites used in applications. Physical ageing is thermo-reversible for all amorphous polymers by heating the polymer above its T_g and subsequently rapidly quenching the material. It is assumed that this thermo-reversible behavior does not occur in thermoset materials due to the tendency for elevated temperature to affect their extent of cross-linking and/or influence chain scission. Operational mean temperature and lifetime thermal history have a strong influence on the rate of physical ageing.

Mechanical degradation mechanisms are irreversible processes that are observable on the macroscopic scale. These degradation mechanisms include matrix cracking, delamination, interface degradation, fiber breaks, and inelastic deformation; and thus have a direct effect on engineering properties such as stiffness and strength. If the stress in a material is too high, its response is no longer elastic, i.e. plastic. This limiting stress level is called the elastic limit. The strain that remains after removal of the stress is called the inelastic strain or the plastic strain. Plastic strain is defined as time-independent although some time-dependent strain is often observed to accompany plastic strain.

Creep is the continuous time-dependent deformation of a material under constant stress. For small strains it is assumed that the cross-sectional area of the specimen remains constant, therefore constant load and constant stress experiments are equivalent. The first or primary stage of creep is associated with increasing strain at a declining strain rate; the secondary creep stage proceeds at a nearly constant strain rate. For most polymers, the tertiary stage is characterized by rapid fracture. Stress relaxation is the gradual decrease in stress (or load) of a material subjected to a constant strain. This stress may asymptotically reach a limiting value over time unless the strain is increased once again.

In some cases, mechanical degradation mechanisms dominate only after chemical or physical ageing mechanisms have altered the polymer properties. For example, thermo-oxidative stability is a problem with many thermoset materials and often leads to matrix cracks along surfaces and edges exposed to the environment. Once these cracks occur, they then serve as initiation sites for extensive crack growth from subsequent mechanical loading. Once the cracks start to grow, the longer cracks provide additional surface area for thermo-oxidative degradation and hence additional sites for new crack growth (Lévêque *et al.*, 2005).

1.2 Background

Physical and chemical ageing are inherently time-dependent processes. This section will focus on the time-dependent response of PMCs, in particular addressing the viscoelasticity of the polymer matrix material and as a result the composite as a whole.

The three basic constituents of advanced PMCs are fiber, interface, and matrix. While the polymer matrix provides many advantages – such as easy processing, low cost, and corrosion resistance – in many circumstances the polymeric matrix is the major constituent contributing to degradation or changes in durability of PMCs. In applications where the matrix material experiences exposure to environments that alter the properties of the matrix or mechanical loads that act over long time periods, the viscoelastic nature of the matrix becomes a dominant factor in the composite performance. Consequently, time-dependent changes in composite stiffness, strength, and fatigue life can all be related to changes in the mechanical properties of the polymer matrix.

Focused studies on the viscoelastic behavior of polymeric composite materials started in the late 1960s and early 1970s. Many of these early studies were driven by the need to describe the temperature- and time-dependent behavior of the PMCs during the fabrication of laminated parts. As composites were used in military and civilian aircraft, long-term reliability became a major concern, sparking interest in the long-term viscoelastic nature of the polymer matrix. More recently, use of composites on supersonic aircraft (Tsotsis *et al.*, 2001; Committee on High Speed Research *et al.*, 2007) and reusable launch vehicles (Freeman *et al.*, 1997; Dragone and Hipp, 1998) that experience elevated temperatures in the substructure and skin during flight has motivated research to explain a range of behaviors related to time-dependent polymer composite response. Additional applications that have motivated interest in time-dependent behavior include use of polymer composites in the civil infrastructure, such as in bridges and pipelines (Liao *et al.*, 1997; Creese and GangaRoa, 1999).

In order to meet exacting performance goals, the aerospace industry has typically relied on the use of expensive, high-performance materials (e.g. high-T_g thermoplastics, thermosets with continuous fiber). Alternatively, mainstream civil infrastructure applications have responded to the need for low-cost structures, forcing use of lower end materials. Although the temperature extremes seen by most civil applications are milder than those of aerospace vehicles, the use of lower T_g materials and less expensive fabrication techniques brings viscoelastic effects into relevancy. Thus, for both aerospace and civil structural applications, environmental effects and the various types of ageing – and their impact on durability and damage accumulation – are prime concerns.

Historically, there have been several types of descriptors used to define the linear and nonlinear time-dependent behavior of materials. For advanced metallics, it has been noted that an 'elastic-viscoplastic' material shows viscous properties in both the elastic and plastic regions while an 'elastic/viscoplastic' material shows viscous properties in the plastic region only. Discussion of this type of behavior can be found in Cristescu and Suliciu (1982). Materials that exhibit initial elastic behavior upon loading followed by creep and initial elastic recovery followed by continuously decreasing strain upon stress removal are termed viscoelastic (Findley *et al.*, 1976). In short, viscoelasticity combines elasticity and viscosity. As discussed in the following sections, the condition for viscoelastic linearity can be expressed mathematically and verified experimentally. For many polymers, particularly those developed for advanced applications, the viscoelastic behavior can easily cross from the linear to the nonlinear range at elevated temperature and/or stress levels. In fact, due to localized states of stress or varying loading histories, composite materials may exhibit attributes of both linear and nonlinear behavior. Hence, constitutive models for nonlinear viscoelasticity are also required to provide a complete description of material response (Drozdoz, 2001).

In a summary article during the mid 1970s, Schapery (1975) provided a comprehensive treatment of the deformation and failure theories for viscoelastic composite materials. For an undamaged composite, he began with the linear viscoelastic response at constant temperature and then systematically addressed response under transient temperatures, and dynamic response. Schapery then analyzed some aspects of viscoelastic fracture mechanics for treatment of crack growth and failure in composites. Applications to composite materials and the effects of microcracking were discussed in a general manner. In the early 1980s, the use of viscoelasticity to describe the nonlinear behavior of PMCs with damage was further investigated by Schapery (1981; 1982). The use of the *J* integral, energy release rate, and correspondence principle for nonlinear viscoelastic media were also developed (Schapery, 1984) to predict crack initiation and growth. Many researchers have successfully used these models to enhance the understanding of short-term viscoelastic composite behavior. Long-term ageing was not explicitly accounted for in these early models.

In 1995, Scott *et al.* (1995) provided a literature review of creep behavior of fiber-reinforced polymeric composites which covered linear and nonlinear theories, accelerated test techniques, and the effects of environment. Among their findings were: Schapery's integral representation lends itself well to numerical techniques; shearing deformations are the dominate modes for time-dependency in PMCs; understanding temperature and moisture effects is critical for predicting long-term behavior; and the time–temperature superposition principle (TTSP) may be useful for predicting long-term performance.

The relationships between elevated temperature and time-dependent behavior of polymer composites have been investigated by many research groups. If one was to catalog the test procedures and analysis methods used, it would be evident that most studies have concentrated on characterizing the creep behavior under isothermal conditions. An example is the creep experiments of Gramoll *et al.* (1990) that were used to quantify the effects of temperature on the nonlinear viscoelastic response of Kevlar-based composites. In that work, the TTSP was employed to predict the time-dependent response of a series of general laminates. In addition to temperature, some studies have focused on the behavior of PMCs under hygrothermal conditions. One example of this is the work of Wang *et al.* (1990) with Kevlar-based composites, which experimentally characterized the transient moisture effects during creep and dynamic mechanical tests. Another example is Han and Nairn (2003), who determined that degradation and changes in microcracking fracture toughness did not change below a threshold value.

1.2.1 Accelerated ageing

Verified accelerated ageing methods are required to provide guidance for materials selection and to assess ageing of new materials accurately. The highly empirical approaches taken for the majority of accelerated ageing studies dictate that the primary objective of an accelerated ageing method is to screen and characterize new material systems. As discussed above in this chapter, materials testing is a costly process that often involves many materials-related disciplines and a wide variety of laboratory equipment. While it is recognized that, real-time testing is required long-term to assess the durability of materials, accelerated ageing may reduce the expense and time involved by significantly narrowing the field of acceptable candidate materials that will go into long-term qualification tests. In addition to materials screening, accelerated ageing may help determine residual service life of existing structures and suggest directions for product improvements (Kim *et al.*, 2002).

In a recent report by Gates (2003), a rational approach to the problem of accelerated testing of high-temperature polymeric composites is presented. The methods provided are considered tools useful in the screening of new materials systems for long-term application to extreme environments that include elevated temperature, moisture, oxygen, and mechanical load. The need for reproducible mechanisms, indicator properties, and real-time data are outlined as well as the methodologies for accelerated ageing of specific time-dependent mechanisms.

1.3 Viscoelasticity

The basic viscoelastic effects, such as creep and relaxation, typically studied for homogeneous polymer systems also appear in polymer composites. For simplicity of presentation, this section will focus on linear response characteristics. It is assumed that the polymeric matrix material alone exhibits viscoelastic response, while the fibers (typically carbon) are elastic and of a much higher modulus than the matrix material. The combined effect in the composite is such that the mechanical responses transverse to the fibers and in shear are significantly impacted by the viscoelasticity of the matrix material, while the response in the fiber direction is constrained by the fibers to be elastic within typical experimental measurement ranges. Looking at a typical two-dimensional (2-D) stiffness matrix for a laminate, this implies that the matrix-dominated terms Q_{22} and Q_{66} are time-dependent.

$$\mathbf{Q} = \begin{bmatrix} Q_{11} & Q_{12} & 0 \\ Q_{12} & Q_{22}(t) & 0 \\ 0 & 0 & Q_{66}(t) \end{bmatrix} \qquad [1.1]$$

In three dimensions, additional terms will be affected but limited experimental data are available for such conditions. In contrast, the 2-D transverse modulus and in-plane shear modulus have been characterized as time-dependent for many material systems.

The two most fundamental concepts in viscoelasticity are creep and relaxation. In order to examine the relaxation response, consider a constant strain applied to a material. For an elastic material, there is a unique stress corresponding to that strain level determined by the modulus of the material. For a constant strain applied to a viscoelastic material however, over time the long-chain molecules in the polymer accommodate that strain by decreasing stress levels as the molecules unwind and disentangle. This stress relaxation (σ(t)) is represented mathematically by a time-dependent modulus for the material, $E(t)$

$$\sigma(t) = E(t - t_0)\varepsilon_0 \qquad [1.2]$$

where ε_0 is the applied initial strain. Similarly, creep in a viscoelastic material can be understood by considering the material subjected to a constant stress. Again, due to the polymer in the matrix of the composite, over time the strain increases as the segments of the long-chain molecules move relative to one another. This is represented mathematically by a time-dependent compliance for the material, D(t)

$$\varepsilon(t) = D(t - t_0)\sigma_0 \qquad [1.3]$$

where σ_0 is the applied initial engineering stress, t is time and t_0 is starting time. Time-dependent moduli and compliances for viscoelastic materials are obtained from tests conducted at constant strain and constant stress, respectively. The initial values of modulus $E(t)$ and compliance $D(t)$ at time $t = 0$ correspond to the initial instantaneous elastic response. The relaxation modulus decays over time – for some materials (e.g. thermosets) this decays to a constant E_∞, known as the rubbery modulus of the material, while for other materials (e.g. thermoplastics) the relaxation modulus decays to zero. The time-dependent moduli and compliances can be represented by a variety of functions such as power laws, exponentials, and series expansions.

1.3.1 Superposition

An extremely important property of linear viscoelastic materials is that of superposition. Superposition principles occur in several domains in viscoelasticity as will be seen shortly (including nonlinear viscoelasticity), but the first one to consider is superposition of responses to stress. For linear viscoelastic materials, the strain responses to two different stress inputs applied separately can be simply superposed to provide the strain response for a combined loading of the two stress inputs superposed: the modulus and compliance ($E(t)$ and $D(t)$) are not functions of stress. This concept is the basis of the Boltzman superposition principle that is used to create a hereditary integral form for a viscoelastic constitutive law. Consider a series of stress steps applied to a viscoelastic material where at time zero an initial stress of σ_0 is applied and at subsequent times, t_i, a stress jump, $\Delta\sigma_i$ (either positive or negative) is applied. Considering each individual stress jump as the applied load and applying the principle of superposition, one can obtain the strain response as

$$\varepsilon(t) = \sigma_0 D(t)H(t) + \sum_i \Delta\sigma_i D(t - t_i)H(t - t_i) \qquad [1.4]$$

where $H(t)$ is the step function. If one now takes the limit as $\Delta t_i \rightarrow 0$, the result is a constitutive law for a viscoelastic material that can be applied to any loading, not limited to discrete step loading:

$$\varepsilon(t) = \sigma_0 D(t)H(t) + \int_0^t (t - t')\frac{\mathrm{d}\sigma(t')}{\mathrm{d}t'}\mathrm{d}t' \qquad [1.5]$$

Equation [1.5] is a Riemann convolution integral and in practice is often written as

$$\varepsilon(t) = \int_0^t D(t - t')\frac{\mathrm{d}\sigma(t')}{\mathrm{d}t'}\mathrm{d}t' \qquad [1.6]$$

where it is assumed that the lower limit really represents $t = 0^-$ and the stress is understood to be expressed with a step function at the origin, $\sigma(t) = \sigma(t)H(t)$.

1.3.2 Linearity

The mathematical representation for linear viscoelastic materials also provides a means to easily test whether one is in the linear range of a given material. In order to sufficiently ensure material linearity, two common techniques are used: the first method is to perform short-term creep tests on the material at several load levels. In the linear viscoelastic range, the compliances, $D(t) = \varepsilon(t)/\sigma_0$, obtained from each creep test will superpose on top of one another. At a certain load level, the compliance response will begin to differ and for successively higher load levels the discrepancies will increase – this defines the nonlinear region of loading for the given material. The second method to verify linearity of a material is to perform a creep and recovery test. The creep portion of the experiment is used to obtain the creep compliance of the material, $D(t)$. This creep compliance is then used to predict the recovery portion of the data via Boltzman's superposition principle. Note that the stress can be written

$$\sigma(t) = \sigma_0 H(t) - \sigma_0 D(t - t_0) \qquad [1.7]$$

Substitution into equation [1.6] yields the prediction for the strain

$$\varepsilon(t) = \sigma_0 D(t)H(t) - \sigma_0 D(t - t_0)H(t - t_0) \qquad [1.8]$$

and the portion evaluated after time t_0 is checked against the experimental data.

1.3.3 Time–temperature superposition

The effect of elevated temperature on a polymer composite is a general increase in matrix-dominated compliance. For most polymeric materials, the quantitative impact of temperature on mechanical properties is embodied in the TTSP (Findley *et al.*, 1976). TTSP, illustrated graphically in Fig. 1.2, indicates that the modulus curves at different temperatures are related to one another by a simple shift on the log time scale. This result implies that the relaxation times for a material, which represent the ease of motion of different segments of the polymer chain, are all scaled by temperature in an identical manner. The relaxation times for a given material are short at high temperatures, long at low temperatures, and can be calculated relative to those at a reference temperature by a simple multiplicative factor

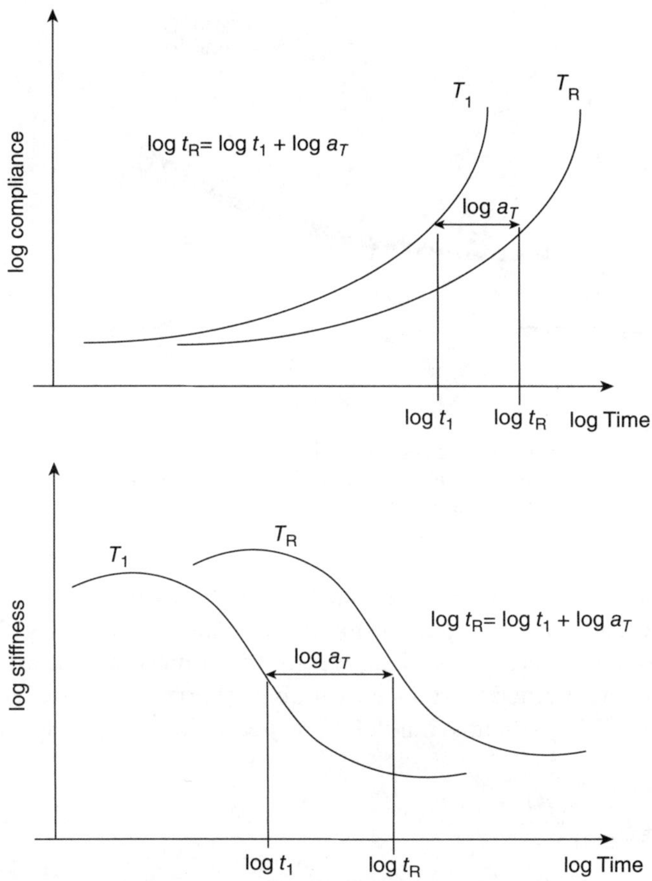

1.2 TTSP indicates that the modulus curves at different temperatures are related to one another by a simple shift on the log time scale. T_1, first temperature; T_R, reference temperature; t_R, reference time; t_1, starting time; a_T, shift time.

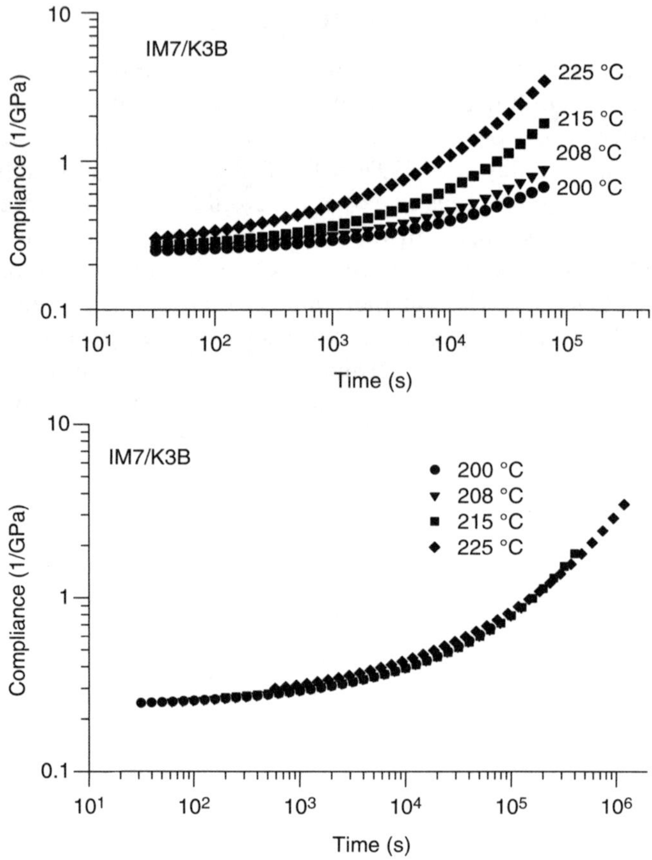

1.3 TTSP shifting process and collapse of the data into a single master curve is illustrated for a series of isothermal creep tests below the T_g of the composite material.

called the temperature shift factor, a_T. The TTSP shifting process and collapse of the data into a single master curve is illustrated in Fig. 2.3 for a series of isothermal creep tests below T_g of the composite material.

Above T_g, the method of reduced variables (Ferry, 1980) can be used to arrive at the WLF (Williams–Landel–Ferry) equation for temperature shift factor

$$\log a_T = \frac{-c_1^0(T-T_0)}{(c_2^0+T-T_0)} \quad [1.9]$$

while below the T_g the Arrhenius equation

$$a_T = \exp\frac{\Delta H}{R}\left[\frac{1}{T}-\frac{1}{T_0}\right] \quad [1.10]$$

should be used where T is the temperature, T_0 is the reference temperature, c_1 and c_2 are constants, and ΔH and R are the activation enthalpy and gas constant, respectively. Note that $a_T = 1$ for $T = T_0$ and T_g is often used as the reference temperature T_0 (Ferry, 1980). One major application of the TTSP is accelerated testing. Accelerated testing based on viscoelasticity typically involves experiments at multiple temperatures where in each test the experimental time range is relatively short. Then long-term response is obtained by use of TTSP.

The vast majority of the work in viscoelastic model development for composites has relied on the use of tensile creep data to provide material constants and verify the predicted time–dependent behavior. Stress relaxation is the analogue of creep in the sense that despite the difference between loading modes, both test types provide a time-dependent response that can be defined in terms of stress, strain, time, and environment. However, the inherent difficulties associated with performing stress relaxation tests with composites has led to a dearth of data describing this type of time-dependent behavior. Considering the anisotropic compliance and modulus matrices for a composite, their interrelation can be written as

$$\sum_j \int_0^t S_{ij}(t-\zeta)Q_{jk}(\zeta)\mathrm{d}\xi = t\delta_{ik} \qquad i, j, k \in [1, 2, 6] \qquad [1.11]$$

where δ_{ik} is the delta function and $S_{ij}(t)$ and $Q_{ij}(t)$ are the compliance and modulus, respectively. Since the time-dependence for composite moduli is limited to the transverse and shear response, the solution to equation [1.11] simplifies considerably and various numerical methods can be used to obtain one function from the other mathematically (Bradshaw and Brinson, 1997; Gates *et al.*, 1999).

1.4 Ageing and effective time

Experimental studies, such as those given in Hastie and Morris (1992), have illustrated that the matrix-dominated properties of continuous fiber-reinforced PMCs, namely the in-plane shear and transverse response, are affected by physical ageing in a manner similar to that observed for polymers. These studies indicated that it was possible to use the general experimental approaches developed by Struik (1978) to isolate the physical ageing component of the time-dependent behavior by performing isothermal creep compliance tests and using linear viscoelasticity with superposition techniques to establish the ageing-related material constants. Struik (1978) was successful in introducing concepts such as effective time and methods for accelerated ageing by determining ageing shift factors from creep tests.

For linear viscoelastic materials, the effect of many factors on material response can be expressed as a simple time shift. The effects of ageing can be quantified by use of the shift factor concept and the associated method of reduced variables. This result gives rise to the *effective time* concept. Considering temperature effects as the example, the modulus E_T at any temperature T can be defined relative to the modulus E_{T_R} at the reference temperature, T_R, by

$$E_T(t) = E_{T_R}(a_T t) \quad [1.12]$$

The 'reduced time' or 'effective time', ξ, is defined by noting that all relaxation mechanisms in time increment dt at temperature T are a_T times slower/faster than those in a time increment $d\xi$ at T_R so that

$$d\xi = a_T dt \rightarrow \xi(t) = \int_0^t a_T(\zeta) d\zeta \quad [1.13]$$

Note that if the temperature T is constant, a_T is also constant and therefore $\xi = a_T t$ and as before

$$E_T(t) = E_{T_R}(\xi(t)) = E_{T_R}(a_T t) \quad [1.14]$$

For the material at temperature T, equation [1.2] can be written in the time domain as

$$\sigma(t) = \int_{0^-}^{t} E_{T_R}(\xi(t) - \xi(t')) \frac{d\varepsilon(t')}{dt'} dt' \quad [1.15]$$

or in the effective time domain as

$$\sigma(\xi) = \int_{0^-}^{\xi} E_{T_R}(\xi - \xi') \frac{d\varepsilon(\xi')}{d\xi'} d\xi' \quad [1.16]$$

For constant temperatures this notation is perhaps cumbersome, but the power of the technique becomes apparent when considering a loading during a variable temperature history, $T(t)$. In this case, again the reduced time can be defined in the same manner as equation [1.13] and the constitutive law by equations [1.15] and [1.16], but now a_T is no longer constant (Bradshaw and Brinson, 1997; Zheng and McKenna, 2003).

1.4.1 Time–ageing time superposition

In order to account for material ageing accurately, a time–ageing time superposition (TASP) procedure must be developed. In order to illustrate this concept, isothermal physical ageing will be considered using creep compliance as the viscoelastic behavior of interest. Assume a polymeric material is quenched from above its T_g to a temperature below its T_g. The time the material exists below the T_g is referred to as the ageing time (t_e). As ageing time progresses, a series of short (in comparison to the elapsed ageing time) creep tests are run to measure the momentary creep compliance of the material. This test procedure is described by Struik (1978) and is illustrated schematically in Fig. 1.4. Experimental data produced from this type of test are provided in Fig. 1.5.

A simple model of momentary creep compliance is the Kohlrausch three-parameter model as described in Brinson and Gates (1995) and given by

$$S(t) = S_0 e^{(t/\tau)^\beta} \quad [1.17]$$

where S_0 is the initial compliance, β is the shape parameter, t is the time, and τ is the relaxation time.

The sequenced creep/ageing curves that result from the testing sequence are collapsed through use of horizontal (time) and vertical (compliance)

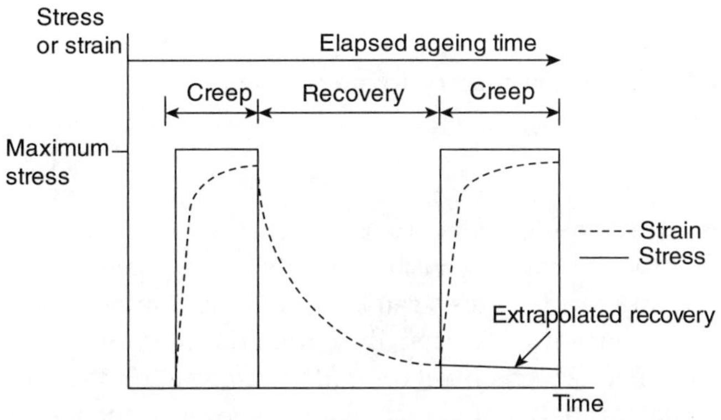

1.4 Schematic representation of a series of short creep tests run to measure the momentary creep compliance of the material.

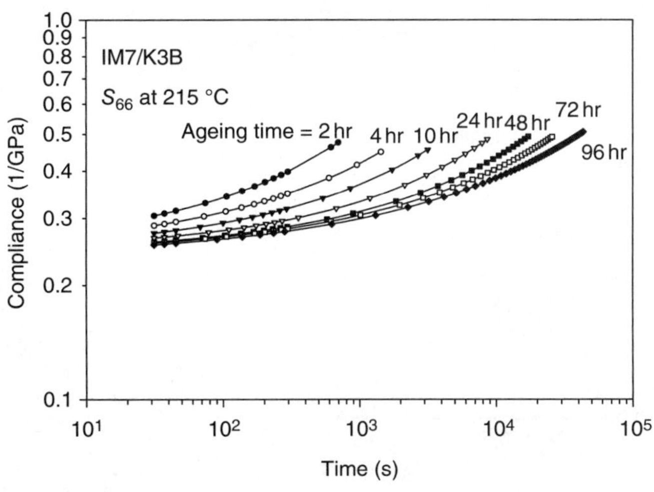

1.5 Experimental data generated by the procedure illustrated schematically in Fig. 1.4.

shifts using any of the curves as the reference curve. The time shifts ($\log a_{te}$) used to collapse the sequenced curves are plotted versus log ageing time and approximated through a linear fit. The slope of this shift factor versus ageing time data is the shift rate (μ). The ageing time shift factor, a_{te}, is defined as the horizontal distance required to shift a compliance curve to coincide with a reference compliance curve. As illustrated in Gates and Feldman (1996), it is possible to shift a series of these momentary curves into a momentary master curve using the ageing shift rate μ. If the ageing time shift factor is plotted as a function of ageing time on a double-log scale, it is found to map a straight line with a slope of μ (Fig. 1.6)

$$\mu = \frac{-\mathrm{d}\log a_{te}}{\mathrm{d}\log t_{\mathrm{e}}} \qquad [1.18]$$

where a_{te} is the ageing shift factor found through test and given as

$$a_{te}\left(\frac{t_{\mathrm{e\,ref}}}{t_{\mathrm{e}}}\right)^{\mu} \qquad [1.19]$$

where $t_{\mathrm{e\,ref}}$ is the reference ageing time. The shift rate usually has values of the order of unity and can be considered to be a material constant. Thus, a series of creep and recovery tests can be used to determine experimentally the value of the shift rate v for any given material.

Since the momentary creep curves collapse through horizontal shifting on the log scale, the only parameter that changes as a function of ageing time is the relaxation time. This allows the relaxation time in equation [1.17] to be given as

$$\tau(t_{\mathrm{e}}) = \tau(t_{\mathrm{e\,ref}})/a_{te} \qquad [1.20]$$

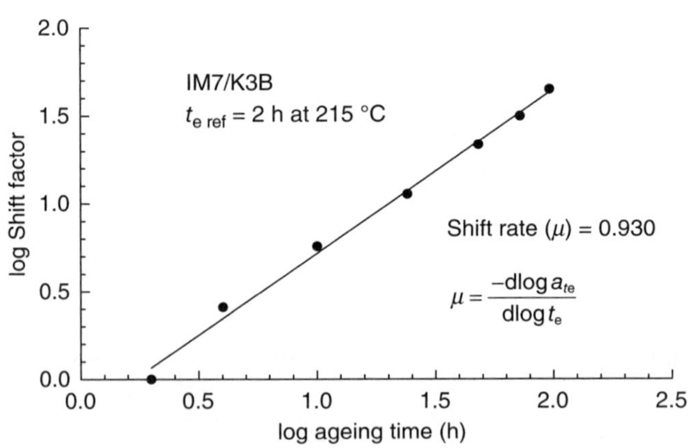

1.6 The ageing time shift factor is plotted as a function of ageing time on a double-log scale.

Therefore, the momentary creep compliance at any ageing time can be found by knowing the initial creep compliance, the shape parameter, the shift rate, and the relaxation time at a reference ageing time. For physical ageing, the shift rate is a function of temperature and will approach zero as test temperature approaches T_g.

Taking the initial ageing time t_e^0 to be the reference ageing time ($t_{eref} = t_e^0$), the shift factor at any instant in time can be defined based on the shift rate ν

$$a_{t_e^0}(t) = \left(\frac{t_e^0}{t_e^0 + t}\right)^{\mu} \tag{1.21}$$

Using the previously introduced concepts of effective time, the effective time increment can then be defined (Struik, 1978)

$$d\lambda = a_{t_e^0}(t)dt \tag{1.22}$$

and the total test time can be reduced to the effective time λ

$$\lambda = \int_0^t a_{t_e^0}(\xi)d\xi \tag{1.23}$$

Integration of equation [1.23] using equation [1.21] gives two distinct expressions for effective time

$$\begin{aligned} \lambda &= t_e^0 \ln(t/t_e^0 + 1) && \text{for } \mu = 1 \\ \lambda &= \frac{t_e^0}{1-\mu}\left[(1 + t/t_e^0)^{1-\mu} - 1\right] && \text{for } \mu \neq 1 \end{aligned} \tag{1.24}$$

Therefore, using effective time in place of real time, equation [1.17] gives

$$S(t) = S^0 e^{(\lambda/\tau(t_e^0))^{\beta}} \tag{1.25}$$

which allows the prediction of long-term behavior based solely on the material parameters determined from short-term or momentary tests. This procedure, along with various enhancements to account for laminated composite plates and automated data reduction schemes, has been used successfully by a number of test programs, details of which can be found in Sullivan *et al.* (1993), Gates *et al.* (1997), Bradshaw and Brinson (1997), and Zheng and Weng (2002).

1.4.2 Time–temperature–ageing superposition

In order to account for both temperature and ageing time effects simultaneously, a combined approach or time–temperature–ageing time superposition can be used. In most cases, this is most easily accomplished by performing isothermal viscoelastic testing (e.g. creep or relaxation) at definite temperature levels below T_g. A general form of the time–temperature–ageing time shift factor is given by Sullivan (1990)

$$\log a = \log a_{t_e} + \log \bar{a}_T \qquad [1.26]$$

where a_{te} is the ageing time shift factor defined by equation [1.19] and $\bar{\alpha}_T$ is the time–temperature shift factor that depends on ageing. As derived in Brinson and Gates (1995), given the ageing shift rate as a function of temperature and the time–temperature shift factor at a single ageing time, $\bar{\alpha}_T$ can be calculated at any ageing time using

$$\frac{\bar{a}_{T_1/T_2}^{t_{e2}}}{\bar{a}_{T_1/T_2}^{t_{e1}}} = \left(\frac{t_{e2}}{t_{e1}}\right)^{\mu(T_2)-\mu(T_1)} \qquad [1.27]$$

where T_1 and T_2 are two different temperatures and t_{e1} and t_{e2} are two different ageing times.

Time–temperature and time–ageing time superposition have been coupled with classical laminated plate theory (CLT) (Jones, 1975) to provide a framework for analysis of laminated composite materials and details can be found in Schapery (1974) and Brinson and Gates (1995). The complete description of this analysis framework is beyond the scope of this text, however the fundamental equations can be easily illustrated.

For a single lamina or ply under plane stress conditions, the stress–strain relations, as given by CLT, are

$$\begin{Bmatrix} \varepsilon_{xx} \\ \varepsilon_{yy} \\ \gamma_{xy} \end{Bmatrix} = [\bar{\mathbf{S}}] \begin{Bmatrix} \sigma_{xx} \\ \sigma_{yy} \\ \sigma_{xy} \end{Bmatrix} \qquad [1.28]$$

where ε_{ij} and σ_{ij} represent the strain and stress in the body axis directions. The transformed compliance matrix $[\bar{\mathbf{S}}]$ is given by

$$[\bar{\mathbf{S}}] = [\mathbf{T}]^{-1}[\mathbf{S}][\mathbf{T}] \qquad [1.29]$$

where $[\mathbf{T}]$ is the transformation matrix and $[\mathbf{S}]$ is the compliance matrix referenced to the material coordinate axis and given by

$$[\mathbf{S}] = \begin{bmatrix} S_{11} & S_{12} & 0 \\ S_{12} & S_{22} & 0 \\ 0 & 0 & S_{66} \end{bmatrix} \qquad [1.30]$$

As demonstrated by Ha[illegible]s (1992), the only time-dependent compliance terms in e[illegible]re the transverse (S_{22}) and shear (S_{66}) terms that are as[illegible]matrix-dominated deformation of a ply. Therefore, based on a[illegible] expression such as given in equation [1.25], the time-dependent compliance terms can be written in a general form as

$$S_{22}(t) = f\left(S_{22}^{0}, \beta_{22}, \tau_{22}(t_{\mathrm{e\,ref}}), \mu_{22}; t\right) \qquad [1.31]$$

$$S_{66}(t) = f\left(S_{66}^{0}, \beta_{66}, \tau_{66}(t_{\mathrm{e\,ref}}), \mu_{66}; t\right) \qquad [1.32]$$

where $t_{\mathrm{e\,ref}}$ is the reference ageing time.

For a laminated composite plate, the laminate compliance is found using CLT. Due to the time-dependence of equations [1.31] and [1.32], the laminate compliance will also be time-dependent. The amount of time-dependence and the effects of ageing will depend on the layup and therefore the relative contributions of equations [1.31] and [1.32] to the total laminate compliance. As an example, if it is assumed that the fiber is linear-elastic, a unidirectional laminate (e.g. $[0]_n$) loaded in the fiber direction will not exhibit viscoelastic behavior. Conversely, the same unidirectional laminate loaded transverse to the fiber direction will have a compliance governed by equation [1.31], and exhibit both viscoelastic and ageing behaviors. Most laminates will have a range of ply and load orientations and consequently will be considered to exhibit some aspects of time-dependent behavior.

In equations [1.5] and [1.6], the hereditary integral constitutive law for a viscoelastic material was developed considering a one-dimensional response; the extension to effective time was shown in equations [1.15] and [1.16]. These results can be applied independently to uniaxial response and shear response for an isotropic viscoelastic material to obtain proper expressions for multiaxial behavior. For an anisotropic material, these expressions can be generalized to

$$\sigma_{ij}(t) = \int_{0}^{t} C_{ijkl}(\xi(t) - \xi(t'))\frac{\mathrm{d}\varepsilon_{kl}}{\mathrm{d}t'}\mathrm{d}t' \qquad [1.33]$$

$$\varepsilon_{ij}(t) = \int_{0}^{t} S_{ijkl}(\xi(t) - \xi(t'))\frac{\mathrm{d}\sigma_{kl}}{\mathrm{d}t'}\mathrm{d}t' \qquad [1.34]$$

where C_{ijkl} and S_{ijkl} are the fourth-order modulus and compliance tensors, respectively, effective time is used in the integration and summation over repeated indices is implied. For the planar analysis used for a thin composite lamina, only four independent constants are relevant in the material property tensors. In this case, these expressions can be simplified to the form

$$\begin{Bmatrix} \varepsilon_1(t) \\ \varepsilon_2(t) \\ \varepsilon_3(t) \end{Bmatrix} = \int_{0}^{t} \begin{bmatrix} S_{11}(\xi(t)-\xi(t')) & S_{12}(\xi(t)-\xi(t')) & 0 \\ S_{12}(\xi(t)-\xi(t')) & S_{22}(\xi(t)-\xi(t')) & 0 \\ 0 & 0 & S_{66}(\xi(t)-\xi(t')) \end{bmatrix} \frac{\mathrm{d}}{\mathrm{d}t'} \begin{Bmatrix} \sigma_1(t') \\ \sigma_2(t') \\ \sigma_6(t') \end{Bmatrix} \mathrm{d}t' \qquad [1.35]$$

where a contracted notation is typically used for a planar formulation in which the index 1 indicates the 11 direction, 2 indicates the 22 direction, and 6 indicates the 12 direction (Jones, 1975).

Note that a single effective time is indicated in equations [1.33] to [1.35]. It has been noted in Gates and Feldman (1996) and Brinson and Gates (1995) that the ageing of a polymeric composite is seen to exhibit different time scales for the shear and transverse response. Fortunately, the shear and transverse response are decoupled in equation [1.35].

1.5 Development of an ageing study

The most important requirement to keep in mind during the development of an ageing program of study, and the required accelerated test methods, is that the accelerated methods must replicate those changes that occur in the real-time, long-term application. The implication of this requirement is that emphasis must be put on the need to understand and reproduce degradation mechanisms associated with each accelerated ageing condition. A mechanistic approach to this requirement would require complete knowledge of all relevant degradation mechanisms and the recognition that competing mechanisms may proceed at different rates as well as interact synergistically. This mechanistic approach is beyond the scope of most test programs. Therefore, an approach is proposed that relies on determination of primary degradation mechanisms that are easy to measure.

The ageing test program must establish a list of indicator properties that will be measured during testing. Examples of these indicator properties include weight, T_g, and damage state (e.g. crack density). These material indicator properties form the basis for development of more economical accelerated ageing schemes for screening for new materials and for evaluating the status of materials in long-term ageing. In order to develop this list of indicator properties, exploratory tests should be run with a wide range of properties investigated using *in-situ* and/or post-test evaluation. Indicator properties should be easy to measure and reliably correlate to changes in residual mechanical properties. Absolute changes and rates of change should be measurable and easily tracked against a firm baseline.

Perhaps the most difficult yet necessary part for any material ageing study is coupling mechanical and physical property data from accelerated ageing with real-time ageing data. The comparison of the accelerated data with real-time data will determine the accelerating factor for any given degradation mechanism. This correlation between real-time and accelerated implies the need to define clearly all environmental degradation factors and determine sensitivity of the factors to variations in parameters (e.g. temperature, humidity). For most commercial, polymeric composite material systems, the cost of developing an extensive real-time database will force laboratories to collaborate and develop national databases. This type of coordinated effort will rely on the use of standardized test methods and data reporting.

For typical environmental degradation factors such as temperature, load, and moisture, it is normal for one or two environmental degradation factors to dominate the ageing of a material system. Therefore, when possible, each ageing mechanism should be investigated by separate real-time, long-term tests in order to determine the dominant environmental degradation factors and provide the critical degradation mechanism of a given material system.

The first step in assessing ageing requires accurate material identification. For this purpose, a thermoset is defined as a cross-linked polymer network that hardens to final shape after cool down from the forming temperature and is incapable of being reshaped. A thermoplastic material is a linear polymer of amorphous, semi-crystalline, or mixed morphology. A thermoplastic softens on heating to a state where the shape may be changed by physical forces and resolidifies on cooling. In principle, the process of softening and solidifying may be repeated indefinitely.

Material performance is defined by a set of indicators that measure a specific property for the material. The indicators will be dependent upon the mechanism of interest. The suggested procedure is as follows.

1 Identify material by class (i.e. thermoplastic, thermoset).
2 Identify mechanism(s) to evaluate (e.g. thermal stability, matrix cracking, etc.).
3 Choose an environmental degradation factor for ageing (e.g. elevated temperature, moisture, etc.).
4 Conduct ageing experiment within limits of the chosen environmental degradation factor using established methods.
5 Perform *in-situ* or post-ageing measurements with indicators sensitive to changes in material performance and compare results to unaged values.

The results of such a study will provide data that help evaluate whether a particular degradation mechanism will be critical for the given application. This procedure can be repeated for all degradation mechanisms of interest for that material. It should not introduce extraneous damage/degradation mechanisms nor should it omit any known degradation mechanisms. In addition, the set of mechanical properties or indicators chosen for screening should be those most critical from the structural performance viewpoint and those most sensitive to degradation. Examples that highlight these criteria are provided in the following sections.

1.5.1 The influence of geometry

The rate and magnitude of most transport processes can be related to the volume and surface area of the material. Therefore, it is worthwhile considering that the geometry of a composite laminate may be used to accelerate the effects of ageing. It is known, for instance, that thermo-oxidative degradation of many materials occurs from the surface inward. Therefore, thinner samples (low volume) should reflect the effects of thermo-oxidative degradation more rapidly than thicker samples (high volume). For continuous fiber composites, these effects are complicated by the transversely isotropic nature of the material, which can lead to degradation rates that vary according to the differences in the laminate surfaces and ply orientation (Boukhoulda *et al.*, 2006).

1.5.2 Thermo-oxidative degradation

In the polyimide material systems, the oxidation reaction occurs through a radical chain process (Schnabel, 1981; David, 1983; White, 1994) and is referred to as autoxidation. Experimentally, thermal methods such as thermogravimetric analysis (TGA) have been used to obtain data related to thermo-oxidative degradation. The TGA relies on weight loss data as a measure of thermo-oxidative stability (TOS) (Kampf, 1986). The experimentalist can utilize several measurements to quantify TOS including the temperature at onset of TOS, the temperature at which half the sample has decomposed, or the apparent activation energy of the reaction from the weight loss data. This latter approach has been used to provide a 'kinetic map' of the reaction as a means of comparing the TOS of similar materials. It should be noted that in a TGA experiment, the reactions are temperature driven and thus occur sequentially as temperature is increased.

It is possible to model the chemical kinetics on a limited basis by coupling the oxygen diffusion to the chemical reaction (Colin *et al.*, 2000). This approach assumes a classical diffusion model such as Fick's law along with a reaction equation that provides the rate of oxygen consumption. The chemical reactions proceed more rapidly at higher temperatures. For the simple case of single activation energy, the reaction may be modeled by the Arrhenius equation (Hsuan and Koerner, 2001). The absolute effects of oxygen exposure during ageing can only be determined if compared against ageing in an inert environment. Some indicators that monitor thermo-oxidative degradation are outlined below.

1 *Weight.* Initial weight loss of a polymer exposed to a thermo-oxidative environment may occur due to loss of moisture and residual volatiles and is not related to polymer breakdown. Usually, after several hundred hours of exposure, the sample weight stabilizes and any additional weight loss is indicative of thermo-oxidative degradation (Colin *et al.*, 2000; Schoeppner *et al.*, 2007). Once polymer degradation occurs, the evolution of gaseous degradation products accompanies weight loss.

2 *Physical changes.* Optical measurements of changes in color, surface texture, and crack density can be indicators of thermo-oxidative degradation (Lafarie-Frenot, 2006).
3 *Glass transition temperature, T_g.* Changes in T_g are frequently observed by dynamic mechanical analysis (DMA). Usually, an increase in T_g suggests chain extension or network cross-linking. A decrease in T_g is normally associated with chain scission. The data in Fig. 1.7 illustrate the experimental values of time-dependent weight loss and increase in T_g for a graphite/bismaleimide composite subjected to isothermal ageing at 204 °C.
4 *Mechanical properties.* Most residual properties such as tension strength, compression strength, and stiffness do not make good indicators of thermo-oxidative degradation early in the ageing process. However, more subtle mechanical properties such as fracture toughness and plasticity are sensitive indicators of short-term ageing owing to their dependence on matrix-dominated behavior.

1.5.3 Thermal degradation

Elevated temperature ageing in polymers in the absence of oxygen can also lead to thermal degradation and changes in material properties due to additional cross-linking and/or chain scission. These changes make an analytical approach based on principles of viscoelasticity and ageing-based superposition quite complicated. For example, if T_g increases during ageing (as is the case with most thermoset materials), then the ageing shift rate is not a constant during ageing and would have to be fully characterized as a

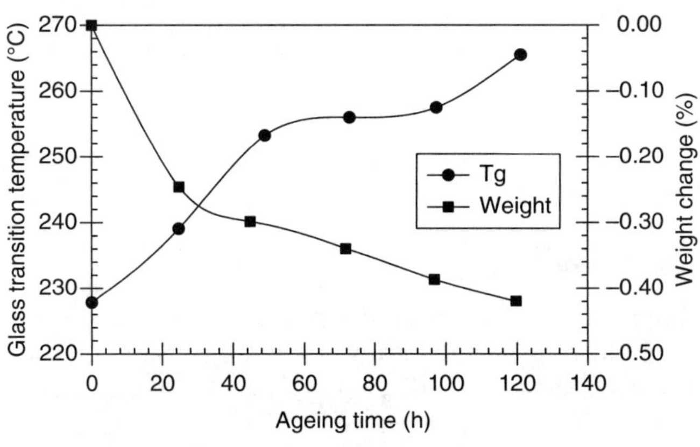

1.7 Experimental values of time-dependent weight loss and increase in T_g for a graphite–bismaliemide composite subjected to isothermal ageing at 204 °C.

function of both time and temperature, unlike physical ageing where the shift rate is strictly a function of temperature.

As a consequence of the difficulties associated with chemical ageing, analytical models capable of predicting the long-term changes due to chemical ageing are rare. An example of a model that correlates change in mechanical properties due to additional cross-linking is given in Zhou (1993). Generally, thermoset systems initially tend to embrittle during purely thermal ageing. Depending upon thermoset chemistry, these systems may either continue to embrittle or start to degrade significantly. Indicators that are useful for tracking thermal degradation are given below and are similar to those for thermo-oxidative degradation. The changes incurred during the elevated temperature fatigue include loss in weight, increase in T_g, and increase in crack density.

1 *Weight.* Initial weight loss of a polymer exposed to a purely thermal environment may occur due to loss of moisture and residual volatiles, just as in thermo-oxidative degradation. However, the magnitude of weight loss for equal exposure times and temperatures is generally much less in purely thermal degradation than it is in thermo-oxidative degradation.
2 *Physical changes.* Optical measurements of changes in color, surface texture, and crack density can be indicators of thermo-oxidative degradation. Frequently, these changes are less noticeable in thermally aged samples compared with those aged under the same conditions with oxygen present. Another measurement that can provide information on physical changes is X-ray photoelectron spectroscopy (Ohno *et al.*, 2000).
3 *Glass transition temperature, T_g.* Changes in T_g are frequently observed by DMA. These changes are considerably smaller for equivalent ageing conditions than for samples aged in the presence of oxygen.
4 *Mechanical properties.* Most residual properties such as tension strength, compression strength, and stiffness do not make good indicators of thermal degradation early in the ageing process. However, more subtle mechanical properties such as fracture toughness are sensitive indicators of short-term ageing.

Hydrolytic degradation

Hydrolytic degradation in polymeric composites is due to diffusion of water into the material leading to moisture uptake and possible plasticization of the polymer matrix. Most mechanical properties are sensitive to hydrolytic degradation. For example, a decrease in fracture toughness has been observed when a polymeric composite was simultaneously exposed to water and mechanical stress (Han and Nairn, 2003). Viscoelastic properties such as creep may also be responsive to moisture-induced degradation and can provide a good method for determining long-term influence on stiffness-related degradation.

The diffusivity of a polymeric composite is found to be a function of specimen geometry, moisture concentration, temperature, and time and is often modeled using the standard Fick's law as described in Springer (1981). The activation energy for water diffusion may be determined from the slope of the natural log of the diffusion coefficient versus the reciprocal immersion temperature. Moisture concentration may reach a plateau level; however the plasticization in a polymer may increase over time. For an inhomogeneous material, diffusion into the polymer can vary which may result in significant differences in water concentration from one region of the sample to the next. This nonuniform concentration or nonFickian behavior can impose stresses on the material. In a composite laminate at constant temperature, this stress can give rise to internal damage in the form of matrix cracks (Roy *et al.*, 2001).

Damage in the form of microcracks may also develop due to the general process of hydrolytic degradation, as shown by Bao and Yee (2002). The occurrence of microcracking will subsequently affect the rates of moisture absorption/desorption during repeated hygrothermal cycling. With internal stresses, microcracks that form in the laminate subsequently allow new pathways for moisture uptake or fiber–matrix debonding (Wang and Hahn, 2007).

Indicators that are useful for tracking hydrolytic degradation are outlined below:

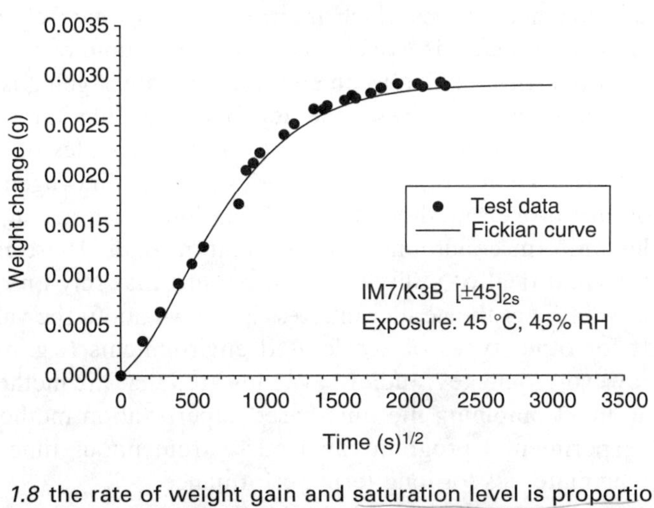

1.8 the rate of weight gain and saturation level is proportional to relative humidity (RH) while the time required to reach saturation is a function of temperature.

1 *Weight.* Exposure to a wet environment will result in weight gain over time. For Fickian behavior, the rate of weight gain and saturation level is proportional to relative humidity, while the time required to reach saturation is a function of temperature. This behavior is illustrated in Fig. 1.8 for $[\pm 45]_{2s}$, IM7/K3B composite test data.
2 *Physical changes.* An increase in crack density may be observed after exposure. Anomalous weight change behavior may be noted during cyclic exposure with the time to saturation and drying shortened by orders of magnitude following microcrack formation.
3 *Mechanical properties.* Fracture toughness, fatigue life, and linear visco-elastic creep are particularly sensitive to hygrothermal degradation (Chateauminois, 2000; Han and McKenna, 2000). Other engineering properties, such as residual tension and compression strength, and stiffness, are also affected to a lesser extent.

1.6 Summary

Ageing is ubiquitous in polymeric composites. However, the rate and degree of degradation, and the retention of properties over time is uniquely defined by material type, environmental and mechanical loading, and the length of the ageing time relative to the expected durability limit. In order to avoid the costly approach of a strictly empirical ageing study, analysis methods and careful experiments must be combined together and managed by a multidisciplinary group of scientists and engineers.

Chemical and physical ageing processes are broadly defined by thermal reversibility. Chemical ageing encompasses a range of environmentally driven degradation mechanisms which are irreversible. Conversely, physical ageing, a reversible process, is associated with free volume evolution and the change in properties relative to an equilibrium state. Ageing is a time-dependent process that requires time-dependent testing and analysis methods. Many analysis methods are based on the principles of polymer viscoelasticity. Time–temperature superposition and time–ageing time superposition are proven methods for assessing long-term ageing performance, under uniform conditions, from short-term data. These methods have shown a great deal of utility in test programs that rely on elevated temperature (sub-T_g) as the accelerant. Less is known about the validity of the methods for other types of accelerated environments (e.g. moisture, solvents). However, some key studies have indicated that the methods may have great utility. Combining the time-based superposition methods with the correct experimental program can lead to tremendous time savings when screening materials for long-term performance.

Increasing the temperature accelerates all thermally activated rate processes and will also reduce the activation energy of chemical bond rupture in the polymer. Elevated temperature (sub-T_g) will also increase the free volume in the polymer, hence decreasing the time needed to age to thermodynamic equilibrium. Increased temperature is also usually associated with decreases in both strength and stiffness in PMCs and will lead to increased ductility and strain to failure. Unfortunately, the use of elevated temperature for acceleration of the ageing mechanism(s) may promote degradation mechanisms that do not occur at use-environment temperatures, or may alter the rates so that degradation may not be accelerated proportionally.

For chemical ageing mechanisms, mechanical load, or stress, may increase the probability of bond rupture within the polymer. Residual stress or the externally applied stress on the chemical bond can accelerate chain scission caused by chemical reaction. It has also been found that stress can alter the effective activation energy for a chemical reaction. Increased stress has traditionally been used as the primary means for accelerating mechanical degradation. The occurrence and growth of microcracks, fiber breaks, and delamination will all be accelerated through the application of increased static or fatigue stress.

Aside from elevated temperature and stress, secondary accelerators such as moisture, partial pressure of oxygen, geometry, and layup should be considered for PMCs. One must also consider that ageing performed in the standard environment may actually represent an accelerated ageing case for material systems that do not operate in the standard environment. As an example, consider real-time isothermal ageing used to establish baseline conditions. For this example, the level of oxygen concentration in laboratory air in ageing ovens exceeds by several orders of magnitude the concentration of oxygen that a laminate on a supersonic aircraft would see when it is at altitude undergoing operational load.

Accelerated testing can reduce the time required and the cost of durability-based material characterization by facilitating material screening and suggesting key degradation mechanisms associated with long-term durability. Accelerated testing can speed up the ageing behavior of the material by influencing the processes of mechanical degradation, chemical ageing, and physical ageing. The rates and degree of interaction of these three processes are dependent on material type, environmental degradation factors, and test methods.

1.7 References

BAO, L.-R. and A. F. YEE (2002). 'Moisture Diffusion and Hygrothermal Aging in Bismaleimide Matrix Carbon Fiber Composites – Part I: Uni-weave Composites.' *Composite Science and Technology* **62**: 2099–2110.

BOUKHOULDA, B. F., E. ADDA-BEDIA, and K. MADANI (2006). 'The Effect of Fiber Orientation Angle in Composite Materials on Moisture Absorption and Material Degradation after Hygrothermal Ageing.' *Composite Structures* **74**: 406–418.

BRADSHAW, R. D. and L. C. BRINSON (1997a). 'Physical Aging in Polymers and Polymer Composites: An Analysis and Method for Time-Aging Time Superposition.' *Polymer Engineering and Science* **37**(1): 31–44.

BRADSHAW, R. D. and L. C. BRINSON (1997b). 'Recovering Nonisothermal Physical Aging Shift Factors via Continuous Test Data: Theory and Experimental Results.' *Journal of Engineering Materials and Technology* **119**(July): 233–241.

BRADSHAW, R. D. and L. C. BRINSON (1997c). 'A Sign Control Method for Fitting and Interconverting Material Functions for Linearly Viscoelastic Solids.' *Mechanics of Time-Dependent Materials* **1**: 85–108.

BRINSON, L. C. and T. S. GATES (1995). 'Effects of Physical Aging on Long Term Creep of Polymers and Polymer Matrix Composites.' *International Journal of Solids and Structures* **32**(6/7): 827–846.

BRISTOW, J. W. (2001). *Ageing Airframes, A Regulatory View from Europe*. 5th Joint NASA/FAA/DoD Conference on Aging Aircraft, Orlando, Florida.

CHATEAUMINOIS, A. (2000). Interactions between moisture and flexural fatigue damage in unidirectional glass/epoxy composites. In *Recent Developments in Durability Analysis of Composite Systems*. A. H. Cardon, H. Fukuda, K. L. Reifsnider, and G. Verchery (Eds). Rotterdam A.A. Balkema: pp. 159–167.

COLIN, X., C. MARAIS, J. L. COCHON, and J. VERDU (2000). Kinetic modeling of weight changes during the isothermal oxidative ageing of bismaleimide matrix. In *Recent Developments in Durability Analysis of Composite Systems*. A. H. Cardon, H. Fukuda, K. L. Reifsnider, and G. Verchery (Eds). Rotterdam, A.A. Balkema: pp. 49–54.

COMMITTEE ON HIGH SPEED RESEARCH, AERONAUTICS AND SPACE ENGINEERING BOARD, COMMISSION ON ENGINEERING AND TECHNICAL SYSTEMS, NATIONAL RESEARCH COUNCIL (2007). *U.S. Supersonic Commercial Aircraft: Assessing NASA's High Speed Research Program*. Washington, DC, National Academy Press.

CREESE, R. C. and H. GANGARAO (Eds) (1999). A Conference on Polymer Composites: Infrastructure Renewal and Economic Development. Lancaster, Pennsylvania.

CRISTESCU, N. and I. SULICIU (1982). *Viscoplasticity*. The Hague, Martinus Nijhoff.

DAVID, A. and D. SIMS (1983). *Weathering of Polymers*. London, Applied Science Publishers.

DRAGONE, T. L. and P. A. HIPP (1998). 'Materials Characterization and Joint Testing on the X-34 Reusable Launch Vehicle.' *Sampe Journal* **34**(5): 7–20.

DROZDOZ, A. D. (2001). 'Physical Aging and Nonlinear Viscoelasticity of Amorphous Glassy Polymers.' *Computational Materials Science* **21**: 197–213.

FERRY, J. D. (1980). *Viscoelastic Properties of Polymers*. New York, John Wiley and Sons, Inc.

FINDLEY, W. N., J. S. LAI, and K. ONARAN (1976). *Creep and Relaxation of Nonlinear Viscoelastic Materials*. Toronto, Canada, North-Holland Publishing Company.

FREEMAN, D. C., T. A. TALAY, and R. E. AUSTIN (1997). 'Reusable Launch Vehicle Technology Program.' *Acta Astronautica* **41**(11): 777–790.

GATES, T. S. (2003). On the Use of Accelerated Test Methods for Characterization of Advanced Composite Materials, NASA.

GATES, T. S., L. C. BRINSON, K. S. WHITLEY, and T. BAI (1999). Stress relaxation of IM7/K3B: the effects of elevated temperature and aging. In *Time Dependent and Nonlinear Effects in Polymers and Composites STP 1357*. R. A. Schapery and C. T. Sun (Eds). West Conshohocken, Pennsylvania, American Society for Testing and Materials.

GATES, T. S. and M. FELDMAN (1996). Effects of physical aging at elevated temperatures on the viscoelastic creep of IM7/K3B. In *Composite Materials: Testing and Design*, vol. 12, ASTM STP 1274. R. B. Deo and C. R. Saff (Eds). Philadelphia, Pennsylvania, American Society for Testing and Materials: pp. 7–36.

GATES, T. S., D. R. VEAZIE, and L. C. BRINSON (1997). 'Creep and Physical Aging in a Polymeric Composite: Comparison of Tension and Compression.' *Journal of Composite Materials* **31**(24): 2478–2505.

GRAMOLL, K. C., D. A. DILLARD, and H. F. BRINSON (1990). Thermoviscoelastic characterization and prediction of kevlar/epoxy composite laminates. In *Composite Materials: Testing and Design*, vol. 9, ASTM 1059. S. P. Garbo (Ed.). Philadelphia, Pennsylvania, American Society for Testing and Materials: pp. 477–493.

HAN, M.-H. and J. A. NAIRN (2003). 'Hygrothermal Aging of Polyimide Matrix Composite Laminates.' *Composites: Part A* **34**: 979–986.

HAN, W. H. and G. B. MCKENNA (2000). The influence of moisture on the physical aging response of epoxy: experimental results and modeling considerations. In *Recent Developments in Durability Analysis of Composite Systems*. A. H. Cardon, H. Fukuda, K. L. Reifsnider, and G. Verchery (Eds). Rotterdam, A.A. Balkema: pp. 153–157.

HASTIE, R. L. and D. H. MORRIS (1992). The effect of physical aging on the creep response of a thermoplastic composite. In *High Temperature and Environmental Effects in Polymer Matrix Composites*, ASTM STP 1174. C. Harris and T. Gates (Eds). Philadelphia, Pennsylvania, American Society for Testing and Materials: pp. 163–185.

HSUAN, Y. G. and R. M. KOERNER (2001). Lifetime prediction of polyolefin geosynthetics utilizing acceleration tests based on temperature. In *Long Term Durability of Structural Materials*. P. J. M. Monteiro, K. P. Chong, J. Larsen-Basse, and K. Komvopoulos. Amsterdam, Elsevier Science Ltd: pp. 145–157.

HUTCHINSON, J. M. (1995). 'Physical Aging of Polymers.' *Progress in Polymer Science* **20**: 703–760.

JONES, R. M. (1975). *Mechanics of Composite Materials*. Washington, DC, Scripta Book Company.

KAMPF, G. (1986). *Characterization of Plastics by Physical Methods.* Munich, Hanser Publishers.

KIM, J., W. I. LEE, and S. W. TSAI (2002). 'Modeling of Mechanical Property Degradation by Short-term Aging at High Temperatures.' *Composites: Part B Engineering* **33**: 531–543.

LAFARIE-FRENOT, M. C. (2006). 'Damage Mechanisms Induced by Cyclic Ply-stresses in Carbon–Epoxy Laminates: Environmental Effects.' *International Journal of Fatigue* **28**: 1202–1216.

LAWFORD, K. (2002). Airbus A300 Extended Service Goals. A summary of the test program used to support the development of widespread fatigue damage analysis methods. 6th Joint FAA/DoD/NASA Aging Aircraft Conference, San Francisco, California.

LÉVÊQUE, D., A. SCHIEFFER, A. MAVEL, and J.-F. MAIRE (2005). 'Analysis of How Thermal Aging Affects the Long-term Mechanical Behavior and Strength of Polymer–Matrix Composites.' *Composites Science and Technology* **65**: 395–401.

LIAO, K., R. I. ALTKORN, S. M. MILKOVICH, J. M. FILDES, J. GOMEZ, C. R. SCHULTHEISZ, D. L. HUNSTON, and L. C. BRINSON (1997). 'Long-term Durability of Glass-fiber Reinforced Composites in Infrastructure Applications.' *Journal of Advanced Materials* **28**(3): 54–63.

MCKENNA, G. B. (1994). 'On the Physics Required for the Prediction of Long Term Performance of Polymers and Their Composites.' *Journal of Research of the National Institute of Standards and Technology* **99**(2): 169–189.

MURRAY, S., C. HILLMAN, and M. PECHT (2003). 'Environmental Aging and Decohesion of Siloxane–Polyimide–Epoxy Adhesive.' *IEEE Transactions on Components and Packaging Technologies* **26**(3): 524–531.

NUSS, M. (2001). The FAA Continued Operational Safety Program for General Aviation. 5th Joint NASA/FAA/DoD Conference on Aging Aircraft, Orlando, Florida.

OHNO, S., M.-H. LEE, K. Y. LIN, and F. S. OHUCHI (2000). 'Thermal Degradation of IM7: BMI5260 Composite Materials: Characterization by X-ray Photoelectron Spectroscopy.' *Materials Science and Engineering* **A293**: 88–94.

ROY, S., W. XU, S. PATEL, and S. CASE (2001). 'Modeling of Moisture Diffusion in the Presence of Bi-axial Damage in Polymer Matrix Composite Laminates.' *International Journal of Solids and Structures* **38**: 7627–7641.

SARGENT, J. P. (2005). 'Durability studies for aerospace applications using peel and wedge tests.' *International Journal of Adhesion and Adhesives* **25**: 247–256.

SCHAPERY, R. A. (1974). Viscoelastic behavior and analysis of composite materials. In *Mechanics of Composite Materials*. G. P. Sendeckyj (Ed.). Academic Press: New York, pp. 85–168.

SCHAPERY, R. A. (1975). Inelastic behavior of composite materials. In *ASME Winter Annual Meeting*. C. T. Herakovich (Ed.). Houston, Texas, ASME: AMD-13, pp. 122–150.

SCHAPERY, R. A. (1981). On viscoelastic deformation and failure behavior of composite materials with distributed flaws. In *Advances in Aerospace Structures and Materials*, The Winter Annual Meeting of the American Society of Mechanical Engineers, Washington, DC.

SCHAPERY, R. A. (1982). Models for damage growth and fracture in nonlinear viscoelastic particulate composites. In Ninth US National Congress of Applied Mechanics, Cornell University, Ithaca, New York. The American Society of Mechanical Engineers.

SCHAPERY, R. A. (1984). 'Correspondence Principles and a Generalized J Integral for Large Deformation and Fracture Analysis of Viscoelastic Media.' *International Journal of Fracture* **25**: 195–223.

SCHNABEL, W. (1981). *Polymer Degradation: Principles and Practical Applications.* New York, Hanser International.

SCHOEPPNER, G. A., G. P. TANDON, and E. R. RIPBERGER (2007). 'Anisotropic Oxidation and Weight Loss in PMR-15 Composites.' *Composites: Part A, Applied Science and Manufacturing* **38**: 890–904.

SCOTT, D. W., J. S. LAI, and A. H. ZUREICK (1995). 'Creep Behavior of Fiber Reinforced Polymeric Composites: A Review of the Technical Literature.' *Journal of Reinforced Plastics and Composites* **14**: 588–617.

SPRINGER, G. S. (Ed.) (1981). *Environmental Effects on Composite Materials*. Westport, Connecticut, Technomic Publishing.

STRUIK, L. C. E. (1978). *Physical Aging in Amorphous Polymers and Other Materials*. New York, Elsevier Scientific Publishing Company.

SULLIVAN, J. L. (1990). 'Creep and Physical Aging of Composites.' *Composites Science and Technology* **39**: 207–232.

SULLIVAN, J. L., E. J. BLAIS, and D. MOUSTON (1993). 'Physical Aging in the Creep Behavior of Thermosetting and Thermoplastic Composites.' *Composites Science and Technology* **47**: 389–403.

TSOTSIS, T. K., S. KELLER, K. LEE, J. BARDIS, and J. BISH (2001). 'Aging of Polymeric Composite Specimens for 5000 hours at Elevated Pressure and Temperature.' *Composites Science and Technology* **61**: 75–86.

WANG, J. Z., D. A. DILLARD, M. P. WHOTT, F. A. KAMKE, and G. L. WILKES (1990). 'Transient Moisture Effects in Fibers and Composite Materials.' *Journal of Composite Materials* **24**(September): 994–1009.

WANG, Y. and T. H. HAHN (2007). 'AFM Characterization of the Interfacial Properties of Carbon Fiber Reinforced Polymer Composites Subjected to Hygrothermal Treatments.' *Composites Science and Technology* **67**: 92–101.

WHITE, J. R. and A. TURNBULL (1994). 'Review – Weathering of Polymers: Mechanisms of Degradation and Stabilization, Testing Strategies and Modeling.' *Journal of Materials Science* **29**: 584–613.

ZHENG, S. F. and G. J. WENG (2002). 'A New Constitutive Equation for the Long-term Creep of Polymers Based on Physical aging.' *European Journal of Mechanics A/Solids* **21**: 411–421.

ZHENG, Y. and G. B. MCKENNA (2003). 'Structural Recovery in a Model Epoxy: Comparison of Responses after Temperature and Relative Humidity Jumps.' *Macromolecules* **36**: 2387–2396.

ZHOU, J. (1993). 'A Constitutive Model of Polymer Materials Including Chemical Aging and Mechanical Damage and its Experimental Verification.' *Polymer* **34**(20): 4252–4256.

2

Ageing of glass–ceramic matrix composites

K. PLUCKNETT, Dalhousie University, Canada

2.1 Introduction

Traditionally, ceramics are viewed as being brittle materials. In this context, they are susceptible to failure from flaws or damage, either surface or internal, and their mechanical performance can be expected to exhibit some degree of variability. The drive to develop advanced ceramics that have more reproducible mechanical properties has progressed on several fronts simultaneously. From the perspective of eliminating processing flaws, new approaches to the forming of ceramic green bodies have been based upon colloidal processing technologies (Lange, 1989; Lewis, 2000). Following this methodology allows the potential elimination of flaws in the form of agglomerates or voids, with the consequent improvement of reliability (Lange, 1989). A second approach to improving reliability involved the generation of an increased understanding of the significance of surface flaws, and the subsequent development of improved ceramic machining processes (Marinescu, 2006). An extension of this line of thought involved the development of materials that could tolerate limited levels of surface damage, due to the generation of some degree of surface compressive stress during processing (e.g. functional gradient structures) which opposes crack growth (Chan, 1997; Hbaieb *et al.*, 2007). Each of these approaches resulted in incremental improvements in the reliability and mechanical performance of advanced ceramics. However, arguably the most significant advance came with the development of fibre-reinforced glasses, and subsequently glass-ceramics, in the 1970s and 1980s. The development of these materials resulted in a significant re-assessment of the design philosophy for advanced ceramics.

2.1.1 Fibre-reinforced glasses and glass-ceramics

The potential benefits of incorporating strong fibres into brittle matrices were first noted more than 30 years ago (Kelly, 1973). Early brittle matrix composites of this type were based upon the incorporation of carbon fibres into SiO_2-based matrices (Crivelli-Visconti and Cooper, 1969; Phillips *et al.*, 1972; Phillips, 1972, 1974; Sambell *et al.*, 1972a, 1972b). From the perspective of potential high-temperature applications, a major advance came in the mid 1970s with the development by Yajima and colleagues of fine-diameter SiC-based fibres (Yajima *et al.*, 1976, 1978a, 1978b; Hasegawa *et al.*, 1980). These fibres were subsequently developed commercially by Nippon Carbon under the tradename Nicalon™. Early grades of Nicalon fibre, such as NLM-102 and NLM-202 contained moderately high oxygen contents, but there are now several manufacturers of such fibres that provide materials with compositions close to stoichiometric SiC. Table 2.1 summarises the various grades of fine SiC-based fibres that have been used in ceramic matrix composite manufacture, along with some of their physical properties.

While the first matrix composites utilised glass matrices, in the early 1980s prototype composites were developed that utilised glasses that could be easily devitrified using the glass–ceramic process. Table 2.2 summarises some typical glass–ceramic compositions that were used in these materials. Invariably, the glass–ceramic matrix compositions are selected such that near-complete crystallisation is possible, and consequently many of the early systems had been developed following studies of monolithic glass–ceramics for more conventional applications. During this period, extensive work at the United Technologies Research Center led to the development of a wide variety of glass and glass–ceramic matrix composites, based on both carbon and fine-diameter SiC-based fibres (Prewo and Brennan, 1980, 1982; Brennan and Prewo, 1982; Prewo, 1989). Some of the first glass–ceramic matrix systems to be developed through that work were based on $Li_2O—Al_2O_3—SiO_2$ (LAS) compositions. Flexure strengths approaching 1000 MPa were achieved for unidirectional Nicalon/LAS-I composites (Brennan and Prewo, 1982), which was close to five times greater than for the unreinforced matrix. The fracture surface of these materials was noted to be highly fibrous, and chemical analysis of the fibre/matrix interface using scanning Auger microscopy showed the formation of a carbon-rich interphase (Brennan, 1986, 1988). This interphase was found to be absent in the case of weaker composites that did not produce fibrous fracture surfaces. Consequently, the presence of a carbon-rich layer was observed to be necessary to allow fibre/matrix debonding and sliding, both of which are features required for pseudo-ductile composite failure behaviour.

2.1.2 Macromechanical behaviour

Figure 2.1(a) demonstrates a schematic 'ideal' load–deflection curve for a unidirectional fibre-reinforced glass or ceramic matrix composite. It is notable that, in many ways, this material behaves in a similar manner to a metal. There is an apparent ductile nature to the material, with a knee in

Table 2.1 The properties of some SiC-based fine-diameter fibres. Adapted from Bunsell *et al.* (1999) and manufacturers' data sheets

Trademark	Manufacturer	Composition (wt%)	Diameter (μm)	Density (g/cm^3)	Strength (GPa)	Elastic modulus (GPa)	Failure strain (%)
Nicalon NLM-202	Nippon Carbide	56.6 Si, 31.7 C, 11.7 O	14	2.55	2.0	190	1.05
Hi-Nicalon	Nippon Carbide	62.4 Si, 37.1 C, 0.5 O	14	2.74	2.6	263	1.0
Hi-Nicalon Type S	Nippon Carbide[1]	69 Si, 31 C, 0.2 O	12	3.10	2.6	420	
Tyranno Lox-M	Ube Industries	54.0 Si, 31.6 C, 12.4 O, 2.0 Ti	8.5	2.37	2.5	180	1.4
Tyranno Lox-E	Ube Industries	54.8 Si, 37.5 C, 5.8 O, 1.9 Ti	11	2.39	2.9	199	1.45
Tyranno ZMI	Ube Industries	56.6 Si, 34.8 C, 7.6 O, 1.0 Zr	11	2.48	3.4	200	1.7
Tyranno ZE	Ube Industries	58.6 Si, 38.4 C, 1.7 O, 1.0 Zr	11	2.55	3.5	233	1.5
Tyranno SA	Ube Industries[2]	67.8 Si, 31.3 C, 0.3 O, <2 Al	10, 7.5	3.10	2.8	380	0.7
HPZ	Dow Corning	59 Si, 28 N, 10 C, 3 Zr	10–12	2.3–2.5	1.7–2.1	180–230	~1
Sylramic	Dow Corning[1]	96 SiC, 3 TiB_2, 1 B_4C, 0.3 O	10	>2.95	>2.75	>310	

[1] Data from http://www.coiceramics.com.
[2] Data from http://northamerica.ube.com.

Table 2.2 Summary of typical glass–ceramic systems examined in the manufacture of advanced fibre-reinforced composites. Adapted from Prewo (1989)

Matrix	Major constituents	Minor constituents	Major crystal phase	Maximum operational temperature (°C)
BMAS	BaO, MgO, Al_2O_3, SiO_2		Barium osumilite	1250
CAS-II	CaO, Al_2O_3, SiO_2	ZrO_2	Anorthite	~1200
LAS-I	Li_2O, Al_2O_3, MgO, SiO_2	ZnO, ZrO_2, BaO	β-Spodumene	1000
LAS-II	Li_2O, Al_2O_3, MgO, SiO_2, Nb_2O_5	ZnO, ZrO_2, BaO	β-Spodumene	1100
LAS-III	Li_2O, Al_2O_3, MgO, SiO_2, Nb_2O_5	ZrO_2	β-Spodumene	1200
MAS	MgO, Al_2O_3, SiO_2	BaO	Cordierite	1200
Ternary mullite	BaO, Al_2O_3, SiO_2		Mullite	~1500
Hexacelsian	BaO, Al_2O_3, SiO_2		Hexacelsian	~1700

the load–deflection curve, and even some form of large-scale yielding prior to failure. In comparison, a conventional brittle material is linear elastic to failure, as shown in Fig. 2.1(a). In order to understand this mechanical response, it is necessary to look at both the macromechanical and micromechanical behaviour of the composite. For the case of a brittle material, even one containing fibres, failure invariably occurs from a single flaw (i.e. a crack), as shown in Fig. 2.1(b). However, for the case of a composite that exhibits the pseudo-ductile behaviour noted above, the failure response is far more complex, as shown in Fig. 2.1(c). In this instance, when the material is strained, load builds up in both the matrix and the fibres until, eventually, cracks form in the matrix. However, unlike the brittle case, these cracks do not penetrate into the fibres, but instead partially debond the interface between the fibre and matrix. Ultimately, the material exhibits an array of micro-cracks, with the majority of the applied load having been transferred to the intact fibres. As loading continues, the fibres will begin to fracture periodically, and then slide within the matrix. All of these processes require

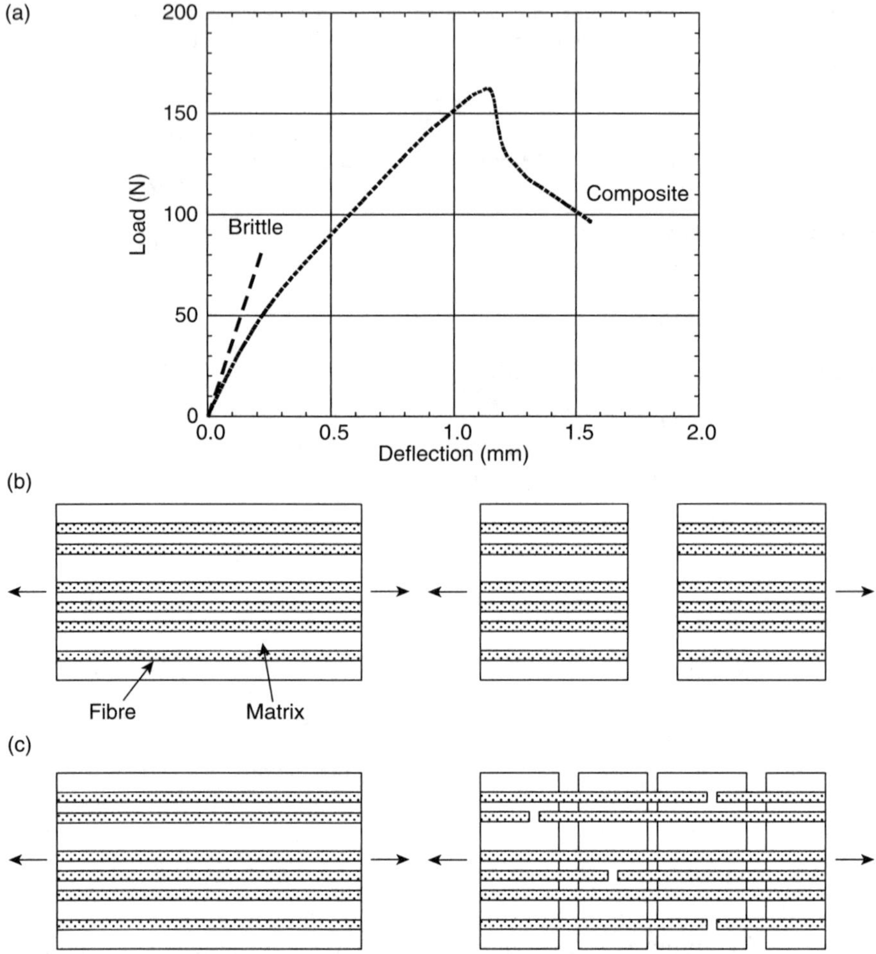

2.1 (a) Schematic load–deflection curves (in flexure) for a fibre-reinforced ceramic showing both composite behaviour and brittle behaviour (e.g. for the case of a strong fibre–matrix bond). (b) Schematic representation of failure for the case of a brittle ceramic matrix composite, showing catastrophic failure from a single crack. (c) Schematic representation of multiple matrix cracking for the case of composite behaviour, where intact fibres bridge the matrix cracks.

additional energy, which in turn leads to increased fracture resistance. This process of matrix micro-cracking and load transfer requires specific interface properties, as discussed in Section 2.1.3 below.

It was noted early on in the study of fibre-reinforced glasses that matrix micro-cracking occurs well before the onset of composite failure. The initial work of Aveston, Cooper and Kelly (1971) developed a simple model, known as the ACK model, to predict the matrix micro-cracking stress, σ_y, based on a number of materials parameters:

$$\sigma_y = \left[\frac{12\tau\gamma_m E_c^2 E_f V_f^2}{E_m^2 V_m R}\right]^{1/3} \quad [2.1]$$

where τ is the fibre/matrix frictional sliding stress, γ_m is the fracture surface energy of the matrix material, R is the fibre radius, and E and V are the elastic modulus and volume fraction of the constituent phases, respectively (in this last instance, subscripts c, m and f represent the composite, matrix and fibre respectively). The ACK model makes several major assumptions in its implementation, notably that (a) the fibre failure strain exceeds that of the matrix, (b) the fibres can effectively take the entire applied load, such that there is no support from the matrix and (c) a frictional interface is present between the fibre and matrix. Using the ACK model, and knowledge of the composite microstructure parameters and matrix fracture energy, it is possible to estimate the fibre/matrix frictional sliding stress, τ, using tensile load–deflection data where the micro-cracking stress can be experimentally measured.

2.1.3 Interfacial micromechanics

As outlined in the previous section (Section 2.1.2), a necessary requirement for obtaining a composite failure mode is that the fibres can both debond and slide during deformation. A critical factor in promoting this behaviour is the fibre/matrix frictional sliding resistance, τ. If τ is too large, then load transfer from the fibre to the matrix is favoured. In this instance the fibres will tend to fail at or near the crack plane. As a consequence, fibre sliding and pull-out will not occur to any significant extent, and the failure mode will be essentially brittle. It can therefore be seen that lower values of τ will tend to promote interfacial failure between the fibre and matrix, allowing fibre/matrix debonding and sliding to occur. It has subsequently been demonstrated that the tendency to either fibre/matrix debonding or fibre failure depends largely on the ratio of the fracture energy of the interface and the fibre, G_i and G_f respectively (Evans *et al.*, 1989; He and Hutchinson, 1989; He *et al.*, 1994). For the promotion of fibre/matrix debonding it is invariably necessary that:

$$\frac{G_i}{G_f} \leq 0.25 \quad [2.2]$$

This simple relationship assumes that the fibre and matrix possess essentially the same elastic modulus, and can be relaxed somewhat when the fibre elastic modulus is significantly higher than that of the matrix (He and Hutchinson, 1989).

Clearly, given the criteria for debonding outlined in equation [2.2] above, a means of accurately determining the interfacial fracture, or debonding, energy is necessary. There are several techniques available for the determination of fibre/matrix interfacial properties (Marshall and Oliver, 1987; Cho *et al.*, 1991; Mackin and Zok, 1992). Probably the most widely applied of these methods is the fibre push-in test, which is based upon a constant shear–stress model developed by Marshall and Oliver (1987). In its most simple form, assuming frictional sliding with no fibre/matrix bond, this model allows the frictional sliding stress, τ, to be determined following the relationship:

$$u = \frac{F^2}{4\pi^2 R^3 E_f \tau} \qquad [2.3]$$

where u is the apparent fibre end displacement (i.e. the actual indentor displacement, less the indentation depth of the indentor tip into the fibre), F is the applied load, R is the fibre radius and E_f is the fibre modulus. In the case of a fibre that is debonded during push-in, this relationship is modified to incorporate the fibre/matrix debond energy, Γ, as follows:

$$u = \frac{F^2}{4\pi^2 R^3 E_f \tau} - \frac{2\Gamma}{\tau} \qquad [2.4]$$

The Marshall and Oliver (1987) treatment benefits from incorporating residual stress effects, but does not account for either the Poisson expansion of the fibre under loading, or the fibre surface roughness. Subsequent to the development of this model, several extensions of this approach have been developed to account for these limitations. Notably, Hseuh (1993) developed a model taking into account Poisson ratio effects, while Parthasarathy *et al.* (1994) developed a comprehensive model to account for fibre surface roughness.

2.1.4 Carbon and boron nitride interphases

The requirement for a low-toughness interface to be present between the fibre and matrix in a ceramic matrix composite system presents certain challenges from the perspective of materials processing. For the case of glass and glass–ceramic composites it was noted in early studies that a carbon-rich layer is often formed between the fibre and matrix during the high-temperature, hot-pressing stage of processing (Brennan, 1986, 1988; Cooper and Chyung, 1987; Benson *et al.*, 1988; Bonney and Cooper, 1990). The presence of this compliant carbon-rich layer leads to fibre/matrix debonding and fibre sliding, both mechanisms that dissipate energy during crack growth. Composites processed in such a way that this carbon-rich layer did not form showed poor mechanical properties and failed in a brittle manner (Prewo, 1989). The growth of this layer has been proposed to occur following one of two mechanisms (Cooper and Chyung, 1987; Benson *et al.*, 1988; Bonney and Cooper, 1990):

$$SiC_{(s)} + O_{2(g)} \rightarrow SiO_{2(s,\,l)} + C_{(s)} \qquad [2.5]$$

or

$$SiC_{(s)} + 2CO_{(g)} \rightarrow SiO_{2(s,\,l)} + 3C_{(s)} \qquad [2.6]$$

Formation of a carbon-rich interphase has been noted in a wide variety of composite systems, including matrices based on LAS, MAS, CAS, BAS, BMAS and YMAS (Y_2O_3-MgO-Al_2O_3-SiO_2) glass–ceramics (Brennan, 1986, 1988; Chaim and Heuer, 1987, 1991; Murthy *et al.*, 1989; Lewis and Murthy, 1991; Bleay and Scott, 1992; Pharaoh *et al.*, 1993; Plucknett *et al.*, 1995a, 1995b, 1995c; Vicens *et al.*, 1995, 2003) and borosilicate glasses (Murthy and Lewis, 1989). It was also shown that the formation of carbon-rich layers in these composite systems was not simply a result of the non-stoichiometry of the Nicalon or Tyranno fibres, as thin carbon layers were also observed to form at the interface between pure α-SiC and a BaO-containing cordierite glass–ceramic (Chaim and Heuer, 1991). While the *in-situ* formation of carbon-rich layers can be viewed as somewhat fortuitous, the relatively poor oxidation of carbon has led to extensive study of alternative interphase materials, and in particular boron nitride (BN)-based coatings deposited by chemical vapour deposition (Sun and Nutt, 1994; Brennan *et al.*, 1995; Sun *et al.*, 1997a). BN possesses a hexagonal layered crystallographic structure that is similar to the graphitic form of carbon, and hence leads to moderately low interfacial debond energies and sliding stresses.

2.1.5 Applications of advanced ceramic matrix composites

When contemplating potential applications of advanced ceramic-based composites, it is clear that their mechanical performance is significantly different from conventional 'monolithic' ceramics. These materials generally exhibit a degree of damage tolerance that cannot be envisaged for conventional materials, which offers the potential for application in scenarios where the mechanical performance variability of monolithic ceramics precludes their use. While glass–ceramic matrix composites have not found widespread commercial usage, the subsequent generations of advanced non-oxide and oxide composite materials are starting to find commercial niche applications. Arguably the most well-publicised uses of these materials are in advanced friction environments, notably as lightweight brake discs in performance automotive and, potentially, aerospace applications (Krenkel *et al.*, 2002; Zhang *et al.*, 2005; Zuber and Heidenreich, 2006). These materials, based primarily on carbon-fibre reinforcement of a silicon carbide/carbon matrix, are produced using relatively low-cost methods such as melt-infiltration (Zuber and Heidenreich, 2006). Similar non-oxide composites, as well as recent generations of all-oxide composites, are also being investigated for aerospace use, most notably as static components in advanced gas-turbine engines (Christin, 2002; Barnard *et al.*, 2004; Naslain, 2004, 2005; Schmidt *et al.*, 2005; van Roode *et al.*, 2007). Examples of such applications include combustor liners, nozzles, flaps and blade tip seals. Stressed, lifetime evaluation tests have been conducted to in excess of 12 000 h, with considerable success, on both SiC/SiC and all-oxide composites (van Roode *et al.*, 2007). It is notable that the ultimate aim of these studies is to develop ceramic matrix composite materials that can safely sustain in excess of 30 000 h operational lifetime (Barnard *et al.*, 2004; van Roode *et al.*, 2007). As will be shown in the subsequent sections, when contemplating the development of composites with one or more non-oxide components, it is necessary to also consider appropriate thermal protection coatings.

2.2 Composite fabrication

One of the major advantages of preparing ceramic matrix composites through the glass–ceramic route is that the matrix material can be processed in the form of a fluid glass, which allows near theoretical density to be achieved through simple processing techniques such as uniaxial hot-pressing. The typical steps that are followed include: (a) glass powder slurry infiltration of the fibre tows, (b) winding of the tows, (c) cutting and lay-up of the impregnated tows, (d) hot-pressing of the laminate to achieve densification, (e) glass-to-ceramic heat treatment and (f) final machining. It is clear that there is a necessary degree of flow of the glass during hot-pressing, and therefore this stage must be conducted at a temperature above which the individual glass constituents have formed a viscous liquid. After hot-pressing an essentially theoretically dense composite can be obtained, which then requires a suitable heat treatment to convert the glass matrix to one that is polycrystalline. Typically this involves a two-stage nucleation and growth heat treatment, and nucleation aids are invariably used to promote this transformation.

2.3 Fast-fracture behaviour

In addition to assessment of room temperature mechanical behaviour, a number of early studies on glass–ceramic matrix composites evaluated the behaviour of the materials at elevated temperature in an oxidising environment (Prewo, 1989; Plucknett *et al.*, 1995a). Figure 2.2 shows the fast-fracture strength as a function of test temperature for a Tyranno/BMAS

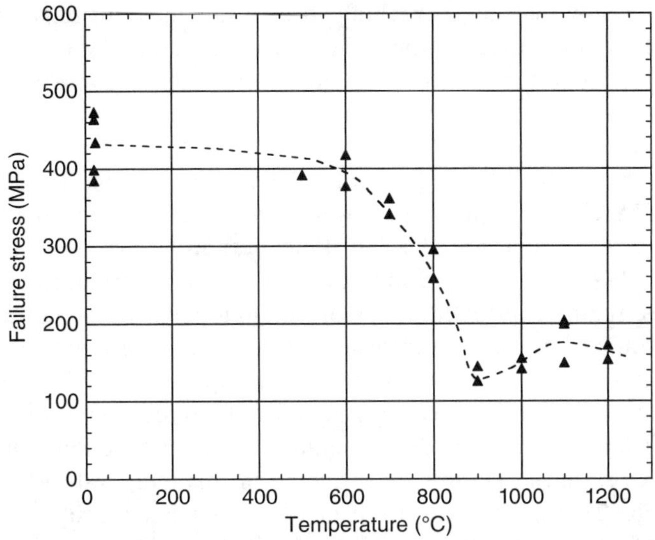

2.2 The effects of testing temperature upon the four-point flexure strength of Tyranno/BMAS tested in air. Adapted from Plucknett *et al.* (1995a).

glass–ceramic composite system. In this instance there is a clear strength reduction at test temperatures above ~800 °C, which is accompanied by a transition to a more brittle failure mode with reduced fibre pull-out. Generally similar observations have been made when testing other glass–ceramics matrix composites at elevated temperatures in air, using either flexure or tensile tests (Luh and Evans, 1987; Mandell *et al.*, 1987; Kahraman *et al.*, 1995; Gyekenyesi and Bansal, 2000; Yasmin and Bowen, 2002). Conversely, similar testing performed in an inert atmosphere such as argon did not show a strength decrease at these lower temperatures, but instead demonstrated plasticity at higher temperatures due to softening and creep of the glass–ceramic matrix phase (Prewo, 1989). It is clear from these early studies that oxidation-induced embrittlement is occurring when strength testing is conducted in air, and this may be attributed to oxidation of the compliant carbon interlayer and the fibre surface.

2.4 Long-term environmental ageing behaviour

While the initial investigation of the high-temperature mechanical behaviour of glass–ceramic matrix composites focused upon fast fracture, especially in oxidising environments, a number of subsequent studies have examined the effects of long-term ageing exposure on the stability of these materials. It has been shown that, under conditions of unstressed ageing, or ageing well below the micro-cracking stress, the behaviour falls into three approximate temperature regimes. These are outlined in the following sub-sections as: (a) high temperature, effectively between 900 and 1200 °C; (b) intermediate temperature, between approximately 600 and 900 °C; and (c) low temperature, between 300 and 600 °C.

Figure 2.3(a) highlights the general behaviour that is observed for both CAS and BMAS composites when aged at various temperatures between 450 and 1200 °C for a period of 100 h (Plucknett and Lewis, 1995; Plucknett *et al.*, 1995a, 1995b, 1995c), and then tested at room temperature; the three temperature regimes outlined above are highlighted on this figure. In particular, it is apparent that the intermediate temperature region is the one that is most severely affected by oxidation exposure. For both of these materials, this region is marked by a significant decrease in strength and a transition to brittle failure. Representative load–deflection curves, for Nicalon/CAS tested in flexure, after ageing in this temperature range, are shown in Fig. 2.3(b). This transition in behaviour is also reflected in the fracture surfaces, with brittle samples exhibiting minimal fibre pull-out in comparison with the as-received material, or the composite when aged above 1000 °C (Fig. 2.4(a) to (d)). Similarly, when examining the interfacial micromechanical properties, it is apparent that both the debonding energy, Γ, and the frictional sliding stress, τ, increase dramatically after intermediate temperature oxidation (Fig. 2.5). In this case data for Tyranno/BMAS are shown (Plucknett *et al.*, 1995a), although generally similar observations were also made for Nicalon/CAS (Plucknett *et al.*, 1995b, 1995c). The following sub-sections describe both the macro- and micromechanical behaviour in greater detail.

2.4.1 High-temperature stability

As demonstrated in Fig. 2.3, extended exposure to high-temperature oxidation (e.g. 900–1200 °C) does not result in a significant strength decrease for these specific composite systems. In this instance, strength is largely retained and the failure mode is one that is comparable with the as-fabricated material (Figs 2.4 and 2.5). Given that oxidation of both the carbon interphase *and* the fibre can be anticipated at such temperatures, it is clear that some form of intrinsic protection mechanism is in operation. Examination of the cross-section of the Nicalon/CAS composite, after unstressed exposure at 1200 °C for 500 h in air shows that the near-surface fibres are partially consumed by oxidation, and that there is a thick silicate oxide scale around these fibres (Fig. 2.6). However, just 30–40 μm in from the surface the fibres appear unaffected, and even demonstrate debonding due to thermal stress crack formation (Fig. 2.6(a)). This indicates the retention of the compliant carbon interlayer, even after such an extreme treatment, and is confirmed

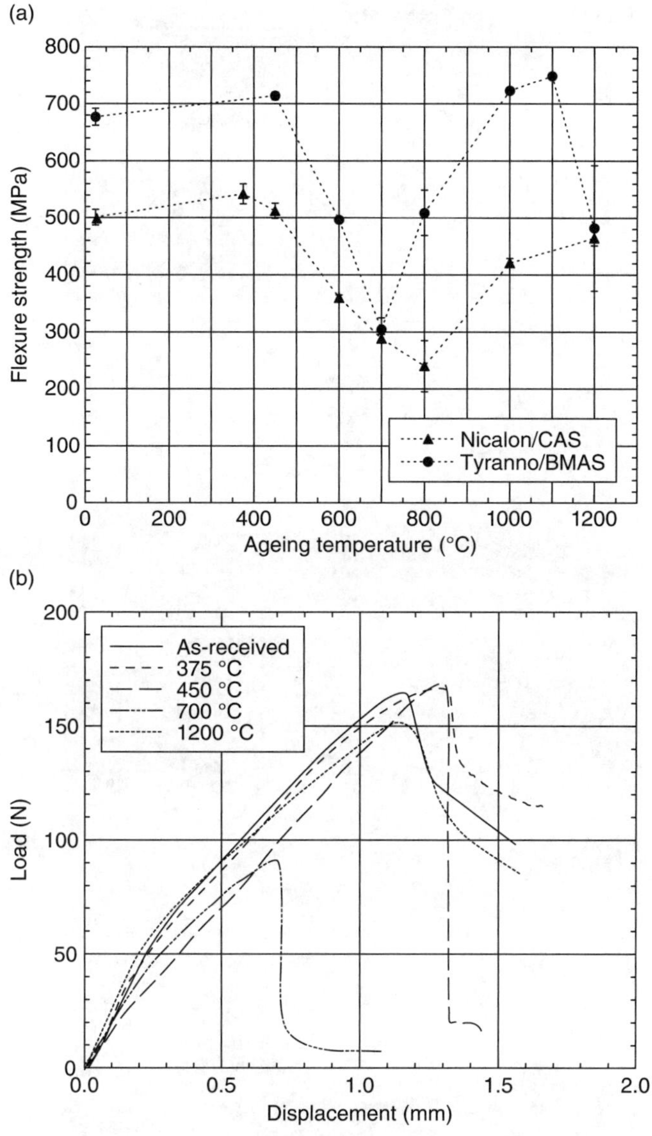

2.3 (a) The effects of ageing temperature, held for a period of 100 h, on the three-point flexure strength of Nicalon/CAS and Tyranno/BMAS. Adapted from Plucknett *et al.*, (1995a, 1995b, 1995c) and Plucknett and Lewis (1995). (b) Typical load–displacement curves obtained using three-point bend flexure tests after ageing at various temperatures for 100 h in air.

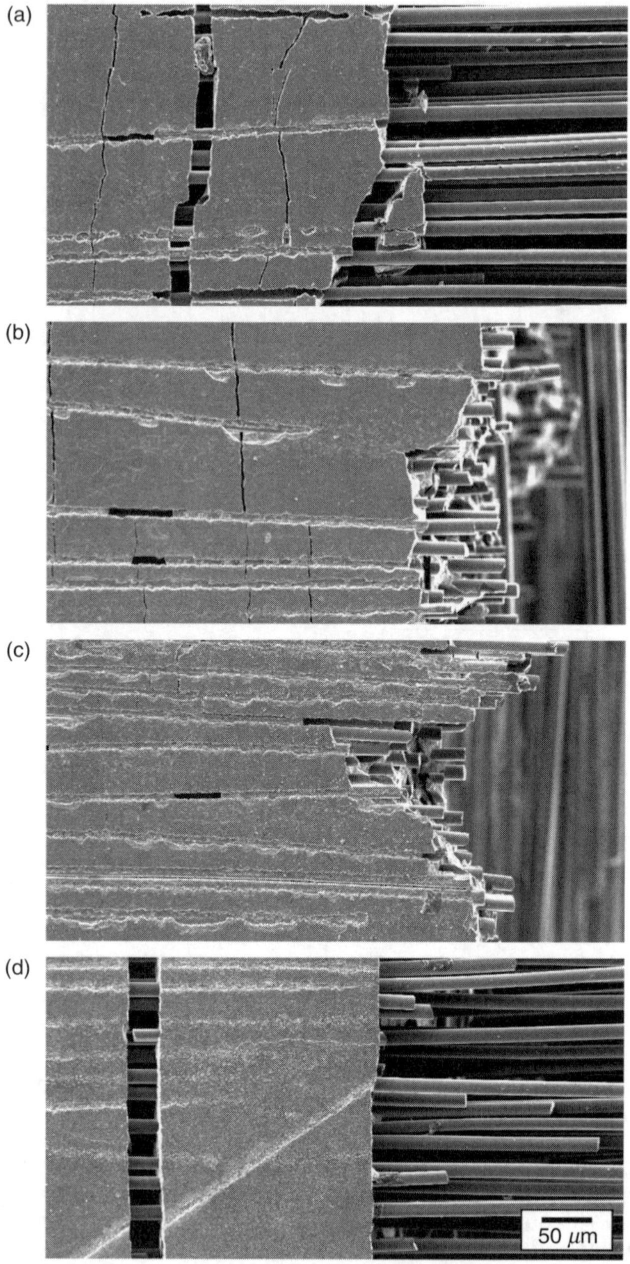

2.4 (a) Scanning electron microscopy (SEM) image of the as-received fracture surface, in comparison with SEM images of the tensile fracture surfaces after ageing at (b) 450 °C for 100 h, (c) 700 °C for 100 h and (d) 1000 °C for 100 h.

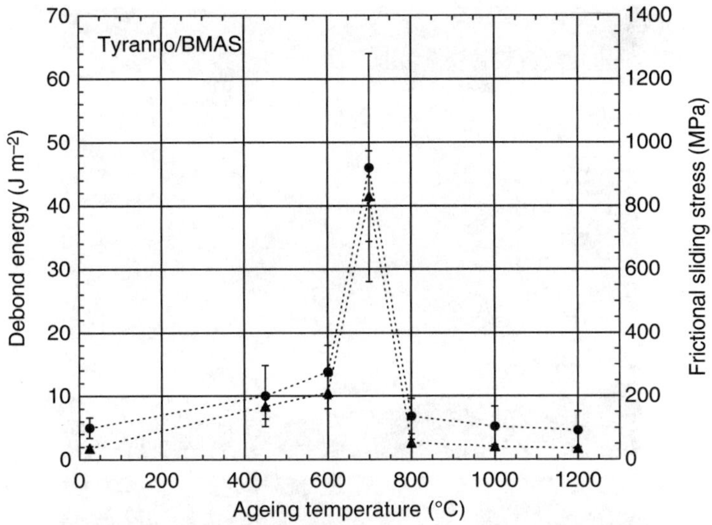

2.5 The effects of ageing temperature, held for a period of 100 h, on the fibre/matrix interfacial micro-mechanical properties of Tyranno/BMAS, as measured using the fibre push-in test (●, debond energy; ▲, frictional sliding stress). Adapted from Plucknett *et al.* (1995a).

by examination of the fracture surface where fibre pull-out is clearly noted just a few microns from the exposed surface (Fig. 2.6(b)). Under these conditions surface sealing occurs rapidly, protecting the underlying composite material. The mechanism of surface sealing is discussed in more detail in Section 2.5.2.

2.4.2 Intermediate temperature degradation

It is apparent from Fig. 2.3(a) that the intermediate temperature range (i.e. 600–900 °C) is the one that exhibits the greatest extent of degradation after oxidation ageing. Firstly, it is clear that strength is significantly reduced after ageing at 600, 700 or 800 °C for 100 h. Secondly, it is also clear that the materials aged in this temperature range are brittle in behaviour (Fig. 2.3(b)), and that they exhibit minimal fibre pull-out after flexure testing (Fig. 2.4(c)). For such behaviour to occur it is clear that the interfacial structure has been severely compromised in some manner. In these particular cases, for Tyranno/BMAS and Nicalon/CAS, transmission electron microscopy and scanning Auger microscopy have been used to demonstrate that the compliant carbon interphase has been partially or completely removed through oxidation (Plucknett *et al.*, 1995a, 1995b, 1995c). In addition, partial or complete bonding of the fibre to the matrix has occurred, through the formation of SiO_2 bridges. The mechanism of carbon layer

(a)

(b)

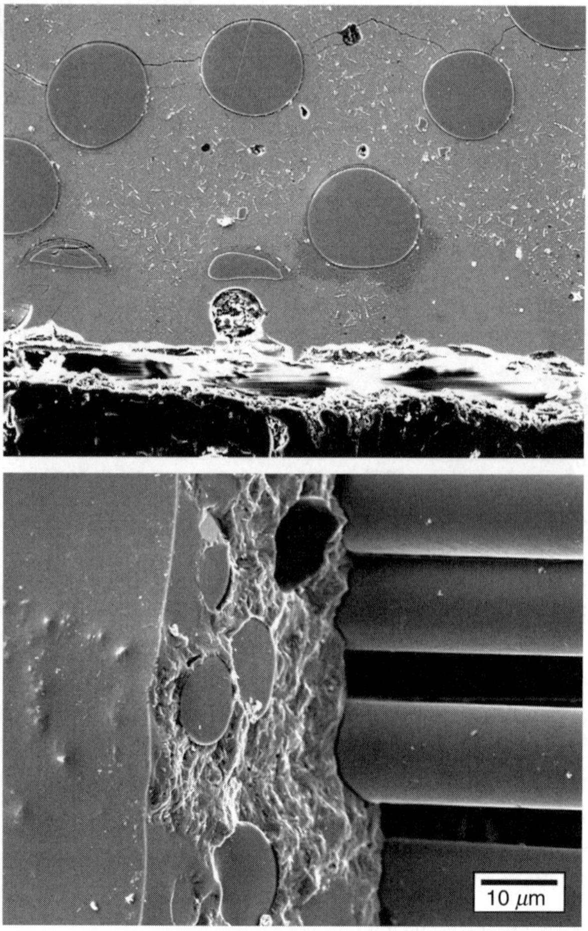

2.6 (a) SEM image of the polished cross-section of Nicalon/CAS after ageing at 1200 °C for 500 h in air. (b) SEM image of the tensile fracture surface of Nicalon/CAS after ageing at 1200 °C for 500 h in air.

removal and oxide bridge formation is discussed in greater detail in Section 2.5.1.

2.4.3 Low-temperature degradation

It is notable from prior studies that very little emphasis has been placed on assessing the low-temperature stability of ceramic matrix composites (CMCs) with carbon-based fibre/matrix interphases (e.g. ageing below 600 °C). The reason for this is not clear, as low-temperature degradation of carbon fibres and carbon/carbon composites is well known (Dhami *et al.*,

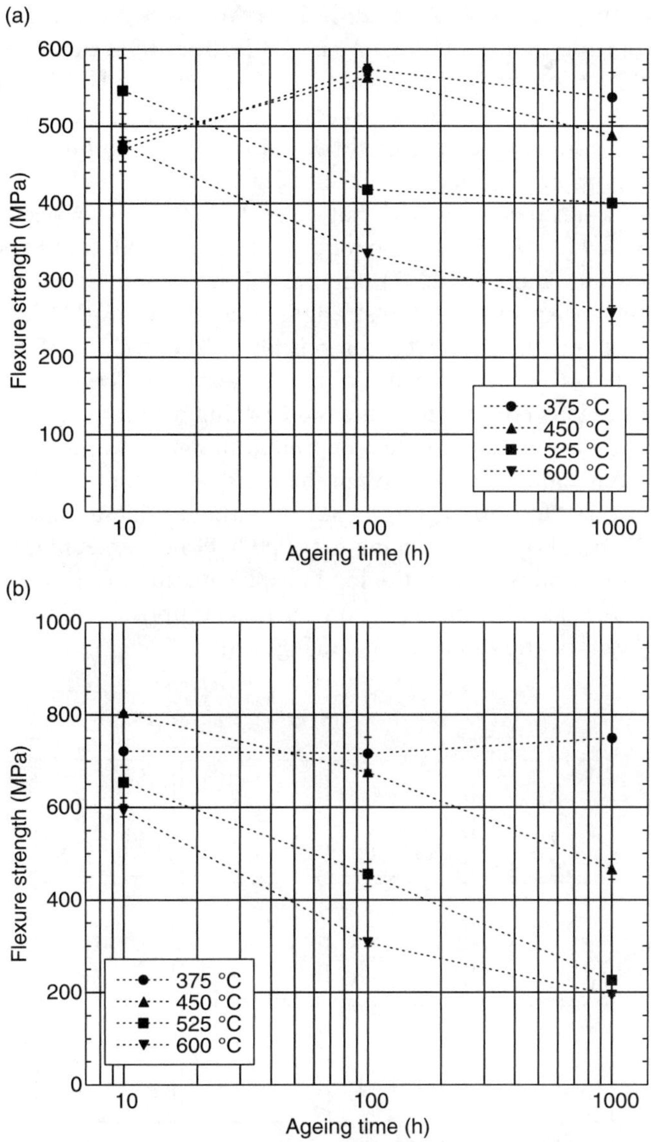

2.7 The effects of low-temperature ageing, for up to 1000 h, on the three-point flexure strength of (a) Nicalon/CAS (adapted from Plucknett and Lin, 2007) and (b) Nicalon/MAS (Plucknett and H.-T. Lin, unpublished data, 2007).

1991; Ismail and Hurley, 1992; Chung, 1994; Westwood *et al.*, 1996). It is apparent from Fig. 2.3(a) that ageing at 450 °C for 100 h does not appear to degrade the strength of either Nicalon/CAS or Tyranno/BMAS. However, examination of typical load–deflection curves for Nicalon/CAS highlights a transition to a nominally brittle mode of failure at this ageing temperature (Fig. 2.3(b)), with minimal fibre pull-out (Fig. 2.4(b)).

Recent work has demonstrated that degradation can even occur at temperatures as low as 375 °C, after extended duration oxidation exposure of 1000 h (Plucknett and Lin, 2007). Figure 2.7 demonstrates the effects of low-temperature exposure on the strength retention of both Nicalon/CAS and developmental Nicalon/MAS composites aged between 375 and 600 °C for up to 1000 h. The former material is a conventional commercial grade glass-ceramic matrix composite, while the Corning prototype Nicalon/MAS composite has a boron-containing component designed to promote surface sealing at lower temperatures through the formation of a borosilicate glass oxidation product. In both instances, severe strength degradation is apparent at 450 °C and above after ageing for 1000 h. However, evidence of degradation is most notable when the load–displacement curves are analysed, as there is a transition from a pseudo-ductile, composite failure mode to one that is more brittle in appearance (Fig. 2.8).

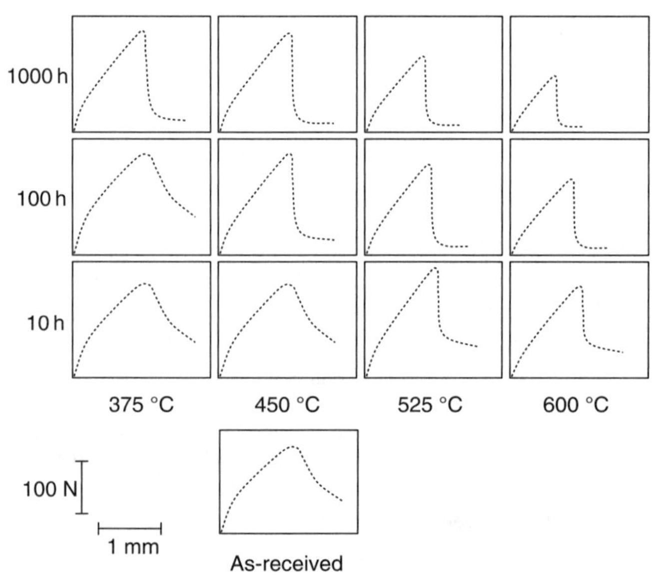

2.8 Schematic representation of the load–deflection curves obtained following low-temperature ageing of Nicalon/CAS. Adapted from Plucknett and Lin (2007).

2.5 Mechanism of oxidation degradation

In the previous section, general observations relating to the effects of long-term ageing in oxidising environments were described. It was apparent from this that there are nominally three distinct regions, namely: (a) high temperature, between 900 and 1200 °C, where there is effectively minimal degradation during unstressed oxidation; (b) intermediate temperature, between 600 and 900 °C, where embrittlement is generally most severe; and (c) low temperature, from 300 to 600 °C, where there is evidence of degradation after long-term exposure (i.e. 1000 h). In the following sub-section (Section 2.5.1), these effects are discussed in more detail with relation to the degradation mechanism(s) operating in each case. The mechanism of high-temperature protection that is noted above ~900 °C is discussed in Section 2.5.2.

2.5.1 Intermediate- and low-temperature degradation

It is clear from the previous discussion that the most severe degradation is generally seen at intermediate temperatures, between approximately 600 and 900 °C. Under such ageing conditions several observations are invariably noted: (a) there is a significant reduction in failure stress relative to the as-fabricated material; (b) there is a large increase in both the debonding energy, Γ, and the interfacial sliding stress, τ; (c) the load–deflection response is one typical of brittle failure; and (d) there is minimal fibre pull-out. A common observation following microstructural assessment of these intermediate-temperature aged materials, for example using transmission electron microscopy or scanning Auger microscopy, is that the compliant carbon interphase that separated the fibre and matrix from direct contact has been removed and replaced by an isolated, or continuous, SiO_2 bridge (Pharaoh *et al.*, 1993; Plucknett *et al.*, 1995a, 1995b, 1995c). The presence of this bridge has the resulting effect of significantly increasing both the debond energy and, to a lesser extent, the frictional sliding stress.

During low- and intermediate-temperature ageing carbon removal occurs via 'pipe-line' oxidation, following one of two reactions:

$$C_{(s)} + O_{2(g)} \rightarrow CO_{2(g)} \quad [2.7]$$

or

$$2C_{(s)} + O_{2(g)} \rightarrow 2CO_{(g)} \quad [2.8]$$

Carbon oxidation has previously been reported to occur at temperatures as low as ~400 °C (Dhami *et al.*, 1991; Ismail and Hurley, 1992; Chung, 1994; Westwood *et al.*, 1996). This process is shown schematically in Fig. 2.9, along with the behaviour for both low and high temperature ranges. In particular, for low-temperature ageing the primary reaction is the loss of carbon

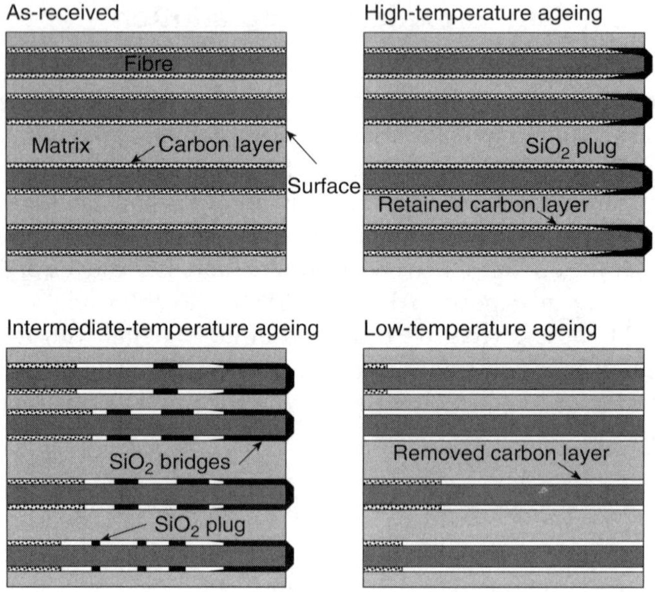

2.9 Schematic representation of the oxidation mechanism in various temperature ranges.

through oxidation. It has been shown that, at low enough temperature (i.e. ageing at 375 °C for 1000 h), essentially complete carbon removal is possible for small samples (e.g. flexure test bars) (Plucknett and Lin, 2007). Conversely, in both intermediate- and high-temperature regimes, the oxidation behaviour is far more complex.

Previously, the oxidation of a carbon interlayer has been assumed to follow sequential linear-parabolic kinetics (Eckel *et al.*, 1995), in a manner similar to silicon (Deal and Grove, 1965), where the rate of carbon recession due to oxidation is described by;

$$t = \frac{x^2}{k_p} + \frac{x}{k_l} \qquad [2.9]$$

where t is time, x the carbon recession distance, k_p the parabolic rate (or recession) constant (m^2/s) and k_l the linear rate (or recession) constant (m/s). Eckel *et al.* (1995) have derived a relationship for k_p based upon a simplified cylindrical pore geometry:

$$k_p = 6.263 \times 10^{10} T^{1/2} \times \ln\left\{\frac{((1+\chi)4.392 \times 10^4\ (Pd/T)+1)}{(4.392 \times 10^4\ (Pd/T)+1)}\right\} \qquad [2.10]$$

where T is the temperature, χ is the fractional partial pressure of the oxidising species (i.e. 1 for pure oxygen or, as in the majority of cases, 0.2 for air),

P is the environmental pressure and *d* is the pore diameter. The pore diameter is here approximated to the annular pore thickness of the carbon interlayer in a 'real' composite. This work compared predicted behaviour with that observed experimentally using a carbon-cored SiC fibre (Textron SCS-6). It should be noted that the minimum dimensions of the carbon core used (~33 μm) were well in excess of the carbon interphase thickness in glass-ceramic matrix composites, and hence the Knudsen diffusion limiting case is not relevant, and the behaviour is based primarily on molecular oxygen diffusion. However, for nanometre-scale carbon layers, and hence pores, it is more likely that Knudsen diffusion effects will become significant. Similar experimental and modelling approaches have been taken by Filipuzzi (Filipuzzi and Naslain, 1994; Filipuzzi *et al.*, 1994), for pore dimensions of 0.1 and 1 μm. It should be noted that both the work of Eckel (Eckel *et al.*, 1995) and Filipuzzi (Filipuzzi and Naslain, 1994; Filipuzzi *et al.*, 1994) set minimum pore diameter limits of 0.1 μm, which is still one order of magnitiude higher than in typical glass–ceramic matrix composites (i.e. 10–20 nm). Their work is primarily focused on the oxidation behaviour of SiC/SiC composites, where the interphase thickness is generally in excess of 0.1 μm. In particular, Eckel *et al.* (1995) highlight the change in the parabolic recession constant, k_p, as a function of pore diameter, which is significant when decreasing from 1 μm to 10 nm. It is clear that further work is necessary in this area, and it is likely that the comprehensive models developed by both Eckel (Eckel *et al.*, 1995) and Filipuzzi (Filipuzzi and Naslain, 1994; Filipuzzi *et al.*, 1994) can be readily adapted to materials with thinner carbon interphases, provided that suitable reaction kinetics data are available.

As previously shown in Figs 2.7 and 2.8, low-temperature ageing can result in the transition to a brittle failure mode, as well as strength reduction. While the effects of such heat treatments on the fibre/matrix interfacial properties have not been studied in detail, Daniel *et al.* (1996) have determined the effects of ageing at low temperatures, for a period of 100 h, on both the debond energy and the frictional sliding stress for Nicalon/CAS (Table 2.3). It was clear from that study that ageing between 450 and 600 °C for 100 h results in a significant increase in the frictional sliding stress, while the debond energy actually tends to decrease slightly relative to the as-fabricated composite. These observations imply that any chemical interfacial bond is effectively destroyed by the oxidation treatment, through oxidative removal of the carbon interlayer, which is confirmed by scanning Auger microscopy of the fibre surfaces after such heat treatments (Plucknett *et al.*, 1995c). In this instance, for Nicalon/CAS, the matrix clamps down onto the fibre (Powell *et al.*, 1993), due to the mismatch in coefficient of thermal expansion (CTE), and the fibre slides directly against the glass-ceramic matrix trough in which it sits. As a consequence, the frictional sliding stress increases significantly, as noted in Table 2.3. This behaviour

Table 2.3 The effects of low-temperature ageing on the fibre/matrix properties of Nicalon/CAS. After Daniel *et al.* (1996)

Heat-treatment condition	Debond energy, G_i (J/m^2)	Sliding stress, τ (MPa)
As-received	8.0 ± 3.0	25 ± 5
375 °C/100 h	1.1 ± 1.0	40 ± 14
450 °C/100 h	4.7 ± 3.7	144 ± 63
525 °C/100 h	6.3 ± 4.3	177 ± 54
600 °C/100 h	9.3 ± 5.5	193 ± 57

contrasts with that observed at intermediate temperatures, where a strong silicate-based oxide bond is formed between the fibre and matrix.

2.5.2 High-temperature sealing

While these materials exhibit intermediate- and even low-temperature degradation, it is readily apparent that at high temperatures they can withstand an oxidation heat treatment of extended duration with minimal degradation in mechanical performance (Fig. 2.3). The reason for this apparent stability relates to the initial oxidation mechanism of the composite at high temperature. Figure 2.6 highlights the surface oxide scale formed on a CAS/Nicalon composite after oxidation at 1200 °C for 500 h. The first feature that is apparent in this instance is that the near-surface fibres are heavily oxidised (Fig. 2.6(a)), and in fact are largely consumed. The oxidation of these fibres can be expected to follow the reaction:

$$\mathrm{Si}_x\mathrm{C}_y\mathrm{O}_{(s)} + \mathrm{O}_{2(g)} \rightarrow \mathrm{SiO}_{2(s,l)} + \mathrm{CO}_{(g)} \qquad [2.11]$$

The process that is occurring in this instance is highlighted schematically in Fig. 2.9.

It is clear that this phenomena leads to very rapid sealing and protection of the exposed fibre/matrix carbon interphase in this material. In fact, as will be noted later (Section 2.9), this effect can be used in a beneficial manner to 'pre-treat' the composite and protect it at lower oxidation temperatures. In this instance only the fibre is being oxidised, as the matrix is an oxide glass ceramic. Based on published oxidation kinetics data (Huger *et al.*, 1993) it is therefore possible to predict the time that will be taken to seal the exposed interphase region at elevated temperature.

A simple geometrical model was developed by Huger *et al.* (1994) to estimate the critical time, t_c, taken for sealing of the exposed fibre ends, such that carbon oxidation ceases, for the case of oxidation of SiC/SiC composites. In this instance *both* the fibre and matrix can oxidise, albeit at different rates. For the case of oxidation of SiC/SiC composites, the critical sealing time, t_c, is then estimated from Huger *et al.* (1994):

$$t_c = \frac{1}{B_f}\left[\frac{H}{\left(1-\frac{1}{\theta_f}\right)+R\left(1-\frac{1}{\theta_m}\right)}\right]^2 \qquad [2.12]$$

where B_f is the parabolic rate constant for oxidation of the fibres in air, H is the carbon interphase thickness, θ_f and θ_m are the respective expansion coefficients of the fibre and matrix due to oxidation (depends upon stoichiometry), and R is the ratio of the SiO_2 thickness on the fibre and matrix. For the case where no oxide product is formed on the matrix, $R = 0$.

The approach taken by Huger can therefore be simplified for the case of composites where only the fibre can oxidise (i.e. $R = 0$), such as those with a glass–ceramic matrix (Plucknett *et al.*, 1995c). For the instance of an oxidising fibre and non-oxidising matrix, the critical sealing time, t_c, then simplifies to the relationship:

$$t_c = \frac{1}{B_f}\left[\frac{H}{(1-1/\theta_f)}\right]^2 \qquad [2.13]$$

For the case of Nicalon NLM-202, the fibre used in many glass–ceramic matrix composites, θ_f is determined to be 1.48 (Huger *et al.*, 1994). An approximation of the critical sealing time can then be determined based on the oxidation kinetics of Nicalon NLM-202 fibres (Huger *et al.*, 1993). Predicted sealing times for the Nicalon NLM-202 fibre in a non-oxidising glass–ceramic matrix, with various carbon layer thicknesses, are shown in Fig. 2.10. At the lowest temperature examined (700 °C), sealing can be predicted to occur in approximately 35 h for a 50 nm thick carbon layer, 5 h for a 20 nm thick layer, and 1 h 20 min for a 10 nm thick layer. Conversely, at 1200 °C, sealing can be predicted to occur in minutes for all but the highest interphase thicknesses examined.

Taking Nicalon/CAS as an example, the residual stress state in the composite is such that the matrix will clamp down onto the fibre if the carbon layer is removed. In this instance the fibre/matrix separation will decrease, and they will potentially come into direct contact. Consequently, sealing times can be expected to be reduced as the fibre/matrix separation decreases. Oxidation kinetics data for the Nicalon NLM-202 are not available for temperatures below 700 °C, but extrapolation of the behaviour observed at higher temperatures indicates that sealing will occur after approximately 20 h at 600 °C for a 20 nm thick fibre–matrix gap. This estimation is in general accordance with weight gain responses observed for Nicalon/CAS during oxidation, where weight gain ceases prior to 100 h exposure at 600 °C (Plucknett and Lin, 2007). A more thorough assessment of the sealing behaviour at these lower temperatures (375–600 °C) will require detailed information on the weight changes as a function of time. However, it is clear

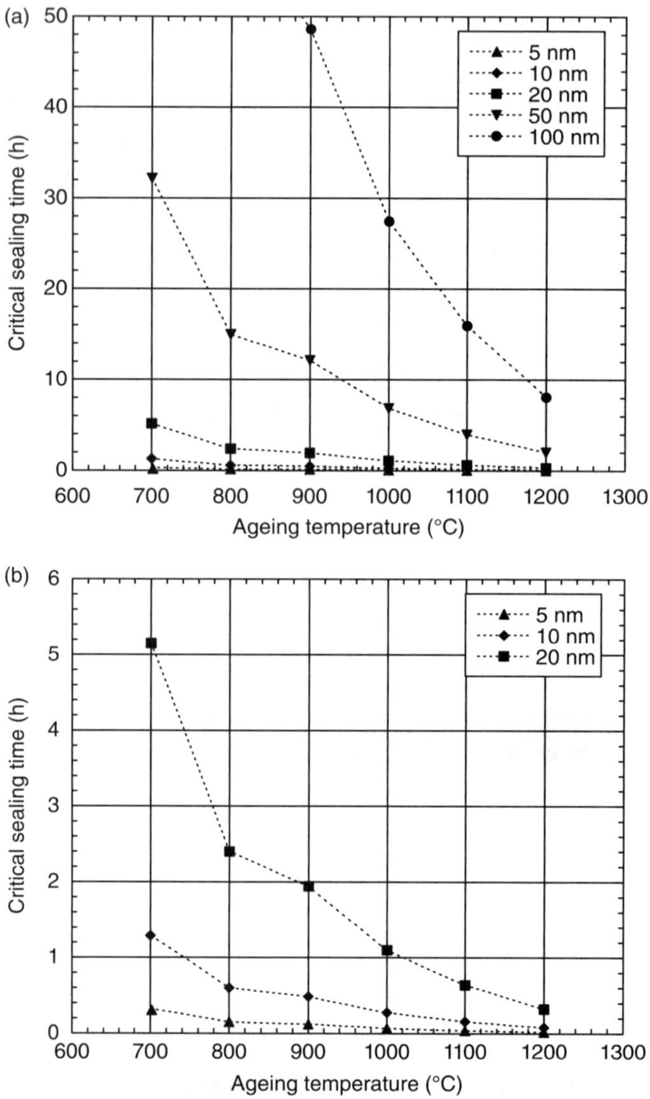

2.10 (a) The critical sealing time as a function of ageing temperature for various carbon interphase thicknesses. (b) Magnified region of (a) for thin carbon interphases.

that after exposure at 450 °C for 100 h there is no measurable oxide product formed on the fibre surfaces (Plucknett *et al.*, 1995c), indicating that sealing is unlikely to be a significant factor at these lower temperatures, and complete carbon layer removal is expected to arise, if oxidation occurs for a sufficient duration.

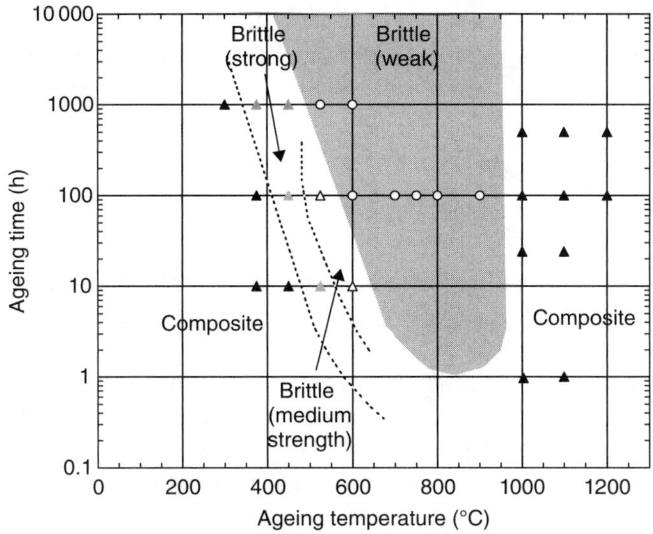

2.11 Failure mechanism map for Nicalon/CAS-II. Adapted from Plucknett and Lin (2007).

2.6 Development of a failure mechanism map

Based on the oxidation response previously outlined (Sections 2.4 and 2.5), it is now possible to generate a simple environmental embrittlement failure mechanism map. An example of such a failure mechanism map is shown in Fig. 2.11, in this case for the Nicalon/CAS-II system. This map highlights the effects of oxidation temperature and duration upon the type of failure mode that will occur. In this case there is a clear region, at extremes of temperature, where the failure mode is largely unaffected; in this instance the composite remains strong and shows a pseudo-ductile stress–strain curve that is largely the same as the material prior to heat treatment. At intermediate temperatures the material's mechanical behaviour is degraded significantly. The influence of ageing time at lower temperatures is also apparent, with a gradual transition from composite to brittle failure occurring at 375 and 450 °C. It can be anticipated that the carbon layer thickness will have a pronounced influence on the size of the degradation zone, such that the upper boundary will be moved to a higher temperature if the carbon layer thickness is increased, and to a lower temperature if it is reduced (simply due to the sealing time that will arise (Fig. 2.10)).

2.7 Oxidation behaviour under applied stress

In most of the previously described work, ageing degradation has occurred due to heat treatment at elevated temperature in an oxidising environment.

However, in many environments it can be expected that the materials will be exposed to a combination of elevated temperature, oxidising atmosphere and applied stress. The following sections briefly discuss the stability of glass–ceramic matrix composites under conditions of static fatigue loading, cyclic fatigue loading and creep loading.

2.7.1 Static fatigue

The low-temperature fatigue behaviour of Nicalon/CAS has been assessed in bending between 450 and 950 °C (Plucknett and Lin, 2007; H.-T. Lin and K. P. Plucknett, unpublished research, 2007). Figure 2.12 demonstrates the combined effects of applied stress and temperature on the lifetime of Nicalon/CAS. It is apparent that the fatigue lifetime is strongly affected by both increasing temperature and applied stress (static fatigue run-out was set at 1000 h in each case). However, even at moderately low temperatures, there is a clear influence of temperature. In the majority of the examples presented in Fig. 2.12, the applied stress is in excess of the matrix micro-cracking stress, and fatigue run-out occurs with decreasing applied stress for increasing temperature. At 950 °C, only samples loaded to 100 MPa, which is below the onset stress for micro-cracking, exhibited static fatigue run-out after 1000 h. In this instance sealing of the exposed fibre ends can be anticipated after ~1 h, and the carbon-based fibre/matrix interphase will

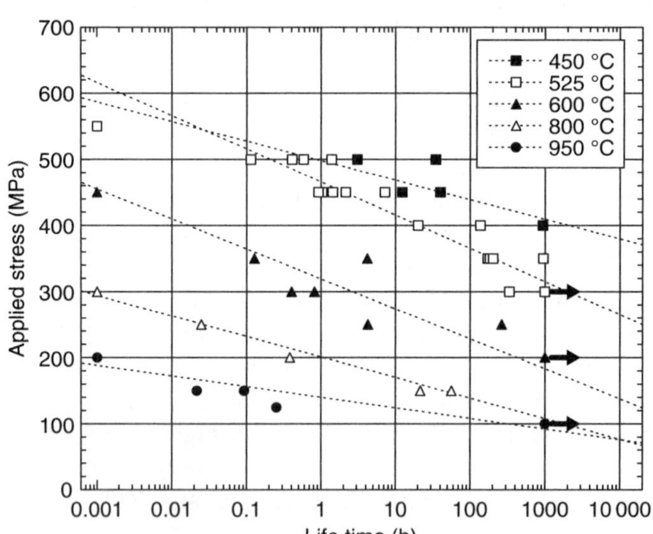

2.12 Static fatigue lifetimes as a function of applied stress for Nicalon/CAS. Adapted from Plucknett and Lin (2007); additional data K. P. Lin and K. P. Plucknett, unpublished observations, 2007. Arrows indicate test samples that exhibited successful fatigue run-out at after 1000 h.

be largely retained. Above the micro-cracking stress, partial or complete removal of the carbon layer is likely, as multiple paths for oxygen ingress will occur. At lower temperatures this effect is less pronounced, but carbon interphase oxidation will still be expected even at 450 °C. It is notable from unstressed oxidation assessment at this low temperature that the strength of the composite is largely retained (Fig. 2.3), while there is a transition from a composite to brittle failure mode, with negligible fibre pull-out (Section 2.4.3). For the case of static fatigue, there appears to be a similar effect with overall composite strength retained, and therefore increased lifetimes relative to the higher temperatures.

Sun *et al.* (1997b) have conducted a similar study on the static fatigue of Nicalon/BMAS composites, with a dual-layer SiC/BN interphase. While only two temperatures were assessed in that earlier study (600 and 950 °C), it was noted that interfacial oxidation occurs at 600 °C throughout the composite, aided by the presence of matrix micro-cracks, while at 950 °C sealing occurs rapidly enough to prevent significant internal oxidation. It was also noted that, at 600 °C, the BN interphase component oxidised by volatilisation rather than the formation of a borosilicate glass.

2.7.2 Cyclic fatigue

There have been a number of comprehensive studies of the room temperature cyclic fatigue behaviour of fibre-reinforced glass–ceramic matrix composites (Holmes and Shuler, 1990; Cho *et al.*, 1991; Zawada *et al.*, 1991; Holmes and Cho, 1992; Vanswijgenhoven *et al.*, 1999; Sørensen *et al.*, 2000, 2002). Generally, there are several features that are consistent across these studies. Firstly, when considering monotonic failure of CMCs, fibre/matrix interfacial debonding and sliding are necessary mechanisms for composite failure to occur (see Sections 2.1.2 and 2.1.3). However, from the perspective of cyclic deformation, even at room temperature, this behaviour gives rise to several problems. At fatigue stresses above the matrix micro-cracking stress, the fibres will be partially debonded, and matrix cracks will be present. Cyclic fatigue results in continued wear of the interface, as the fibre slides back and forth against the matrix. Under such conditions a decrease in the fibre/matrix frictional sliding coefficient, τ, is invariably noted (Holmes and Shuler, 1990; Holmes and Cho, 1992). In extreme cases, the sliding stress can be observed to decrease by more than a factor of four during the early stages of fatigue (i.e. from >20 to ~5 MPa), after which it remains essentially constant (Cho *et al.*, 1991; Holmes and Cho, 1992). Under high cycle fatigue conditions (e.g. 200 Hz), when the composite is stressed in excess of the proportional limit (i.e. such that micro-cracking and fibre sliding occurs), the extent of frictional heating that occurs within the composite can be significant (Holmes and Shuler, 1990; Cho *et al.*, 1991; Homes and Cho,

1992). During such fatigue testing, even under run-out conditions, microstructural damage is still evolving in the composite after 10^8 cycles (Sørensen *et al.*, 2002). As a consequence, it has been demonstrated that the fatigue life decreases with increasing cyclic frequency (Holmes *et al.*, 1994).

While room temperature cyclic fatigue behaviour has received considerable attention, there have been relatively few studies of the effects of temperature on cyclic fatigue of glass–ceramic matrix composites. In recent work, Yasmin and Bowen (2004) have compared the cyclic fatigue of Nicalon/CAS at both room temperature and 800 °C. It was observed that at room temperature the number of fatigue cycles to failure decreased with increasing applied stress. For example, at a maximum applied stress of 400 MPa, with a stress ratio $R = 0.1$, failure occurred after 190 cycles, while run-out (successful completion of 10^6 cycles) occurred at 200 MPa with the same stress ratio, $R = 0.1$. Conversely, at 800 °C, only 566 cycles were successfully completed at an applied stress of 130 MPa ($R = 0.1$). Only fatigue cycling at stresses below the micro-cracking stress (tested at 100 MPa, $R = 0.1$) resulted in consistent cycling to run-out at 10^6 cycles. At room temperature, a fibrous fracture surface was retained, even after thousands of cycles, and the degradation was attributed to fibre/matrix interfacial shear stress degradation through repeated sliding. Conversely, at 800 °C, the fracture surfaces were largely absent of fibrous failure at the higher stresses, indicating a strong fibre/matrix bond. However, the samples fatigued at 100 MPa for 10^6 cycles exhibited better retained strength *after* fatigue, when compared with those fatigued at room temperature, in combination with fibrous fracture surfaces. In this instance, sealing of the exposed fibre ends occurs, protecting the carbon-based interphase.

2.7.3 Creep

As outlined in the previous sections (Sections 2.7.1 and 2.7.2), static and cyclic fatigue studies are inevitably performed at temperatures where matrix plasticity is minimal, and the materials can micro-crack in a conventional manner. However, at high enough temperatures matrix plasticity can become appreciable, and the materials are able to undergo creep deformation. Invariably the temperature at which this occurs is sufficiently high that surface protection mechanisms can operate, such as those outlined in Section 2.5.2, and the compliant carbon or boron nitride interlayer is protected. This is particularly the case for typical glass–ceramic matrix materials, where negligible creep is expected below ~900 °C, and the matrix can be viewed as elastic in behaviour.

Wu and Holmes (1993) have investigated the tensile creep behaviour of both unidirectional and cross-ply Nicalon/CAS at 1200 °C. It was noted that the creep rate of the unidirectional composite increased by a factor of 5 when the applied stress was increased from 60 to 200 MPa. Significant creep recovery was noted in these materials after unloading, with approximately 50% of the creep strain recovered for a cross-ply material tested at 60 MPa. Under these conditions, fibre fracture was rarely noted, although void formation was observed in the matrix, even at moderately low applied stresses. The high-temperature compression creep behaviour of a similar cross-ply Nicalon/CAS has been assessed by Nair *et al.* (2001), between 1275 and 1325 °C, for applied stresses between 15 and 50 MPa. It was shown that the 'on-axis' cross-ply material behaved in an intermediate manner to the unidirectional composite, with fibres either parallel (most creep resistant) or perpendicular to the applied stress axis. Fibre creep was found to be the rate-limiting process in the 'on-axis' cross-ply material. 'Off-axis' studies were also conducted on the cross-ply composite, and it was postulated that the behaviour of such materials could be modelled based on a thorough knowledge of the response of 'on-' and 'off-axis' unidirectional materials.

Sutherland *et al.* (1995) assessed the tensile creep behaviour of two continuous fibre-reinforced composites, namely a unidirectional Nicalon/Pyrex glass–matrix system and a cross-ply Tyranno/BMAS glass-ceramic matrix system. It was found that the glass–matrix material exhibited behaviour with a creeping matrix but elastic fibres in the temperature range of interest (400–560 °C). Conversely, for elevated temperature testing of the Tyranno/BMAS composite (between 1125 and 1200 °C), both the matrix and fibre exhibited creep. A simple load-partitioning model was developed to estimate the load transfer to the fibres in both systems, with a time constant, θ, calculated based on an exponential response. It was shown that for both materials, after a duration of $\sim 5\theta$, the fibres carried in excess of 99% of the applied load.

The flexure creep behaviour of a developmental Nicalon/YMAS composite has been assessed at moderately low temperatures (e.g. 800–1100 °C), in both air and vacuum, by Vicens and colleagues (Vicens *et al.*, 1997; Chermant *et al.*, 2002). It was shown that matrix micro-cracking occurs, in a manner similar to static fatigue. The apparent creep strain rate increases significantly at applied stresses greater than 150 MPa, which is likely to be close to the micro-cracking stress, indicative of the fact that matrix micro-cracking is the primary creep strain mechanism. Under such conditions, testing in air results in the removal of the carbon-based interphase, and the formation of SiO_2 ligaments between the fibre and matrix, in a manner similar to stress-free ageing. Sun and colleagues have investigated the creep behaviour of a prototype Nicalon/BMAS, with a BN/SiC double-layer interphase (Sun *et al.*, 1995, 1996; Widjaja *et al.*, 1999). It was demonstrated that with a highly crystalline BMAS matrix, creep was limited below 1130 °C ($\sim 10^{-9}$ s^{-1}) (Sun *et al.*, 1995). Furthermore, a creep-strengthening process could be employed where the composite proportional limit was increased after creep pre-treatment (Widjaja *et al.*, 1999).

2.8 Thermal shock cycling

It can be anticipated that in most applications a CMC component will be thermally cycled many times during its service life, and that therefore thermal cycling studies are important to assess potential long-term degradation. Zawada and Wetherhold (1991) conducted an early investigation into the thermal cycling fatigue behaviour of Nicalon fibre-reinforced aluminosilicate glass–matrix composite. It was noted that repeated cycling through intermediate temperature ranges (either 250–700 °C or 250–800 °C), for typically 500 cycles in air, results in significant strength degradation and a transition to brittle fracture. Similar strength degradation was also noted following an isothermal hold at 650 °C for 16 h. It was proposed in this work that oxidative removal of the carbon interlayer occurs, and a strong SiO_2 bond is subsequently formed between the fibre and matrix. Recently there has been renewed interest in the effects of thermal cycling upon the mechanical behaviour of selected CMCs, from both an experimental and modelling perspective (Blissett *et al.*, 1998; Kastritseas *et al.*, 2005, 2006). It is generally shown in these studies that cyclic thermal loading results in significant accumulation of micro-cracking within the matrix. Increasing the extent of the thermal transition, ΔT, resulted in an increase in the concentration of matrix micro-cracks (Kastritseas *et al.*, 2006).

It is clear from both these and other investigations that an understanding of thermal shock response is necessary when contemplating the use of these materials in real-world applications, such as those outlined in Section 2.1.5. It is notable that prior work has focused on composites in their as-fabricated form. However, ultimately it is probable that coatings will be applied to these materials, and extensive thermal shock assessment will be required in the future on such materials, in order to assess the durability of the coating/composite combination.

2.9 Composite protection methods

In Section 2.5.2, a surface-sealing, self-protection mechanism is outlined for glass–ceramic matrix composites with non-oxide fibres. Based on this understanding of high-temperature sealing phenomena, it is possible to develop simple high-temperature sealing treatments that allow subsequent use of the composites at lower temperatures, where intermediate-temperature degradation may occur (Plucknett and Lewis, 1995). It was shown in this work that through the use of a short-duration heat treatment in air, at 1100 °C for 1 h, intermediate-temperature degradation could be avoided (with samples subsequently aged at 700 °C for 100 h). Under these conditions the fibre surface rapidly oxidises, forming SiO_2, which helps to protect the fibre/matrix interface from degradation. Compared with samples that had not been pre-treated, those that had undergone a surface sealing cycle at 1100 °C showed both high strength retention and a composite failure mode after intermediate-temperature ageing. In a similar manner, Wetherhold and Zawada (1991) utilised an isothermal heat treatment at 800 °C to protect a Nicalon fibre-reinforced, alkaline-earth aluminosilicate glass-matrix composite. In this instance viscous flow of the glass was believed to occur, rather than fibre oxidation, resulting in sealing of any exposed fibre–matrix interfacial region. Subsequent ageing at lower temperature (650 °C for 16 h) resulted in significantly reduced degradation in comparison with the untreated material.

However, while these two approaches may be successful for unstressed intermediate-temperature ageing, under stresses greater than the matrix micro-cracking stress it can be envisaged that environmental embrittlement will still occur. As a consequence, alternative approaches are required for protection of the composites at such temperatures. Ferraris *et al.* (2004) have outlined a simple approach, where a glass–ceramic coating is applied from a slurry through a three-stage dip coating, consolidation and crystallisation process. The zinc borosilicate glass used softens at approximately 600 °C, and provides good oxidation protection at the same temperature, with a potential maximum usage temperature of 700 °C.

It is important to note that for any coating procedure to be successful in operation, the coating must to able to operate over a wide temperature range and withstand the strains that can be observed in CMCs without cracking. Given the potential operational temperature range of these materials where oxidation-induced degradation may occur (i.e. 375–1200 °C), this can be seen as a major challenge in materials design. The approaches outlined earlier in this section all suffer from one or more drawbacks in operation. For example, they either cannot withstand the strains observed in the underlying composite in use or, for coatings that show some viscous flow or plastic flow sealing behaviour, the functional mechanism cannot operate over the entire temperature range outlined earlier. This current limitation creates a significant barrier to the successful implementation of CMCs with non-oxide components in oxidising operating environments.

2.10 Conclusions and future trends

The development of fibre-reinforced glasses and ceramics during the past three decades has led to significant renewed interest in these materials for use in advanced engineering applications such as gas turbines. The early generation of glass–ceramic matrix composites offered much promise in such environments, but the stability issues present with carbon and, to a lesser extent, boron nitride as a fibre/matrix interphase material have limited the commercial acceptance of these materials. However, through their development and testing, and the attainment of an in-depth understanding of their micro- and macromechanical behaviour, new generations of non-oxide and all-oxide composites have been developed in recent years. These materials, especially the all-oxide systems, are now the focus of much attention for application as static and, potentially, dynamic components in gas turbines.

It has been shown in this chapter that composites with carbon-based fibre/matrix interphases exhibit both intermediate- and low-temperature degradation phenomena that relate to the intrinsic oxidation stability of the interphase. Degradation of these materials can occur at temperatures as low as 375 °C in an oxidising environment, after several hundred hours of exposure, while significantly more rapid degradation occurs in the intermediate temperature range (e.g. 600–800 °C). In both cases, pipeline oxidation of the compliant interphase material occurs, leading to its partial or complete removal, and potentially to the formation of oxidation products that strongly bond the fibre to the matrix. These studies have significant implications for more advanced non-oxide composites developed with multiple carbon/SiC layers, as degradation can still occur at low temperatures. The adoption of surface sealing or coating procedures can, in part, mitigate this issue of environmental stability. However, it is clear the demands on such a coating system are significant, as it must be capable of protecting the material over a very wide temperature range (e.g. 350–1200 °C) at stresses above the micro-cracking stress of the matrix itself.

2.11 References

AVESTON J, COOPER G A and KELLY A (1971), 'Single and multiple fracture', in *Properties of Fibre Composites*, Conference Proceedings of the National Physical Laboratory, Surrey, IPC Science and Technology Press, Ltd, pp. 15–26.

BARNARD P, HENDERSON M B and RHODES N (2004), 'CMC Integration and Demonstration for Gas Turbine Engines (CINDERS)', *Appl Thermal Engng*, **24** (11–12), 1755–1764.

BENSON P M, SPEAR K E and PANTANO C G (1988), 'Interfacial characterization of glass matrix/Nicalon SiC fibre composites: A thermodynamic approach', *Ceram Engng Sci Proc*, **9** (7–8), 663–670.

BLEAY S M and SCOTT V D (1992), 'Effect of heat-treatment in air on the structure and properties of barium osumilite reinforced with Nicalon fibres', *J Mater Sci*, **27** (3), 825–838.

BLISSETT M J, SMITH P A and YEOMANS J A (1998), 'Flexural mechanical properties of thermally treated unidirectional and cross-ply Nicalon-reinforced calcium aluminosilicate composites', *J Mater Sci*, **33** (16), 4181–4190.

BONNEY L A and COOPER R F (1990), 'Reaction-layer interfaces in SiC-fibre-reinforced glass-ceramics: a high resolution scanning transmission electron microscopy study', *J Am Ceram Soc*, **73** (10), 2916–2921.

BRENNAN J J and PREWO K M (1982), 'Silicon carbide fibre reinforced glass-ceramic matrix composites exhibiting high strength and toughness', *J Mater Sci*, **17** (8), 2371–2383.

BRENNAN J J (1986), 'Interfacial characterization of glass and glass-ceramic matrix Nicalon SiC fibre composites', in Tressler R E, Messing G L, Pantano C G and Newnham R E (Eds), *Tailoring Multiphase and Composite Ceramics*, New York, Plenum Press, pp. 549–560.

BRENNAN J J (1988), 'Interfacial chemistry and bonding in fibre-reinforced glass-ceramic matrix composites', in Pask J A and Evans A G (Eds), *Ceramic Microstructures: The Role of Interfaces*, New York, Plenum Press, pp. 387–400.

BRENNAN J J, NUTT S R and SUN E Y (1995), 'Interfacial microstructure and stability of BN-coated Nicalon fibre/glass–ceramic matrix composites', in Evans A G and Naslain R (Eds), *High Temperature Ceramic Matrix Composites II: Manufacturing and Materials Development, Ceramic Transactions*, Volume 58, Westerville, The American Ceramic Society, pp. 53–64.

BUNSELL A R, BERGER M H and KELLY A (1999), 'Fine ceramic fibers', in Bunsell A R and Berger A H (Eds), *Fine Ceramic Fibers*, New York, Marcel Dekker Inc., pp. 1–62.

CHAIM R and HEUER A H (1987), 'The interface between Nicalon SiC fibers and a glass–ceramic matrix', *Adv Ceram Mater*, **2** (2), 154–158.

CHAIM R and HEUER A H (1991), 'Carbon interfacial layers formed by oxidation of SiC in SiC/Ba-stuffed cordierite glass–ceramic matrix reaction couples', *J Am Ceram Soc*, **74** (7), 1663–1667.

CHAN H M (1997), 'Layered ceramics: Processing and mechanical behavior', *Ann Rev Mater Sci*, **27**, 249–282.

CHERMANT, J L, BOITIER G, DARZENS S, FARIZY G, VICENS J and SANGLEBOEUF J C (2002), 'The creep mechanism of ceramic matrix composites at low temperature and stress, by a material science approach', *J Eur Ceram Soc*, **22**, 2443–2460.

CHO C, HOLMES J W and BARBER J R (1991), 'Estimation of interfacial shear in ceramic composites from frictional heating measurements', *J Am Ceram Soc*, **74** (11), 2802–2808.

CHRISTIN F (2002), 'Design, fabrication and application of thermostructural composites (TSC) like C/C, C/SiC, and SiC/SiC composites', *Adv Engng Mater*, **4** (12), 903–912.

CHUNG D D L (1994), *Carbon Fibre Composites*, Newton, Massachusetts, Butterworth-Heinemann.

COOPER R F and CHYUNG K (1987), 'Structure and chemistry of fibre matrix interfaces in silicon carbide fibre-reinforced glass–ceramic composites – an electron microscopy study', *J Mater Sci*, **22** (9), 3148–3160.

CRIVELLI-VISCONTI I and COOPER G A (1969), 'Mechanical properties of a new carbon fibre material', *Nature*, **221** (5182), 754–755.

DANIEL A M, MARTÍN MEIZOSO A, PLUCKNETT K P and BRASKI D N (1996), 'Interface modification during oxidation of a glass ceramic matrix/SiC fibre composite, *Ceram Engng Sci Proc*, **17** (4), 280–287.

DEAL B E and GROVE A S (1965), 'General relationship for the thermal oxidation of silicon', *J Appl Phys*, **36** (12), 3370–3378.

DHAMI T L, MANOCHA L M and BAHL O P (1991), 'Oxidation behavior of pitch based carbon-fibers', *Carbon*, **29** (1), 51–60.

ECKEL A J, CAWLEY J D and PARTHASARATHY T A (1995), 'Oxidation-kinetics of a continuous carbon phase in a non-reactive matrix', *J Am Ceram Soc*, **78** (4), 972–980.

EVANS A G, HE M Y and HUTCHINSON J W (1989), 'Interface debonding and fibre cracking in brittle matrix composites', *J Am Ceram Soc*, **72** (12), 2300–2303.

FERRARIS M, SALVO M, MATEKOVITS I and BOCCACCINI A R (2004), 'Oxidation protective glass-ceramic coating for SiC fibre reinforced glass matrix composites', *Adv Engng Mater*, **6** (11), 910–914.

FILIPUZZI L, CAMUS G, NASLAIN R and THEBAULT J (1994), 'Oxidation mechanisms and kinetics of 1D-SiC/C/SiC composite-materials: 1, An experimental approach', *J Am Ceram Soc*, **77** (2), 459–466.

FILIPUZZI L and NASLAIN R (1994), 'Oxidation mechanisms and kinetics of 1D-SiC/C/SiC composite-materials: 2, Modeling', *J Am Ceram Soc*, **77** (2), 467–480.

GYEKENYESI J Z and BANSAL N P (2000), 'High temperature tensile properties of unidirectional Hi-Nicalon/celsian composites in air', NASA Technical Report, NASA/TM-2000-210214.

HASEGAWA Y, IIMURA M and YAJIMA S (1980), 'Synthesis of continuous silicon-carbide fibre. 2: Conversion of polycarbosilane into silicon carbide fibers', *J Mater Sci*, **15** (3), 720–728.

HBAIEB K, MCMEEKING R M and LANGE F F (2007), 'Crack bifurcation in laminar ceramics having large compressive stress', *Int J Solids Struct*, **44** (10), 3328–3343.

HE M Y and HUTCHINSON J W (1989), 'Crack deflection at an interface between dissimilar elastic materials', *Int J Solids Struct*, **25** (9), 1053–1067.

HE M Y, EVANS A G and HUTCHINSON J W (1994), 'Crack deflection at an interface between dissimilar elastic materials – role of residual stresses', *Int J Solids Struct*, **31** (24), 3443–3455.

HOLMES J W and SHULER S F (1990), 'Temperature rise during fatigue of fibre-reinforced ceramics', *J Mater Sci Lett*, **9** (11), 1290–1291.

HOLMES J W and CHO C D (1992), 'Experimental observations of frictional heating in fibre-reinforced ceramics', *J Am Ceram Soc*, **75** (4), 929–938.

HOLMES J W, WU X and SØRENSEN B F (1994), 'Frequency dependence of fatigue life and internal heating of a fibre-reinforced ceramic matrix composite', *J Am Ceram Soc*, **77** (12), 3284–3286.

HSUEH C H (1993), 'Interfacial debonding and fibre pull-out stresses of fibre-reinforced composites: 9, A simple treatment of Poisson effect for frictional interfaces', *Mater Sci Engng*, **A161** (1), L1–L6.

HUGER M, SOUCHARD S and GAULT C (1993), 'Oxidation of Nicalon fibres', *J Mater Sci Lett*, **12** (6), 414–416.

HUGER M, FARGEOT D and GAULT C (1994), 'Ultrasonic characterization of oxidation mechanisms in Nicalon/C/SiC composites', *J Am Ceram Soc*, **77** (10), 2554–2560.

ISMAIL I M K and HURLEY W C (1992), 'Modeling carbon-fibre oxidation in air at constant heating rates', *Carbon*, **30** (3), 419–427.

KAHRAMAN R, MANDELL J F and DEIBERT M C (1995), 'High-temperature mechanical behaviour of multidirectional Nicalon/CAS-II composite', *J Mater Sci*, **30** (24), 6329–6338.

KASTRITSEAS C, SMITH P A and YEOMANS J A (2005), 'Thermal shock fracture in unidirectional fibre-reinforced ceramic-matrix composites', *Comp Sci Technol*, **65** (11–12), 1880–1890.

KASTRITSEAS C, SMITH P A and YEOMANS J A (2006), 'Damage characterisation of thermally shocked cross-ply ceramic composite laminates', *J Mater Sci*, **41** (3), 951–962.

KELLY A (1973), *Strong Solids*, 2nd edition, Oxford, Oxford University Press.

KRENKEL W, HEIDENREICH B and RENZ R (2002), 'C/C-SiC composites for advanced friction systems', *Adv Engng Mater*, **4** (7), 427–436.

LANGE F F (1989), 'Powder processing science and technology for increased reliability', *J Am Ceram Soc*, **72** (1), 3–15.

LEWIS J A (2000), 'Colloidal processing of ceramics', *J Am Ceram Soc*, **83** (10), 2341–2359.

LEWIS M H and MURTHY V S R (1991), 'Microstructural characterization of interfaces in fibre-reinforced ceramics', *Comp Sci Technol*, **42** (1–3), 221–249.

LUH E Y and EVANS A G (1987), 'High-temperature mechanical properties of a ceramic matrix composite', *J Am Ceram Soc*, **70** (7), 466–469.

MACKIN T J and ZOK F W (1992), 'Fibre bundle pushout – a technique for the measurement of interfacial sliding properties', *J Am Ceram Soc*, **75** (11), 3169–3171.

MANDELL J F, GRANDE D H and JACOBS J (1987), 'Tensile behavior of glass ceramic composite-materials at elevated temperatures', *J Engng Gas Turbines Power*, **109** (3), 267–273.

MARINESCU I D (2006), *Handbook of Advanced Ceramic Machining*, Boca Raton, Florida, CRC Press.

MARSHALL D B and OLIVER W C (1987), 'Measurement of interfacial mechanical-properties in fibre-reinforced ceramic composites', *J Am Ceram Soc*, **70** (8), 542–548.

MURTHY V S R and LEWIS M H (1989), 'Interface structure and matrix crystallisation in SiC (Nicalon)-Pyrex composites', *J Mater Sci Lett*, **8** (5), 571–572.

MURTHY V S R, LEWIS M H, SMITH M E and DUPREE R (1989), 'Structure and degradation of Tyranno fibers', *Mater Lett*, **8** (8), 263–268.

NAIR B G, COOPER R F and PLESHA M E (2001), 'High temperature creep of a bi-directional, continuous-SiC-fibre-reinforced glass–ceramic composite', *Mater Sci Engng A*, **300** (1–2), 68–79.

NASLAIN R R (2004), 'Design, preparation and properties of non-oxide CMCs for application in engines and nuclear reactors: an overview', *Comp Sci Technol*, **64** (2), 155–170.

NASLAIN R R (2005), 'SiC-matrix composites: Nonbrittle ceramics for thermostructural application', *Int J Appl Ceram Technol*, **2** (2), 75–84.

PARTHASARATHY T A, MARSHALL D B and KERANS R J (1994), 'Analysis of the effect of interfacial roughness on fibre debonding and sliding in brittle matrix composites', *Acta Metall Mater*, **42** (11), 3773–3784.

PHARAOH M W, DANIEL A M and LEWIS M H (1993), 'Stability of interfaces in calcium aluminosilicate matrix nicalon SiC fibre composites', *J Mater Sci Lett*, **12** (13), 998–1001.

PHILLIPS D C (1972), 'The fracture energy of carbon-fibre reinforced glass', *J Mater Sci*, **7** (10), 1175–1191.

PHILLIPS D C, SAMBELL R A J and BOWEN D H (1972), 'Mechanical properties of carbon-fibre reinforced Pyrex glass', *J Mater Sci*, **7** (12), 1454–1464.

PHILLIPS D C (1974), 'Interfacial bonding and the toughness of carbon fibre reinforced glass and glass-ceramics', *J Mater Sci*, **9** (11), 1847–1854.

PLUCKNETT K P and LEWIS M H (1995), 'Inhibition of intermediate temperature degradation of calcium aluminosilicate/Nicalon by high temperature pretreatment', *J Mater Sci Lett*, **14** (17), 1223–1226.

PLUCKNETT K P, SUTHERLAND S, DANIEL A M, CAIN R L, WEST G, TAPLIN D M R and LEWIS M H (1995a), 'Environmental ageing effects in a silicon carbide fibre-reinforced glass–ceramic matrix composite', *J Microscopy*, **177** (3), 251–263.

PLUCKNETT K P, CAIN R L and LEWIS M H (1995b), 'Interface degradation in CAS/Nicalon during elevated temperature ageing', in *Ceramic Matrix Composites: Advanced High-Temperature Structural Materials* (Materials Research Society Symposium Proceedings Vol. 365), Pittsburgh, Pennsylvania, Materials Research Society, pp. 421–426.

PLUCKNETT K P, LIN H-T, BRASKI D N and BECHER P F (1995c), 'Environmental aging degradation in continuous fibre ceramic composites', in *Characterisation and Ceramic Matrix Composites* (Proceedings of the 10th International Conference on Composite Materials, Volume IV), Cambridge, UK, Woodhead Publishing Limited, pp. 803–810.

PLUCKNETT K P and LIN H-T (2007), 'Low-temperature oxidation embrittlement of SiC (Nicalon^{TM})/CAS ceramic matrix composites', *J Am Ceram Soc*, **90** (12), 4050–4054.

POWELL K L, SMITH P A and YEOMANS J A (1993), 'Aspects of residual thermal-stresses in continuous-fibre-reinforced ceramic–matrix composites', *Comp Sci Technol*, **47** (4), 359–367.

PREWO K M and BRENNAN J J (1980), 'High strength silicon carbide fibre-reinforced glass matrix composites', *J Mater Sci*, **15** (2), 463–468.

PREWO K M and BRENNAN J J (1982), 'Silicon carbide yarn reinforced glass matrix composites', *J Mater Sci*, **17** (4), 1201–1206.

PREWO K M (1989), 'Fibre reinforced glasses and glass–ceramics', in Lewis M H (Ed.), *Glasses and Glass-Ceramics*, London, Chapman and Hall, pp. 336–368.

SAMBELL R A J, PHILLIPS D C and BOWEN D H (1972a), 'Carbon fibre composites with ceramic and glass matrices, Part 1 – Discontinuous fibres', *J Mater Sci*, **7** (6), 663–675.

SAMBELL R A J, BOWEN D H, BRIGGS A and PHILLIPS D C (1972b), 'Carbon fibre composites with ceramic and glass matrices, Part 2 – Continuous fibres', *J Mater Sci*, **7** (6), 676–681.

SCHMIDT S, BEYER S, IMMICH H, KNABE H, MEISTRING R and GESSLER A (2005), 'Ceramic matrix composites: A challenge in space propulsion technology applications', *Int J Appl Ceram Technol*, **2** (2), 85–96.

SØRENSEN B F, HOLMES J W and VANSWIJGENHOVEN E (2000), 'Rate of strength decrease of fibre-reinforced ceramic matrix composites during fatigue', *J Am Ceram Soc*, **83** (6), 1469–1475.

SØRENSEN B F, HOLMES J W and VANSWIJGENHOVEN E (2002), 'Does a true fatigue limit exist for continuous fibre-reinforced ceramic matrix composites?', *J Am Ceram Soc*, **85** (2), 359–365.

SUN E Y and NUTT S R (1994), 'Interfacial microstructure and chemistry of SiC/BN dual-coated Nicalon-fibre-reinforced glass–ceramic matrix composites', *J Am Ceram Soc*, **77** (5), 1329–1338.

SUN E Y, NUTT S R and BRENNAN J J (1995), 'Flexural creep of coated SiC-fibre-reinforced glass–ceramic matrix composites', *J Am Ceram Soc*, **78** (5), 1233–1239.

SUN E Y, NUTT S R and BRENNAN J J (1996), 'High temperature tensile behavior of a boron nitride-coated silicon carbide-fibre-reinforced glass–ceramic matrix composite', *J Am Ceram Soc*, **79** (6), 1521–1529.

SUN E Y, NUTT SR and BRENNAN J J (1997a), 'Fibre coatings for SiC-fibre-reinforced BMAS glass–ceramic composites', *J Am Ceram Soc*, **80** (1), 264–266.

SUN E Y, LIN H-T and BRENNAN J J (1997b), 'Intermediate-temperature environmental effects on boron nitride-coated silicon carbide-fibre-reinforced glass–ceramic composites', *J Am Ceram Soc*, **80** (3), 609–614.

SUTHERLAND S, PLUCKNETT K P and LEWIS M H (1995), 'High temperature mechanical and thermal stability of silicate matrix composites', *Comp Engng*, **5** (10–11), 1367–1378.

VAN ROODE M, PRICE J, KIMMEL J, MIRIYALA N, LEROUX D, FAHME A and SMITH K (2007), 'Ceramic matrix composite combustor liners: a summary of field evaluations', *J Engng Gas Turbines Power*, **129** (1), 21–30.

VANSWIJGENHOVEN E, WEVERS M L and VAN DER BIEST O (1999), 'Influence of the laminate lay-up on the fatigue behaviour of SiC-fibre/BMAS–matrix composites', *Composites: Part A*, **30** (5), 623–635.

VICENS J, DOREAU F and CHERMANT J L (1995), 'The microstructure of experimental SiC-fibre-reinforced yttrium magnesium aluminosilicate (SiC-YMAS) materials', *J Microsc*, **177** (3), 242–250.

VICENS J, DOREAU F and CHERMANT J L (1997), 'Microstructure and creep characteristics of experimental SiC-YMAS composites', *J Microsc*, **185** (2), 168–178.

VICENS J, FARIZY G and CHERMANT J L (2003), 'Microstructure of ceramic composites with glass–ceramic matrices reinforced by SiC-based fibres', *Aerospace Sci Technol*, **7** (2), 135–146.

WESTWOOD M E, WEBSTER J D, DAY R J, HAYES F H and TAYLOR R (1996), 'Oxidation protection for carbon fibre composites', *J Mater Sci*, **31** (6), 1389–1397.

WETHERHOLD R C and ZAWADA L P (1991), 'Heat-treatments as a method of protection for a ceramic fibre-glass matrix composite', *J Am Ceram Soc*, **74** (8), 1997–2000.

WIDJAJA S, JAKUS K, RITTER J E, LARA-CURZIO E, WATKINS T R, SUN E Y and BRENNAN J J (1999), 'Creep-induced residual strengthening in a Nicalon-fibre-reinforced BMAS glass–ceramic matrix composite', *J Am Ceram Soc*, **82** (3), 657–664.

WU X and HOLMES J W (1993), 'Tensile creep and creep-strain recovery behavior of silicon-carbide fibre calcium aluminosilicate matrix composites', *J Am Ceram Soc*, **76** (10), 2695–2700.

YAJIMA S, HIYASHI J, OMORI M and OKAMURA K (1976), 'Development of a silicon-carbide fibre with high-tensile strength', *Nature*, **261** (5562), 683–685.

YAJIMA S, HASEGAWA Y, OKAMURA K and MATSUZAWA T (1978a), 'Development of high-tensile strength silicon-carbide fibre using an organosilicon polymer precursor', *Nature*, **273** (5663), 525–527.

YAJIMA S, HASEGAWA Y, HAYASHI J and IIMURA M (1978b), 'Synthesis of continuous silicon-carbide fibre with high-tensile strength and high Young's modulus. 1: Synthesis of polycarbosilane as precursor', *J Mater Sci*, **13** (12), 2569–2576.

YASMIN A and BOWEN P (2002), 'Fracture behaviour of cross-ply Nicalon/CAS-II glass–ceramic matrix composite laminate at room and elevated temperature', *Composites*, **A33** (9), 1209–1218.

YASMIN A and BOWEN P (2004), 'Fatigue behaviour of cross-ply Nicalon/CAS-II glass–ceramic matrix composite laminate at room and elevated temperature', *Composites*, **A35** (1), 83–94.

ZAWADA L P, BUTKUS L M and HARTMAN G A (1991), 'Tensile and fatigue behavior of silicon carbide fibre-reinforced aluminosilicate glass', *J Am Ceram Soc*, **74** (11), 2851–2858.

ZAWADA L P and WETHERHOLD R C (1991), 'The effects of thermal fatigue on a SiC fibre/aluminosilicate glass composite', *J Mater Sci*, **26** (3), 648–654.

ZHANG Y, XU Y, LOU J, ZHANG L, CHENG L and CHEN Z (2005), 'Braking behavior of C/SiC composites prepared by chemical vapor infiltration', *Int J Appl Ceram Technol*, **2** (2), 114–121.

ZUBER C and HEIDENREICH B (2006), 'Development of a net shape manufacturing method for ventilated brake discs in single piece design', *Mat-wiss U Werkstoff-tech*, **37** (4), 301–308.

3
Chemical ageing mechanisms of glass fibre reinforced concrete

H. CUYPERS, Vrije Universiteit Brussel, Belgium; and
J. ORLOWSKY, Institut für Bauforschung der RWTH Aachen, Germany

3.1 Introduction

3.1.1 Scope

Although most people would probably spontaneously associate 'chemical attack' with rapid degradation of a material under an aggressive environment, it can also occur under environmental conditions that give the impression of being undisruptive. If, in addition, the rate of chemical attack is relatively low, its effect in the long term can easily be underestimated or even completely overlooked. However, if ageing of a material leads to an apparent loss of mechanical performance during the service life of a structure, its effects should be taken into account at the design stage. The main focus in this chapter is thus the modelling of ageing due to chemical attack, which is occurring slowly but surely during the service life of the structure. Unfortunately, the design of models for simulating ageing under chemical attack is still awkward. Firstly, it is not always easy to identify the chemical attack mechanisms, to determine their relative importance and to decide which effects need to be implemented into the mathematical description of the ageing process. Secondly, once a durability model is chosen – based on literature and/or own experiments – it is not easy to calibrate this model in the laboratory within a reasonable time span. This problem is usually solved through the use of accelerated ageing techniques. This should, however, be done with care. Thirdly, a well-controlled and conditioned laboratory environment does not always represent a realistic and variable environment and the simulated loss of performance can be highly dependent on the method and hypotheses used in this stage of extrapolation. The three above-mentioned stages, needed to predict ageing of composites under chemical attack, will be discussed in more detail in this chapter.

3.1.2 Focus and limitations

Rather than focusing on the description of detailed chemical reactions for several material combinations, this chapter aims to present a global methodology for the modelling of ageing due to chemical attack from an engineering point of view. In order to illustrate the stages of identification of mechanisms, calibration of models and development of predictive tools, one specific material combination is chosen and will be used throughout this chapter: glass fibre reinforced concrete (GRC). It was soon generally recognised that in the case of GRCs the fibres were aged by the matrix material itself (e.g. Budd, 1961; Majumdar and Ryder, 1968; El-Shamy *et al.*, 1972). The reasons for using GRC as an illustrative example in this chapter are: (a) chemical attack in this type of composite has been widely discussed and documented; (b) GRC composites are subjected to chemical degradation, even in a seemingly non-aggressive environment – including everyday outdoor weathering; (c) many researchers have been able to improve the GRC materials, based on knowledge of the ageing mechanisms. Owing to the improvement of GRC material combinations, other fields of applications became possible. One of the more recent developments in GRC will be discussed in more detail this chapter, i.e. textile reinforced composites. Whereas the main aim of the introduction of fibres into a concrete matrix is usually to enhance the toughness and to obtain better crack control, textile reinforced concrete (TRC) is a relatively new load-bearing material that combines glass fibres and a concrete matrix (Brameshuber, 2006 (RILEM TC 201-TRC)). If TRC composites are produced one can obtain a tensile strength that is as high as or even higher than the compressive strength (Cuypers, 2002).

Although glass fibres are subject to static fatigue and although this static fatigue is also highly influenced by the chemical environment, this chapter will focus on the chemical attack of glass fibres in the absence of mechanical load. Since Chapter 4 in this book will be dedicated to static fatigue (also referred to as stress corrosion if the chemical environment clearly plays an important role in the degradation under mechanical load), this subject will not be studied thoroughly in this chapter. Still, some observations in this chapter will refer to specimens under constant load, but this will be mentioned explicitly, if necessary.

3.2 Problem identification

3.2.1 Introduction

Ageing of materials due to chemical attack under service conditions can be globally determined in two ways: (a) from specimens that were subjected to real environmental conditions and (b) from results obtained on similar specimens stored in a well-controlled environment in the laboratory. Unfortunately, when the studied processes are slow the first method is not a viable option for studying 'new' material combinations. Basic knowledge of the degradation mechanisms can however be used to study the materials of interest under 'accelerated' ageing, meaning that the nature of the attack is not modified and processes are only accelerated within a laboratory environment. For GRCs this method is commonly used (e.g. Litherland *et al.*, 1981) and will be discussed in detail. Since ordinary glasses (such as soda lime–silica or borosilicate compositions) were found to lose strength at an unacceptable rate – from an engineering point of view – in an alkaline environment (Budd, 1961; El-Shamy *et al.*, 1972; Paul, 1982), alkali-resistant (AR)-glass fibres were developed in the early days of GRC research (Dimbleby and Turner, 1926; Majumdar and Ryder, 1968; Majumdar and Tallentire, 1973). A high amount (~16 wt%) of zirconium (Zr) is incorporated into the glass network in order to improve the resistance towards ageing. Although loss of strength is still reported for GRCs with AR-glass fibres, the degradation mechanisms slow down considerably. Recent results on ageing mentioned in this chapter are thus obtained on AR-glass fibres, unless mentioned otherwise.

3.2.2 Possible damage mechanisms

When composite materials are subject to ageing in general, several mechanisms can be attributed to the loss of strength. In GRCs, for example, the following possible damage mechanisms are usually recognised:

- direct chemical attack of the fibres by the alkaline pore solution of the matrix;
- increasing transverse pressure on the fibres due to the deposition of reaction products from the matrix;
- embrittlement of the matrix–fibre bond;
- static fatigue (under permanent mechanical load).

Although the introduction of damage might occur due to mechanical action for some of the above-mentioned mechanisms (e.g. transverse pressure on fibres by reaction products), a slowly progressing chemical reaction is usually at the source of this effect (e.g. deposition of reaction products at the matrix–fibre interface).

3.2.3 Identification of external sources influencing the chemical attack

Many authors have discussed the chemical attack of various glass compositions under water and/or submerged in alkaline solutions and found that the rate of chemical attack increases with increasing temperature (e.g. Hillig and Charles, 1965; Wiederhorn, 1972; Litherland *et al.*, 1981) and with increasing partial pressure of water (e.g. Wiederhorn, 1967; Freiman, 1980) or increasing water content in the concrete (e.g. Orlowsky *et al.*, 2004). Obviously, ageing of the studied composites is thus a function of the climate in which the composite has to fulfil its load-bearing function. If any predictive tool is to be developed for the chemical attack of glass fibres in GRCs, it should thus take into account the dependency on temperature and humidity.

3.3 Experimental methods

Loss of performance of composite materials, which will for now be restricted to loss of strength, can be due to several mechanisms, which can influence one another and are usually not easily identified individually within full-scale composite testing. As already mentioned, loss of strength might occur due to several mechanisms in the fibres, in the matrix or at the interface between both fractions. These mechanisms can be chemical, mechanical or a combination of both. In this chapter a multi-scale approach will thus be used to explain the individual identification of several possible effects and to discuss their relative importance. The following stages, with increasing complexity, will be discussed: (a) direct chemical attack of the fibres, (b) matrix–fibre interface effects and (c) full composite action, including pull-out effects and redistribution of stresses. Since the main focus of this section is chemical attack, this subject will be discussed in more detail than the other topics. In this section a global overview of the testing methods that are commonly used to determine the nature or evolution of ageing of GRCs will be given.

3.3.1 Single fibres in a pore solution

Direct chemical attack of fibres can be studied if fibres are stored in solutions representing or resembling the same chemical environment as the solid matrix. The advantage of this technique is that direct chemical attack can be studied without the influence of mechanical damage, occurring at the interface between the fibres and the solid matrix. The relative importance of chemical attack can thus be properly assessed. On the one hand, the magnitude and evolution of chemical attack can be determined from analysis of the solution in order to determine which elements are leached out of the glass structure with time and at various temperatures (e.g. Larner *et al.*, 1976; Paul, 1982). On the other hand, glass fibres can be studied using a large variety of methods, including weight loss determination (Scarinci *et al.*, 1986), microscopic examination of the evolution of fibre sections, surface roughness profiling with atomic force microscopy (Gao *et al.*, 2003a), scanning electron microscopy (SEM) (Orlowsky *et al.*, 2004), X-ray photoelectron spectroscopy (Koshizaki, 1988), nano-indentation (Gao *et al.*,

2003b) and single filament tensile testing (Orlowsky *et al.*, 2004). Since the loss of strength of single glass fibres is the result of the combined effects of attack of the glass structure and linear elastic fracture mechanics, as will be explained later, one should be careful when, for example, drawing a straightforward relationship between the loss of fibre section and remaining strength. Since even at the scale of one single fibre several mechanisms could interact (see Section 3.3.2), several experimental techniques should be combined in order to understand the relationship between chemical attack and subsequent loss of load-bearing performance.

3.3.2 Fibres in a matrix block

Although the observation of fibres in solutions, simulating the environment in the adjacent proximity of the fibres, can provide some useful information on the existence and magnitude of loss of strength of the fibres due to chemical attack, one cannot be sure that the studied aqueous solutions correspond to the real environment in the composite at all times. In real composites, ongoing reactions within the matrix and – even more – at the interface between matrix and fibres might lead to changes of the surroundings of the fibres. Moreover, microstructural changes in the interface might induce mechanical effects such as transverse pressure on the fibres. If one wishes to study the combined effect of chemical attack of the fibres and densification of the interface, a bundle of fibres can be embedded into a matrix block for a limited length. One can then determine the remaining strength of the fibres after ageing through a direct tensile test on the fibre bundle. For GRCs this type of test, called the strand-in-cement (SIC) test has been used by many researchers (Litherland *et al.*, 1981; Proctor *et al.*, 1982) and is currently described by the standard EN 14649:2005. The global specimen set-up is shown in Fig. 3.1.

Apart from the residual strength, the evolution of single filament or fibre bundle pull-out curves can be used to determine the mechanical effects of modifications at the interface (e.g. Nammur and Naaman, 1989; Li and Chan, 1994; Banholzer *et al.*, 2006). Apart from pull-out, petrography and image analysis (Purnell *et al.*, 2000), laser scanning microscopy or SEM (Banholzer, 2004) can also provide useful information on the global mecha-

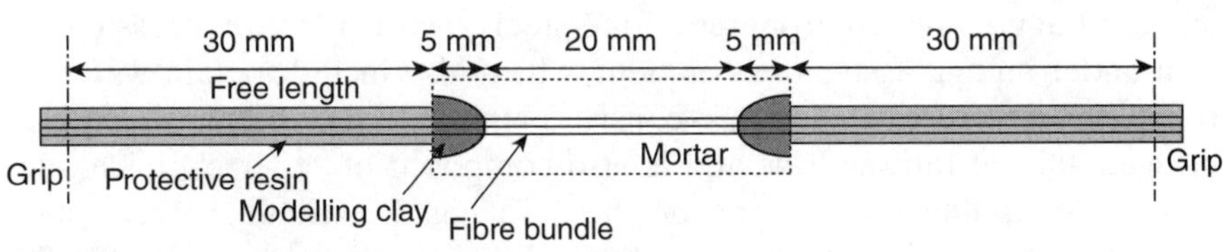

3.1 Strand-in-cement (SIC) specimen.

nisms at the interface. It should, however, be mentioned that the complexity of these evolutions at the interface is extreme. Since mechanical effects and chemical and/or microstructural analysis at the interface have been studied separately, chemical and mechanical results are usually not linked numerically to each other.

3.3.3 Full composite testing

The last, but from engineering point of view most interesting, stage is the study of full composite specimens. Basically, the experimental methods discussed for the matrix–fibre bundle unit cell discussed above can also be applied on full composite specimens. For GRCs and especially for TRC, however, care should be taken when comparing results obtained under direct tensile loading and results obtained under bending. Since the behaviour under tensile stress is clearly non-linear (Aveston *et al.*, 1971; Aveston *et al.*, 1974; Cuypers, 2002; Cuypers and Wastiels, 2006), there will be a continuous redistribution of stresses in specimens under bending. The evolution of the 'apparent' failure stress due to ageing obtained under bending or tensile testing might therefore be considerably different (Blom *et al.*, 2007) if linear elasticity is used to determine 'material stresses' from measured forces.

3.4 Modelling of the chemical attack of fibres

In order to explain how the loss of strength of composites can be modelled in general, and more specifically for GRC composites, firstly a literature overview of the possible mechanisms will be given; secondly, previous findings on the influence of temperature and humidity will be discussed. Based on this overview, a global methodology will be presented, which can be used to validate the durability of new material combinations and to discuss the usefulness of several possible models, as presented in the literature. This methodology will be illustrated in detail on an experimentally obtained series of data.

3.4.1 Literature overview

Figure 3.2 shows a Si–O–Si structure under a combination of mechanical load and aggressive environment. Like steel, glass can suffer stress corrosion under an aggressive environment, which also includes plain water. In reality, the structure of silica glass consists of interconnected rings of oxygen-bridged silicate tetrahedrals and is also composed of other chemical elements (depending on the type of glass) so Fig. 3.2 should be seen as a two-dimensional simplification of a three-dimensional structure. Within the theory of stress corrosion for glass, it is assumed that the glass structure is

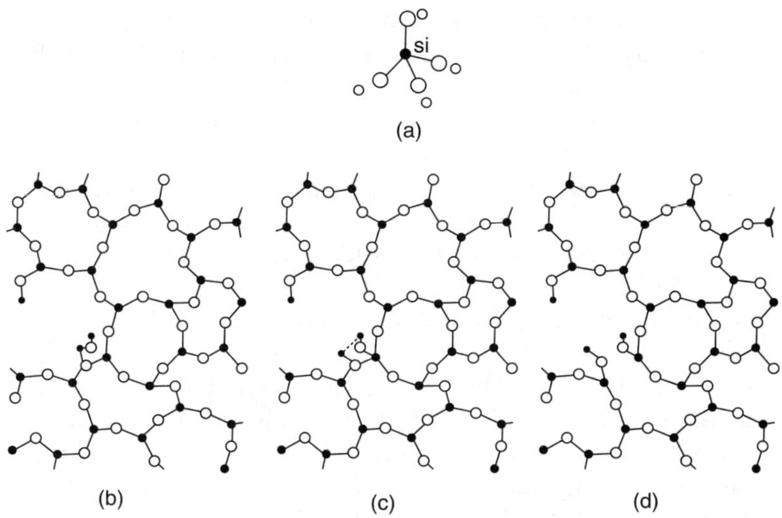

3.2 (a) Silicate tetrahedron and (b)–(d) two-dimensional simplification of chemical attack of Si—O—Si structure by water molecule. (b) formation of hydrogen; (c) breaking of Si—O bond (in the glass) and an O—H bond (in the water molecule); (d) the weak hydrogen bond is broken. Figures adapted from Wright (1993) and Michalske and Bunker (1987).

preferentially chemically attacked at those points where stress concentrations are existing and where the chemical bonds are therefore already strained (e.g. Wiederhorn, 1967; Freiman, 1980). Stress-enhanced attack by water molecules therefore occurs preferentially at existing crack tips. When water molecules react with the glass structure at an existing crack tip, this stress corrosion usually comprises three stages (following, for example, Michalske and Bunker (1987)): (1) a water molecule adsorbs to the crack tip bond (through a hydrogen bond formation with the bridging oxygen); (2) at the crack tip both an Si–O bond (in the glass) and an O–H bond (in the water molecule) are then broken and two new silanol groups are formed; finally (3) the weak hydrogen bond is broken. When glass fibres are studied, the above-mentioned cracks will occur in the form of flaws. Flaws are small defects, with an order of magnitude of 10–100 nm, that are inherently present in the glass fibres and are introduced during the production process. These flaws were studied using atomic force microscopy (AFM) by Gao *et al.* (2003a) and using SEM by Orlowsky *et al.* (2004) and both research groups noted that for AR-glass fibres the fibre surface was not attacked within an alkaline environment as a whole, but that only a limited number of flaws were deepened with time. Owing to the brittle behaviour of glass however, the growth of a limited number of glass flaws is sufficient to lead to a considerable decrease in load-bearing capacity.

More generally, according to Michalske and Bunker (1987) every molecule that has a labile cation and a lone electron pair could theoretically serve as a stress-corrosion agent for glass. However, since this process preferentially occurs at crack tips and since the crack itself has limited opening, only molecules with a limited molecular size will be able to reach the crack tip. From comparison of the acid–base properties of several test chemicals and the corrosive effect of these chemicals on silica glass, Michalske and Bunker (1987) decided that the order of magnitude of steric hindrance is about 0.5 nm.

In a cementitious environment, however, the described process becomes more complex. Several alkalis are present in the pore solution of a Portland cement-based matrix, which is the most commonly used cement in the concrete industry. Various authors have indicated the presence of soluble silica in water after ageing of various glass fibre types in different solutions. From the study of solutions with different pH values (e.g. El-Shamy *et al.*, 1972; Paul, 1982; Scholze, 1982; Adams, 1984) it can be seen that the solubility of silica is low in acidic and neutral solutions, but increases rapidly with alkalinity. The main corrosion mechanism in alkaline solutions is thus slightly different from that presented above, and can be globally described as (Charles, 1958; Simhan, 1982):

$$\text{Si—O—Si} + \text{OH}^- \rightarrow \text{Si—OH (silanol group in glass structure)}$$
$$+ \text{Si—O}^- \text{ (in solution)}$$

Depending on the pH of the studied solution, the form in which the soluble silica is present might differ. Following Paul (1982), the soluble silica is mainly present as H_2SiO_3 when the pH equals or is lower than 10, as $HSiO_3^-$ between pH 10 and 12, and as SiO_3^- when the pH is higher than 12. According to Charles (1958) and Simhan (1982), the Si—O^- in solution further reacts with water to form silanol (Si—OH) and hydroxide (OH^-). Furthermore, if sodium exists within the glass structure (e.g. AR-glass contains around 15% Na_2O), several authors have noted, from analysis of the solutions in which the glass fibres were stored, that this sodium is also clearly leached out of the glass structure (Douglas and Isard, 1949; Larner *et al.*, 1976; Simhan, 1982).

The presence of $Ca(OH)_2$ in concrete complicates the evolution of the corrosion of glass fibres in concrete even more. Accelerating as well as decelerating effects have been assigned to the precipitation of $Ca(OH)_2$. Although all reports (e.g. Scarinci *et al.*, 1986; Koshizaki, 1988; Yilmaz and Glasser, 1991) have noted that a surface layer, which is rich in calcium, is built at the glass surface, the effect of this layer is interpreted in different ways. On the one hand, Scarinci *et al.* (1986) reports that the relative permeability of this layer limits further corrosion and leads to a diffusion-controlled rate of chemical attack. Yilmaz and Glasser (1991), on the other hand, suggested that $Ca(OH)_2$ could precipitate into the existing flaws of the glass fibres and that the internal pressure of $Ca(OH)_2$ crystals could notch the glass fibres. There are both older and more recent interesting papers, however, that disagree with the idea that this mechanism could be the main source of the loss of strength of fibres in a concrete matrix: Majumdar (1980) concluded that $Ca(OH)_2$ is typically softer than glass and can thus not lead to notching and Orlowsky *et al.* (2004) showed that glass fibres also lose strength in simulated concrete pore solutions when calcium is left out. In fact, the loss of strength in the simulated pore solutions showed the same order of magnitude as the loss of strength of the benchmark composite specimens on which the pore solution composition was based (Orlowsky *et al.*, 2004).

To summarise, even when only the direct chemical attack of glass fibres in concrete matrices is discussed, there is still quite some discussion on the nature of the chemical attack of a glass fibre in a concrete environment and on the relative importance of chemical mechanisms on the loss of strength. Of course the vast number of glass fibre compositions, coatings and concrete matrix compositions that have been investigated make it difficult, if not impossible, to define one single degradation mechanism that holds globally for all GRC. Even so, nowadays the majority of authors (including the authors of this chapter) agree on some important conclusions that can be used to model the loss of strength of glass fibres due to chemical attack in simulated pore solutions as a function of temperature, humidity and time.

1 The remaining strength of the glass (and thus also of glass fibres) is determined by fracture mechanics: the loss of strength of the fibres is thus not due to global loss of fibre diameter, but due to the growth of the flaws (small defects) inherently present in the fibre. The order of magnitude of these flaws is about 10–100 nm (e.g. Gao *et al.*, 2003a; Orlowsky *et al.*, 2004). The rate of loss of strength is directly linked to the rate of growth of the flaws.
2 The temperature dependency of the rate of chemical attack is expressed according to an Arrhenius relationship (Hillig and Charles, 1965; Wiederhorn, 1972; Litherland *et al.*, 1981; Purnell *et al.*, 2001; Orlowsky *et al.*, 2004). Usually it is assumed that only one (Hillig and Charles, 1965; Wiederhorn, 1972; Litherland *et al.*, 1981; Purnell *et al.*, 2001), or two (Orlowsky *et al.*, 2004) mechanisms occur as a function of the studied time–temperature frame. This Arrhenius relationship is thus usually used to express the dependency of a maximum of two rate constants as a function of temperature.
3 The rate of chemical attack of glass is a function of the humidity. For example, Wiederhorn (1967) and Freiman (1980) noted that the rate of

chemical attack of glass increases with increasing partial pressure of water within the surrounding environment. Orlowsky (2005) describes an increasing rate of ageing of full composite TRC specimens with increasing relative humidity.

The three above-mentioned conclusions will be further used to develop various models for the description of loss of strength of GRCs due to ageing under chemical attack.

3.4.2 Direct chemical attack of fibres: modelling the evolution with time

Following the assumptions given in the previous sections, an expression can be constructed for the loss of strength of the fibres under chemical attack. The expression and calibration technique that will be presented in this section are globally independent from the detailed chemical reactions. Since it is assumed that the evolution of strength with ageing is a function of the growth of surface flaws, the bulk failure stress of glass fibres can be calculated as a function of the flaw size and the geometry of the largest flaw according to classical linear elastic fracture mechanics (LEFM):

$$\sigma_{\mathrm{bf}} = \frac{K_{1\mathrm{c}}}{A\sqrt{\pi a}} \qquad [3.1]$$

where:

σ_{bf} = bulk tensile strength of the fibre (= force at failure/fibre area) ($\mathrm{MN/m^2}$)
$K_{1\mathrm{c}}$ = critical stress intensity factor (mode I) ($\mathrm{MN/m^{-3/2}}$)
A = shape factor (—)
a = flaw depth (m)

The critical stress intensity factor is a function of the material used and it is usually assumed that this parameter is invariable with time. Although chemical attack could lead to a change in the shape of flaws, it is quite often assumed that this parameter is also constant. Effective changes of the shape factor (A) are then included into the assumed variation of the flaw depth (a) during calibration of the model. If it is thus assumed that only the variation of the depth of the largest flaw leads to loss of strength of the glass fibre, the relative loss of strength L of fibres, stored under a stable environment for a certain time t, can be written as:

$$L = \frac{\sigma_{\mathrm{bf}}(t=0) - \sigma_{\mathrm{bf}}(t)}{\sigma_{\mathrm{bf}}(t=0)} = \frac{\dfrac{K_{1\mathrm{c}}}{A\sqrt{\pi a_0}} - \dfrac{K_{1\mathrm{c}}}{A\sqrt{\pi a}}}{\dfrac{K_{1\mathrm{c}}}{A\sqrt{\pi a_0}}} = 1 - \frac{1}{\sqrt{1 + \dfrac{X}{a_0}}} \qquad [3.2]$$

where:

a_0 = flaw depth at time $t = 0$
X = increase of flaw depth

This means that the evolution of the loss of strength is function of the evolution of the local chemical attack of the glass fibre. Unfortunately, several points of view can be found in the literature concerning the rate-dependent mechanisms of the growth of the crack (X) and therefore several models have been proposed for the evolution of X. An overview of these mechanisms is given in Table 3.1 and a more detailed explanation of the models is presented below.

Some authors have noted that the growth of the largest flaw is linear with time (Purnell *et al.*, 2001) and that the ion exchange at the crack tip is the limiting factor for the total rate (see mechanism 1 in Table 3.1).

Other authors have indicated that at a certain point the chemical attack of glass fibres becomes diffusion controlled owing to the precipitation of reaction products from the glass itself (Orlowsky *et al.*, 2004), as a result of the precipitation of a $Ca(OH)_2$-rich layer (Scarinci *et al.*, 1986), due to the fact that aggressive ions have to diffuse towards the crack tip through a small flaw (Wiederhorn, 1967; Orlowsky *et al.*, 2004) or due to the increasing Zr/Si ratio at the glass surface of AR-glass fibres when Si is leached out of the glass (Simhan, 1982). This diffusion-controlled rate is shown as mechanism 2 in Table 3.1.

The previously mentioned mechanisms can also be combined (mechanism 3) and it is then assumed that the rate of degradation is initially determined by the ion exchange at the crack tip and that the reaction becomes diffusion controlled at a later stage (Orlowsky *et al.*, 2004). In some cases, chemical attack might change the shape of cracks or flaws and even lead to a more blunted crack tip with time. Although flaw depths are thus increasing, the stress concentrations at the crack tip do not increase linearly with the depth of the crack tip due to the modified shape factor (A). Since in this case the apparent loss of strength L seems to occur at a slower rate than would be predicted if only the growth of the largest flaw X is introduced, a factor n is introduced to express the non-linear dependency of X and L. This effect, which leads to an apparent decelerating effect, is shown as mechanism 4 in Table 3.1. Larner (1976) noted that the corrosion velocity of glass fibres in simulated pore solutions decreases with time, but proposed a logarithmic relationship between the growth and time based on his experimental observations; this is shown as mechanism 5 in Table 3.1.

3.4.3 Modelling the influence of temperature

Rather than expressing any prejudice on any of the above-mentioned evolutions with time, it will now be discussed further how these models can be

Table 3.1 Overview of models to describe the growth of flaws in glass fibres under chemical attack

Mechanism no.	Mechanism name	Model	Reference
1	Ion exchange	$\frac{dX}{dt} = k_1(T, RH)$	Purnell *et al.* (2001)
2	Diffusion controlled	$\frac{dX}{dt} = \frac{k_2(T, RH)}{X}$	Simhan (1983) Scarinci *et al.* (1986) Wiederhorn (1967)
3	Combined model	$\frac{dX}{dt} = \frac{1}{\frac{1}{k_1(T, RH)} + \frac{X}{k_2(T, RH)}}$	Orlowsky *et al.* (2004) Cuypers and Wastiels (2006)
4	*n*-Model	$\frac{dX}{dt} = \frac{k_3(T, RH)}{X^n}$	Based on Freiman (1980)
5	Logarithmic	$X = a_0 + k_4(T, RH)\log(t)$	Larner *et al.* (1976)

RH, relative humidity.

calibrated, based on short-term, well-controlled accelerated test series, and how these results can then be used to predict the lifetime of a similar material under less stringent, long-term environmental conditions. As already mentioned in the literature overview, many authors agree that the dependency of the rate of the flaw growth on the temperature can be described by an Arrhenius relationship. Globally, one can thus write for a rate parameter:

$$k_i = k_{0i} e^{-\frac{\Delta H_i}{RT}} \quad [3.3]$$

where:

k_i = rate parameter of the degradation mechanism under consideration (see Table 3.1)
R = gas constant
ΔH_i = activation energy of process I
k_{0i} = material constant, linked to process I
T = temperature (in Kelvin)

For each rate constant used within the description of the flaw growth, two model constants are thus introduced to represent the dependency of the rate on the temperature. If the evolution of the strength of single filaments, SIC specimens or full composites is thus obtained for several temperatures under a constant relative humidity (usually immersed in water), the proposed model(s) can be calibrated for temperature dependency. If a new material combination is to be modelled for durability, the steps to be followed are as shown below.

1 Determination of the loss of strength at several time intervals for various elevated temperatures.
2 Determination of the experimental evolution of the crack growth for various temperature series (see equation [3.2]).
3 Determination of the rate constants as function of temperature for the chosen model (following Table 3.1).
4 Determination of the Arrhenius parameters (see equation [3.2]).

One of the dangers of the above-mentioned methodology is the inadequate use of ageing under elevated temperatures. The Arrhenius equation parameters are usually obtained from test series at higher temperatures in order to obtain useful results within a limited time schedule (typically weeks or months instead of years). For traditional cementitious matrices, this methodology is usually valid, since ongoing reactions are indeed only accelerated and not altered when the temperature is increased. Several authors, however, have already indicated that this technique should be used with care when modified – more durable – matrices are developed (Alshaer, 2006; Van

Itterbeeck *et al.*, 2007). It was noted in this case that the relative importance of various chemical reactions alters considerably with increasing temperature. In this case the resulting loss of strength at higher temperatures and shorter times cannot be linked to long-term variations at lower everyday temperatures. The predictive extrapolative value of the accelerated ageing technique is thus lost.

3.4.4 Modelling the influence of humidity

In contrast to the vast number of papers published on the effect of temperature on the chemical attack of fibres, only a few authors mention the effect of humidity – although it is also an important parameter. Based on the measurement of crack velocities (Wiederhorn, 1967) in glass under load and under varying relative humidity levels, the overall crack growth could be divided into regions according to the stress intensity at the crack tip, which is function of the crack size (among other factors). Initially, as long as the stress intensity was relatively small at the crack tip, the velocity of the crack growth was noted to be a function of the relative humidity. Freiman (1980) specified that the rate of crack growth is influenced by the partial pressure rather than the absolute quantity of water. If one takes a closer look at the curves representing the rate of corrosion versus stress intensity factor, as presented by Wiederhorn (1967), it can be seen that the logarithm of the rate of corrosion is a linear function of relative humidity. Figure 3.3 shows the evolution of the corrosion rate as a function of the relative humidity at one fixed stress intensity value (extrapolated towards zero force in this case), as determined indirectly on the resulting curves, published by Wiederhorn (1967). This would mean that an exponential relationship can therefore be used to link the rate of corrosion to the relative humidity (*RH*):

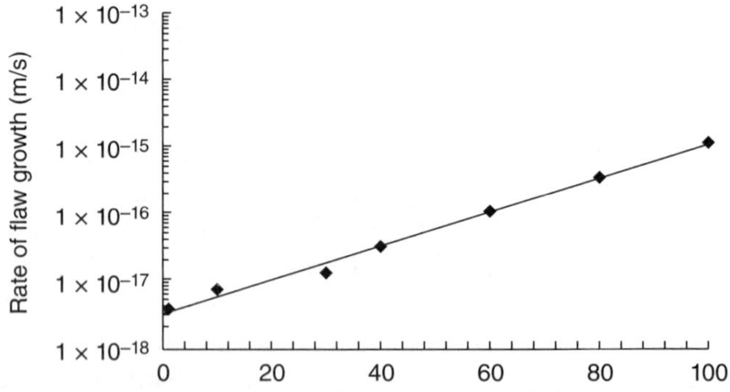

3.3 Effect of relative humidity on crack velocity. Figure adapted from Wiederhorn (1967) and Freiman (1980).

$$k_{i,RH} = k_{i,100} e^{C_i(RH-100)} \qquad [3.4]$$

where:

$k_{i,RH}$ = rate parameter, function of relative humidity for constant temperature, linked to process I
C_i = material constant
RH = relative humidity
$k_{i,100}$ = rate parameter as measured at 100% relative humidity, linked to process i

Orlowsky (2005) studied the corrosion of GRC under varying relative humidity and also concluded that there is a noticeable relationship between the water content in the concrete matrix and the corrosion rate.

3.4.5 Calibration example and discussion

The theoretical background explained in the previous section will be used in this section on a series of recently obtained experimentally results and thus for illustrative purposes. Several series of SIC specimens, which consisted of AR-glass fibre bundles in a fine-grained concrete matrix, were tested for remaining strength, after they were stored under water at various temperatures. AR-glass fibres with a diameter of 14 μm were bundled into bundles of 320 tex (1 tex = 1 g/km), meaning that around 800 fibres were used per bundle.

The composition of the concrete matrix is given in Table 3.2. In this case normal Portland cement is used without any modifications in order to obtain a high alkalinity and thus create a potential aggressive chemical reaction. The pH of the solution in the pores was determined and was found to equal 13.5. Three series of SIC specimens (see Section 3.3.2) were prepared. All series were stored under water: the first series at 20 °C, the second series at 50 °C and the third series at 80 °C. After specific periods of

Table 3.2 Composition of the fine-grained concrete matrix used

	Additives						
Cement	Silica fume	Fly ash	Super plasticiser	Total binder content (cement + additives)	Water/ binder ratio	Quartz sand 0.2–0.6 mm	Quartz flour 0–0.2 mm
(kg/m³)	(kg/m³)	(kg/m³)	(kg/m³)	(kg/m³)	—	(kg/m³)	(kg/m³)
490	35	175	1[1]	700	0.4	714	500

[1] Mass percentage relative to cement content.

Table 3.3 Loss of strength of AR-glass fibre bundles in concrete matrix A, after storage under water at elevated temperature: results of series stored at 20, 50 and 80 °C; SIC specimens (adapted from Orlosky (2005))

20 °C		50 °C		80 °C	
Time	Loss	Time (days)	Loss	Time (days)	Loss
5	0.003	5	0.142	7	0.571
7	0.036	7	0.213	14	0.665
45	0.08	14	0.28	28	0.663
56	0.115	20	0.323	56	0.656
80	0.12	25	0.337	90	0.675
112	0.158	28	0.403		
168	0.185	56	0.471		
180	0.119	112	0.498		
365	0.234	180	0.492		

Table 3.4 Calculated growth of largest flaw (nm)

20 °C		50 °C		80 °C	
Time (days)	Growth (nm)	Time (days)	Growth (nm)	Time (days)	Growth (nm)
5	0.241 084 34	5	14.335 7187	7	177.342 875
7	3.043 336 03	7	24.581 8567	14	316.426 821
45	7.258 979 21	14	37.160 4938	28	312.208 789
56	11.070 8928	20	47.273 5524	56	298.020 552
80	11.652 8926	25	50.998 2278	90	338.698 225
112	16.420 3542	28	72.230 6115		
168	20.220 5578	56	102.938 311		
180	11.535 6994	112	118.727 639		
365	28.171 4375	180	115.000 31		

storage, ten specimens were removed from the storage tank, dried for 7 days and subsequently tested in tension up to failure. Table 3.3 gives an overview of the evolution of the remaining strength of the SIC specimens as a function of time.

Section 3.3.5 explained briefly in four steps how experimental results can be used to calibrate a durability model. This methodology will be shown here in detail, assuming that ion exchange at the crack tip is the main mechanism that controls the rate of loss of strength. The experimentally obtained results are used to calculate the associated evolution of the growth of the largest flaw X, using equation [3.2]. The value of the initial flaw depth is set to 40 nm in this case (Orlowsky, 2005). Table 3.4 depicts the calculated evolution of X. For each temperature series, a value of k_i can now be determined. This value is determined in such a way that the model representing the growth of the largest flaw associated to the ion exchange mechanism (see Table 3.1, mechanism 1) finds a best fit with the experimental results. In this case the slope of the best-fitting linear line going through the points of Table 3.4 is thus withheld. This means that:

$k_1 = 0.0894$ nm/day at 20 °C
$k_1 = 0.894$ nm/day at 50 °C
$k_1 = 5.0212$ nm/day at 80 °C

The relation between the applied temperature and rate constant can be described through the Arrhenius relationship (equation [3.3]). If now for the three temperature series the logarithm of the rate constant is depicted as a function of $1/T$ (where T is temperature in Kelvin), Fig. 3.4 can be constructed. The values of the model parameters are thus determined, since according to Fig. 3.4, $k_{01} = 1.35 \times 10^9$ nm/day and the value of $\Delta H/R = 6950$ K. It should however be noted that, although Fig. 3.4 seems to show that the assumed Arrhenius dependency was confirmed, serious discrepancies were found between the assumed linear evolution of flaw growth with time and the evolution as determined from the loss of strength.

Figure 3.5 shows, for example, the experimental and theoretical growth of the flaws of the series stored at 50 °C. It can be seen that the experimentally observed growth seems to decelerate with time. Since the other four models, depicted in Table 3.5, describe a decelerating trend with time, they are now calibrated on the same experimental results. The proposed meth-

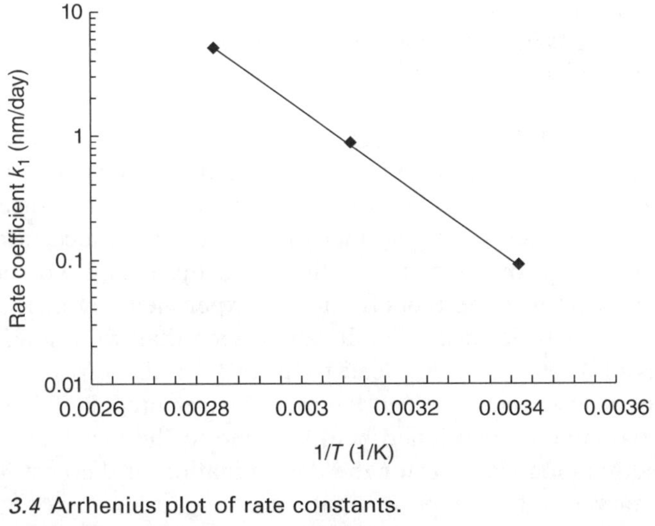

3.4 Arrhenius plot of rate constants.

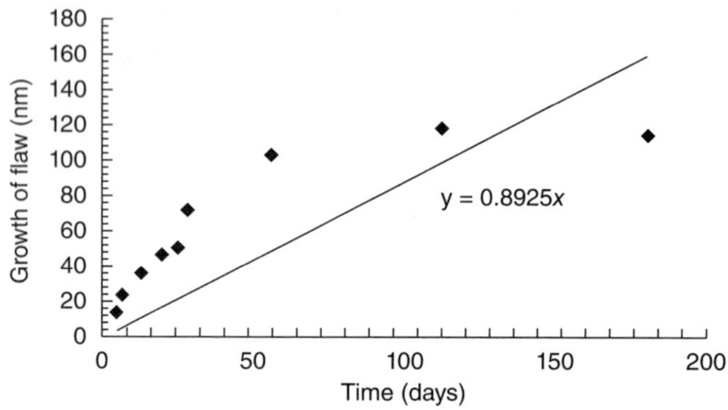

3.5 Theoretical (straight line) versus experimental (◆) evolution of growth of flaw at 50 °C.

odology cannot, however, be used in a straightforward way for the fourth model, depicted in Table 3.1, since this model comprises a parameter that influences the rate, but is constant for all tested temperature series, whereas all other rate constants (k_i) can be determined per temperature series. For this reason, the determination of the model constants is performed in a slightly different way: the model parameters of this model are determined at the same time, using all experimental data together. The parameters are determined in such a way that the total discrepancy between all experimental results and theoretical predictions is minimised. This minimum is expressed by the total least square coefficient value as follows:

$$LSC = \frac{\sum (L_{theoretical} - L_{experimental})^2}{\text{Number of points}} \qquad [3.5]$$

In Table 3.5, the resulting values of the model parameters and minimal values of the LSC are depicted for all models.

As can be seen from Table 3.5, the diffusion-controlled model, the combined model and the *n*-model seem to provide the lowest discrepancy between experimental results and theoretical predictions. The fact that model parameter *n* seems to be close to 1 for the *n*-model indicates that the ageing of the tested series was indeed diffusion controlled rather than being controlled by ion exchange at the crack tip. Figure 3.6 shows the worst- and best-performing models and the experimentally observed loss of strength for the series discussed. It can be seen that, with time, the first mechanisms of Table 3.1 would lead to an over-prediction of the loss of strength. The discrepancy between the model parameters in Table 3.5 and the model parameters mentioned before is due to the fact that a slightly different methodology is used for the determination of the best fit of the theoretical curves and the experimental results.

Table 3.5 Model coefficients and goodness-of-fit of several models, determined on series of experimental results obtained from SIC specimens

Mechanism no.	Mechanism name	Model	Material constants	LSC
1	Ion exchange	$\frac{dX}{dt} = k_1(T, RH)$	$k_{01} = 6.31 \times 10^{11}$ nm/day $\Delta H_1/R = 8600$ K	0.0058
2	Diffusion controlled	$\frac{dX}{dt} = \frac{k_2(T, RH)}{X}$	$k_{02} = 3.16 \times 10^{19}$ nm^2/day $\Delta H_2/R = 13\,200$ K	0.0012
3	Combined model	$\frac{dX}{dt} = \frac{1}{\frac{1}{k_1(T, RH)} + \frac{X}{k_2(T, RH)}}$	$k_{01} = 2.34 \times 10^{18}$ nm^2/day $\Delta H_1/R = 12\,790$ K $k_{02} = 4.72 \times 10^{18}$ nm^2/day $\Delta H_2/R = 12\,617$ K	0.0012
4	*n*-Model	$\frac{dX}{dt} = \frac{k_3(T, RH)}{X^n}$	$k_{03} = 1.58 \times 10^{19}$ nm$^{(n+1)}$/day $\Delta H_3/R = 13\,000$ K $n = 0.98$	0.0012
5	Logarithmic	$X = a_0 + k_4(T, RH)\log(t)$	$k_{04} = 5.01 \times 10^9$ nm/day $\Delta H_4/R = 6000$ K	0.0019

LSC, least square coefficient.

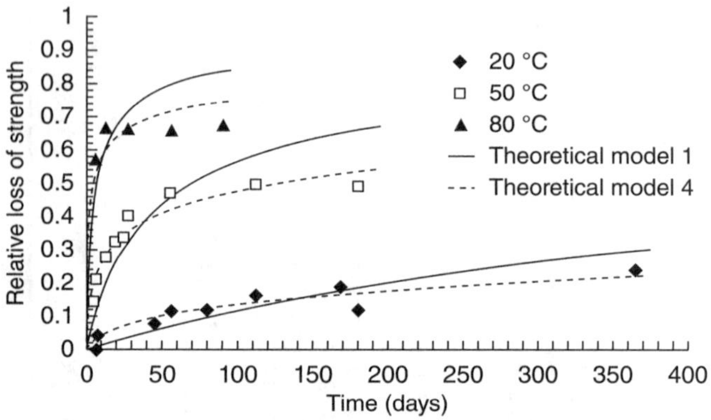

3.6 Results obtained on SIC specimens together with the 'best-fitting' and 'worst-fitting' model of Table 3.5.

3.5 Interface effects

Until now, no straightforward experimental methods have been available for the determination of the interface properties. Many different alternative test set-ups and experimental techniques have been developed in order to understand and interpret the bond properties and bond evolution. Moreover, the phenomena occurring at the interface can be studied at several length scales: one can find in literature results obtained on single filament pull-out, fibre bundle pull-out and composite specimens with textiles (e.g. Banholzer, 2004). As already mentioned, the interpretation of the behaviour of the interface is not straightforward. The results of single fibre pull-out are global load–displacement curves. If one wants to determine the local relation between slip and shear stress, a proposal is usually made considering the shape of the shear strain versus slip curve before the back calculation is performed. Figure 3.7 shows some of these proposed theoretical shapes. Usually, at low load levels, the interface can be assumed to behave elastically. Furthermore, it is often assumed that interface failure occurs and debonding takes place when the stress at the interface reaches a critical value. After reaching this critical value stress transfer between matrix and fibres only occurs due to friction.

Recently (Banholzer, 2004) developed a methodology that can be used to determine the slip versus bond stress from global load–displacement curves, without any preliminary hypothesis on the global shape of the curve of slip versus shear stress. From knowledge of the load versus displacement relation $P(\omega)$, the embedded length L, the fibre stiffness E_f and cross-section A_f, the slip versus bond stress $\tau(s)$ can be determined piece-wise with the help of an iterative inverse boundary value method. For more detailed information on this methodology and an in-depth overview and

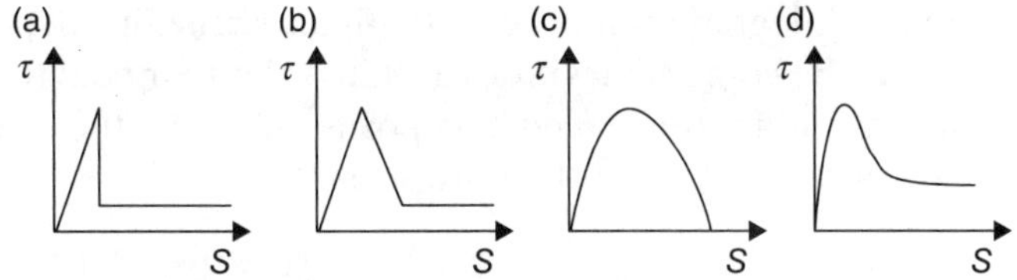

3.7 Proposed slip (s) versus bond shear stress (τ) relationships. (a) Linear elastic part followed by a sudden stress drop and constant residual friction; (b) linear elastic part followed by gradual softening and constant residual friction; (c) non-linear relationship; (d) schematically shown global shape of slip versus bond stress as obtained from an iterative inverse boundary value method. Figures adapted from Banholzer *et al.* (2006).

discussion of other theories the reader is referred to the work of, for example, Banholzer (2004).

If fibre bundles are used instead of single fibres, the evolution of the bond with time as determined from single fibre pull-out does not always correspond to the evolution of the interface when a fibre bundle is used. Many authors indicated that scatter on the results is usually extremely high when fibre bundle pull-out is used. Moreover, one cannot be sure that the penetration of the matrix into a fibre bundle for pull-out specimens can be compared with penetration of the fibre bundle within a composite, since production techniques may be quite different. It is therefore advisable to determine the evolution of the interface on composite specimens directly, if possible. For GRCs a fairly simple technique exists that can be used to obtain a global idea of the magnitude of the global shear stress. Since in GRC and TRC composites the matrix shows a much lower tensile strength than the fibres, matrix multiple cracking usually occurs at stresses well below the actual failure stress of the composite. Based on the initial theories for this cracking phenomenon, measurement of the distance between neighbouring cracks, fibre volume fraction and average matrix strength can be used to obtain an order of magnitude for the quality of the interface. For load-bearing TRC specimens, the changes at the interface are however usually limited and do not affect the global durability greatly (Banholzer, 2004; Orlowsky, 2005).

3.6 Composite loading effects

Once the main degradation mechanisms are identified within the multi-scale approach, a model can be proposed for the ageing of full composites, based on previously obtained knowledge. Although the relative importance of the basic mechanisms for ageing of GRC is still under discussion, most authors agree that a chemical degradation process (with an Arrhenius dependence on temperature) can be combined with LEFM in order to model the ageing of this type of glass fibre reinforced composite. Usually, the composite strength is written as a function of the fibre strength as follows:

$$\sigma_c = \eta_0 \eta_l V_f \sigma_f \qquad [3.6]$$

where V_f is the fibre volume fraction, and η_0 and η_l efficiency factors for fibre orientation and length respectively. Some authors therefore suggest (Purnell *et al.*, 2001) that the strength of the composite can be written as a function of the fibre strength as follows:

$$\sigma_c = \frac{\eta_0 \eta_1 V_f K_{1c}}{A\sqrt{\pi\left(a_0 + k_0 e^{-\frac{\Delta H}{RT}} f(t)\right)}} \qquad [3.7]$$

and that the relative loss of strength can thus be modelled as:

$$S = \frac{1}{\sqrt{1 + \frac{k_0}{a_0} e^{-\frac{\Delta H}{RT}} f(t)}} \qquad [3.8]$$

This would thus indicate that the evolution of the composite strength would be the same function as that used for the evolution of the fibre strength, as measured for example on SIC specimens. If, however, it is assumed that loss of strength of the fibres within a matrix is mainly due to transverse pressure of reaction products at the interface, the efficiency factors η_0 and η_l are not necessarily constant any more with time. Moreover, if the composite is loaded in bending, one should be extra careful when GRCs are used and even more so when TRCs are used. Since the composite shows a non-linear behaviour due to the introduction of matrix multiple cracking, a redistribution of stresses might be expected if the composite is loaded, as mentioned previously in Section 3.3.3 (Blom *et al.*, 2007).

3.7 *In situ* degradation of composites due to chemical attack

As shown in the previous sections, moisture and temperature are significant parameters with respect to the durability of glass fibre reinforced composites. Usually, laboratory tests are performed under a well-controlled and steady-state hydrothermal regime. However, real weather comprises continuous transient states of humidity, temperature and precipitation. Unfortunately, the amount of data on the evolution of moisture in composites measured in an outside variable environment is limited. Although it can reasonably be assumed that the evolutions of the internal and external temperature are globally equal and can be easily linked to each other, this is not the case for the internal humidity. Unfortunately, only a limited number of papers have been dedicated to the effects of variable natural weather on durability. For GRCs specifically, Purnell (2004) mentioned that the effect of the variability of the temperature could be significant in predicting the strength loss of GRCs within a certain lifetime. It was shown that not only a high average temperature, but also large variations in temperature could accelerate ageing. Orlowsky *et al.* (2004) noted that the same conclusion is valid for the influence of humidity. Orlowsky *et al.* (2004) measured the internal humidity in TRC composite specimens for 2.5 years. This internal humidity was measured by means of a miniature multi-ring electrode. The amount of water in the concrete, present due to natural evolutions of the weather, was linked to the measured electrical resistance. Based on calibration measurements in a constant and well-controlled laboratory environment, it was noted that specimens, which were stored in a climate chamber with a relative humidity of 80% and higher showed clear loss of strength with time, while this ageing was less clear for series of specimens that were stored at lower relative humidities. The resistance that was measured inside the calibration specimens stored in a constant environment at 80% relative humidity, was around 12 kΩm. The temperature and electrical resistance of several concrete specimens were then measured over 2.5 years. Each hour, one data point was saved and used further in the predictive models. The composition of the specimens was similar to that used in Section 3.4.5 in this chapter. The production technique however was slightly different and full composite specimens were prepared. Some of the specimens were subjected to outdoor weather conditioning (in Aachen, Germany) and other specimens were produced to calibrate the material model through series of tests as described in Section 3.4.5 (Orlowsky *et al.*, 2004). This time it was noted that the combined model seemed to fit the results best and the values of the model parameters, which will therefore be used further for extrapolation purposes, are:

$$k_{01} = 1.31 \times 10^{11} \text{ nm/day}$$
$$\Delta H_1/R = 7832 \text{ K}$$
$$k_{02} = 1.71 \times 10^{24} \text{ nm}^2\text{/day}$$
$$\Delta H_2/R = 16\,393 \text{ K}$$

The prediction of the evolution of the strength loss was performed in three different ways:

1 It is assumed that the influence of the humidity can be neglected, since this influence is not easily introduced.

2 It is assumed that temperature and humidity can change. Both effects are measured inside the specimens. For the humidity, it is assumed that no corrosion occurs as long as the humidity stays below a certain limit and full corrosion occurs (as measured under water) as soon as the internal humidity increases beyond this limit. The humidity was measured through the electrical resistance of the concrete, as explained above.
3 It is assumed that temperature and humidity can change. Both effects are measured outside the specimens. It is assumed that the internal temperature equals the external temperature. Since the specimens were only 6 mm thick, it was shown that this assumption could hold. Unfortunately, the internal humidity could not be linked directly to the external conditions. A simplified method, which seemed to be legitimate globally for the specific test conditions and climate, was however developed. This methodology was based on findings obtained on concrete under natural weathering (Andrade *et al.*, 1999; Andrade and Castillo, 2003):
   (a) if precipitation is measured within one hour, it is assumed that during this hour chemical attack occurs as if the specimen is immersed.
   (b) if no precipitation is measured during an hour, it is assumed that the partial pressure inside the specimen equals the partial pressure outside the specimen.

These simplifications usually do not describe the internal humidity within a concrete construction very well. It should, however, be noted that the specimens were quite different from normal concrete constructions since they were very thin (6 mm total thickness) and no concrete cover was used to protect the reinforcement. Secondly, measurement of the pore size distribution revealed that about 15 vol.% of pores was present (Orlowsky, 2005) in the fine-grained matrix. This means that the internal humidity inside the specimens changed rapidly according to the outside environment.

At time t_{i+1}, the depth of the largest flaw X_{i+1} was calculated from integration of the chosen model, shown in Table 3.1, between t_i and t_{i+1}. For the combined model for example (mechanism 3), this means that, since:

$$\frac{\mathrm{d}X}{\mathrm{d}t} = \frac{1}{\dfrac{1}{k_1(T, RH)} + \dfrac{X}{k_2(T, RH)}} \qquad [3.9]$$

and since we assume that, within one hour, temperature and humidity are constant:

$$\frac{(X_{i+1})^2}{2k_2(T, RH)} + \frac{X_{i+1}}{k_1(T, RH)} = \frac{(X_i)^2}{2k_2(T, RH)} + \frac{X_i}{k_1(T, RH)} + t_{i+1} - t_i \qquad [3.10]$$

we can now calculate two possible values of X_{i+1}. One of the solutions is negative and is thus discarded. The other value is implemented in equation [3.2] to calculate the loss of strength. The influence of the humidity is limited for the first and second methodology, since it is assumed that the humidity is 100% or 0%. This is not the case for the third methodology. As long as it rains, it is also assumed that the concrete is immersed. In periods without rain, however, it is assumed that the relative humidity can vary between 0 and 100%. The relationship between relative humidity and corrosion rate was established according to Section 3.3.3. Several series of specimens were stored for 180 days at a constant relative humidity and temperature and then tested for remaining strength. All series were aged at the same temperature level (50 °C) but at various levels of external relative humidity. The corrosion rate, defined as the rate of growth of the largest flaw, was calculated with the help of equation [3.2] from the loss of strength. Figure 3.8 shows the logarithm of the calculated rate as function of the external relative humidity. As can be seen from this figure a linear relationship could also be found for the tested material combination. The straight line depicted in Fig. 3.8 was used to calibrate the relationship between the relative humidity and corrosion rate for the second methodology.

Figure 3.9 shows that the calculated loss of strength leads to an overprediction of the loss of strength if it is assumed that the concrete is always immersed. If, however, the internal humidity is measured and used as an on/off switch to decide if corrosion occurs or not, the predicted value of the loss of strength seems to be more accurate. If only the measurement of the outside temperature/relative humidity and precipitation is used, this predic-

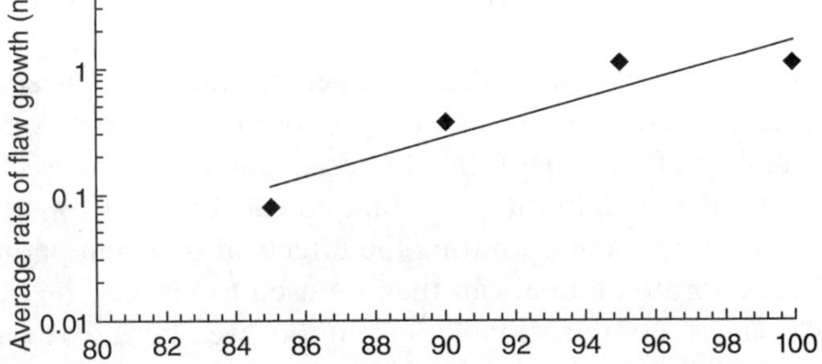

3.8 Rate of corrosion as a function of the relative humidity of the environment.

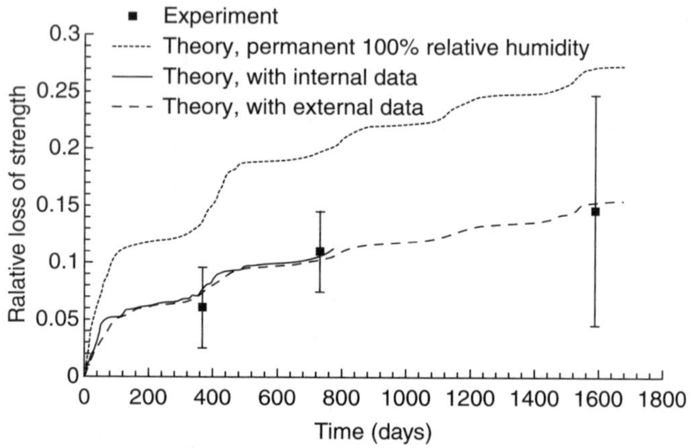

3.9 Evolution of the strength loss due to outdoor weathering.

tion seems to coincide rather well with the previous curve. The advantage of the latter technique, however, is that weather data are usually widely available, which is not the situation for internal measurements. In this case, for example, the internal resistance was not measured any more after 2.5 years, while the external weather data were still available. Figure 3.9 shows the predicted value after 5 years for the first and third methodology. The second methodology could not be extrapolated due to lack of necessary data. Some specimens were still left outside for 5 years and the results will soon be published in full (Cuypers *et al.*, 2008).

3.8 Conclusions

In this chapter, ageing of composites is discussed in terms of calibration and extrapolation methodologies rather than in terms of well-defined and thoroughly described chemical reactions. It has been shown from a literature overview – especially for GRC, which was used for illustrative purposes throughout this chapter – that for ageing, which is induced through chemical attack, several models can be proposed. The choice of the best model is usually a function of the material combination used. It was, however, shown that a calibration technique can be used for a material combination of interest, incorporating the effects of time, temperature and humidity. The calibrated model can then be used to predict the behaviour of the same material combination under outdoor weathering. It was shown within an exemplary calculation that the effects of variable temperature, humidity and precipitation can be integrated into the calculations so that the prediction of strength loss under real weathering conditions becomes possible.

3.9 Acknowledgements

For the investigation at MeMC Vrije Universiteit Brussel, the support of the post-doctoral position of the first author and of the project G.0047.05 'Analysis of the durability of cementitious composites with glass fibre reinforcement for building applications', both sponsored by the Research Foundation–Flanders (FWO, Fonds Wetenschappelijk Onderzoek – Vlaanderen), is gratefully acknowledged.

The investigations at ibac, RWTH Aachen University, are part of the Collaborative Research Center 532 project 'Textile reinforced concrete – Basics for the development of a new technology' and are sponsored by the Deutsche Forschungsgemeinschaft (DFG); the support is gratefully acknowledged.

3.10 References

ADAMS PB (1984), 'Glass corrosion', *Journal of Non-Crystalline Solids*, **67**, 193–205.

ALSHAER M (2006), *Optimization of properties of Inorganic Phosphate Cement (IPC) for construction and high-temperature applications*. PhD thesis, VUB (available online on http://wwwtw.vub.ac.be/memc/website/index.htm).

ANDRADE C, SARRIA J, ALONSO C (1999), 'Relative humidity in the interior of concrete exposed to natural and artificial weathering', *Cement and Concrete Research*, **29**, 1249–1259.

ANDRADE C, CASTILLO A (2003), 'Evolution of reinforcement corrosion due to climatic variations', *Materials and Corrosion*, **54**, 379–386.

AVESTON J, COOPER GA, KELLY A (1971), 'Single and multiple fracture. The properties of fibre composites', *Proceedings of the National Physical Laboratories Conference*, London, November 1971, p. 15, IPC Science & Technology Press, Guildford, UK.

AVESTON J, COOPER GA, KELLY A (1974), 'Fibre reinforced cements – scientific foundations for specifications', *Composites – Standards, Testing and Design, Proceedings of the National Physical Laboratories Conference*, London, April 1974, IPC Science & Technology Press, Guildford, UK.

BANHOLZER B (2004), *Bond behaviour of a multi-filament yarn embedded in a cementitious matrix*. PhD thesis, RWTH-Aachen (available through the online library of the RWTH Aachen).

BANHOLZER B, BRAMESHUBER W, JUNG W (2006), 'Analytical evaluation of pull-out-tests – the inverse problem', *Cement and Concrete Composites*, **28**, 564–571.

BLOM J, CUYPERS H, VAN ITTERBEECK P, WASTIELS J (2007), 'Modelling the behaviour of Textile Reinforced Cementitious composites under bending', *Proceedings of Fibre Concrete 2007*, Prague, 12–13 September 2007, pp. 205–210.

BRAMESHUBER W ed. (2006), *State-of-the Art Report of RILEM technical committee 201-TRC: Textile Reinforced Concrete*, RILEM, Bagneaux, France.

BUDD SM (1961), 'The mechanisms of chemical reaction between silicate glass and attacking agents', *Physics and Chemistry of Glasses*, **2**, 111–114.

CHARLES RJ (1958), 'Static fatigue of glass I', *Journal of Applied Physics*, **29**(11), 1549–1553.

CUYPERS H (2002), *Analysis and design of sandwich panels with brittle matrix composite faces for building applications*. PhD thesis, VUB (available online on http://wwwtw.vub.ac.be/memc/website/index.htm).

CUYPERS H, WASTIELS J (2006), 'Stochastic matrix-cracking model for textile reinforced cementitious composites under tensile loading', *Materials and Structures*, **39**, 777–786.

CUYPERS H, BUETTNER T, ORLOWSKY J, RAUPACH M (2008), 'Durability of textile reinforced concrete under variable environment', *Proceedings of Challenges for Civil Construction*, 16–18 April 2008, Porto, Portugal (CD-Rom).

DIMBLEBY V, TURNER WES (1926), 'The relationship between chemical composition and the resistance of glasses to the action of chemical reagents. Part I', *Society of Glass Technology*, **10**, 304–358.

DOUGLAS RW and ISARD JO (1949), 'The flow of glass', *Journal of the society of Glass Technology*, **33**, 138–163.

EL-SHAMY TM, LEWINS J, DOUGLAS RW (1972), 'The dependence on the pH of the decompositions of glasses by aqueous solutions', *Glass Technology*, **13**(3), 81–87.

FREIMAN SW (1980), 'Fracture mechanics of glass', in *Elasticity and Strength in Glass 5*, Uhlmann DR, Kriedl NJ eds, Academic Press, New York, pp. 21–78.

GAO SL, MÄDER E, ABDKADER A, OFFERMANN P (2003a), 'Environmental resistance and mechanical performance of alkali-resistant glass fibers with surface sizings', *Journal of Non-Crystalline Solids*, **325**, 230–241.

GAO SL, MÄDER E, ABDKADER A, OFFERMANN P (2003b), 'Sizings on alkali-resistant glass fibres: environmental effects on mechanical properties', *Langmuir*, **19**, 2496–2506.

HILLIG WB, CHARLES RJ (1965), *High Strength Materials*, Zackay VF ed., Wiley, New York, pp. 682–704.

KOSHIZAKI N (1988) 'Interfacial analysis between zirconia-containing glass and cement by X-ray photoelectron', *Journal of Materials Science Letters*, **7**(11), 1190–1192.

LARNER LJ, SPEAKMAN K, MAJUMDAR AJ (1976), 'Chemical interactions between glass fibres and cement', *Journal of Non-Crystalline Solids*, **20**, 43–74.

LI VC, CHAN YW (1994), 'Determination of interfacial debond mode for fiber-reinforced cementitious composites', *Journal of Engineering Mechanics*, **120**(4), 707–719.

LITHERLAND KL, OAKLY DR, PROCTOR BA (1981), 'The use of accelerated ageing procedures to predict the long term strength of GRC composites', *Cement and Concrete Research*, **11**(3), 455–466.

MAJUMDAR AJ, RYDER JF (1968), 'Glass fibre reinforcement of cement products', *Glass Technology*, **9**, 78–86.

MAJUMDAR AJ, TALLENTIRE AG (1973), 'Glass fibre reinforced cement', *International Symposium on Applications of Fibre Reinforced Concrete*, Ottawa, Canada, 11 October 1973, American Concrete Institute, Farmington Hills, Michigan.

MAJUMDAR AJ (1980), 'Some aspects of glass fibre reinforced cement research', in *Advances in* Cement–Matrix Composites, Roy DM, Majumdar AJ, Shah SP, Manson JA eds, Material Research Society, Boston, pp. 37–59.

MICHALSKE TA, BUNKER BC (1987), 'Steric effects in stress corrosion fracture of glass', *Journal of the American Ceramic Society*, **70**, 780–784.

NAMMUR G, NAAMAN AE (1989) 'Bond stress model for fiber reinforced concrete based on bond-stress–slip relationship', *ACI Materials Journal*, **89**(1), 45–57.

ORLOWSKY J, RAUPACH M, CUYPERS H, WASTIELS J (2004), 'Durability modelling of glass fibre reinforcement in cementitious environment', *Materials and Structures*, **28**, 155–162.

ORLOWSKY J (2005), *Zur Dauerhaftigkeit von AR-glasbewehrung in Textilbeton*, Deutscher Ausschuss für Stahlbeton, Heft 558 (in German).

PAUL A (1982), *Chemistry of Glasses*, Chapman & Hall Ltd, London.

PROCTOR BA, OAKLEY DR, LITHERLAND KL (1982) 'Developments in the assessment and performance of GRC over 10 years', *Composites*, **13**(2), 173–179.

PURNELL P, BUCHANAN AJ, SHORT NR, PAGE C, MAJUMDAR AJ (2000), 'Determination of bond strength in glass fibre reinforced cement using petrography and image analysis,' *Journal of Materials Science*, **35**(18), 4653–4659.

PURNELL P, SHORT NR, PAGE CL (2001), 'A static fatigue model for the durability of glass fibre reinforced cement', *Journal of Materials Science*, **36**, 5385–5390.

PURNELL P (2004), 'Interpretation of climatic temperature variations for accelerated ageing models', *Journal of Materials Science*, **39**, 113–118.

SCARINCI G, FESCA D, SORARU G, GRASSI G, STAFFERRI L, BADINI C (1986), 'Corrosion behaviour of a ZrO_2–containing glass in aqeous acid and alkaline media and in hydrating cement paste', *International Journal of Cement Composites,* **1**(3), 103–109.

SCHOLZE H (1982), 'Chemical durability of glasses', *Journal of Non-Crystalline Solids*, **52**, 91–103.

SIMHAN RG (1983), 'Chemical durability of ZrO_2 containing glasses', *Journal of Crystalline Solids*, **54**, 335–343.

VAN ITTERBEECK P, CUYPERS H, ORLOWSKY J, WASTIELS J (2007) 'Evaluation of the strength in cement (SIC) test for GRCs with improved durability', *Materials and Structures*, in press.

WIEDERHORN S (1967), 'Influence of water vapor on crack propagation in soda-lime glass', *Journal of the American Ceramic Society*, **50**, 141–407.

WIEDERHORN (1972), 'A chemical interpretation of static fatigue', *Journal of the American Ceramic Society*, **55**, 81–85.

YILMAZ VT, GLASSER FP (1991), 'Reaction of alkali-resistant glass fibres with cement. Part 1. Review, assessment, and microscopy', *Glass Technology*, **32**, 91–98.

4
Stress corrosion cracking in glass reinforced polymer composites

A. CHATEAUMINOIS, Ecole Supérieure de Physique et Chimie Industrielles (ESPCI), France

4.1 Introduction

In many structural applications, fibre reinforced composite materials are exposed to the long-term action of both mechanical stresses and environmental ageing processes. In the case of glass fibre reinforced polymers (GFRPs), it has long been recognized that the strength of the fibre reinforcement is very sensitive to humidity. Under the combined action of moisture and applied stress, a delayed failure of the glass reinforcement is induced which can significantly alter the durability of GFRP structures. Termed 'stress corrosion cracking' (SCC), this phenomenon is a concern in many applications, such as filament wound pipes for the oil industry, including aggressive offshore environments (Frost and Cervenka 1994; Ghotra 1999; Hale *et al.* 2000), tanks in chemical plants (Myers *et al.* 2007) or electrical insulators (Kumosa *et al.* 2001; Kumosa *et al.* 2005). Despite a considerable research effort into this area, there is still a lack of a predictive durability model that can account for the reduction of the fatigue life of GFRP materials under SCC conditions.

The purpose of this chapter is to show how well-established concepts for the delayed failure of bulk glasses can be applied to the fatigue behaviour of GFRP laminates under hygrothermal ageing conditions. As an introduction, Section 4.2 presents a review of experimental evidence for SCC processes in GFRP materials. The different time and length scales associated with the coupled interactions between fatigue damage and physico-chemical water ageing processes are also considered. Sections 4.3 and 4.4 are devoted to the application of SCC concepts to the prediction of the early stages of fatigue damage development in unidirectional glass reinforced composite materials. In such composites, fatigue damage is largely controlled by the progressive accumulation of broken fibres by delayed failure (Talreja 1987). How scaling laws describing the kinetics of fibre failure as a function of applied stress and environment can be derived from a knowledge of subcritical crack propagation rates in the glass filaments will be detailed. In addition, how an SCC approach can also be used as a framework for stiffness loss prediction under static and cyclic fatigue loadings in wet environments will be reviewed.

4.2 Overview of stress corrosion cracking in glass reinforced polymer matrix composites

4.2.1 Experimental evidence for stress corrosion cracking processes in glass fibre reinforced polymer laminates

Early experimental evidence of SCC in glass/epoxy and glass/polyester laminates was provided by Hogg *et al.* (Hogg and Hull 1980; Hogg and Hull 1982), Jones *et al.* (Jones *et al.* 1983a; Jones *et al.* 1983b; Jones 1989) and Aveston *et al.* (Aveston *et al.* 1980; Aveston and Sillwood 1982). In these works, the SCC behaviour of unidirectional and cross-ply laminates was investigated under static and/or cyclic fatigue loading in wet environments including acidic solutions (sulphuric and hydrochloric acids). SCC under acid conditions is found to be enhanced and the associated failures are characterized by typical planar fracture surfaces (Fig. 4.1) which form at 90° to the applied stress and normally within that part of the composite in direct contact with the environment. These failures occur over increasingly longer time scales as the initial applied stress is reduced. In their study of the acidic stress corrosion of wound GFRP pipes, Hogg and Hull (1982) also provided some strong evidence that the SCC

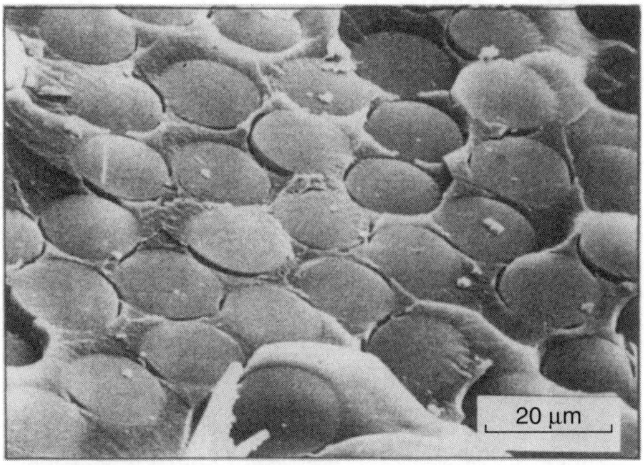

4.1 Scanning electron micrograph showing the smooth fracture surface induced by subcritical crack growth in a unidirectional glass fibre/polyester composite exposed to hydrochloric acid (from Hogg and Hull (1982)).

processes can be activated by water diffusion through the matrix in the absence of any pre-existing matrix micro-cracking. Interestingly, stress corrosion fractures were also reported by Jones *et al.* (1983a) to occur in unimmersed parts of the specimens. According to Jones *et al.*, these failures involved the transportation of glass degradation products (metallic ions leached from the glass surface) and a corrosive medium along the interface, which itself is subjected to stress corrosion, to the part of the specimen out of contact with the environment. These observations show that diffusion at the fibre/matrix interface can play a significant role in SCC mechanisms.

It has long been established that SCC in glass materials is associated with the breakage of the Si–O–Si siloxane bonds that form the silicate network under the action of mechanical stresses and water molecules (Wiederhorn 1978b):

$$-\mathrm{Si}-\mathrm{O}-\mathrm{Si}- + \mathrm{H_2O} \rightarrow -\mathrm{Si}-\mathrm{OH} + \mathrm{HO}-\mathrm{Si}-$$

The contribution of stress to this chemical reaction is attributed to the extension of Si–O bonds, which enhances the chemical bond breakage reaction rate.

In the case of E-glass fibre reinforcement, an exchange reaction is also known to occur between superficial ions (typically Na^+, Ca^{2+} or Al^{3+}) and protons that are present in the surrounding environment (Douglas and El-Shamy 1967):

$$-\mathrm{Si}-\mathrm{O}-\mathrm{M} + \mathrm{H^+} \rightarrow -\mathrm{Si}-\mathrm{O}-\mathrm{H} + \mathrm{M^+}$$

This reaction results in a progressive increase in the pH of the crack tip environment (Wiederhorn 1978b), which can have a catalytic effect on the hydrolysis of the siloxane bonds (Charles 1958b), according to the following reaction:

$$-\mathrm{Si}-\mathrm{O}-\mathrm{Si}- + \mathrm{OH^-} \rightarrow -\mathrm{Si}-\mathrm{O^-} + \mathrm{HO}-\mathrm{Si}-$$

Another complication can arise from this ion exchange mechanism in a corrosive medium. When the glass fibres are exposed to an acid environment, the amount of extracted alkaline ions is much more important. As a consequence of the volume changes associated with ion exchange on the glass surface, tensile stresses are induced within the fibres' superficial layers.

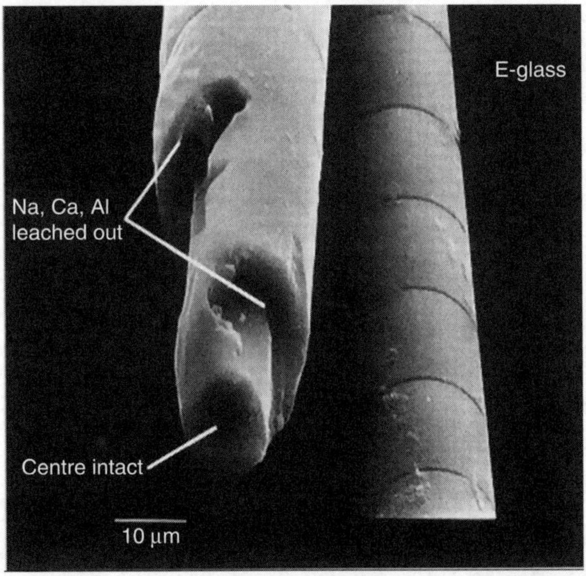

4.2 Scanning electron micrograph of E-glass fibres showing the occurrence of spiral cracking induced by ion exchange mechanisms (courtesy of Owens Corning).

These processes can result in the spontaneous spiral cracking of the outer sheath of the fibres without any external applied stress (Fig. 4.2). Such failures have been extensively reported in the case of E-glass exposed to H_2SO_4 (Aveston and Sillwood 1982; Jones *et al.* 1983b; Rodriguez 1987). In alkaline media, the degradation of the fibres occurs by an etching process that involves hydration followed by total dissolution of the glass (Rodriguez 1987).

The chemical reaction between glass and aqueous solutions depends on the composition of the glass. It is well recognized that the higher the SiO_2 content of the glass, the better its chemical resistance in acidic media. Accordingly, some studies (Kawada and Srivastava 2001; Mizoguchi *et al.* 2001; Myers *et al.* 2007) indicate that the stress corrosion resistance of composites reinforced by C-glass, ECR®-glass or S-2 glass® are improved as compared with that of general purpose fibres with a lower SiO_2 content, such as E-glass or A-glass. As shown by Kawada *et al.* (Kawada and Srivastava 2001; Kawada *et al.* 2001), an increase in the acid concentration of the environment (HCl and H_2SO_4) results in a lowering of the crack propagation threshold of GFRP and in enhanced propagation velocities above this threshold. However, it can be noted in Fig. 4.3 that crack propagation in water is promoted almost as much as that in acid solution in comparison with the propagation rate in air. Temperature effects can also

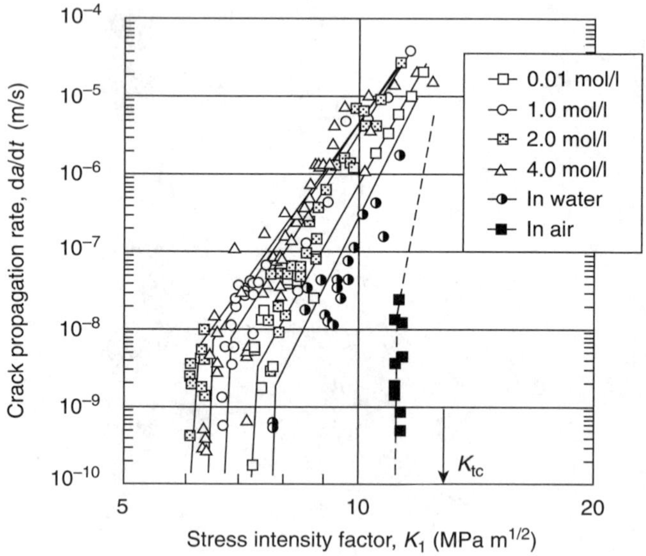

4.3 Subcritical crack propagation rate as a function of the stress intensity factor for a glass fibre/vinylester composite in HCl solutions with different concentrations; K_{IC} is the material toughness (from Kawada *et al.* (2001)).

be different depending on the nature of the acid. For woven C-glass/vinyl ester laminates in 1 M HCl, Kawada *et al.* (2001) observed that the threshold decreased with increased temperature while the reverse trend was observed for 1 M H_2SO_4.

4.2.2 Basic mechanisms for the interaction between environment and fatigue damage

During SCC of GFRP, the interactions between glass fibres and moist environments can be schematically dissociated into two mechanisms involving different time and length scales.

The first mechanism corresponds to localized interactions between the ageing media and bare portions of glass fibres at the tip of macroscopic cracks. Such situations can typically result from early transverse matrix micro-cracking in cross-ply and angle-ply laminates. Similar processes can also occur in unidirectional composites after the nucleation of matrix cracks perpendicular to the fibre direction. In both cases, the characteristic time scales corresponding to the occurrence of SCC processes can be very short as they involve capillary diffusion of the corrosive medium along microcracks. From an experimental point of view, such a situation can be simulated using fracture mechanics tests where the crack velocity of notched composite specimens is monitored in various environments.

A second mechanism for the activation of SCC is the diffusion of water through the polymer matrix or along the fibre/matrix interface. In such a situation, SCC crack propagation can be viewed as a diffusion-controlled fracture mechanism. Water diffusion within thermoset matrix composites is a very well-documented topic, which has been the subject of a considerable amount of work since the pioneering work of Springer and co-workers (Schen and Springer 1981). These processes can conveniently be investigated by means of gravimetric measurements of the relative moisture uptake, M_t, of specimens exposed to liquid water or humidity at various temperatures. From these data, it is now widely established that water diffusion is a thermally activated process with characteristic times that vary as a function of the square of the thickness of the structure; typical values of diffusion coefficient for glass/epoxy composites at room temperature are of the order of 10^{-6} to 10^{-8} $mm^2\ s^{-1}$. Depending on the resistance of the fibre surface treatment to hydrolysis, accumulation of water at the interface can result in drastic changes in the physico-chemical environment of the fibres, which in turn may affect the SCC behaviour. Such effects have been clearly identified in the case of dicyandiamide (DICY)-hardened epoxy matrix, where the leaching and the hydrolysis of unreacted fractions of the DICY hardener result in the extensive formation of ammonia (Kasturiarachi and Pritchard 1984; Vauthier *et al.* 1996). The associated increase in the pH (up to 9–10) strongly decreases the resistance of the glass reinforcement to SCC due to etching. As a consequence of matrix hydrolysis and swelling stresses, micro-cracks can also be nucleated within the aged composite, which promotes further direct attack of the glass fibres by the environment. Extensive hydrolysis of the epoxy matrix has often been reported in the case of resins hardened by aliphatic amines or anhydrides (Bonniau and Bunsell 1981; Dewimille and Bunsell 1982; Dewimille and Bunsell 1983). On the other hand, resins hardened by aromatic amines (such as the diglycidyl ether of Bisphenol A (DGEBA)/diaminodiphenylmethane (DDM) and DGEBA/diaminodiphenylsulfone (DDS) systems used in aeronautical applications) are much less prone to hydrolysis (Schen and Springer 1981; Chateauminois *et al.* 1994).

SCC processes activated by moisture diffusion can be investigated from the fatigue testing of water-saturated composite specimens. An example is shown in Fig. 4.4 in the case of the static fatigue of a unidirectional glass/epoxy composite aged in water at different temperatures. The lifetimes at a given strain level are strongly dependent on the ageing temperature, but they were not found to be related to any significant changes in the moisture content of the material. The temperature effect can rather be related to the nucleation of additional defects at the surface of the fibres during ageing

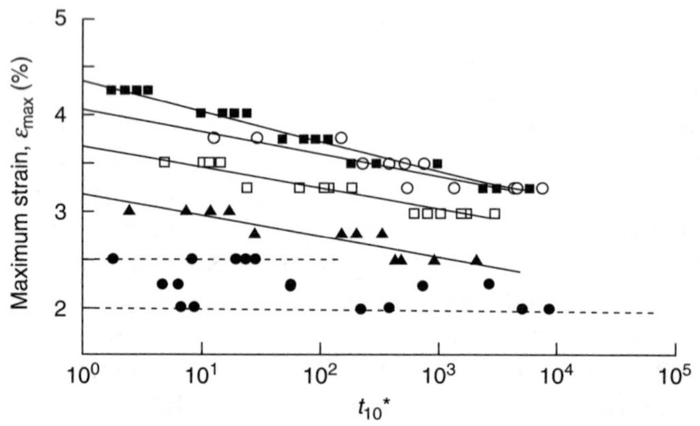

4.4 Fatigue curves of an S-glass/epoxy composite under static fatigue conditions (constant imposed strain, three-point bending). Unaged, ■ aged for 100 days at 30 °C (○), 50 °C (□), 70 °C (▲) and 90 °C (●). t_{10}* is the time to 10% stiffness loss normalized with respect to the loading time. Fatigue tests carried out in water at R.T.

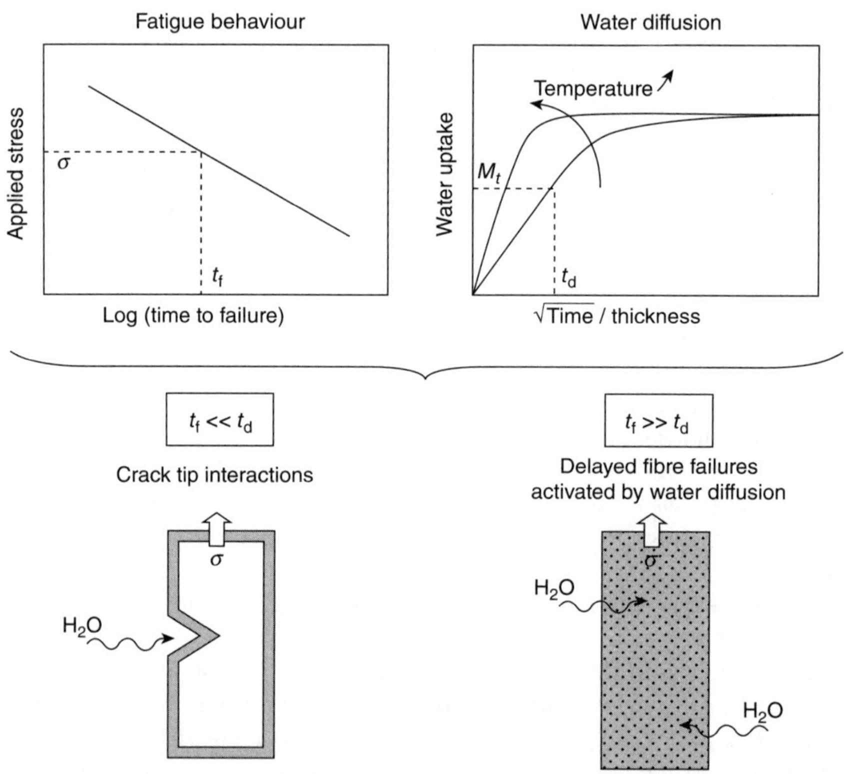

4.5 Schematic of the interactions between water ageing and fatigue damage in GFRPs.

at elevated temperature, a process that is enhanced by capillary diffusion along the fibre/matrix interface (Chateauminois *et al.* 1994).

As shown schematically in Fig. 4.5, the relative contributions of localized crack tip interactions and diffusion-controlled mechanisms to SCC depend on the applied stress level and on the water diffusion kinetics. In the case of high stresses and/or slow water diffusion (i.e. low temperatures or thick composite parts), localized interactions between environment and crack tip will account for most of the SCC damage. On the other hand, low stress levels and/or rapid water saturation (at high temperature or in thin parts) will promote SCC damage by water diffusion through the bulk composite materials. Such a situation is especially relevant to the prediction of the fatigue limit of composite parts. The latter relies mostly on the resistance of the composite to the accumulation, at the microscopic scale, of delayed fibre failures.

4.3 Stress corrosion cracking of glass fibres

4.3.1 Physical and mechanical processes involved in the delayed failure of glasses

The delayed failure of glasses is associated with the progressive propagation of cracks from pre-existing surface defects under the action of static or dynamic mechanical stresses. This phenomenon is known to occur at stress levels that can be well below the short-time strength of the material. There is a long history behind the study of delayed glass failures which is recalled in reviews by Maugis (1985), Wiederhorn (1978a) and Lawn (1993). Early studies by Holland and Turner (1940) on glasses have shown the existence of a threshold stress for the occurrence of delayed failure which is approximately 20–30% of the short-time failure stress. The influence of moisture on delayed fracture was first emphasized by Baker and Preston (1946), followed by Wiederhorn (1967) who showed that orders of magnitude variations in the crack growth rate can result from changes in the relative humidity of the surrounding environment. Although the presence of humidity is of primary importance, some evidence for subcritical crack propagation in glasses and ceramics has also been reported (Pukh *et al.* 1970; Wiederhorn *et al.* 1974; Chevalier *et al.* 2002).

Subcritical crack propagation in brittle materials such as glasses is usually accounted for within the framework of linear elastic fracture mechanics. In such an approach, the relevant parameter for describing crack velocity, v, as a function of applied stress and crack length is the stress intensity factor, K, defined as:

$$K = \sigma Y \sqrt{a} \qquad [4.1]$$

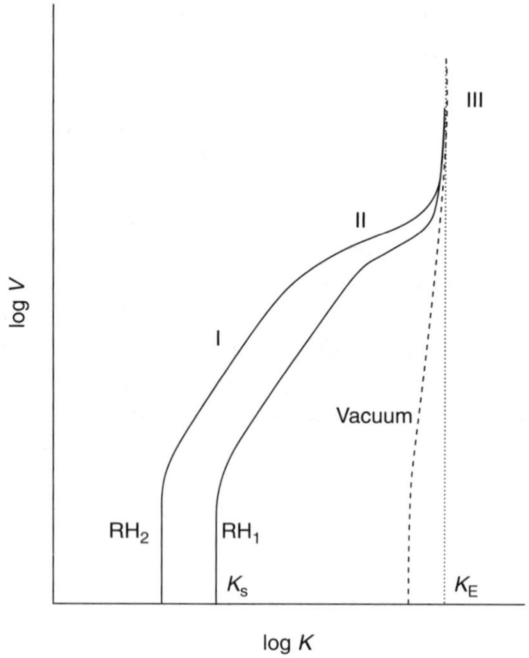

4.6 Schematic diagram for subcritical crack growth in glass. RH_1, and RH_2 denote two different relative humidity levels ($RH_2 > RH_1$).

where *a* is the crack length, σ is the applied stress and *Y* is a geometrical factor. When the crack velocity is plotted as a function of *K*, three regions are usually identified in the curves (Fig. 4.6).

1 A 'stage I' domain which is located just above a propagation threshold, K_s. Charles (1958a) has shown that the *v*–*K* curve can be fitted in a large part of this region by the following empirical equation:

$$v = \frac{da}{dt} = AK^n \qquad [4.2]$$

which can be used for lifetime predictions. In this expression, *A* and *n* are two empirical constants depending on the chemical composition of the glass and on the environment (temperature, humidity, pH). The threshold, K_s, corresponds to the Griffith criterion for an equilibrium crack (Maugis 1985). Accordingly, a crack will heal if $K < K_s$ and propagate (either in a stable or unstable manner) if $K > K_s$. This propagation threshold can be related to the intrinsic surface energy, γ, of the glass by the relation $K_s = \sqrt{2\gamma E/1-v^2}$ under plane strain conditions (*E* is the Young's modulus and *v* is the Poisson's ratio). Typical values of K_s for soda lime or borosilicate in water are close to 0.25 MPa $m^{1/2}$ (Wiederhorn and Bolz 1970).

2 A 'stage II' zone where the crack velocity remains nearly constant when K is increased. In this region, crack motion has often been claimed to be limited by the arrival rate of vapour or the viscous drag of water at the crack tip.

3 'Stage III' corresponds to a sharp increase in the crack velocity which returns to the vacuum values as water or moisture do not have sufficient time to reach the crack tip during propagation. This domain is associated with the catastrophic failure of the specimen in a very short time. It is often used to define a critical stress intensity factor or material toughness, K_c, which should not be confused with the Griffith equilibrium. For soda lime glass, K_c was found to be equal to 0.76 MPa $m^{1/2}$ by Wiederhorn and co-workers (Wiederhorn 1969; Wiederhorn *et al.* 1974).

Crack propagation rates in stages II and III are elevated, as compared with stage I. As a result, stages II and III are of minor importance regarding the long-term strength of glass materials. Most of the durability models for glass materials therefore concentrate on stage I, which is the relevant domain regarding the long-term behaviour. As schematized in Fig. 4.6, the $\log v$–$\log K$ curve is shifted to lower values of K in the presence of humidity. This can be related to two different effects. The first one is the lowering of the surface energy, γ, which is clearly emphasized by the decrease in the propagation threshold K_s. In addition, chemical effects associated with SCC are also involved. According to the mechanisms detailed in Section 4.2.1, the subcritical propagation rate of cracks under stage I conditions is dictated by the kinetics of the breakage of the siloxane bonds. As in many chemical reactions, these processes are affected by the temperature and by other physico-chemical factors such as pH. The effects of these environmental parameters on subcritical crack propagation rates can be described by considering the associated changes in the parameters, A and n, of the power law relation between crack velocity and stress intensity factor (equation [4.2]). It has been widely established that parameter A of this relation can be considered as the product of a thermal activation term and a hygrometry-dependent term (Metcalfe and Schmitz 1972; Wiederhorn 1978b):

$$A = A_{[H_2O]} \exp(-E_a/RT) \qquad [4.3]$$

where E_a is activation energy, R is the gas constant and T is absolute temperature. For silicate glasses, the activation energy is of the order of 130 kJ mol^{-1} (Wiederhorn 1978b). In the case of glass fibres in water, E_a ranges from 60 to 100 kJ mol^{-1} depending on the amount of alkaline ions (Metcalfe and Schmitz 1972; Ritter *et al.* 1988).

The stress corrosion parameter, n, is usually reported to be relatively insensitive to the amount of water in the environment (Wiederhorn 1978b).

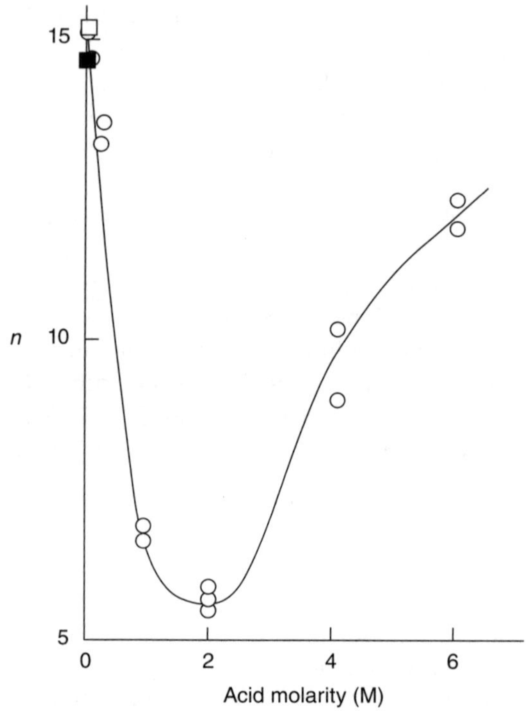

4.7 Stage I crack propagation parameter, *n,* plotted against acid molarity for stress corrosion of E-glass fibres in water and aqueous HCl at 20 °C (○), resin impregnated E-glass strands in water at 20 °C (■), and bulk soda lime glass in water at 20 °C (□) (data taken from Aveston *et al.* (1980), Cowking *et al.* (1991b) and Wiederhorn and Bolz (1970)).

On the other hand, it is strongly affected by the pH of the surrounding media (Cowking *et al.* 1991a; Cowking *et al.* 1991b). As an example, the effects of HCl at various molar ratios on the static fatigue behaviour of bundles of unimpregnated E-glass fibres are reported in Fig. 4.7. The stress corrosion parameter, n, is about 16 in deionized water and decreases with increasing acid strength until a minimum ($n = 5.7$) occurs at acid molarity ≈2 M. The fact that concentrated acids are less aggressive than diluted acids is believed to be due to the larger dissociation constants of diluted acids.

4.3.2 Determination of subcritical crack propagation velocities in glass fibres

From an experimental point of view, v–K curves are usually obtained from the tensile testing of notched bulk specimens where crack velocity can be continuously monitored. Such experiments are much more difficult, if not impossible, to perform using single glass filaments which have a diameter ranging from 10 to 25 μm. Using larger silica optical fibres of about 200 μm in diameter, Muraoka *et al.* (1993) were, however, able to monitor by optical observation the subcritical propagation of a crack from a pre-existing notch induced by Vickers indentation. These experiments confirmed the existence of stage I propagation, which was found to be markedly sensitive to relative humidity. More recently, indirect techniques based on the tensile testing of fibre bundles associated with acoustic emission (AE) detection of fibre failure have emerged as tools to assess subcritical crack velocity in E-glass fibres (Pauchard *et al.* 2000). The basis of this approach is to consider that the lifetime distribution under a constant applied strain contains the crack propagation law in an integrated form.

According to equations [4.1] and [4.2], the lifetime, t_f, under a constant applied strain, ε, can be expressed as:

$$t_f = \int_{a_0}^{a_f} \frac{da}{v} = \frac{2}{\varepsilon^2 E^2 Y^2} \int_{K_0}^{K_f} \frac{K dK}{v(K)} \qquad [4.4]$$

where a_0 and $K_0 = Y\sigma(a_0)^{1/2}$ are the crack length and the initial values of the stress intensity factor, respectively. K_f and a_f denote the values of the stress intensity factor and the crack length when catastrophic failure occurs. Without any hypothesis regarding the form of $v(K)$ relation, this integral equation can be derived with respect to K_0:

$$\frac{dt_f}{dK_0} = -\frac{2K_0}{\varepsilon^2 E^2 Y^2 v(K_0)} \qquad [4.5]$$

Noting that $K/K_c = \varepsilon/\varepsilon_i$, where ε_i is the strain to failure of the fibre in an inert environment, i.e. in the absence of SCC processes, equation [4.5] can be rewritten in the following form:

$$v(K_0) = \frac{2K_c^2}{\varepsilon_i^2 E^2 Y^2} \frac{d\varepsilon_i}{dt_f} \qquad [4.6]$$

This equation indicates that it is sufficient to determine the relation between ε_i and t_f in order to determine, after derivation, the v–K curve. However, it is obviously impossible to obtain t_f and ε_i on the same fibre. Following a suggestion by Fett and Munz (1985), these data can be generated from a combination of monotonic and static fatigue tests carried out using two separate populations of fibres. If the number of fibre specimens in each population is great enough, they can be considered as statistically identical in terms of surface defect distributions. The distribution of fibre strength determined using one set of fibres can thus be associated to the lifetime distribution given by the static fatigue testing of a second set. AE techniques such as those described by Huguet *et al.* (2002) and Cowking *et al.* (1991a)

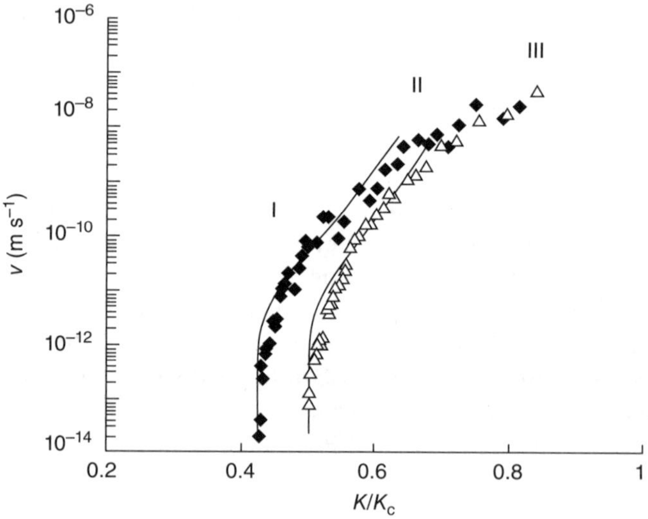

4.8 Subcritical crack propagation rate in ECR-glass® fibres at 10% (△) and 50% (◆) relative humidity and at 23 °C (from Pauchard *et al.* (2001)). Solid lines correspond to theoretical predictions (equation 4.3).

provide a convenient way to establish such data. During the tensile testing of a bundle containing thousands of glass monofilaments, individual fibre failure events can be monitored continuously. As a result, bundle testing can conveniently be used to generate the statistical sets of strength or lifetime data that are required to establish the *v*–*K* relationship.

An example of a *v*–*K* curve obtained using this approach is shown in Fig. 4.8 for E-glass fibres exposed to two different levels of relative humidity. Both curves exhibits evidence of SCC mechanisms, namely parallel stage I domains above a well-defined threshold. As expected, the value of the threshold is lowered when the relative humidity is increased. At high values of the stress intensity factor, the two curves merge, which allows stage III to be identified. Between stages I and III, stage II is more difficult to identify: while very short for 50% relative humidity, it is not really distinguishable for 10% relative humidity. It can be noted that the crack velocity associated with stage II at 50% relative humidity (about 10^{-8} m s^{-1}) is more than three orders of magnitude lower than that measured on bulk glasses (Wiederhorn and Bolz 1970). This difference could be attributed to the fact that the glass surface is not directly exposed to the surrounding environment which has to diffuse through the organic coating deposited on the fibres during manufacturing. From the data reported in Fig. 4.8, the exponent, *n*, of the power law expression for stage I crack velocity (equation [4.2]) is estimated to be about 20, which is close to the value obtained by

Cowking *et al.* (1991a) for E-glass immersed in water. It can also be noted that the slope of the v–K curve in stage I, i.e. the value of n, is independent of the level of humidity, according to literature data for bulk glasses.

These experiments therefore support the relevance of SCC concepts in describing the delayed failure of unimpregnated glass fibres. In the following section, it will be shown how the stage I crack propagation law can be used for lifetime prediction as a function of stress level.

4.3.3 Lifetime prediction of unimpregnated glass fibres under SCC conditions

Lifetime of a single fibre

The time to failure of a single fibre specimen corresponds to the time for a surface crack to grow from its initial size, a_0, to a critical size, a_f, which corresponds to the occurrence of catastrophic failure. This problem will be addressed in the general case of a time-dependent applied stress, $\sigma(t)$. If the contribution of stages II and III is neglected, the lifetime, t_f, can be deduced from the integration of equation [4.2] with $K = \sigma(t)Y\sqrt{a}$:

$$\int_0^{t_f} \sigma(t)^n \mathrm{d}t = \int_{a_0}^{a_f} \frac{\mathrm{d}a}{AY^n a^{n/2}} \qquad [4.7]$$

if $n \gg 1$ (n is indeed reported to be greater than 10 for glasses) and $a_0 \ll a_c$, the above integral equation can be approximated by:

$$\int_0^{t_f} \sigma(t)^n \mathrm{d}t \approx \frac{2a_0^{(2-n)/2}}{AY^n(n-2)} \qquad [4.8]$$

If the occurrence of subcritical crack propagation is neglected in an inert environment, it comes out that $K_c = \sigma_i Y\sqrt{a_0}$ (σ_i is the initial strength of the fibre), which after substitution in equation [4.8] provides the following general integral equation for the lifetime:

$$\int_0^{t_f} \sigma(t)^n \mathrm{d}t \approx \frac{2K_c^{2-n}\sigma_i^{n-2}}{AY^2(n-2)} \qquad [4.9]$$

In the case of any periodic stress with $t_f >> 1/\upsilon$ (υ is the loading frequency), an explicit expression for t_f is obtained from equation [4.9] in the form:

$$t_f \approx \frac{2K_c^{2-n}\sigma_i^{n-2}}{AY^2(n-2)\lambda}\sigma_{max}^{-n} \qquad [4.10]$$

where σ_{max} is the maximum value of the applied stress and λ is a stress-dependent parameter defined by:

$$\lambda = \frac{1}{t}\int_0^t \left(\frac{\sigma(t)}{\sigma_{max}}\right)^n dt \qquad [4.11]$$

which can readily be calculated from a knowledge of the stress history.

In the case of a constant applied stress, σ_a, λ is a constant equal to unity and the lifetime is thus simply given by:

$$t_f \approx \frac{2K_c^{2-n}\sigma_i^{n-2}}{AY^2(n-2)}\sigma_a^{-n} \qquad [4.12]$$

Equation [4.12] indicates that the logarithm of the lifetime under constant stress (or strain) conditions is linearly related to the logarithm of the applied stress (or strain). The slope of the curve is thus a measurement of the stress corrosion parameter, n. This result has been largely validated for bulk glass and notched optical fibres, although the reported values of the stress corrosion parameter are widely scattered (e.g. Wang *et al.* 1979).

Delayed failure of a statistical population of fibres under stress corrosion cracking conditions

In a composite material containing several thousands of glass fibres, the lifetimes of the glass filaments are distributed due to the statistical nature of the surface defects. As a starting point, it can be assumed that the statistical distribution of surface flaw sizes (in the nanometer range for glass fibres) is correlated to the distribution of fibre strength. The latter can be conveniently described by means of a 'weakest link' theory which yields the following Weibull expression (Weibull 1951) for the survival probability, P_s, of a fibre at a given applied strain, ε_i:

$$P_s(\varepsilon_i) = \exp\left[-\left(\frac{\varepsilon_i}{\varepsilon_0}\right)^m\right] \qquad [4.13]$$

where ε_0 is a scaling factor. The parameter m is the so-called 'Weibull modulus' which is an indication of the breath of the strength distribution. For glass fibres, m is typically of the order of 2–6. It is known experimentally that the strength of fibres is length dependent: for short lengths, the probability of finding a critical defect is reduced and the average strength of the fibre is thus increased as compared with that of longer glass filaments. In the above expression for P_s, this size effect is implicitly embedded in the scaling parameter, ε_0. The parameters m and ε_0 of the Weibull law can be determined experimentally from the tensile testing of fibre bundles in a dry environment. Under such conditions, the survival probability at a given applied strain is simply calculated from the ratio of the number of unbroken fibres to the initial number of fibres. Using such techniques, the validity of equation [4.13] in describing the distribution of the strength properties of glass fibres has been established (Cowking *et al.* 1991b; Zinck *et al.* 1999; Pauchard *et al.* 2000).

If it is assumed that the distribution of lifetimes, t_f, and strain to failure, ε_i, have the same statistical nature, equations [4.10] and [4.13] can be combined to provide the following scaling for the logarithm of the survival probability, P_s, of a statistical population of fibres under periodic loading:

$$\ln P_s(t) = -kt^{m/(n-2)}\varepsilon_{\max}^{mn/(n-2)}\lambda^{m/(n-2)} \tag{4.14}$$

with

$$k = \left[\frac{AY^2(n-2)E^2}{2K_c^{2-n}\varepsilon_0^{n-2}}\right]^{m/(n-2)}$$

where $\varepsilon_{\max}$ is the maximum applied strain. In equation [4.14] the prefactor k may simply be assimilated to an empirical constant to be determined experimentally for the considered glass material and environmental condition. In so doing, it turns out that equation [4.14] provides a very simple scaling relationship for the fibre survival probability as a function of time, applied strain and frequency. It relies only on the determination of the Weibull modulus, m, and of the stress corrosion parameter, n. The validity of this approach with both unimpregnated and matrix-impregnated glass strand bundles has been verified by Aveston *et al.* (1980) and Pauchard *et al.* (2002a). The above expression for the survival probability P_s constitutes the basis of the SCC model which will be extended below in the context of water-aged unidirectional GFRP composites.

4.4 Stress corrosion cracking in unidirectional glass fibre reinforced polymer composites

4.4.1 Micromechanical analysis of delayed fibre failure within water-aged glass fibre reinforced polymers

The validity of the SCC model introduced above was considered further by Pauchard *et al.* (2002a) from an *in situ* micromechanical analysis of delayed failure within unidirectional glass/epoxy composites. Aged and unaged composite specimens were analyzed under static fatigue conditions at the imposed strain. Experiments were restricted to the initial stages of fatigue damage, i.e. before the nucleation of macroscopic matrix cracks. Within this domain, damage mostly consists of the progressive accumulation of delayed glass fibre failures due to SCC. Moreover, the use of three-point bending conditions allowed these fibre failures to be concentrated within a restricted material volume located on the tensile side of the specimens and beneath the loading span, where strain and ageing conditions can be regarded as roughly uniform. Taking advantage of the transparency of the matrix and the reinforcement, broken fibres can be detected in this elementary composite volume by optical microscope observations (Vauthier *et al.* 1998).

Compared with unimpregnated fibre bundles, delayed fibre failures within a composite is potentially complicated by fibre/matrix interface stress transfer. These processes can manifest into two different ways. The first is the fragmentation of the glass filaments into fragments whose length depends on the strength properties of the reinforcement and on the interface shear strength (Pigott 1987). Secondly, stress redistribution at the vicinity of a broken fibre can induce the failure of adjacent fibres. These two situations are clearly not accounted for in the fibre bundle approximation embedded in the above-detailed SCC model. However, *in situ* observation indicated that such processes were limited below a strain of about 2.0% (Pauchard *et al.* 2002a). Below this threshold, the broken fibres detected may thus be considered as essentially non-interactive defects. As a first-order approximation, the fibre portions enclosed within an elementary composite volume can thus be treated as isolated glass filaments in a bundle.

Figure 4.9(a) shows an example of the increase in the number of broken fibres as a function of time and applied strain, where the kinetics of fibre failure is clearly shown to be non-linearly activated by the applied strain. A progressive saturation of the number of broken fibres is also observed, except at the highest strain level ($\varepsilon_{max} = 1.9\%$). At this latter strain, *in situ* optical observations revealed that stress concentrations at the vicinity of some of the broken fibres induced subsequent localized failures of adjacent fibres, which in turn resulted in the early formation of matrix cracks. At lower strains, the saturation of delayed fibre failure can be interpreted by considering that the distribution of the fibres' strength is associated with a distribution of the initial values of the stress intensity factors, K^{ini}, at $t = 0$, i.e.

$$K^{ini} = \frac{\varepsilon K_s}{\varepsilon_i} \qquad [4.15]$$

where ε is the applied strain and ε_i is the catastrophic failure strain under an inert environment. Within the fibre population, some specimens will break (if $\varepsilon_i < K_s\varepsilon/K^{ini}$) while others will not (if $\varepsilon_i > K_s\varepsilon/K^{ini}$). A limit value of the survival probability, P_s^{∞}, can thus be defined by substituting the failure strain of the last broken fibres in the expression giving the strength distribution (equation [4.13]):

$$P_s^{\infty} = \exp\left[-\left(\frac{\varepsilon K_c}{\varepsilon_0 K_s}\right)^m\right] \qquad [4.16]$$

Before this limit survival probability is reached, delayed fracture should obey equation [4.14]. In order to validate this model, the following data are needed:

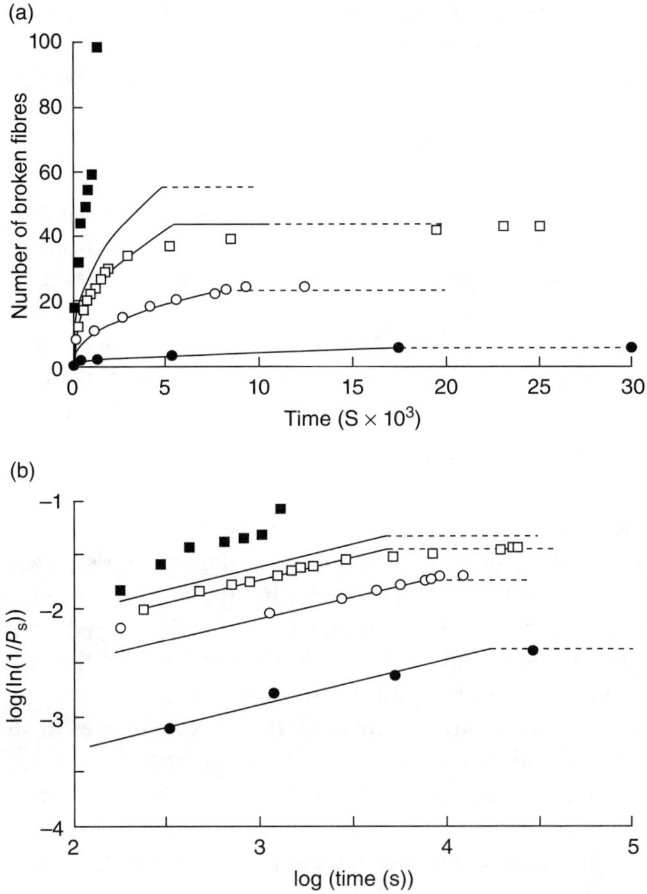

4.9 In situ monitoring of delayed fibre failures in a water-aged glass/epoxy unidirectional composite under static fatigue conditions (three-point bending, imposed strain ε_{max}). (a) Number of broken fibres as a function of time and applied strain. (b) Representation of the experimental data in panel (a) in the form of a survival probability, P_s. Solid lines correspond to the theoretical prediction of the SCC model (equations [4.14]), dashed lines to the number of broken fibres at saturation (equation [4.16]). Symbols represent: ●, ε_{max} = 1.0%; ○, ε_{max} = 1.5%; □, ε_{max} = 1.8%; ■, ε_{max} = 1.9%.

- the values of the parameters ε_0 and m corresponding to the statistical distribution of initial fibre strength. As detailed in Pauchard *et al.* (2001) and (2002a), these values can be determined from the *in situ* observation of composite specimens during both monotonic and static fatigue loading. Alternatively, these values can also be obtained from independent tensile testing of unimpregnated fibre bundles, provided that additional defects are not induced at the surface of the fibres during the manufacturing of the composite.

- the value of the exponent, *n*, of the power law relation describing stage I subcritical crack propagation. If the Weibull parameter, *m*, is known, *n* can be obtained from the slope, $m/(n - 2)$, of a log–log plot of $1/P_s$ versus time. Alternatively, *n*, can be obtained from the *v*–*K* curve corresponding to the glass fibres embedded within the aged composite matrix. The latter can be obtained from *in situ* observations of delayed fibre failure using a procedure similar to that described in Section 4.3.2.
- the prefactor, *k*, in equation [4.14] corresponds to the only fitting parameter of the model. It can be determined from the distribution of the lifetimes in a reference test under known strain and ageing conditions.

An example of the prediction of the SCC model is given by the solid lines in Fig. 4.9(b). At $\varepsilon = 1.9\%$, the experimental kinetics of fibre failure is clearly underestimated by the model due to the early nucleation of macroscopic matrix cracks. On the other hand, the theoretical predictions agree well with the theoretical data at lower strains, including the number of broken fibres at saturation. This *in situ* micromechanical analysis of delayed fibre fracture therefore supports the validity of SCC concepts to describe the initial stages of fatigue damage accumulation in water-aged unidirectional GFRPs. However, information regarding the density of broken fibres or the probability of failure of individual glass filaments are of little use for lifetime prediction of composite structures. In the following section, it will be shown that, in some situations, the theoretical SCC approach can be extended to the prediction of macroscopic mechanical parameters such as stiffness.

4.4.2 Lifetime prediction for unidirectional glass/epoxy composite beams under stress corrosion cracking conditions

A theoretical description of the relationship between a microscopic damage parameter such as the density of broken fibres and a macroscopic property such as stiffness would require a complete modelling of the composite behaviour, including stress transfer phenomena. This difficulty can be circumvented if some empirical relationship can be assumed between these variables. This was shown to be the case for the three-point bending fatigue behaviour of unidirectional glass/epoxy composite beams (Pauchard *et al.* 2002b). According to *in situ* microscopic observations and to finite element simulations, it was established that the relative stiffness loss, S/S_0, is linearly related to the number of broken fibres, N_f, within the considered material volume:

$$\frac{S}{S_0} = 1 - \alpha N_f \qquad [4.17]$$

where S_0 is the initial stiffness of the beam and α is an empirical parameter depending on the nature of the composite and on the loading configuration. Taking into account that $P_s = 1 - N_f/N_t$ (where N_t is the total number of fibres), equation [4.17] can be rewritten as follows:

$$\frac{S}{S_0} = 1 - \alpha N_t + \alpha P_s N_t \qquad [4.18]$$

In the case of the monotonic loading of an unaged material, equation [4.13] can be substituted in equation [4.18] to give an expression of the stiffness loss as a function of the applied strain, ε:

$$\frac{S}{S_0}(\varepsilon) = 1 - \alpha N_t + \alpha \exp\left[-\left(\frac{\varepsilon}{\varepsilon_0}\right)^m\right] N_t \qquad [4.19]$$

Similarly, equations [4.14] and [4.18] provide the stiffness loss under a periodic ($\lambda < 1$) or static ($\lambda = 1$) loading:

$$\frac{S}{S_0}(t) = 1 - \alpha N_t + \alpha \exp[-kt^{m/(n-2)} \varepsilon_{\max}^{mn/(n-2)} \lambda^{m/(n-2)}] N_t \qquad [4.20]$$

the limiting value of the relative stiffness loss being given by:

$$\frac{S_\infty}{S_0}(t) = 1 - \alpha N_t + \alpha \exp\left[-\left(\frac{\varepsilon_{\max} K_c}{\varepsilon_0 K_s}\right)\right] N_t \qquad [4.21]$$

It can be shown that expressions [4.19] and [4.20] can be approximated to a good level of accuracy by the following scaling relations:

$$\ln \frac{S}{S_0}(\varepsilon) \approx -\left(\frac{\varepsilon}{\varepsilon_0^*}\right)^m, \quad \text{monotonic loading} \qquad [4.22]$$

$$\ln \frac{S}{S_0}(t) \approx -k^* t^{m/(n-2)} \varepsilon_{\max}^{mn/(n-2)} \lambda^{m/(n-2)}, \quad \text{fatigue loading} \qquad [4.23]$$

with

$$\ln \frac{S_\infty}{S_0}(t) \approx -\left(\frac{\varepsilon_{\max}}{\varepsilon^*}\right)^m \qquad [4.24]$$

where k^*, ε_0^* and ε^* are empirical constants to be determined from experiments. As will be shown below, these scaling laws for $\ln(S/S_0)$ against time and applied strain can be verified experimentally for both static and cyclic fatigue loading in order to provide a consistent basis for the prediction of stiffness loss under the combined action of mechanical stresses and water ageing.

Static fatigue behaviour (R = 1)

Constant strain static fatigue experiments under three-point bending were reported by Pauchard *et al.* (2002b) for glass/epoxy beams aged in water at room temperature. In this study, the SCC analysis of stiffness loss was restricted to the initial stages of fatigue damage development (up to about 5–10% stiffness loss), i.e. before the development of significant macroscopic cracking. It is noteworthy that current practice for the fatigue design of composite beams is often based on a 5% or 10% stiffness loss criterion, which is therefore consistent with the validity domain of the SCC approach. In Fig. 4.10, it can be seen that $\ln(S_0/S)$ is linearly related to the loading time in a log–log plot, according to the prediction of equation [4.23]. Whatever the applied strain level, the slope of regression lines is roughly constant and yields a value of the ratio $m/(n-2)$. Equation [4.23] also indicates that, when the applied strain is varied, the stiffness loss lines should be shifted along the stiffness loss axis by a factor $mn/(n-2)\log\varepsilon_{max}$. From the experimental values of the $m/(n-2)$ and $mn/(n-2)$ ratios, a realistic estimate of the Weibull modulus, m, and of the stress corrosion parameter, n, can be obtained ($m = 4.1 \pm 0.1$ and $n = 12.6 \pm 1.5$) (Pauchard *et al.* 2002b).

It can be noted in passing that the parameters m and n can also be estimated by other independent approaches. According to equation [4.22], the representation, in a Weibull plot, of the relative stiffness loss of a dried composite beam during monotonic loading should give a line with a slope corresponding to m. This is indeed the case (Fig. 4.11) and the corresponding value of m (4.3 ± 0.15) is very consistent with that determined independently from the static fatigue experiments. In addition, the value of n can

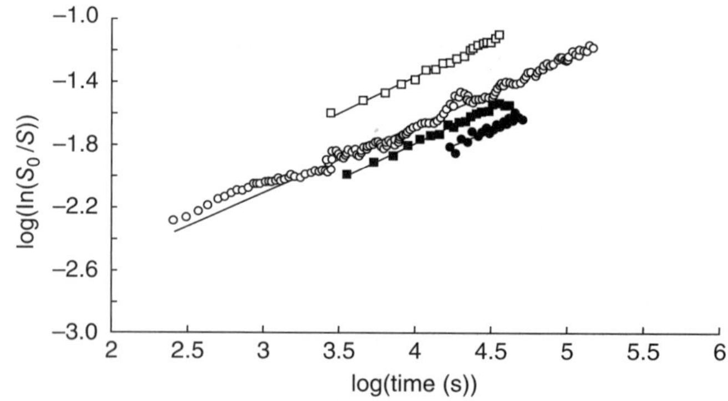

4.10 Log–log plot of the logarithm of the reverse of the relative stiffness loss, S/S_0, against time for a water-aged unidirectional glass/epoxy beam under static fatigue (three-point bending). Symbols represent: ●, $\varepsilon_{max} = 1.2\%$; ■, $\varepsilon_{max} = 1.4\%$; ○, $\varepsilon_{max} = 1.6\%$; □, $\varepsilon_{max} = 1.9\%$ (from Pauchard *et al.* (2002b)).

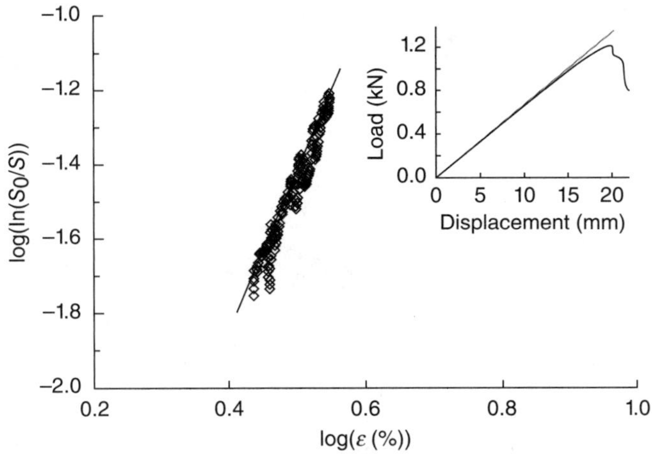

4.11 Log–log plot of the logarithm of the reverse of relative stiffness loss, S/S_0, as a function of applied strain, ε during the monotonic flexural bending of a unidirectional glass/epoxy composite. Insert: load plotted against displacement (data taken from Pauchard *et al.* (2001)).

also be estimated independently from a v–K curve established using stiffness loss data under both monotonic and static fatigue loadings. Such a v–K curve can be obtained using the same procedure as that described in Section 4.3.2, since the stiffness loss data can be related to the amount of broken fibre by relation [4.17]. This approach was found to provide a value of the parameter n that is consistent with that obtained from the analysis of static fatigue data (Pauchard *et al.* 2001).

At this stage, it is also interesting to consider the ability of an SCC model to take into account the effects of temperature. The data reported in Fig. 4.12 correspond to the stiffness loss behaviour of specimens aged in water at 20 °C and subsequently submitted to a static loading in water at the same applied strain but at different temperatures. Within the considered range of temperature (from 20 to 60 °C), the water content at saturation was not found to vary substantially with temperature. All the testing parameters except the temperature can thus be considered as constant during the mechanical loading. The static fatigue results indicate that increasing the temperature results in a substantial activation of the stiffness loss kinetics. This phenomenon can be related to the thermal activation of the subcritical crack velocity. From equations [4.2], [4.3] and [4.23], the following modified expression for the stiffness loss under static fatigue conditions can be derived:

$$\ln\frac{S}{S_0}(t) \approx -k^{**}t^{m/(n-2)}\varepsilon_{\max}^{mn/(n-2)}\lambda^{m/(n-2)}\exp(-Ea/RT)^{m/(n-2)} \quad [4.25]$$

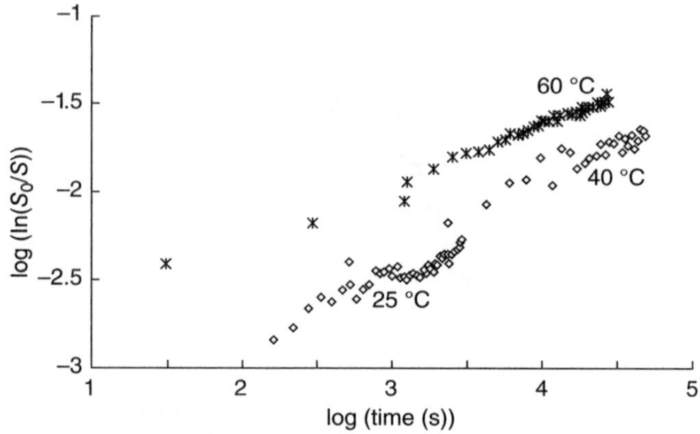

4.12 Log–log plot of the logarithm of the reverse of the relative stiffness loss, S/S_0, as a function of time for an aged unidirectional glass/epoxy composite under static fatigue in water at different temperatures (three-point bending, $\varepsilon_{max} = 1.2\%$) (from Pauchard *et al.* (2001)).

where k^{**} is another empirical prefactor derived from k^*. Accordingly, the shift of the $\log(\ln(S/S_{0)}))$–$\log t$ curve as a function of temperature provides a value of the activation energy which is of the order of 50 kJ mol^{-1} for E-glass/epoxy composites (Pauchard *et al.* 2001).

Cyclic fatigue behaviour ($R \neq 1$)

In the context of SCC behaviour, the delayed failure of fibres should be independent of the loading frequency. This means that the fibre lifetimes can be deduced from the integration of the subcritical crack propagation rate over the loading period, irrespective of the loading rate. Evidence of such a frequency-independent behaviour is provided from a comparison of lifetime data under static and cyclic fatigue loading. Aveston and Sillwood (1982) showed that the lifetimes under static and square-wave cyclic fatigue are the same if only the time at maximum load is taken into account during cyclic loading. Similar investigations by Pauchard *et al.* (2002b) under static and sinusoidal fatigue also showed that the relevant parameter for the fatigue life of water-aged unidirectional composites is the time spent under a given load rather than the number of cycles. As an example, Fig. 4.13 shows the stiffness loss curves obtained for water-aged glass/epoxy composite beams at different frequencies. As the frequency is increased, a shift of the stiffness loss curve to increasing numbers of cycles can clearly be observed (Fig. 4.13(a)). For a given number of cycles, more time is spent at a given strain level at low frequency than at high frequency. As a result, the

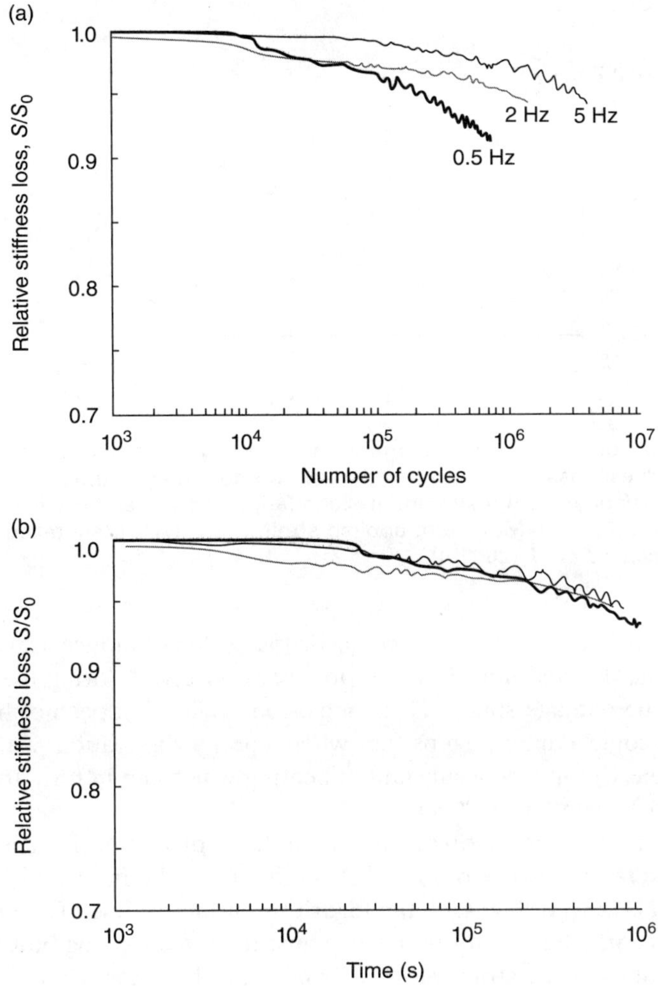

4.13 Relative stiffness loss of a water-aged unidirectional glass/epoxy composite against (a) the number of cycles and (b) time under fatigue at different frequencies (three-point bending, $\varepsilon_{max} = 1.4\%$, strain ratio $R = \varepsilon_{min}/\varepsilon_{max} = 0.7$) (data taken from Pauchard *et al.* (2002b)).

fibres' surface defects have more time to grow during each loading cycle and the total number of cycles to failure is reduced. Accordingly, the different stiffness loss curves can be reduced, within the experimental scatter, to a single curve if a time scale is considered (Fig. 4.13(b)). This result demonstrates that the relevant parameter regarding the frequency effect is the time spent at a given stress level, in accordance with the hypothesis of a fatigue response dominated by SCC mechanisms. Within the frequency range under investigation, it transpires that the dissipative processes associ-

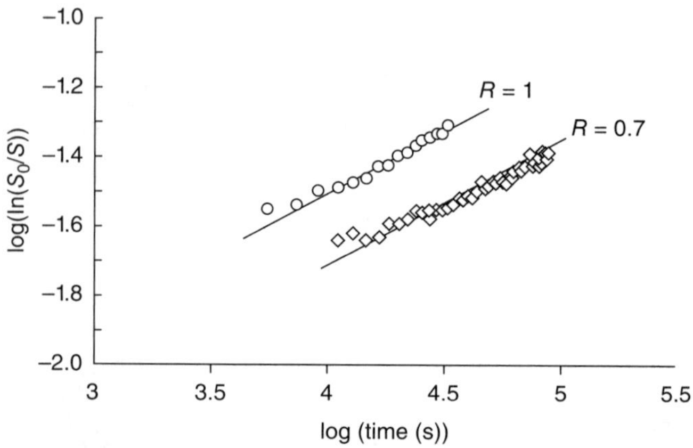

4.14 Log–log plot of the logarithm of the reverse of the relative stiffness loss, S/S_0, as a function of time for an aged unidirectional glass/epoxy composite under static fatigue ($R = 1$) and cyclic fatigue ($R = 0.7$, 5 Hz). Maximum applied strain $\varepsilon_{max} = 1.4\%$ (data taken from Pauchard *et al.* (2002b)).

ated with matrix and interface viscoelastic losses do not induce a significant change in the delayed fibre fracture processes observed during the microscopic fatigue damage stages. This conclusion obviously does not hold for the macroscopic damage steps, i.e. when energy dissipation within the cracked areas can induce a substantial heating which can in turn affect the temperature-dependent SCC processes.

Another parameter involved in the lifetime of composite specimens under cyclic fatigue is the ratio of the minimum to the maximum applied strain, $R = \varepsilon_{min}/\varepsilon_{max}$. In Fig. 4.14, the logarithm of the reverse of the relative stiffness loss, S/S_0, has been reported against time in a log–log plot for two different values of the strain ratio (0.7 and 1). In both cases, a linear relationship is observed, the value of the slope being independent of the strain ratio. This observation is consistent with the SCC model which predicts that a change in the strain ratio must only induce a shift of the $\log(\ln(S_0/S)$ versus $\log(t)$ regression lines by a factor $m/(n-2)\log\lambda$ along the stiffness axis (equation [4.23]). It has been shown that the experimental shift factor is consistent with the theoretical predictions (Pauchard *et al.* 2002b).

4.5 Concluding remarks and future trends

SCC is one of the major damage mechanisms involved in the durability of glass reinforced polymer matrix composites exposed to the combined action of fatigue loading and water ageing. It is especially relevant in the case of unidirectional composites, where most of the fatigue life is controlled by the delayed failure of the glass reinforcement. In such systems, SCC is involved at all stages of fatigue damage development. During the early stages of the fatigue life, the progressive accumulation, at the microscopic scale, of broken fibres is strongly activated by SCC processes resulting from water diffusion through the polymer matrix. As the density of broken fibres increases, macroscopic cracks are nucleated and SCC can subsequently proceed by localized interactions, at the crack tip, between the glass fibres and the environment. The control of the kinetics of delayed fibre failure during the early stages of damage development is especially relevant to the prediction of the fatigue limit of GFRP composites. As reviewed in this chapter, much experimental evidence shows that well-established concepts for the delayed fracture of bulk glasses can be transposed to GFRPs in order to derive predictive statistical models for fibre lifetime. Using this approach, scaling laws are established that predict the changes in the density of broken fibres as a function of time, applied strain and environmental conditions (relative humidity, temperatures etc.). These SCC models can also be extended to macroscopic mechanical parameters such as the stiffness of composite beams. From a calculation of the probability of failure of glass fibres within the matrix, realistic estimates for the stiffness loss under fatigue loading are obtained as a function of frequency, applied strain and strain ratio. The approach basically relies on the determination of two kinds of materials characteristics. The first corresponds to the Weibull statistical distribution of initial fibre strength within the unaged composite. In addition, a knowledge of the fibre subcritical crack propagation law is required in the ageing state to be considered for lifetime prediction. Both kinds of information can conveniently be obtained from simple mechanical tests under monotonic or static fatigue loading conditions.

As mentioned above, this approach is restricted to the initial stages of fatigue life, when the damage mostly consists of the accumulation of delayed fibre failure in the absence of macroscopic cracks. The associated stiffness losses (of the order of 5–10%) are, however, consistent with the lifetime criterion used in engineering applications. Some deviations from the SCC approach to GFRP durability are also expected at high frequency and strain ratio, when viscoelastic effects in the matrix or at the interface can come into play. Experimental observations on aged unidirectional GFRPs tend to indicate that SCC behaviour is still observed up to a frequency of a few Hertz. This range encompasses many applications, such as offshore composite structures under the action of a wave spectrum.

One of the challenges for the application of SCC-based durability models in engineering applications is the control of the physico-chemistry of the water-ageing processes occurring within the polymer matrix and at the interface. Subcritical crack propagation rates are known to be strongly dependent not only on water concentration and temperature, but also on parameters such as pH. In the case of a polymer matrix sensitive to hydrolysis and/or leaching of low molecular weight products, it becomes clear that the evolving physico-chemical environments of the fibres will require delicate and time-consuming identification of the SCC parameters as a function of water concentration, ageing time and temperature. These complexities could probably be circumvented by the selection of a polymer matrix with improved resistance to hydrolysis and leaching, where the changes in subcritical crack growth rate will essentially be driven by water sorption kinetics.

4.6 References

AVESTON, J., KELLY, A. and SILLWOOD, J. M. (1980). Long term strength of glass reinforced plastics in wet environments. In *Advances in Composite Materials*. A. R. Bunsell, C. Bathias, A. Martrenchar, D. Menkes and G. Verchery (Eds). New York, Pergamon Press, vol. 1, pp. 556–568.

AVESTON, J. and SILLWOOD, J. M. (1982). 'Long term strength of glass reinforced plastics in dilute sulphuric acid.' *Journal of Materials Science* **17**: 3491–3498.

BAKER, T. C. and PRESTON, F. W. (1946). 'Fatigue of glass under static loads.' *Journal of Applied Physics* **17**: 170.

BONNIAU, P. and BUNSELL, A. R. (1981). 'A comparative study of water absorption theories applied to glass epoxy composites.' *Journal of Composite Materials* **15**: 272–293.

CHARLES, R. J. (1958a). 'Dynamic fatigue of glasses.' *Journal of Applied Physics* **29**(12): 1657–1662.

CHARLES, R. J. (1958b). 'Static fatigue of glass-I.' *Journal of Applied Physics* **29**(11): 1549.

CHATEAUMINOIS, A., CHABERT, B., SOULIERT, J. P. and VINCENT, L. (1994). 'Interfacial degradation during hygothermal ageing. Investigations by sorption/desorption experiments and viscoelastic analysis.' *Polymer* **35**(22): 4765–4774.

CHEVALIER, J., GREMILLARD, L., ZENATI, R., JORAND, Y., OLAGNON, C. and FANTOZZI, G. (2002). Slow crack growth in zirconia ceramics with different microstructures. In *Fracture Mechanics of Ceramics*. R. C. Bradt, D. Munz, M. Sakai, V. Y. Shevchenko and K. White (Eds). New York, Kluwer Academics, vol. 13, p. 287.

COWKING, A., ATTOU, A., SIDDIQUI, A. M. and SWEET, A. S. (1991a). 'An acoustic emission study of failure by stress corrosion in bundles of E-Glass fibres.' *Journal of Materials Science* **26**: 301–306.

COWKING, A., ATTOU, A., SIDDIQUI, A. M., SWEET, M. A. S. and HILL, R. (1991b). 'Testing E glass fiber bundles using acoustic emission.' *Journal of Materials Science* **26**(5): 1301–1310.

DEWIMILLE, B. and BUNSELL, A. R. (1982). 'The modeling of hydrothermal aging in glass-fiber reinforced epoxy composites.' *Journal of Physics D – Applied Physics* **15**(10): 2079–2091.

DEWIMILLE, B. and BUNSELL, A. R. (1983). 'Accelerated aging of a glass fiber-reinforced epoxy-resin in water.' *Composites* **14**(1): 35–40.

DOUGLAS, R. W. and EL-SHAMY, T. M. M. (1967). 'Reactions of glasses with aqueous solutions.' *Journal of the American Ceramic Society* **50**: 1.

FETT, T. and MUNZ, D. (1985). 'Determination of v-KI curves by a modified evaluation of lifetime measurements in static bending tests.' *Journal of the American Ceramic Society* **68**(8): C213–C215.

FROST, S. R. and CERVENKA, A. (1994). 'Glass fibre-reinforced epoxy matrix filament-wound pipes for use in the oil industry.' *Composites Manufacturing* **5**(2): 73–81.

GHOTRA, J. S. (1999). 'Oilfield FRP liners face corrosion and high temperatures.' *Polymers & Polymer Composites* **7**(3): 143–164.

HALE, J. M., SHAW, B. A., SPEAKE, S. D. and GIBSON, A. G. (2000). 'High temperature failure envelopes for thermosetting composite pipes in water.' *Plastics Rubber and Composites* **29**(10): 539–548.

HOGG, P. J. and HULL, D. (1980). 'Micromechanisms of crack growth in composite materials under corrosive environments.' *Metal Science* **14**: 441–449.

HOGG, P. J. and HULL, D. (1982). Role of matrix properties on the stress corrosion of GRP. In *13th Reinforced Plastics Congress*, Brighton. The British Plastics Federation, London.

HOLLAND, A. J. and TURNER, W. E. S. (1940). 'Effect of sustained loading on breaking strength of sheet glass.' *Journal of the Society of Glass Technology* **24**: 47–57.

HUGUET, S., GODIN, N., GAERTNER, R., SALMON, L. and VILLARD, D. (2002). 'Use of acoustic emission to identify damage modes in glass fibre reinforced polyester.' *Composites Science and Technology* **62**(10–11): 1433–1444.

JONES, F. R., ROCK, J. W. and BAILEY, J. E. (1983a). 'The environmental stress corrosion cracking of glass fibre-reinforced laminates and single E-glass filaments.' *Journal of Materials Science* **18**: 1059–1071.

JONES, F. R., ROCK, J. W. and BAILEY, J. E. (1983b). 'Stress corrosion cracking and its implications for the long term durability of E-glass fibre composites.' *Composites* **14**(3): 262–269.

JONES, F. R. (1989). 'The role of moisture diffusion and matrix plasticisation on the environmental stress corrosion of GRP.' *Journal of Strain Analysis* **24**(4): 223–233.

KASTURIARACHI, K. A. and PRITCHARD, G. (1984). 'Free dicyandiamide in crosslinked epoxy resins.' *Journal of Materials Science Letters* **3**: 283–286.

KAWADA, H., MIZUNO, M., KATSUNO, H., TOGE, K. and TSUBOI, T. (2001). Characteristic of crack propagation and threshold in woven GFRP laminates under acid stress environment. In *13th International Conference on Composite Materials*, Beijing, China.

KAWADA, H. and SRIVASTAVA, V. K. (2001). 'The effect of an acidic stress environment on the stress-intensity factor for GRP laminates.' *Composites Science and Technology* **61**(8): 1109–1114.

KUMOSA, L., ARMENTROUT, D. and KUMOSA, M. (2001). 'An evaluation of the critical conditions for the initiation of stress corrosion cracking in unidirectional E-glass/polymer composites.' *Composites Science and Technology* **61**(4): 615–623.

KUMOSA, L. S., KUMOSA, M. S. and ARMENTROUT, D. L. (2005). 'Resistance to brittle fracture of glass reinforced polymer composites used in composite (nonceramic) insulators.' *IEEE Transactions on Power Delivery* **20**(4): 2657–2666.

LAWN, B. (1993). *Fracture of Brittle Solids.* Cambridge, Cambridge University Press.

MAUGIS, D. (1985). 'Subcritical crack growth, surface energy, fracture toughness, stick-slip and embrittlement.' *Journal of Materials Science* **20**(9): 3041–3073.

METCALFE, A. G. and SCHMITZ, G. K. (1972). 'Mechanism of stress corrosion in E-glass fibres.' *Glass Technology* **13**(1): 5.

MIZOGUCHI, M., MORII, T., FUJII, Y. and HAMADA, H. (2001). Study on acid stress corrosion in E-glass reinforced vinylester composites. In *13th International Conference on Composite Materials*, Beijing, China.

MURAOKA, M., EBATA, K. and ABÉ, H. (1993). 'Effect of humidity on small-crack growth in silica optical fibers.' *Journal of the American Ceramics Society* **76**(6): 1545–1550.

MYERS, T. J., KYTOMAA, H. K. and SMITH, T. R. (2007). 'Environmental stress-corrosion cracking of fiberglass: lessons learned from failures in the chemical industry.' *Journal of Hazardous Materials* **142**(3): 695–704.

PAUCHARD, V., BROCHADO, S., CHATEAUMINOIS, A., CAMPION-BOULHARTS, H. and GROSJEAN, F. (2000). 'Measurements of sub-critical crack growth rates in glass fibres by means of acoustic emission.' *Journal of Materials Science Letters* **19**(23): 2141–2143.

PAUCHARD, V., CHATEAUMINOIS A., BOULHARTS-CAMPION, H., GROSJEAN, F. and ODRU, P. (2001). 'Développement d'un modèle de durabilité de poutres composites unidirectionnelles renforcées par des fibres de verre.' *Oil & Gas Science and Technology* **56**(6): 581–595.

PAUCHARD, V., CHATEAUMINOIS, A., GROSJEAN, F. and ODRU, P. (2002a). 'In situ analysis of delayed fibre failure within water aged GFRP under static fatigue conditions.' *International Journal of Fatigue* **24**: 447–454.

PAUCHARD, V., GROSJEAN, F., CAMPION-BOULHARTS, H. and CHATEAUMINOIS, A. (2002b). 'Application of a stress corrosion cracking model to the analysis of the durability of glass/epoxy composites in wet environments.' *Composite Science and Technology* **62**: 493–498.

PIGOTT, M. R. (1987). *Load Bearing Composites.* Oxford, Pergamon Press.

PUKH, V. P., LATERNER, S. A. and INGAL, V. N. (1970). *Soviet Physics – Solid State* **12**: 881.

RITTER, J. E., SERVICE, T. H. and JAKUS, K. (1988). 'Predicted static fatigue behaviour of specially coated optical glass fibers.' *Journal of the American Ceramic Society* **71**(11): 988–992.

RODRIGUEZ, E. L. (1987). 'Corrosion of glass fibres.' *Journal of Materials Science Letters* **6**: 718-720.

SCHEN, C. H. and SPRINGER, G. S. (1981). Moisture absorption and desorption of composites materials. In *Environmental Effects on Composite Materials*. G. S. Springer (Ed.). Basel, Technomic Publications. vol. 1, p. 15.

TALREJA, R. (1987). *Fatigue of Composites Materials*. Basel, Technomic Publishing Co. Inc.

VAUTHIER, E., CHATEAUMINOIS, A. and BAILLIEZ, T. (1996). 'Fatigue damage nucleation and growth in a unidirectional glass/epoxy composite subjected to hygrothermal ageing.' *Polymer and Polymer Composites* **4**(5): 343–351.

VAUTHIER, E., ABRY, J. C., BAILLIEZ, T. and CHATEAUMINOIS, A. (1998). 'Interactions between hygrothermal ageing and fatigue damage in unidirectional glass/epoxy composites.' *Composites Science and Technology* **58**: 687-692.

WANG, T. T., NAZIRANI, H. N., SCHONHORN, H. and ZUPKO, H. M. (1979). 'Effects of water and moisture on strength of optical (silica) fibers coated with a UV cured epoxy acrylate.' *Journal of the American Ceramic Society* **23**: 887-892.

WEIBULL, W. (1951). 'A statistical distribution function of wide applicability.' *Journal of Applied Mechanics* **18**: 293–296.
WIEDERHORN, S. M. (1967). 'Influence of water vapor on crack propagaition in soda-lime glass.' *Journal of the American Ceramic Society* **50**: 407.
WIEDERHORN, S. M. (1969). 'Fracture surface energy of glass.' *Journal of the American Ceramic Society* **52**: 99.
WIEDERHORN, S. M. (1978a). *Fracture Mechanics of Ceramics*. New York, Plenum Press.
WIEDERHORN, S. M. (1978b). Mechanisms of subcritical crack growth in glass. In *Fracture Mechanics of Ceramics*. R. C. Bradt (Ed.). New York, Plenum Press, vol. 4, pp. 549–580.
WIEDERHORN, S. M. and BOLZ, L. H. (1970). 'Stress corrosion and static fatigue of glass.' *Journal of the American Ceramic Society* **53**: 543–548.
WIEDERHORN, S. M., JOHNSON, H., DINESS, A. M. and HEUER, A. H. (1974). 'Fracture of glass in vacuum.' *Journal of the American Ceramic Society* **57**: 336.
ZINCK, P., PAYS, M. F., REZAKHANLOU, R. and GERARD, J. F. (1999). 'Extrapolation techniques at short gauge lengths based on the weakest link concept for fibres exhibiting multiple failure modes.' *Philosophical Magazine A* **79**(9): 2103–2122.

5
Thermo-oxidative ageing of composite materials

T. TSOTSIS, The Boeing Company, USA

5.1 Introduction

This chapter discusses some of the issues related to thermo-oxidative ageing in polymer-matrix composite materials (PMCs). In general, all PMCs are limited by the range of temperatures where they may be used. In the simplest cases, the glass-transition temperature (T_g) or a heat-distortion temperature will govern the maximum temperature at which a given PMC may be successfully used. However, for longer-term usage, the degradation of the matrix polymer due to thermal and thermally activated processes (mainly oxidation) defines the maximum use temperature of PMCs. From here forward, characterization of oxidation resistance will be described as thermo-oxidative stability.

5.1.1 Importance of thermo-oxidative ageing in composites development

Polymeric composites are generally used to produce lighter-weight structures for a particular application than an alternative materials system could provide. In many of these cases, the application requires long-term exposures to elevated temperatures. Many of the earliest applications of PMCs were in aerospace, mainly for aircraft uses as the lifetimes required for aircraft components are typically in the tens of thousands of hours vs. only tens or a few hundred hours for space-vehicle components and structures. Some of the earliest applications of PMCs at elevated temperatures were for supersonic aircraft as described in reference 1, which summarizes the first large-scale, long-term ageing study of a variety of composite systems.

The effects of thermal or thermally activated degradation processes typically only become apparent after long periods of time due to the physical and chemical processes involved. During matrix degradation, several processes occur simultaneously: diffusion of oxygen into the matrix, diffusion of degradation products out of the matrix, reaction of oxygen with the

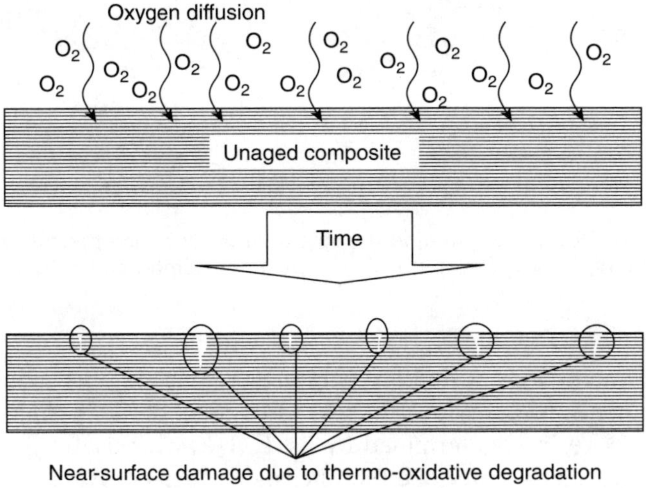

5.1 Schematic of oxygen diffusion into a composite and resulting near-surface damage after long-term, thermo-oxidative exposure.

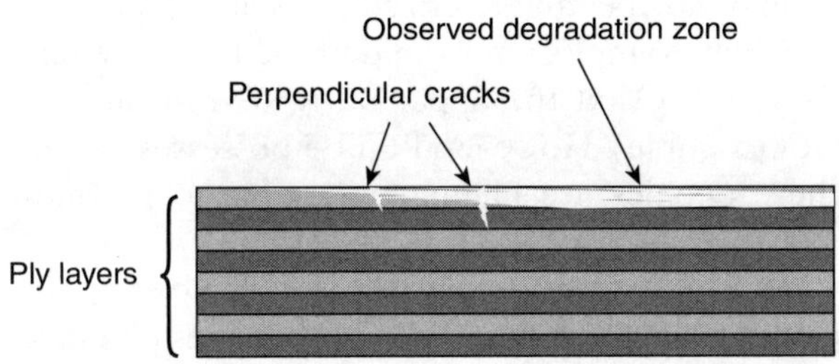

5.2 Schematic of surface degradation due to thermo-oxidative ageing.

matrix, thermal degradation reactions within the matrix, and reaction of degradation and oxidation byproducts with the matrix, with each other, and with oxygen. A schematic of this process is shown in Fig. 5.1. Furthermore, because of the multiple reactions and mass transfers into and out of the matrix, this is a dynamic problem, making it very difficult to characterize. Some of these concerns are described in references 2 and 3. After significant amounts of time, the degradation may create even greater damage than shown in Fig. 5.1 and start to wear away some of the composite surface layers. Figure. 5.2 shows a depiction of increased surface damage.

5.1.2 Application areas and relevance to matrix chemistry

As mentioned above, most applications requiring long-term, thermo-oxidative stability have been for aircraft, where the combination of long-

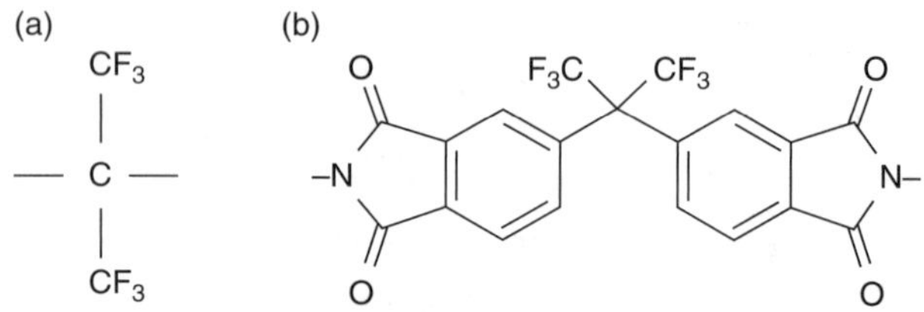

5.3 (a) Fluorine-containing group shown to enhance thermo-oxidative stability; (b) 4,4′-(hexafluoroisopropylidene) diphthalic anhydride (6FDA).

term requirements at elevated temperatures requires stable materials. In addition, as mentioned previously, a base assessment of the thermal capability of a matrix resin is determined by its T_g. However useful the T_g data may be, T_g is not an indicator of either thermal or thermo-oxidative stability. Indeed, T_g is principally a measure of a polymer's stiffness at temperature (as will be described further in Section 1.3) and, as such, is only marginally related to a material's thermo-oxidative stability. Thus, a material's thermo-oxidative stability may not be characterized with any degree of accuracy using T_g data alone. For example, a polyimide-based resin system, PETI-5, has a reported T_g of 235 °C,[4] at the upper range of T_g values for epoxies (see reference 5, for example). However, PETI-5 possesses excellent stability at 177 °C,[6] where epoxies are highly susceptible to thermo-oxidative degradation.[3]

Other factors that affect thermo-oxidative stability besides the generic chemistry (e.g. epoxy, bismaleimide (BMI), polyimide, phenolic, cyanate ester, etc.) of a matrix resin are the monomeric constituents that make up the resin. For example, polymers, especially polyimides, containing the fluorinated group shown in Fig. 5.3 have been shown to have improved (a) thermo-oxidative stability, often with (b) 4,4′-(hexafluoroisopropylidene) diphthalic anhydride, more commonly known at 6FDA, as a key resin component.

In addition to the effects of the matrix resin's backbone, endcap materials have also been found to have a strong effect on thermo-oxidative stability. Many high-temperature polyimides are based on nadic endcaps as shown in Fig. 5.4, such as NASA's PMR-15, PMR-II-50, RP-46, etc. More recently, polyimides (NASA-developed PETI-5, PETI-330, HFPE-52-II, and Air Force-developed AFR-PE-4) have been formulated with phenylethynyl endcaps (also shown in Fig. 5.4), which have been found to impart a high degree of thermo-oxidative stability to composites based on these materials.[4]

Although the examples listed above relate to polyimides, analogous cases are easily found for other polymer types. Molecular features such as the degree of crosslinking, the presence of and types of pendant groups along

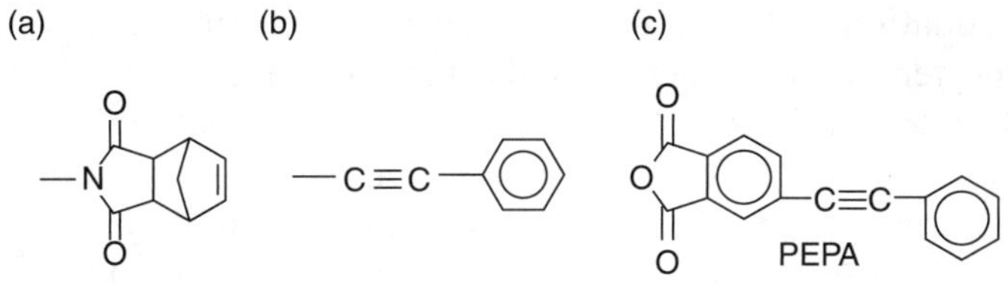

5.4 (a) Nadic endcap; (b) phenylethynyl endcap; (c) phenylethynyl phthalic anhydride (PEPA), incorporating phenylethynyl endcap.

the polymer chain, and stereoisomeric differences (e.g. *meta*, *para*, or *ortho* orientations of groups around an aromatic ring) can also have significant effects on the thermo-oxidative stability of matrix resins. An excellent summary of various matrix-resin chemistries may be found in Bader *et al.*[7]

5.1.3 Key factors in characterization

In determining the suitability of a matrix resin for long-term application at a given temperature, it is necessary to generate data at the relevant temperature because, as will be discussed in more detail in Section 5.2, reliable predictive methodologies are lacking. Because thermo-oxidative degradation in PMCs is almost entirely dependent on the stability of the matrix resin, characterization methods need to focus on resin-dominated properties.

In the simplest terms, oxidation of a matrix may be described by the simple relation

$$\mathrm{rA} + \mathrm{sO_2} \rightarrow \mathrm{tC} + \mathrm{uD} + \ldots \qquad [5.1]$$

where A represents the matrix polymer or a subcomponent of it and C, D and possibly other components are the reaction products. Although this is a highly oversimplified view of the actual mechanisms involved in thermo-oxidative degradation, it is useful for helping to understand some basic physical and chemical effects. If one simplifies the reaction shown in equation [5.1] to a single or single-equivalent reaction product, then first-order reaction kinetics may be used to evaluate the relative effects of the different constituents by the rate equation shown in equation [5.2], where K_{eq} is the equilibrium coefficient:

$$Q = \frac{[\mathrm{C}]}{[\mathrm{A}][\mathrm{O_2}]} \qquad [5.2]$$

when $Q = 1$, the reaction is at equilibrium; when $Q < K_{eq}$, a forward reaction occurs; when $Q > K_{eq}$, a reverse reaction occurs. Thus, for the relationship shown in equation [5.1], the reaction can be accelerated in several ways. First, as the temperature increases, the reactivity increases. Furthermore, and not depicted in the simplistic model shown in equation [5.1], reaction mechanisms can change with increasing temperature, especially near or above a material's T_g.[2] Secondly, the reaction is governed by the amount of oxygen present. The amount of oxygen may be increased in two ways: first, by increasing the pressure[8] and second by allowing more oxygen to penetrate into the matrix. The latter effect can occur by an increase in the diffusivity of oxygen into the matrix, which can occur when there is an increase in the total surface area of the composite – such as can occur when transverse cracks are induced from ageing, damage, or mechanical loading.[9] The toughness of the matrix resin will affect the amount of matrix cracking and thus also the thermo-oxidative stability.[9]

Another factor in the rate of degradation is the presence of reaction products. If the reaction products form a layer protecting the unaffected matrix resin from oxygen, then the reaction will slow, according to the relationship in equation [5.2]. In contrast, if the degraded material is removed by spalling, vaporization, etc., then the simple kinetics analysis shows that the reaction will proceed at least as fast as it did initially. Another feature that needs to be mentioned is that the initial oxidation of the polymer may *increase* the weight of the overall system if the amount of oxygen that reacts into the matrix and any reaction byproducts are less than the amount of material that is volatilized. Typically, this is only observed early on during oxidation, but, because these reaction-related weight gains continue throughout thermo-oxidative degradation, weight-loss measurements need to take into account the competing weight-gain mechanism of oxidation and the weight-loss mechanisms associated with volatiles and other degraded material that is removed from the composite during ageing.

It is important to note that crosslinking and chain extension – (the basic reaction mechanisms in polymerization), as well as chain scission (one of the principal mechanisms in thermo-oxidative degradation), are thermally activated processes. This distinction is necessary because, when performing accelerated ageing tests, attempting to accelerate degradation through the use of elevated temperature alone will probably result in changes in mechanisms and changes in the relative rates of different degradation mechanisms due to ageing at temperatures higher than the ultimate-use temperature of interest.

5.1.4 Test methodologies

In order to perform studies of long-term ageing of composites properly, particularly in the absence of reliable accelerating methods, a large test program is required to assess mechanical-property degradation. There are many considerations in setting up such a program, not unlike any large-scale environmental-testing program. Parameters such as specimen lay-up, thickness, and overall panel size need to be carefully considered. Because oxidation is a diffusion-related phenomenon, using specimens that are too thick will delay the onset of observable changes in many properties because there may be sufficient remaining undamaged material to carry mechanical loads, such that the loss or reduction in properties of a material's outer layers masks the degradation. Different lay-ups will affect the interlaminar stresses, which will affect diffusion rates.[10] Additionally, edge effects can play a role. Making ageing panels larger can mitigate edge effects, but, because any aged panels will have to be machined into smaller test coupons, damage due to machining may have an unwanted effect on test results, such that desired trends may be difficult to determine.

Tests should be selected to maximize the role of the matrix in the material properties. This requirement leads to specimens for compression, shear, and fracture properties. Specimens that are highly dependent on tensile properties that are fiber-dominated will not exhibit any significant reductions in strength until the load transfer between fibers via the matrix is reduced to a sufficient extent that strengths are reduced. Some general discussion of many different test methods is given in reference 11.

In addition to mechanical tests of composites, it is sometimes of interest to try to determine degradation methods via thermal analyses; mainly of the neat resin, but composite test specimens may be used in certain instances. The T_g is the most common of the thermal-analysis properties of interest and may be obtained using various methods including: dynamic mechanical analysis (DMA), which may be performed in torsion or flexure, depending on the test apparatus; differential scanning calorimetry (DSC); or thermo-mechanical analysis (TMA). In using DMA, glass transitions may be determined from the inflection point of the torsional or flexural stiffness (in-phase response) or by the peak of the ratio of the out-of-phase to the in-phase response, generally referred to as tan δ. For the purposes of understanding mechanical behavior, only the change in stiffness is of practical interest. T_g data from TMA are closely related to the DMA change in stiffness, but T_g data from DSC are not necessarily correlated to mechanical changes and, thus, are of limited value for the present discussion.

Thermogravimetric analysis (TGA) is used to measure the rate of change of weight vs. time and temperature in a controlled-gas environment. Typical gases are air, oxygen, and nitrogen, though others may certainly be used. TGA is highly dependent on the available surface area of the specimens placed into the test chamber, thus whether a specimen is monolithic or pulverized has a large effect on the results. Moreover, the sizes of the particles in a powder will also affect results accordingly. TGA may be combined with other methods such as gas chromatography (GC), mass spectroscopy

(MS), or Fourier transform infrared (FTIR) analysis to try to evaluate the degradation products in the effluent gas stream. For detailed discussions regarding the thermal-analysis techniques mentioned above, the reader is advised to consult the general literature.

5.2 Developments in understanding thermo-oxidative ageing

Since the 1970s when polymer-based composite materials first began to be used in a significant way, data and understanding of their long-term durability have become increasingly important. Unlike other widely used structural materials like metal, concrete, and wood, composites do not possess either the large historical database or the level of characterization of these other materials. Thus, the requisite data have had to be generated in parallel with the implementation of new materials.

Another factor that has affected the composite-materials database is the lack of the product standards that are available for metal alloys, etc. All matrix-resin formulations are trade secrets and their fabrication methods are additionally proprietary. This means that most studies have had to rely on whatever commercially available materials existed at the time of the study and not on model compounds. Because of the aforementioned relationships between resin backbone and endcap chemistries and thermo-oxidative stability, it is generally difficult to draw definitive conclusions from the dataset of long-term, composite-materials' ageing studies. Lastly, there is no standard testing protocol to determine or assess thermo-oxidative stability so that most studies are either stand-alone or remain proprietary to the end user.

Most high-temperature PMCs are based on nadic-endcapped polyimide derived from chemistry originally developed by Lubowitz[12] and then modified by Serafini *et al.*[13] into polymerization-of-monomeric-reactants (PMR) chemistry, of which PMR-15 is the best-known matrix resin, By and large, high-temperature PMCs and studies of long-term, thermo-oxidative ageing have disproportionately been with PMR-based resins, though there are many highly relevant studies that deal with lower-temperature PMCs in order to obtain basic understandings regarding the long-term behavior of composite matrices. Thus, materials such as epoxies,[14–16] BMIs,[17,18] etc. have also been studied in this context.

5.3 Initial studies – Kerr and Haskins

The first large-scale ageing study of composites was performed by Kerr and Haskins and is summarized in their final report in reference 1. In this study, $[0°]_6$ and $[0°/\pm45°]_s$ coupons of carbon- and boron-reinforced epoxy

(A-S/3501–5 and Rigidite 5505/4, respectively), boron/aluminum (5.6-mil boron/6061 Al), and polyimide (Celion 6000/LARC-160) materials were exposed to elevated temperatures and various air pressures for up to 50 000 hours. Different air pressures were used because, at the altitudes at which supersonic transport would operate, the ambient air pressure would be substantially lower than at sea level and because increasing air (oxygen) pressure is known to increase degradation rate.[19–23]

The materials above were downselected after an intensive pre-screening phase that exposed specimens for up to 10 000 hours. Tensile test specimens were all 6 plies in thickness using lay-ups of $[0°]_6$ and $[0°/\pm 45°]_s$ except for the LARC-160 for which no unidirectional specimens were used. Additional tests were also performed on 6-ply $[0°/\pm 45°]_s$ and 24-ply $[0°/\pm 45°]_{4s}$ lay-ups to measure room-temperature compressive, interlaminar shear, fatigue resistance, and T_g changes, and weight change. Additionally, some of the LARC-160 coupons were coated with a solution of polyphenylquinoxaline. All epoxy-based specimens were aged at 122 °C (250 °F) and 177 °C (350 °F) for tensile coupons and at 100 °C (212 °F) and 122 °C (250 °F) for compression and shear. The polyimide-based specimens were aged at 177 °C (350 °F) for tensile testing and at 177 °C (350 °F) and 232 °C (450 °F) for compression and shear. Additional tests were performed to study the effects of moisture and creep. The boron/aluminum specimens will not be discussed further as the following discussion will focus on polymer responses.

The goal of this program was to characterize advanced composite systems before and after exposures to simulated supersonic-cruise environments for times of up to 50 000 hours. Fatigue data for specimens tested at various temperatures and stress levels were also gathered. Tests were performed to measure the changes in mechanical properties, and metallography and fractography were used to examine post-ageing specimens to help identify the degradation mechanisms during high-temperature ageing.

The times and temperatures used for most thermo-oxidative ageing studies are based on the skin temperatures at different Mach numbers from aerodynamic heating due to friction with the air. A National Materials Advisory Board publication[24] summarizes the requirements for supersonic aircraft quite well based on prior studies, many related to the Concorde, which are also cited in this reference. At Mach 2.0, the skin temperature remains well below 100 °C but rises to 120 °C at Mach 2.2 and rises further to 150 °C at Mach 2.4.

Kerr and Haskins[1] noted edge cracking and severe property degradation at 177 °C and 0.1 MPa after 5000 hours of ageing. However, at the lower pressure of 0.014 MPa at 177 °C such degradation was not observed until after 25 000 hours. Additionally, degradation in tension-dominated specimens (unidirectional $[0°]_6$) was not seen until well after it had been detected in specimens with stronger matrix dependence (orthotropic $[0°/\pm 45°]_s$). It should also be noted that significant reductions in mechanical properties of the orthotropic lay-ups were not observed until the 0° layer had experienced significant degradation.

For epoxy-based specimens, ageing effects were readily observable after only 1000–5000 hours of ageing at 177 °C (350 °F), whereas it took between 10000 and 25000 hours for such effects to be seen at 121 °C (250 °F). Because the ageing at 177 °C occurs very near the T_g of the epoxy matrix, it is highly likely that the mechanisms of degradation, in addition to being accelerated compared with those at 121 °C, are indeed different as well.[11] At 177 °C, degradation was sufficiently pervasive – matrix pulverization or crumbling on the specimen surfaces – that even ageing at the lower pressure did not eliminate this mechanism; although, for the reasons given above in Section 5.1.3, the rate of degradation was slower.

Due to their inherently higher thermo-oxidative stability, the polyimide-based specimens proved to be far more durable than the epoxy specimens, even after 25000 hours exposure at 232 °C (450 °F). At 232 °C, tensile properties were observed to decrease after 50000 hours exposure even though no visible damage was seen. Clearly, if tensile properties decreased, matrix strength had to be severely compromised to reduce the ability of the matrix to transfer load between the 0° fibers. No independent measurement of the matrix strength was made and, indeed, very few studies have even tried to measure losses in *in situ* matrix properties due to thermo-oxidative ageing. At the higher temperature of 288 °C (550 °F), tensile-strength degradation was observed after only 10000 hours. Continued exposure up to 25000 hours created visible damage in the form of delaminations. Large weight losses were also observed after 25000 hours of ageing. Although degradation of epoxy-based systems at 121 °C (250 °F) could be quite severe, ageing at 100 °C (212 °F) showed far less degradation. It was postulated that epoxy materials may be suitable for 50000 hours at this lower temperature, but a lifetime of only 25000 hours was recommended at this temperature.

The Kerr and Haskins study created the base test methodology that has been used ever since for the assessment of long-term ageing in composites and identified many of the materials and testing issues that remain important today. No study of the thermo-oxidative stability of polymeric composites should be performed without first considering the results and conclusions of this study.

5.4 Overview of other studies

During the performance of the more-than-one-decade-long Kerr and Haskins study, the use of composite materials continued to expand and both high-temperature materials and supersonic transports continued to be investigated. Out of the multiple efforts to develop high-temperature mate-

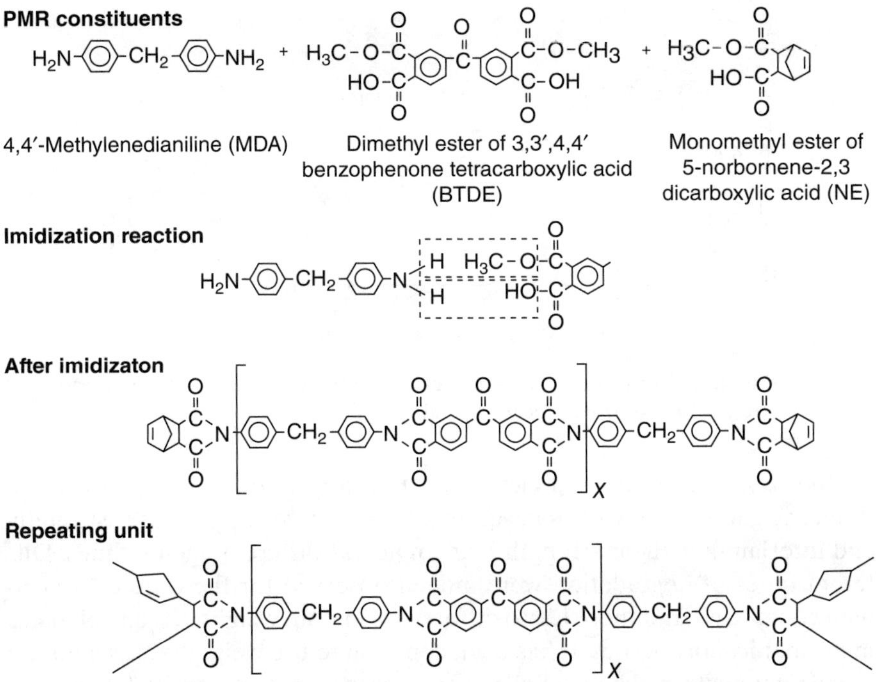

5.5 PMR-15 and the reaction route wherein the molar ratio of reactants is 2.000 NE: 3.087 MDA: 2.087 BTDE.[25]

rials, only one has had significant commercial success to date: PMR-15. A schematic of PMR-15 and its reaction route is shown in Fig. 5.5. The advantages of PMR-15 were its low cost and good high-temperature properties. These features helped to overcome PMR-15's processing difficulties and brittleness. As it began to be adopted on to multiple US government aircraft, specifications were created to support it and, as the *de facto* high-temperature materials standard, it became the subject of many studies.

5.4.1 PMR-15

With the development of PMR-15 and its introduction into production parts, there was a great incentive to study the effects of thermo-oxidative ageing on this material. Many of these studies were conducted by NASA at their Lewis (now Glenn) Research Center, with the earliest by Alston[20] and Cavano and Winters[25]; however, most were led by Kenneth Bowles.[21,26–36]

In order to better understand the long-term behavior of PMR-15, a study[26] examined the effects of specimen geometry on the thermo-oxidative stability and the mechanical-properties retention of Celion 12000/PMR-15

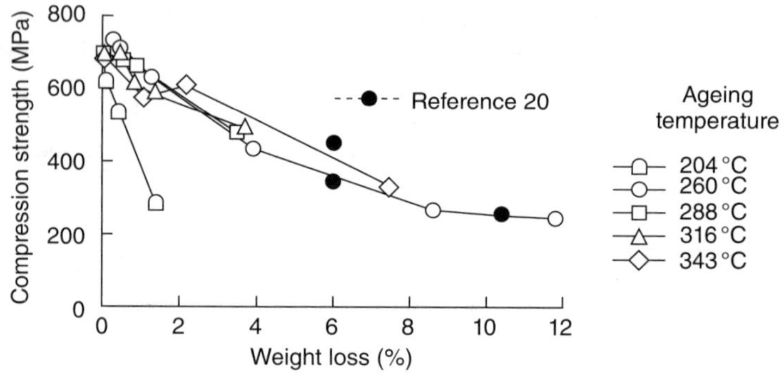

5.6 Reduction in compression strength for PMR-15 vs. percentage weight loss at different temperatures.[27]

composites. Ageing was conducted at 316 °C for up to 1639 hours for three different geometries with measurements of weight loss, flexural strength, and interlaminar shear strength being made at different ageing times. Different types of degradation were indeed observed for the different specimen geometries and these led to differences in both weight loss and changes in mechanical properties. It has been common to use weight loss as a metric to correlate with mechanical-property degradation (references 1, 2, and 27 *inter alia*). Generally, as weight loss increases, mechanical properties decrease, as shown in Fig. 5.6. This is, of course, complicated by weight gains due to initial oxidation as previously pointed out in Section 5.1.3. Despite this, very large strength losses are seen for even small (<2%) losses in weight. The effects of surface treatments on thermal stability were examined[28,29] on bare fibers, but in spite of the difference observed on these, they had little discernible effect on PMR-15 composite results. A later study[30] did show evidence of different degradation rates with different sizings on AS-4 fibers.

Degraded resin properties were measured at different areas of aged test coupons as the thermo-oxidative ageing of PMR-15 was observed to produce nonuniform degradation, especially on the surface where a distinct layer, usually attributed to oxidative degradation, was seen to form having different properties from the bulk matrix away from the surface.[31] Tests were performed on aged material at the surface and in the interior with the thickness of the oxidized surface layer measured along with the shrinkage and coefficient of thermal expansion (CTE). Four-point-bend tests were performed to determine the differences in modulus of both the oxidized surface layer and the interior material, which were found to indeed be substantially different.

Another study[32] indicated that simple, linear relationships exist between the compression properties of graphite-fiber-fabric/PMR-15 composites

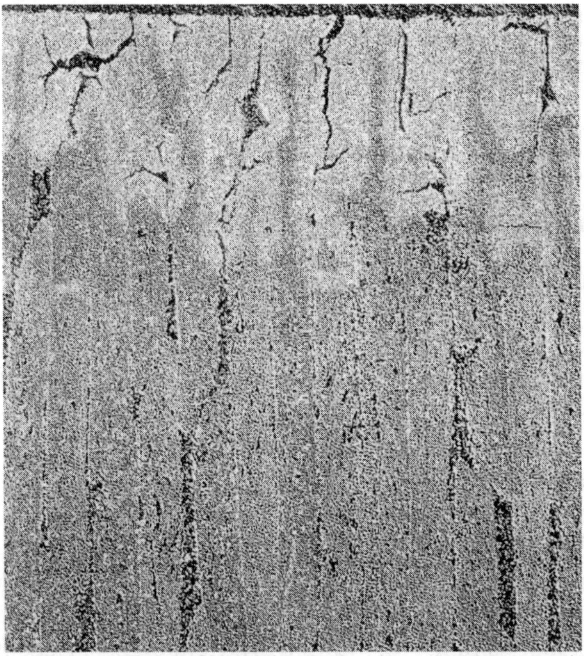

5.7 Photomicrograph of thermo-oxidatively aged PMR-15 coupon showing greater degradation at the specimen edge (right) with the presence of cracking and whitening where the material has oxidized.[33] Each layer is approximately 0.15–0.20 mm in thickness.

and the depth of the surface layer that develops and grows during periods of ageing at elevated temperatures. Interestingly, although the depth of the surface layer was found to be indicative of the decrease in strength, the remaining undamaged central core volume of matrix was determined to the principal factor behind this observation. Thus two contributing factors were attributed to property loss: surface degradation and reactions in the material beneath the surface that tend to degrade the mechanical properties away from the surface.

In addition to the surface effects described above, edge effects were important and indeed were visible, as shown in Fig. 5.7 where cracks are present alongside oxidized material that appears 'whiter' relative to the unoxidized matrix.

The effects of material toughness were also examined[34] and it was found that modifications to PMR-15 that decreased crosslinking led to reduced thermo-oxidative stability. Other results[35] strongly suggested that compression-property changes in PMR-15 composites can be correlated with changes in the surface layer, dimensions, and weight loss induced by thermal ageing; work described in reference 36 reiterates that interfacial effects between the fiber and matrix can affect thermo-oxidative stability.

Nam and Seferis[37] described the anisotropy of thermo-oxidative ageing in carbon-fiber-reinforced composites and Salin and Seferis[38] showed how this anisotropy affects thermogravimetry. Later, Schoeppner *et al.*[39] examined the effects of anisotropy on thermo-oxidative ageing through different specimen geometries that were chosen such that different surface-area ratios (i.e. ratios of surface area perpendicular to the fibers to surface area parallel to the fibers) were obtained. Measurements of unidirectional composite oxidation rates in the axial fiber direction and the transverse direction showed that the axial oxidation rate was an order of magnitude greater than the rate in the transverse direction. Specimen geometry was found to not influence weight loss in argon when the specimen weight loss was normalized to volume; however, a strong dependence of the weight loss on the surface-area ratios was observed for specimens aged in air. In contrast, Marceau and Hilaire[40] found no effect of fibers on degradation kinetics, a result that surprised them.

Intuitively it would seem that protecting a material's surface from oxygen diffusion would improve its thermo-oxidative stability and, indeed, some studies verify this supposition. Hurwitz and Whittenberger[41] used aluminum foil to coat Celion 6000/PMR-15 composites and did find that a thin coating provided significant protection from oxidation. Miller *et al.*[42] found that the use of an expanded graphite barrier reduced the rate of oxidative degradation of PMR-15 resin such that the thermo-oxidative stability increased by up to 25%.

A direct corollary to the effect of coatings is the fact that edge effects are also important in ageing and, most importantly, in the testing of composites for thermo-oxidative ageing. A previous example was shown in Fig. 5.7 where oxidation and cracks were clearly visible. Nelson[43] observed large differences between larger panels and individually aged specimens. Degradation was seen on the outer parts of the large panels with only minor degradation in the center, but the smaller coupons exhibited large amounts of degradation throughout.

5.4.2 Other materials

Although the largest portion of work on thermo-oxidative ageing has been performed on PMR-15, there are many useful studies of other materials. Lee and Levi[15] performed one of the early studies on epoxies. In their work, they studied the effects of varying the cure temperature on thermal degradation using gravimetric techniques (TGA) and found, for the materials that they had evaluated, that higher cure temperatures generally led to more stable epoxies. Greer[14] examined ten carbon/epoxy combinations for

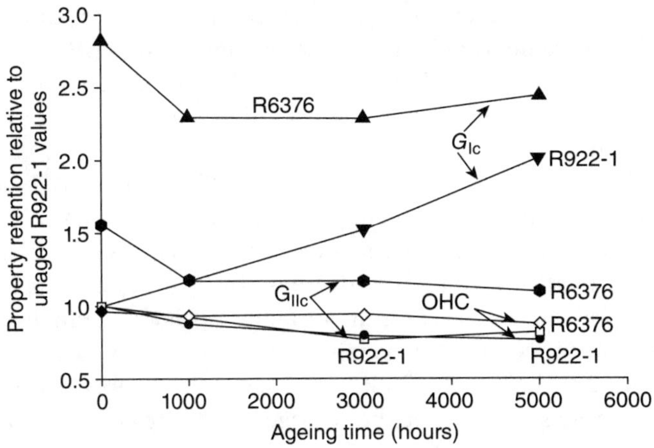

5.8 Relative property retention of a toughened composite (R6376/Celion G30-500) vs. an untoughened one (R922-1/Celion G30-500) after ageing at 177 °C for open-hole compression (OHC), Mode I fracture toughness (G_{Ic}), and Mode II fracture toughness (G_{IIc}) showing higher property retention for the toughened system. Adapted from reference 2.

their capabilities for long-term service at 177 °C and measured changes in tension, compression, flexure, and interlaminar-shear properties. Not surprisingly, based on the discussion presented earlier, severe degradation was noted at this temperature with visible acute degradation on all specimen surfaces. Testing at 150 °C showed much less degradation.

Tsotsis[2,3,8,9,11] performed a series of studies on epoxies. It was found that ageing too near or above a materials' T_g produced thermo-oxidative ageing behavior that was not consistent with that at more likely usage temperatures. Moreover, because material behavior becomes highly nonlinear as T_g is approached, it is difficult, if not impossible, to make accurate life predictions based on data generated at too high a temperature. It was recommended that all long-term ageing be performed at temperatures significantly below T_g.

In one study, Tsotsis[3] also showed that a material's toughness directly affects its long-term thermo-oxidative stability, as shown in Fig. 5.8. Increased toughness, if the material providing the toughness is not itself much more susceptible to degradation than the matrix resin, tends to increase property retention after ageing because the types of cracks noted in Fig. 5.2 are less likely to occur. Thus, the rate of generation of degraded surface and hence the overall degradation rate is suppressed with increases in toughness. This study also pointed out that previously used test matrices that focused mainly on tensile and other fiber-dominated properties probably overestimated useful lifetimes for the materials they studied. It was also noted that the degradation and ageing mechanisms observed in the epoxies of the studies of Tsotsis[2,3,8,9,11] were nearly identical to those observed in other (e.g. BMIs, cyanate esters, polyimides) PMCs. Thus, conclusions regarding test methodologies (not necessarily test conditions) could be generalized for all PMCs to a large extent.

Parvatareddy *et al.*[44] reported results on the effects of up to 9 months of ageing at 150 °C in nitrogen, ambient air, and 13.8 kPa air on the transverse tensile strength of carbon-fiber-reinforced cyanate esters and thermoplastics. Bending strengths were seen to decrease by 30–40% after ageing, with corresponding 40–60% decreases in ultimate strain and an up to 20% increase in moduli. Neat-resin weight losses were only observed to be between 1 and 2%, however. In all cases, ageing was most severe for the ambient air and least severe for the nitrogen, thereby showing the importance of oxidation vs. thermal degradation in thermo-oxidative ageing.

During NASA's High-Speed Research (HSR) program, studies were conducted that focused mainly on high-temperature polymers. As a spin-off of the HSR program, much work was done targeting High-Speed Civil Transport (HSCT) aircraft with a speed target between Mach 2.0 and 2.5 with property retention after 60000–120000 hours. At Mach 2.4, the continuous-use temperature required for structural composites is approximately 175 °C.

In one study to evaluate potential HSCT materials, Morgan *et al.*[23] performed a test program to evaluate the synergistic effects of stress, temperature, moisture, time, radiation, and oxygen level on the properties of BMIs. Thermal cycling, creep rupture, and isothermal stress/air flow were used to simulate in-service operating conditions. Creep-rupture results showed steady declines in strength with time at both ambient (23 °C) and elevated (250 °C) temperatures with decreases in properties much higher for the elevated temperature after 1000 hours of ageing. Additionally, the higher-temperature ageing was shown to give a T_g 100–225 °C greater than for an isothermal cure at 121 °C and 70–200 °C greater than for cure at 180 °C – the two temperatures being in the range of long-term-exposure temperatures of a potential HSCT. As mentioned previously, ageing at such a high temperature, in addition to yielding unrealistically high T_g values, led to anomalous effects that make it difficult to extend the conclusions of this work more broadly to other materials and other ageing conditions.

5.4.3 Gravimetric methods

One of the primary difficulties in characterizing the effects of ageing on changes in structural properties is the correlation of 'ageing'-related data such as weight loss, changes in color, etc. to changes in material mechanical properties such as strength, stiffness, and toughness. As discussed previously

(Section 5.4.2), weight loss is dependent on two dynamic parameters: reaction kinetics and mass-transfer rates, though the kinetics may largely be controlled by the mass-transfer rates, also as previously mentioned. In order to separate the reaction kinetics from the mass-transfer effects, the distance for reactants to travel – dependent on volume and thickness – must be very small. Experimentally, in TGA, this means that very small (typically <10 μm) particles are needed. The use of weight-loss or gravimetric methods to determine the thermal stability of polymers was first established by Flynn and Wall.[45] In this method, thermogravimetric data using a TGA were used to calculate activation energies to determine kinetics models of material degradation. In this model, measurements are taken at several different heating rates (β) and then calculations elicit the temperatures (T in K) at which different weight-loss points ($1 - C$, where C is the fraction of the original mass remaining) are reached. Points representing different $1 - C$ values are plotted on a log β vs. $1/T$ graph. By connecting points that represent constant $1 - C$ values, straight lines are produced with slopes that are proportional to the reaction activation energy (E) as calculated by

$$E \propto \frac{\mathrm{d}(\log \beta)}{\mathrm{d}(1/T)}$$

This model may be generalized for different reaction orders.

A similar model was developed by Friedman[46] in which reaction order is also included. Papazian[47] performed analyses of polymer degradation under vacuum and showed how TGA data could be used to accurately predict isothermal degradation kinetics. Henderson and Tant[48] also used Flynn and Wall's methodology in their determination of the thermal stability of glass- and talc-filled phenolic composites. Grayson and Fry[49] used Flynn and Wall's approach[45] to develop a kinetic map for defining the detection limit of a TGA experiment and to aid in the determination of useful lifetimes of polymers vs. temperature from TGA data. The validity of this TGA-based approach is questioned by Arnold *et al.*[50] who maintain that such methods provide unreliable estimates of activation energies. MacCallum[51] also discusses the use of TGA for kinetic analyses and also cautions on the accuracy of kinetic parameters obtained from TGA. In particular, both Arnold *et al.*[50] and MacCallum[51] warn of mass-transfer effects and their influence on kinetic data. Sheppard[52] pointed out that TGA predictions generally indicated more severe degradation than the bulk composite, further reinforcing this point. Salim and Seferis[38] showed that TGA results may be affected by material anisotropy, making the analysis of TGA-based data even more complex.

Tsotsis[3] later concluded that gravimetric methods are not useful predictors of long-term performance. Tsotsis especially cautioned against macroscopic weight-loss measurements where even small measured weight losses

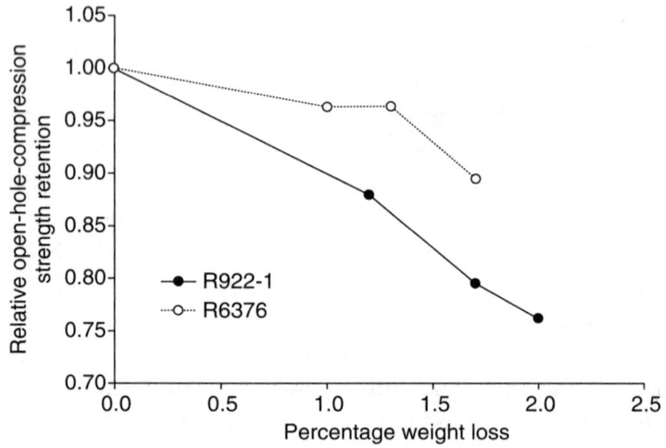

5.9 Differences in open-hole-compression property retention vs. percentage weight loss for a toughened (R6376/Celion G30-500) and an untoughened (R922-1/Celion G30-500) system, adapted from reference 2. Note the large difference in property retention for the same weight loss.

could lead to correspondingly large decreases in some mechanical properties, as shown in Fig. 5.9. Thus, even small errors in the determination of weight loss could lead to large errors in the estimates of mechanical properties. Lastly, it was stated that even weight-loss data that are gathered with corresponding mechanical data do not represent a good indicator of property degradation. Weight loss is thus only useful as a negative indicator in that it can only tell someone if a material is unacceptable. It is not useful for estimating whether a material is suitable for long-term usage at a given temperature for a given time.

5.4.4 Effects of pressure

In the work of Kerr and Haskins,[1] tests were performed at ambient pressure, at sea-level pressure (~1 atm), and at simulated, high-altitude pressure. In all cases, degradation at the lower pressure was found to be less severe than at 1 atm. Others have noted similar behavior. As noted previously, attempting to reduce the testing time needed for assessing long-term thermo-oxidative stability by using temperature alone is problematic since degradation mechanisms and relative degradation rates of different mechanisms will change with temperature. Thus, the use of accelerated ageing using elevated pressure can avoid the pitfalls of testing at temperatures above the ultimate end-use temperature. In this case, all thermally activated processes will occur at the same rate as at the end-use temperature irrespective of the pressure. However, diffusion-related phenomena such as the diffusion of oxygen into a matrix or the diffusion out of the matrix of chain-scission byproducts, for example, would be altered by the elevated pressure. In this case, the diffusion of oxygen may be enhanced, while the diffusion of reaction byproducts may be retarded. Because the chief cause of thermo-oxidative degradation in polymeric composites has been observed to be oxidation, which is diffusion controlled, accelerating ageing by using elevated pressure, as shown by Tsotsis *et al.*,[8] offers an opportunity for accelerated ageing that is scalable in a way that accelerating at temperatures higher than the ultimate end-use temperature is not.

In a diffusion-controlled process, if the rate-limiting step in thermo-oxidative ageing is the diffusion of oxygen to reactive sites, as some studies suggest, then the rate of degradation will largely be governed by this. However, the diffusion rate may change as a material degrades by the formation of a barrier layer or by the growth of cracks that increase the surface area available for both oxygen and chain-scission-product diffusion. In order to fully characterize diffusion-related degradation, lay-up considerations, such as those pointed out by Nam and Seferis,[37] and by Salin and Seferis,[38] need to be taken into account. Nam and Seferis considered the effects of lay-up on diffusion, but the lay-up also influences the mechanical properties of a composite material and its degradation rate because changes in strength, stiffness, etc. are directly related to which plies within a laminate become degraded. For example, whether load-bearing (e.g. 0°) or non-load-bearing (e.g. 90°) plies are closer to the surface will influence the rate of property degradation as those nearest to the surface will tend to degrade first.

Ciutacu *et al.*[53] accelerated the ageing of glass/epoxy composites by using varying oxygen pressure in a manner similar to Tsotsis *et al.*[7] They found a relationship between mechanical-property (flexural strength) degradation and degradation kinetics. They further developed a relationship to show how the flexural failure strain (ε) depended on oxygen pressure by the following:

$$\varepsilon \propto A e^{-BP^n}$$

where A is a constant related to the initial strain-to-failure value, B is a function of exposure time and temperature, P is pressure and n is a constant. Although good agreement was obtained by Ciutacu *et al.*[53] in their experiments, the dependence on material anisotropy cited by Nam and Seferis[37] was not considered, thereby making generalization of their results impossible without further verification of these effects.

Tsotsis *et al.*[54] performed tests up to 5000 hours at four different pressures (0.101, 0.345, 1.03, and 1.72 MPa) for carbon/epoxy specimens aged at 121 °C on open-hole compression and tensile-shear properties. The results

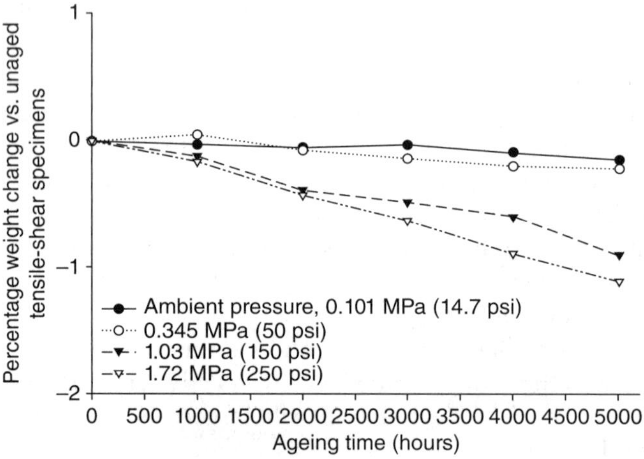

5.10 Percentage weight change in tensile-shear, $[\pm 45°]_{2s}$, specimens vs. ageing time in air at various pressures and at 121 °C for AS4/3501-6.[54]

did show a distinct accelerating effect with the use of elevated pressures, particularly for tensile-shear specimens. In most cases, the accelerating effect of pressure on mechanical-property degradation did not become fully evident until after 1000–2000 hours of exposure for tensile-shear coupons and until after 3000–4000 hours for open-hole-compression coupons. Despite these observations, weight loss was clearly accelerated by elevated temperature, even after 1000 hours of exposure. An example of this discrepancy is seen in Figs 5.10 and 5.11 where, despite clear weight-loss trends with increasing pressure, the dependence of mechanical-property change on pressure does not become evident until after significant ageing times.

The benefits of using elevated pressure to accelerate thermo-oxidative ageing were clearly shown in this study, although no formal means for extrapolating property changes with time in the manner of Ciutacu *et al.*[53] was developed. Accelerated-ageing methods are clearly needed as, in order to simulate a 60 000-hour lifetime at temperature, real-time testing of nearly 7 years would be required.

5.4.5 Modeling

In all cases where thermo-oxidative ageing has been studied, the ultimate goal has been to determine the useful lifetimes for composite structures. In the data of Kerr and Haskins,[1] real-time data were gathered. Since this study, no such large-scale endeavor has been undertaken to develop a reliable database to allow designers to use PMCs for very long times at elevated temperatures. Efforts have begun, however, to try to model

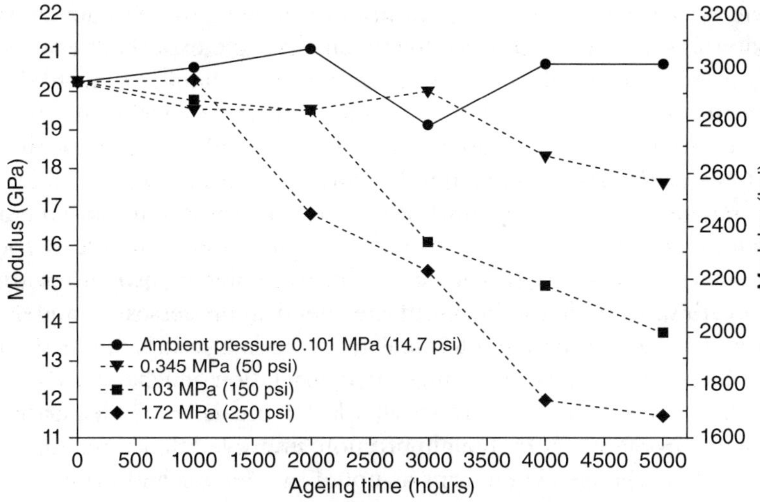

5.11 Modulus of tensile-shear, $[\pm 45°]_{2s}$, specimens vs. ageing time in air at various pressures and at 121 °C for AS4/3501-6.[54]

thermo-oxidative degradation, despite the inherent problems in the basic understanding of degradation mechanisms and how to observe and measure them.

Tandon *et al.*[55] and Pochiraju and Tandon[56] have begun such efforts. Tandon *et al.*[55] used a multidisciplinary approach to try to predict the performance and life expectancy of PMR-15. Emphasis was on the implementation and extension of hierarchical models to represent the polymer behavior/properties as a function of the degradation state. Both neat-resin and composite PMR-15 specimens were evaluated for various ageing conditions to try to develop a constitutive law for use in micromechanical analysis to predict the behavior of PMR-15-reinforced composites. Thermo-oxidative ageing was simulated with a diffusion-reaction model similar to that discussed above in which temperature, oxygen concentration, and weight loss are all considered. The modeling approach did not consider anisotropy explicitly, as a one-dimensional simulation was used.

It was observed that the oxidation growth process was clearly diffusion controlled, and that the diffusivity of the oxidized region was the controlling parameter, in the manner described when considering the simple model in Section 5.1.3. It was further pointed out that any modeling must take into account physical and chemical ageing in addition to thermo-oxidative ageing as all three will occur simultaneously. McManus *et al.*[57] devised a computational method to predict transverse matrix cracks in a composite laminate subjected to cyclic thermal loading. Shear-lag stress approximations and simple energy-based failure criteria were used to predict crack density as a function of temperature by assuming that fatigue degrades the material's inherent resistance to cracking. Simple experiments were used to provide data on progressive cracking of a laminate with decreasing temperature, and on cracking induced by thermal cycling, to enable some correlation to model predictions and good correlation was obtained.

One of the chief difficulties in modeling is obtaining reliable material properties and mass-transport parameters. This is due to both inherent difficulties in measuring *in situ* properties of a dynamic chemical process and the fact that, because the processes are dynamic, the properties are changing with time. Trying to back-out the chemical processes to model material properties without having to measure them is also difficult as the reaction products that might give clues to degradation mechanisms will readily oxidize as well in most cases.[3] This last point is of particular interest as changes in chemistry that might promote improved thermo-oxidative stability, such as changes in endcaps as noted by Meador and Frimer,[58,59] cannot be assessed *a priori* due to the lack of detailed understanding of underlying key degradation mechanisms and kinetics.

Colin and Verdu[60] developed a strategy for reducing the amount of testing required in lifetime prediction. This strategy was based on a model that couples the oxygen diffusion and consumption kinetics and takes into consideration the thickness distribution of all the chemical changes involved during thermo-oxidative ageing. The goal was to have a model in which all the parameters have a physical meaning that can be determined experimentally. The Arrhenius law was used to characterize degradation rates as it is well-suited for diffusion-based phenomena. As for the other work mentioned above, no considerations of anisotropy were included.

Lehrle *et al.*[61] suggest the use of pyrolysis-gas chromatography (GC) to obtain quantitative data to help to obtain mechanistic information about degradation processes. The dependence of the rate constants on volatiles evolution and initial molecular weight and pyrolysis temperature were used to obtain specific information regarding the mechanisms of initiation and termination in chain scission. Sample-thickness effects were also evaluated to provide information regarding diffusion-dependent degradation effects. A framework for determining degradation mechanisms was provided based on the pyrolysis-GC approach.

5.5 Areas for future study

5.5.1 General observations

There have been several attempts to outline methods for developing test programs suitable for characterizing thermo-oxidative degradation. Gates and Grayson[62] outline a framework for consideration of the various ageing mechanisms (thermo-oxidative, physical, chemical) in the context of a large-scale test program. Only a framework and no new data were included in this work. Gates further discusses this approach in a later work.[63] One of the main problems with thermo-oxidative testing is the large number of specimens and the long times needed to generate the data necessary to use composite materials for high-temperature service. This makes the tests quite expensive and, because of the long times required, difficult to qualify within production-qualification deadlines. These are the reasons why PMR-15 has been the primary material of choice – it has the large database necessary, even though it has many deficiencies, as noted widely in the literature. Thus, most 'new' high-temperature materials systems have only found (limited) success in applications that do not require long exposure times, such as space or re-entry vehicles. There are instances where some new materials have been tested for long-term exposure, but these are proprietary and, as such, cannot be discussed here.

5.5.2 Focus areas

Modeling and characterization

Any accelerated testing method must be coupled with a means to correlate the data to long-term ageing; therefore, it is currently required that both be done in parallel to develop the database necessary for the correlations to be validated. Correlations imply the development of models, too, and these must also be validated by testing in order to be implemented. In addition, in all of these areas, it is imperative to determine what types of tests need to be performed both to generate a database that is useful, and to perform correlations and validations.

Taking this into consideration, the following need to be established:

(a) mechanical-property test methods – determine what key properties need to be evaluated from the end-use perspective;
(b) accelerated-ageing, mechanical-property tests – determine the desired test coupons, ageing times, temperatures, and pressures;
(c) methods for understanding material degradation – develop improved analytical test methods or models to fundamentally understand thermo-oxidative degradation.

Of these, the first is generally straightforward as all vehicle systems have specific tests that are used for qualification and allowables testing. The second point is quite contentious, as the main discussion in this chapter has made clear. It appears that increased pressure offers a path forward, but the paucity of data makes this more conjectural than desired at present. The third point brings up some of the most complex problems facing researchers studying thermo-oxidative stability.

It has been seen that trying to detect degradation products via analytical chemical techniques (e.g. TGA-gas chromatography or TGA-mass spectroscopy) poses serious practical difficulties[3] as trying to measure degradation products in air or oxygen generally results in further oxidation of the degradation products in the gas stream, such that only methane and water are ultimately detected. Unless new methods are discovered to solve this problem, it is likely that this will remain a fundamental roadblock to gas-based detection techniques. Some work has been done in analyzing the degraded layers of a composite material (reference 20, for example), but this, too, is experimentally very difficult and time-consuming. Moreover, it is very difficult, if not impossible in most cases, to determine the mechanical properties of degraded zones within composites.

It is likely that the ultimate solution to understanding the chemical mechanisms behind thermo-oxidative degradation and how to combat it will be found analytically. At the time of the writing of this chapter, such computational tools are not yet available for this to be done practically. It is believed that as computational tools and computing power both continue to become ever more capable, this will some day be feasible.

Materials development

The studies cited herein, among others, reveal a large amount of general understanding regarding thermo-oxidative stability, although, as pointed out above, this has not translated into an ability to predict long-term behavior. This understanding, however, has revealed ways to improve polymer performance through innovative formulations. Methods that have been studied include, but are not limited to, the following: interpenetrating networks (e.g. reference 64); modified polymer-chain backbone structures (references 65–67, among others); modified endcaps (e.g. references 68 and 69); and surface coatings (e.g. references 11, 42, and 70). In addition to these the use of inorganic–organic blends is another resin-formulation option. It is clear that there is much work to be done in all of these areas and that they may be complementary. Indeed, work in all of these areas is ongoing.

One area that has not received much attention for high-temperature polymeric composites, but one which is quite active in general, is the use of nanotechnology to improve thermo-oxidative stability. Nanomaterials may be those that act to reduce the diffusivity of polymers and, thus, act similarly to a barrier or surface coating. They may also be functionalized to provide *in situ* property improvements, such as for polyhedral oligomeric silsesquioxane (POSS) (shown in Fig. 5.12), which combines both nanotechnology and inorganic–organic blending, as some recent work[71–73] has found. Many functionalized or unfunctionalized nanomaterials may also potentially be used to improve thermo-oxidative stability.

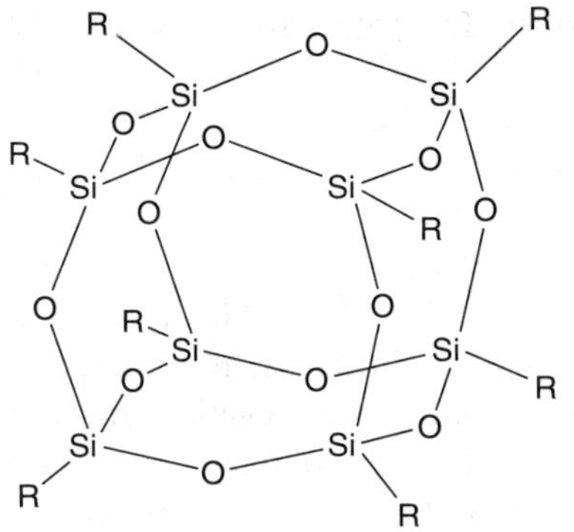

5.12 Schematic of polyhedral oligomeric silsesquioxane (POSS) molecule where R groups may be tailored to provide desired functionality.

5.6 Conclusions and recommendations

Although certain understandings about the effects of ageing on mechanical properties have been reached previously, the exact mechanisms controlling such degradation are still not well understood. From the standpoint of predicting long-term performance, a clearer understanding of the roles of the constituent materials in a composite – fiber, matrix, interface – is needed in order to better assess and predict laminate-level performance when exposed to thermo-oxidative ageing if real-time-ageing tests are to be replaced with accelerated-ageing methods.

It has been shown that weight loss and T_g changes are not clearly related to discrete changes in mechanical properties. Moreover, these changes do not *directly* reflect different chemical and physical mechanisms that control mechanical-property degradation. Indeed, the various composite properties that are affected by thermo-oxidative ageing-induced changes in fiber, matrix, interface properties, or a combination of these, will vary in different ways with ageing, as shown above. Despite this, mechanical-property degradation is clearly tied to changes in molecular structure that are manifested in weight loss, changes in T_g, or visibly through cracks and discoloration. Moreover, the molecular changes that affect the various materials variables (both mechanical and physical) affect these in different ways.

Test methods should focus on properties most affected by changes in the matrix, as it is the principal area where thermo-oxidative degradation takes place. It is likely that the interface also contributes, but separating out interface effects is difficult. Studies (e.g. reference 39) have shown that the interface affects the anisotropy of degradation and that different behavior is observed with different fiber sizings (e.g. reference 30), but no rigorous work has been performed to determine exact cause-and-effect relationships. These should also be studied.

Thermo-oxidative ageing is generally performed using unstressed specimens exposed for long times at a constant temperature. Thus, the interactions between ageing and mechanical loading – discrete and cyclic – are not captured. Furthermore, the effects of rapidly driven-off moisture, such as from a rapid rise in temperature or 'thermal spike',[74] have not been examined extensively. As stress can enhance diffusion, it certainly must be accounted for. In-service temperature loadings are cyclical and it would be instructive to examine if cyclic temperature excursions – in addition to the 'thermal spike' effect – cause different behavior from the constant-temperature tests. In this case, it would be interesting to observe whether the cyclic-temperature loadings would promote fatigue crack growth along the lines of the Paris crack-growth law.[75] The most important issue in thermo-oxidative ageing will remain the development of accurate means for performing accelerated ageing. This will enable much more widespread usage of new materials as well as lowering the risk for new applications of high-temperature polymeric composites.

5.7 References

1 KERR J R and HASKINS J F, "Time-Temperature-Stress Capabilities of Composite Materials for Advanced Supersonic Technology Application," NASA CR-178272, May, 1987.

2 TSOTSIS T K, "Thermo-Oxidative Aging of Composite Materials," *Journal of Composite Materials*, 1995, **29** (3), 410–422.

3 TSOTSIS T K, "Long-Term Thermo-Oxidative Aging of Composite Materials: Experimental Methods," *Journal of Composite Materials*, 1998, **32** (11), 1115–1135.

4 CONNELL J W, SMITH JR J G, and HERGENROTHER P M, US patent 6,441,099, "Phenylethynyl Containing Reactive Additives," August 27, 2002.

5 NEWMAN-EVANS R H, US patent 4,721,799, "Epoxy Resins Based on Tetraglycidyl Diamines," January 26, 1988.

6 HERGENROTHER P M, CONNELL J W, and SMITH J G, "Phenylethynyl Containing Imide Oligomers," *Polymer*, 2000, **41** (13), 5073–5081.

7 BADER M G, SMITH W, ISHAM A B, ROLSTON J A, and METZNER A B, "*Processing and Fabrication Technology*," Volume 3, in *Delaware Composites Design Encyclopedia*, Carlsson L A and Gillespie J W, Eds, Technomic, Lancaster, PA, 1990.

8 TSOTSIS T K, KELLER S, BARDIS J, and BISH J, "Preliminary Examination of the Use of Elevated Pressure to Accelerate Thermo-Oxidative Aging in Composites", *Polymer Degradation and Stability*, 1999, **64** (2), 207–212.

9 TSOTSIS T K and LEE S M, "Long-Term Thermo-Oxidative Aging in Composite Materials: Failure Mechanisms," *Composites Science and Technology*, 1998, **58**, 355–368.

10 WEITSMAN Y, "Stress-Assisted Diffusion in Elastic and Viscoelastic Materials," *Mechanics and Physics of Solids*, 1987, **35** (1), 73–93.

11 TSOTSIS T K, "Long-Term Thermo-Oxidative Aging in Composite Materials: Experimental Methods," *Journal of Composite Materials*, 1998, **32** (11), 1115–1121.

12 LUBOWITZ H R, "New Polyimide Polymers and Composites," in *High-Temperature Polymers, Reinforcements, and Composites*, Polymer Conference Series, University of Utah, June 8–13, 1970.

13 SERAFINI T T, DEIVIGS P, and LIGHTSEY R, "Preparation of Polyimides from Mixtures of Monomeric Diamines and Esters of Polycarboxylic Acids," US patent 3,745,149, July 10, 1973.

14 GREER R H, "Thermal Aging of Contemporary Graphite/Epoxy Materials," in *National SAMPE Symposium Exhibition Proceedings*, Book 2, 1979, pp. 1039–1051.

15 LEE H T and LEVI D W, "Effect of Curing Temperature on the Thermal Degradation of an Epoxide Resin," *Journal of Applied Polymer Science*, 1969, **13**, 1703–1705.

16 ROUQUIE S, LAFARIE-FRENOT M C, CINQUIN J, and COLOMBARO A M, "Thermal Cycling of Carbon/Epoxy Laminates in Neutral and Oxidative Enviroments," *Composites Science and Technology*, 2005, **65**, 403–409.

17 COUSTUMER P L, LAFDI K, and OBERLIN A, "Aging of Carbon-Fiber-Reinforced Bismaleimide-Matrix Composites in Oxidative Conditions," *Composites Science and Technology*, 1994, **52**, 433–437.

18 PEDERSON C L, GILLESPIE JR J W, MCCULLOUGH R, ROTHSCHILDS R J, and STANEK S L, "The Effect of Isothermal Aging on Transverse Crack Development in Carbon Fiber Reinforced Cross-Ply Laminates," *Polymer Composites*, 1995, **16** (2), 154–160.

19 KERR J R, HASKINS J F, and STEIN B A, "Program Definition and Preliminary Results of a Long-Term Program of Advanced Composites for Supersonic Cruise Aircraft Applications," in *Environmental Effects on Advanced Composite Materials*, ASTM STP 602, American Society for Testing and Materials, Philadelphia, PA, 1976, pp. 3–22.

20 ALSTON W B, "Characterization of PMR-15 Polyimide Resin Composition in Thermo-Oxidatively Exposed Graphite Fiber Composites," NASA Technical Memorandum 81565, 1980.

21 BOWLES K J, JAYNE D, and LEONHART T A, "Isothermal Aging Effects on PMR-15 Resin," *SAMPE Quarterly*, 1993, **24** (2), 2–9.

22 PARVATAREDDY H, WANG J Z, DILLARD D A, and WARD T C, "Environmental Aging of High-Performance Polymeric Composites: Effects on Durability," *Composites Science and Technology*, 1995, **53**, 399–409.

23 MORGAN R J, SHIN E, DUNN C, FOUCH E, JUREK B, and JUREK A, "The Durability of Composites for Potential High Speed Civil Transport Applications," in *Proceedings of the 39th International SAMPE Symposium*, April 11–14, 1994, pp. 1564–1575.

24 COMMITTEE ON EVALUATION OF LONG-TERM AGING OF MATERIALS and STRUCTURES USING ACCELERATED TEST METHODS, NATIONAL MATERIALS ADVISORY BOARD, "Accelerated Aging of Materials and Structures," Publication NMAB-479, National Academy Press, Washington DC, 1996.

25 CAVANO P J and WINTERS W E, "Fiber Reinforced PMR Polyimide Composites," NASA CR-13537, 1978.

26 BOWLES K J and MEYERS A, "Specimen Geometry Effects on Graphite/PMR-15 Composites During Thermo-Oxidative Aging," in *Proceedings of the 31st International SAMPE Symposium*, April 7–10, 1986, pp. 1285–1299.

27 BOWLES K J, MCCORKLE L, and INGRAHM L, "Comparison of Graphite Fabric Reinforced PMR-15 and Avimid N Composites After Long-Term Isothermal Aging at Various Temperatures," NASA Technical Memorandum TM-19980107529, February 1998.

28 BOWLES K J, JAYNE D, LEONHARDT T A, and BORS D, "Thermal Stability Relationships Between PMR-15 Resin and Its Composites," NASA Technical Memorandum 106285, 1993.

29 BOWLES K J, "Thermo-Oxidative Stability Studies of PMR-15 Polymer Matrix Composites Reinforced with Various Continuous Fibers," NASA Technical Memorandum 102439, 1990.

30 BOWLES K J, MADHUKAR M, PAPADOPOULOS D S, INGHRAM L, and MCCORKLE L, "The Effects of Fiber Surface Modification and Thermal Aging on Composite Toughness and its Measurement," *Journal of Composite Materials*, 1997, **31** (6), 552–579.

31 TSUJI L C, MCMANUS H L, and BOWLES K J, "Mechanical Properties of Degraded PMR-15 Resin," NASA TM-1998-208487, 1998.

32 BOWLES K J, "Thermal and Mechanical Durability of Graphite-Fiber-Reinforced PMR-15 Composites," NASA TM-1998-113116 REV1, 1998.

33 BOWLES K J, PAPADOPOULOS D S, INGHRAM L L, MCCORKLE L S, and KLAN O V, "Long-time Durability of PMR-15 Matrix Polymer at 204, 260, 288, and 316 °C," NASA-TM-2001-210602, July 2001.

34 VANUCCI R D and BOWLES K J, "Graphite/PMR Polyimide Composites with Improved Toughness," *SAMPE Quarterly*, January 1986, 12–18.

35 BOWLES K J, TSUJI L, KAMVOURIS J, and ROBERTS G D, "Long-Term Isothermal Aging Effects on Weight Loss, Compression Properties, and Dimensions of T650-35 Fabric-Reinforced PMR-15 Composites – Data," NASA-TM – 2003-211870, 2003.

36 BOWLES K J, JAYNE D, LEONHARDT T A, and BORS D, "Thermal Stability Relationships Between PMR-15 Resin and its Composites," NASA Tech Briefs, LEW-16202, 1994.

37 NAM J and SEFERIS J C, "Anisotropic Thermo-Oxidative Stability of Carbon Fiber Reinforced Polymeric Composites," *SAMPE Quarterly*, 1992, **24** (1), 10–18.

38 SALIN I M and SEFERIS J C, "Anisotropic Effects in Thermogravimetry of Polymeric Composites," *Journal of Polymer Science: Part B: Polymer Physics*, 1993, **31**, 1019–1027.

39 SCHOEPPNER G A, TANDON G P, and RIPBERGER E R, "Anisotropic Oxidation and Weight Loss in PMR-15 Composites," *Composites: Part A*, 2007, **38**, 890–894.

40 MARCEAU C and HILAIRE B, "Thermal Aging of PMR15 Polyimide Matrix," *Polymer*, 1993, **34** (11), 2458–2459.

41 HURWITZ F I and WHITTENBERGER J D, "The Effect of a Coating on the Thermo-Oxidative Stability of Celion 6000 Graphite Fiber/PMR 15 Polyimide Composites," *Composites Technology Review*, 1983, **5** (4), 109–114.

42 MILLER S, PAPADOPOULOS D, HEIMANN P, INGHRAM L, and MCCORKLE L, "Graphite Sheet Coating for Improved Thermal Oxidative Stability of Carbon Fiber Reinforced/PRM-15 Composites," *Composites Science and Technology*, 2007, **66**, 2183–2190.

43 NELSON J B, "Long-Term Thermal Aging of Two Graphite/Polyimide Composite Materials," NASA Technical Paper 2369, 1984.

44 PARVATAREDDY H, WANG J Z, DILLARD D A, WARD T C, and ROGALSKI M E, "Environmental Aging of High-Performance Polymeric Composites: Effects on Durability," *Composites Science and Technology*, 1995, **53**, 399–409.

45 FLYNN J H and WALL L A, "A Quick, Direct Method for the Determination of Activation Energy from Thermogravimetric Data," *Polymer Letters*, 1966, **4**, 323–328.

46 FRIEDMAN H L, "Kinetics of Thermal Degradation of Char-Forming Plastics from Thermogravimetry," *Journal of Polymer Science: Part C*, 1965, **6**, 183–195.

47 PAPAZIAN H A, "Prediction of Polymer-Degradation Kinetics at Moderate Temperatures from TGA Measurement," *Journal of Applied Polymer Science*, 1972, **16**, 2503–2510.

48 HENDERSON J B and TANT M R, "A Study of the Kinetics of High-Temperature Carbon-Silica Reactions in Ablative Polymer Composite," *Polymer Composites*, 1983, **4** (4), 233–237

49 GRAYSON M A and FRY C G, "On the Use of a Kinetic Map to Compare the Thermal Stability of Polymeric Materials Undergoing Weight Loss," in *Proceedings of the 21st North American Thermal Analysis Society*. Atlanta, Georgia, September 13–16, 1992, p. 194.

50 ARNOLD M, VERESS G E, PAULIK J, and PAULIK F, "Problems of the Characterization of Thermoanalytical Processes by Kinetic Parameters, Part 1," *Journal of Thermal Analysis*, 1979, **17**, 507–528.

51 MACCALLUM J R, "Thermogravimetric Analysis of Polymers for Assessing Thermal Degradation," *Thermochimica Acta*, 1985, **96**, 285–281.

52 SHEPPARD, C H, "Thermal and oxidative stability of carbon fibers and composites," *SAMPE Quarterly*, 1987, **18** (2), 14–17.

53 CIUTACU S, BUDRUGEAC P, and NICULAE I, "Accelerated Thermal Aging of Glass-Reinforced Epoxy Resin Under Oxygen Pressure," *Polymer Degradation and Stability*, 1991, **31**, 365–372.

54 TSOTSIS, T K, KELLER S, LEE K, BARDIS J, and BISH J, "Aging of Polymeric Composite Specimens for 5000 Hours at Elevated Pressure and Temperature," *Composites Science and Technology*, 2001, **61**, 117–127.

55 TANDON G P, POCHIRAJU K V, and SCHOEPPNER G A, "Modeling of Oxidative Development in PMR-15 Resin," *Polymer Degradation and Stability*, 2006, **91**, 1861–1869.

56 POCHIRAJU K V and TANDON G P, "Modeling Thermo-Oxidative Layer Growth in High-Temperature Resins," *Journal of Engineering Materials and Technology*, 2006, **128**, 107–116.

57 MCMANUS H L, BOWLES D E, and TOMPKINS S S, "Prediction of Thermal Cycling Induced Matrix Cracking," in *American Society for Composites 8th Technical Conference on Composite Materials*, Cleveland, OH, Technomic, Lancaster, PA, 1993.

58 MEADOR M A B and FRIMER A A, "End Caps Increase Thermo-Oxidative Stability of Polyimides," NASA Technical Support Package LEW-16987, 2001.
59 MEADOR M A B and FRIMER A A, "Better End-Cap Processing for Oxidation-Resistant Polyimides," NASA Technical Support Package LEW-17429, 2004.
60 COLIN X and VERDU J, "Strategy for Studying Thermal Oxidation of Organic Matrix Composites," *Composites Science and Technology*, 2005, **65**, 411–419.
61 LEHRLE R, ATKINSON D, COOK S, GARDNER P, GROVES S, HANCOX R, and LAMB G, "Polymer-Degradation Mechanisms: New Approaches," *Polymer Degradation and Stability*, 1993, **42**, 281–291.
62 GATES T S and GRAYSON M A, "On the Use of Accelerated Aging Methods for Screening High Temperature Polymeric Composite Materials," in *Proceedings of the 40th AIAA/ASME/ASCE/AHS/ASC Structures, Structural Dynamics, and Materials Conference,* Volume 2, 1998, AIAA Paper 99–1296, pp. 925–935.
63 GATES T S, "On the Use of Accelerated Test Methods for Characterization of Advanced Composite Materials," NASA Technical Publication TP-2003-212407, May 2003.
64 PATER R H, LOWTHER S E, SMITH J Y, CANNON M S, WHITEHEAD F M, and ELY R M, "High-Performance, Semi-Interpenetrating Polymer Network," NASA Report LAR-14339, 1992.
65 DEZEM J F, "Polyimides Containing Amide And Perfluoroisopropyl Links," NASA Report LAR-14608, 1993.
66 PATER R H, ELY R M, STANFIELD C E, DICKERSON G E, SNOHA J J, SRINIVASAN K, and HOU T, "Low-Toxicity PMR Polyimide," NASA Report LAR-14639, 1994.
67 VANUCCI R D, MALARIK D S, PAPADAPOULOS D S, and WATERS J F, "Autoclavable Addition Polyimides for 371 °C Composite Applications," NASA Technical Memorandum 103233, 1990.
68 CHUANG K C and WATERS J E, "Effects of Endcaps on the Properties of Polyimide/Carbon Fiber Composites," in *Proceedings of the 40th International SAMPE Symposium*, 1995, pp. 1113–1123.
69 HERGENROTHER P M, CONNELL J W, and SMITH JR J G, "Phenylethynyl-Containing Imide Oligomers," *Polymer*, 2000, **41**, 5073–5081.
70 HANSON M P and SERAFINI T T, "Surface Protection of Graphite Fabric/PMR-15 Composites Subjected to Thermal Oxidation," in *High Temperature Polymer Matrix Composites*, Serafini T T, Ed., Noyes Data Corporation, Park Ridge, NJ, 1987, pp. 282–292.
71 ZHANG Z, GU A, LIANG G, REN P, XIE J, and WANG X, "Thermo-oxygen degradation mechanisms of POSS/epoxy nanocomposites," *Polymer Degradation and Stability*, 2007, **92** (11), 1986–1993.
72 MU J, LIU Y, and ZHENG S, "Inorganic–organic interpenetrating polymer networks involving polyhedral oligomeric silsesquioxane and poly(ethylene oxide)," *Polymer*, 2007, **48** (5), 1176–1184.
73 NI Y, ZHENG S, and NIE K, "Morphology and thermal properties of inorganic–organic hybrids involving epoxy resin and polyhedral oligomeric silsesquioxanes," *Polymer*, 2004, **45** (16), 5557–5568.

74 FEDDERLY J J and AUGL J M, "Microscopic Evaluation of a Polyimide (PMR-15)-Graphite Composite," in *High Temperature Polymer Matrix Composites*, Serafini T T, Ed., Noyes Data Corporation, Park Ridge, NJ, 1987, pp. 293–308.
75 PARIS P and ERDOGAN F, "A critical analysis of crack propagation laws," *Transactions of the American Society of Mechanical Engineers, Journal of Basic Engineering*, 1963, December, 528–534.

6
Fourier transform infrared photoacoustic spectroscopy of ageing composites

R. W. JONES and J. McCLELLAND, Iowa State University, USA

6.1 Introduction

Photoacoustic spectroscopy (PAS) is a spectroscopic technique that makes use of the acoustic response produced when a gaseous or condensed-phase sample absorbs radiation. Whenever modulated or pulsed radiation is absorbed, the deposited energy produces a thermal expansion that results in an acoustic signal. This photoacoustic effect was discovered by Alexander Graham Bell (1880). He used it not only for wireless communication via his 'photophone' but also for PAS using his 'spectrophone' (Bell, 1880, 1881). Despite this early start, PAS did not receive much attention until the 1970s when instrumentation improvements made it practical in the visible and ultraviolet regions (Rosencwaig, 1973a, 1973b, 1975), and its use for solid samples was placed on a firm theoretical footing (Parker, 1973; Rosencwaig and Gersho, 1976; McDonald and Wetsel, 1978). For infrared spectroscopy, the most important advance came in the 1980s when Fourier transform infrared (FT-IR) spectrometers became widespread. The high throughput and multiplex advantage of the FT-IR approach made mid-infrared PAS without laser sources possible. These and the intensity-modulated infrared beam of FT-IR spectrometers make FT-IR and PAS a natural pairing. Photoacoustic spectroscopy is sometimes called optoacoustic spectroscopy, but the photoacoustic name is far more common.

The principal advantage of PAS over other analysis techniques is that it can be applied to nearly every type of solid or semisolid material – whether crystalline, amorphous, powder, gel, or smear (Rosencwaig and Gersho,

Note: This manuscript has been authored by Iowa State University of Science and Technology under contract no. DE-AC02-07CH11358 with the US Department of Energy. The United States Government retains and the publisher, by accepting the article for publication, acknowledges that the United States Government retains a non-exclusive, paid-up, irrevocable, worldwide licence to publish or reproduce the published form of this manuscript, or allow others to do so, for United States Government purposes.

1976). Usually these materials can be analyzed with little or no sample preparation, so PAS is nondestructive. Unlike conventional spectroscopy, it is applicable to dark, opaque, and highly light-scattering materials (Rosencwaig and Gersho, 1976). Conventionally, spectroscopy is done by either passing radiation through the sample under test or reflecting it off the sample, and then measuring the fraction that made it through or was reflected back. The amount of radiation absorbed is then calculated from this fraction. This places strict limits on the surface morphology or transparency of the sample or both, and it often means substantial sample preparation is necessary before a solid can be analyzed. Conversely, PAS measures the absorption directly by detecting the heat deposited when radiation is absorbed near the sample surface. This reduces the complications arising from very dark materials or highly light-scattering samples. It is also possible to determine some depth profiling information about structured specimens because the probe depth of PAS is adjustable.

For composites, PAS offers not only nondestructive analysis but also the ability to analyze the matrix even when carbon or other strongly light-absorbing fibres are present. In the next section, we review the basic theory behind the photoacoustic effect that gives rise to its unusual capabilities and the proper protocols for PAS analysis. After that, we survey the applications involving composite ageing and degradation to which FT-IR PAS has been applied.

6.2 Theory and practice of photoacoustic spectroscopy

6.2.1 Theory

The concept behind PAS is illustrated in Fig. 6.1. The sample material is sealed inside a small chamber with a microphone and a window in one wall. Intensity-modulated radiation enters through the window and strikes the sample. If the radiation is absorbed, the sample is heated, and the heat diffuses to the surface of the sample, where it heats the surrounding air or other gas. Because the incident radiation is modulated, the heating of the gas is also modulated. The heating causes the pressure inside the chamber to oscillate, which the microphone detects. If the absorption coefficient of the sample is α, then the intensity of the radiation within the sample is proportional to $e^{-\alpha x}$, where x is the depth within the sample. Very roughly speaking the radiation is absorbed within a distance $1/\alpha$ of the surface. If the thickness, l, of the sample is much greater than $1/\alpha$, then the sample is opaque, but it will not necessarily appear black in a photoacoustic measurement because only a portion of the absorbed radiation may contribute to the photoacoustic signal. Roughly, only the energy deposited within a

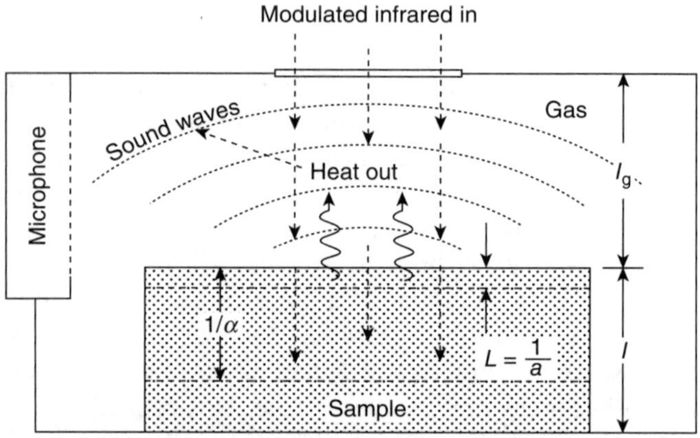

6.1 Schematic of a photoacoustic detector.

thermal diffusion length, L, of the sample surface is able to diffuse to the surface and contribute to the photoacoustic signal. L depends on the modulation frequency of the incoming radiation, f:

$$L = \frac{1}{a} = \left(\frac{D}{\pi f}\right)^{1/2} \qquad [6.1]$$

where a is the thermal diffusion coefficient of the sample, D is the thermal diffusivity of the sample, and $D = k/\rho C$, where k is the thermal conductivity of the sample, C is the heat capacity of the sample, and ρ is the density of the sample. The dependence of L on f makes it user adjustable. If $L \ll 1/\alpha$, then the amount of radiation absorbed within L of the surface is proportional to α, and the photoacoustic spectrum is equivalent to a conventionally acquired spectrum of a sample approximately L thick.

An FT-IR spectrometer modulates its beam of infrared radiation, with each wavenumber (the inverse of wavelength) modulated at a different frequency, so FT-IR and PAS make a natural combination. Most FT-IR spectrometers use a Michelson interferometer, which divides the beam of infrared radiation into two paths of differing lengths and then recombines them so that the two halves mutually interfere (Griffiths and de Haseth, 2007). The spectrometer scans by changing the path difference (or retardation) linearly with time, usually by moving a mirror in one of the two separated paths that reflects the beam back upon itself. As a result, the speed at which the path difference changes is twice the mirror velocity, v. The changing retardation modulates the amplitude of the radiation as the beams in the two paths go in and out of phase with one another. For radiation at a specific wavenumber, $\tilde{\upsilon}$, the frequency of modulation is given by $f = 2v\tilde{\upsilon}$. The pitches of the sound waves produced in photoacoustic spectroscopy

Table 6.1 Dependence of the thermal diffusion length on scanning speed and wavenumber

Scanning speed (cm/s)	Thermal diffusion length			
	Polymer ($D = 0.001\ cm^2/s$)		Helium ($D = 1.8\ cm^2/s$)	
	400 cm^{-1}	4000 cm^{-1}	400 cm^{-1}	4000 cm^{-1}
1	2.0 μm	6.3 μm	0.1 mm	0.3 mm
0.1	6.3 μm	20 μm	0.3 mm	0.8 mm
0.01	20 μm	63 μm	0.8 mm	2.7 mm

Note: the listed scanning speed is the mirror velocity, which is assumed to be half the retardation velocity.

are therefore linear functions of both the wavenumber and the spectrometer scanning speed.

For many polymers and composites, the thermal diffusivity is within a factor of two of $0.001\ cm^2/s$. Table 6.1 lists L for several spectrometer scanning speeds based on this value for D. Although L is an estimate of the effective probe depth, PAS cannot probe any deeper than the spectrometer beam can penetrate into the sample, so the PAS probe depth is ultimately limited by the absorption coefficient of the sample. An opaque buried layer, such as a carbon-fibre mat, would block photoacoustic analysis from viewing deeper into the sample.

The detailed theory behind PAS for homogeneous samples has been developed by Rosencwaig and Gersho (1976). They derived a general equation for $\boldsymbol{Q}$, the complex envelope of the sinusoidal pressure variation that constitutes the photoacoustic signal, which is valid for all sample thicknesses. Their equation can be greatly simplified for composites because we can safely assume that the sample is optically thick ($l \gg 1/\alpha$) and thermally thick ($l \gg L$). For an optically and thermally thick sample, their equation becomes

$$\boldsymbol{Q} = \frac{\alpha I_0 \gamma P_0}{2^{3/2} k l_g a_g T_0 (\alpha^2 - \boldsymbol{\sigma}^2)} \left(\frac{\boldsymbol{r} - 1}{g + 1} \right) = \frac{Y\alpha}{k a_g (\alpha^2 - \boldsymbol{\sigma}^2)} \left(\frac{\boldsymbol{r} - 1}{g + 1} \right) \qquad [6.2]$$

where I_0 is the incident light flux, γ is the heat capacity ratio of the gas, P_0 and T_0 are the ambient pressure and temperature in the photoacoustic cell respectively, l_g is the gas thickness, $a_g = (\pi f/D_g)^{1/2}$ is the thermal diffusion coefficient for the gas, D_g is the thermal diffusivity of the gas, $\boldsymbol{\sigma} = (1 + j)a$, k_g is the thermal conductivity of the gas, $g = k_g a_g/(ka)$, and $\boldsymbol{r} = ½(1 - j)\alpha/a$, with j being the square root of -1. $\boldsymbol{Q}$, $\boldsymbol{\sigma}$, and $\boldsymbol{r}$ are all vectors, having both magnitude and phase. $Y = I_0 \gamma P_0/(2^{3/2} l_g T_0)$, and it does not depend on any property of the sample or on the modulation frequency, so Y can usually be considered a constant.

When the absorption coefficient is small enough or the modulation frequency is high enough, then $\alpha \ll a$ (the case illustrated in Fig. 6.1), and Equation [6.2] simplifies to

$$\boldsymbol{Q} = \frac{Y}{ka_g\boldsymbol{\sigma}^2}\left(\frac{1}{g+1}\right)\alpha = \frac{-jY}{2ka_g a^2}\left(\frac{1}{g+1}\right)\alpha \qquad [6.3]$$

so $\boldsymbol{Q}$ is a linear function of the absorption coefficient, α. Both a_g and $\boldsymbol{\sigma}$ are proportional to $f^{1/2}$, so the photoacoustic signal is proportional to $f^{-3/2}$. At the other extreme, when $\alpha \gg a$,

$$\boldsymbol{Q} = \frac{Y(1-j)}{2aka_g(g+1)} \qquad [6.4]$$

so the signal is independent of the absorption coefficient. The signal is saturated with respect to α, but a and a_g are both proportional to $f^{1/2}$, so the signal is inversely proportional to the modulation frequency.

In the intermediate range, where $\alpha \approx a$, Equation [6.2] does not simplify, so the signal depends on the absorption coefficient but is less than proportional to it. The signal is partially saturated. Because a is proportional to $f^{1/2}$, this partial-saturation zone can be shifted to higher α by increasing the modulation frequency. Of course, the signal-to-noise ratio decreases with increasing frequency because $\boldsymbol{Q}$ is proportional to f^{-x}, where x is between 1 and 1.5, so there is a trade-off between the noise level and the onset of saturation. In practice, the tops of the strongest peaks in the spectra of most composites will be at least partially saturated at all practical modulation frequencies or scanning speeds. This is usually not a problem because there are many other, smaller peaks available for quantitative measurements. When it is a problem, techniques have been developed for either linearizing a spectrum (Carter, 1992; Michaelian, 2007) or adjusting the degree of saturation to be the same in a series of spectra taken at different scanning speeds (Jones and McClelland, 2002).

The above discussion is based on the entire photoacoustic signal being generated by thermal waves within the sample transferring heat to the surrounding gas. In fact, there is another route by which signal is generated. When the sample is heated by the absorption of infrared radiation, it expands slightly, causing a pressure change in the sealed photoacoustic cell. Put another way, the infrared absorption produces acoustic waves within the sample that transfer to the surrounding gas. McDonald and Wetsel (1978) have developed a 'composite piston' model that takes both signal-generation routes into account. The acoustic-wave contribution is dependent on the total absorbance within the sample, so if the sample is optically thick ($\alpha l \gg 1$), then the acoustic-wave contribution is independent of α. In practical terms, the acoustic-wave contribution is usually a small, constant background signal that can be ignored. It becomes important only when the thermal-wave contribution is small, such as when α is small (but αl is not), or when the sample is inhomogeneous, with a nearly transparent surface layer and a much stronger absorption coefficient below the surface, causing the infrared to be absorbed well below the surface.

The quantity $\boldsymbol{Q}$ derived by Rosencwaig and Gersho (1976) is the complex envelope of the sinusoidal pressure variation in the photoacoustic cell, not the actual pressure variation itself (i.e. the photoacoustic signal). The pressure variation, ΔP, is given by

$$\Delta P = Q_1 \cos(2\pi ft - \pi/4) - Q_2 \sin(2\pi ft - \pi/4) \qquad [6.5]$$

where Q_1 and Q_2 are the real and imaginary components of $\boldsymbol{Q}$ and t is time, so the photoacoustic signal lags behind $\boldsymbol{Q}$ by $\pi/4$ radians. The phase of the photoacoustic signal depends on the absorption coefficient, α, even for a homogeneous sample. Conceptually, the greater α is, the nearer to the sample surface the absorption occurs, and so the quicker the deposited heat reaches the sample surface and generates a signal. Mathematically, at the weak absorption limit, Equation [6.3] applies, and the phase of $\boldsymbol{Q}$ is $-j$ relative to the modulation of the incident radiation, so the photoacoustic signal lags behind the incident radiation by 135°. At the strong absorption limit, Equation [6.4] applies, and the phase of $\boldsymbol{Q}$ is $1 - j$, so the signal lags the incident radiation by only 90°. For thermally thick, homogeneous samples, the photoacoustic phase should always fall within this 45° range.

When a sample is heterogeneous, the photoacoustic signal becomes more complicated, but this affords the possibility of determining something about the depth profile of the sample from the photoacoustic signal. The simplest, most qualitative approach is to acquire a series of spectra at a variety of different scanning speeds. The thermal diffusion length, as given by Equation [6.1], increases with decreasing scanning speed, and the slower-speed spectra reveal features from deeper within the sample (Dittmar *et al.*, 1991; McClelland *et al.*, 1993). As shown in Table 6.1, the effective probe depth generally ranges from a few micrometres to a few tens of micrometres for most materials. The variation in signal saturation with scanning speed can complicate the depth profiling interpretation, but a method for equalizing saturation has been developed (Jones and McClelland, 2002). More advanced methods of depth profiling capitalize on the fact that the signal phase grows later as the depth within the sample increases. A model for discretely layered samples has been developed (Jiang *et al.*, 1995). In this model, each layer acts according to the theory of Rosencwaig and Gersho (1976), but the diminution of radiation reaching a buried layer and the phase delay induced by the deposited heat having to diffuse through overlying layers must be added in. The situation becomes even more complex for samples in which the composition changes continuously with depth. Unfortunately, quantitatively determining the absorption depth profile that gives rise to an observed photoacoustic signal from the photoacoustic signal when no *a priori* information is available about the depth profile is very difficult. It is what mathematicians call an ill-posed problem (Power and Prystay, 1995). Small changes in the photoacoustic signal may reflect much larger changes in the depth profile, so modest noise or any other inaccuracies in the photoacoustic signal can prevent recovery of the depth profile. Various approaches to this problem have been developed (Mandelis *et al.*, 1991; Power and Prystay, 1995), and depth profiling remains an area of research.

Although it is outside the scope of FT-IR PAS, it should be noted that depth profiling and imaging using laser ultrasound probes based on the photoacoustic effect are active areas of research, particularly for medical imaging. With such probes, a laser pulse generates ultrasonic acoustic waves within or on the surface of an object. The approach has been applied to composites to evaluate their porosity (Karabutov *et al.*, 2006), to image stresses near the tips of cracks (Muratikov *et al.*, 2002), and to image the matrix–fibre interface (Déom *et al.*, 1990).

6.2.2 Instrumentation and practice

The principal manufacturer of FT-IR photoacoustic detectors for analysis of solid and liquid samples is MTEC Photoacoustics, Inc. (Ames, IA, USA). The description herein refers to its products, but it is generally applicable to any photoacoustic detector used for the same purpose. Commercial PAS instruments (not just FT-IR units) were summarized in 2001 (Handley, 2001). A detector is mounted in the sample compartment of the FT-IR spectrometer and consists of the sample chamber with a window and microphone (as illustrated schematically in Fig. 6.1), a mechanism for sealing the sample in the detector, plumbing to purge the chamber, a mirror for focusing the spectrometer beam onto the sample, and a microphone preamplifier. The microphone preamplifier allows the detector to be plugged directly into the spectrometer. From the point of view of the spectrometer, the photoacoustic detector is like any other detector, and the signal is processed just as that from a conventional detector is processed. Modern detectors are reasonably immune to air-borne noise, but a soft, compliant mount is required to isolate it from mechanical vibration either produced by the spectrometer as it scans or transmitted through the spectrometer from the bench on which it sits. Units designed for field use where vibration may be hard to avoid often use a pair of microphones facing each other across the chamber so that the internally generated photoacoustic signal moves the microphone diaphragms in opposite directions and the signals add, while external vibration moves the diaphragms in the same direction and the signals cancel. Custom units for analyzing large objects nondestructively use an open-ended chamber that seals against the surface of the test object (Fishman and Bard, 1981; Takamoto *et al.*, 1992).

The sample chamber is usually purged with helium prior to spectrum acquisition. Helium increases the photoacoustic signal over air by a factor of two to three because of its superior thermal properties (k_g, a_g, and γ in Equation [6.2]). Purging also eliminates interfering mid-infrared absorptions from carbon dioxide and water vapour. Many sample materials slowly evolve moisture in the sample chamber when warmed by the spectrometer beam. It is therefore often advantageous to put a small amount of desiccant in the sample chamber. Typically in the MTEC detectors, magnesium perchlorate is placed in a sample cup at the bottom of the chamber, a spacer is placed atop that cup, and then the sample in its cup goes on top of that.

This stacking up of elements also fills up some of the empty space in the chamber, which is beneficial. $\boldsymbol{Q}$ is inversely proportional to l_g in Equation [6.2], so the detected signal increases as the volume the acoustic waves must fill decreases. Raising the sample up towards the window at the top of the cell by filling in the space below also reduces the gas column the spectrometer beam passes through and reduces any interfering absorption peaks from vapours. This can be carried too far, however. When the space between the sample and the window is too small, the window acts as a heat sink and reduces the signal by damping the heating of the gas. The signal reduction becomes noticeable when l_g drops to two to three times the thermal diffusion length of the gas (Aamodt *et al.*, 1977). The thermal diffusion length in a gas is given by Equation [6.1], with the thermal diffusivity of the gas replacing that of the sample. D for helium is 1.8 cm^2/s, so the diffusion length can be a millimetre or two at low scanning speeds, as listed in Table 6.1. At typical scanning speeds, maintaining a sample-to-window spacing of at least 2 mm is sufficient.

Transmission spectroscopy uses a zero-absorption blank as a reference and raw sample spectra are normalized by ratioing with the blank spectrum to remove wavenumber-dependent variations of the spectrometer and detector. In PAS the reference is the exact opposite of that, a 100% absorber. Specifically, the reference should be a strong absorber that will absorb all of the radiation within L of its surface so that all of the beam intensity contributes to the photoacoustic signal. MTEC detectors include a reference consisting of a carbon-black coating on a thermally thin, free-standing polymer film. This produces the largest signal possible because heat can transfer to the gas from both sides of the film. Thicker reference materials lose half of the potential signal from heat diffusing away from the surface. Carbon-black powder may be used as a reference, but it tends to pack somewhat differently each time it is placed in a cup, so it is not a highly reproducible reference. Neither a film nor a powder makes a good reference when the phase of the signal is important. Because they are not thermally thick, their signal phases do not behave as Equation [6.2] predicts. When phase is important, either a heavily carbon-black-filled polymer or glassy carbon should be used. These two have been found to be good phase references, with glassy carbon being slightly better than the filled polymer (Jones and McClelland, 2001).

It was difficult to access the phase of the signal using early FT-IR spectrometers. Instrument makers assumed that any variation in phase was an artefact of the instrument and required correction. This is less of a problem with modern instruments. Research-grade FT-IR spectrometers include a second method of scanning called phase modulation or, less appropriately, step scanning, which makes phase much more accessible and data interpretation easier for PAS. In phase modulation, the interferometer mirror does not move at constant speed. Instead it steps through a series of stopping points, and at each of these stops the mirror oscillates a short distance centred about the stop. The signal from this oscillation is then demodulated to determine the signal magnitude and phase. The magnitudes and phases from all the stopping points constitute the interferogram to be Fourier transformed. For PAS, phase modulation has the advantages of substantially easier access to signal phase, a constant modulation frequency over the whole spectrum so L is not wavenumber dependent, and better quality low-frequency modulation (Smith *et al.*, 1988; Manning *et al.*, 1991). Spectrometer makers have also extended phase modulation to higher frequencies than could be reached by conventional scanning by using a square-wave oscillation and demodulating its harmonics (Drapcho *et al.*, 1997).

6.2.3 Further reading on Fourier transform infrared photoacoustic spectroscopy

There were few publications on photoacoustic spectroscopy prior to the 1970s (Rosencwaig, 1973a, 1973b, 1975; Rosencwaig and Gersho, 1976; McDonald and Wetsel, 1978). That early period produced two books that survey the general field of photoacoustics well, but they predate the widespread adoption of FT-IR spectroscopy, so FT-IR PAS receives little mention (Pao, 1977; Rosencwaig, 1980). With the adoption of FT-IR spectroscopy in the 1980s, FT-IR PAS blossomed, and three reviews from that period focus on FT-IR PAS of condensed-phase samples (Vidrine, 1982; McClelland 1983; Graham *et al.*, 1985).

A series of approximately biennual conferences on photoacoustic and photothermal phenomena was founded in 1979 by one of the present authors (J.F.M.). The conference, whose name has evolved over time, is now called the International Conference on Photoacoustic and Photothermal Phenomena. It covers both spectroscopic and nonspectroscopic applications of photothermal phenomena. Almost all of the conferences have published proceedings volumes. The three most recently published are for the 10th, 12th, and 13th conferences (Scudieri and Bertolotti, 1999; Macrander, 2003; Vargas, 2005).

During the past 15 years, there have been several reviews of FT-IR PAS worth noting. Palmer and Jiang (1994) reviewed FT-IR PAS of solid samples with emphasis on phase modulation spectroscopy. The present authors and their co-workers have written a series of chapters on FTIR-PAS of solids, which have emphasized sample handling and basic techniques (McClelland *et al.*, 1993), using sampling depth and the basics of signal phase and phase modulation (McClelland *et al.*, 1998), and more advanced phase modulation and depth profiling (McClelland *et al.*, 2002). The most recent review is that by Schmid (2006), which covers all forms of PAS of both gases and condensed phases. In the second edition of their classic text on FT-IR spectroscopy, Griffiths and de Haseth (2007) devote Chapter 20 to FT-IR PAS and provide an excellent tutorial. The only recent book-length monograph on photoacoustic spectroscopy is by Michaelian (2003), which is a thorough, up-to-date description.

Most of the applications of photoacoustics to composites have been in a nonspectroscopic mode, but there have been spectroscopic uses beyond the ageing-related work discussed in the next section. Noda *et al.* (2002) have used FT-IR PAS to study the structure of nylon at the matrix–fibre interface in glass fibre/nylon 66 composites. Quintanilla and Pastor (1994) and Quintanilla *et al.* (1994) have used FT-IR PAS to study the structure of the polymer in glass fibre/polyamide and glass fibre/polyester composites. Dubois *et al.* (1993, 1994) have used it to determine absolute optical absorptions and optical penetration depths in graphite/epoxy composites. As far as the authors have been able to determine, the present work is the first review of FT-IR PAS applied to composites, ageing or otherwise.

FT-IR PAS has seen far more use in studies of unfilled polymers than of composites, including studies of ageing. Because of its surface sensitivity and adjustable probe depth, FT-IR PAS has been especially favoured for studying photodegradation. Delprat and Gardette (1993) used it to measure the surface degradation of photo-oxidized polypropylene. Gonon *et al.* (1999, 2001) used various scanning speeds to determine depth profiles for photo-oxidized styrene–isoprene copolymers. Kim and Urban (2000) also used a set of scanning speeds to determine depth information about photo-oxidized epoxy and urethane films. Carter *et al.* (1989) and Carter and McCallum (1994) have used FT-IR PAS to follow the weathering, both natural and accelerated, of acrylonitrile–butadiene–styrene (ABS) rubber and of polyester urethane coatings. Thermal degradation studies include one by Xie and Qu (2001) on the 300 and 430 °C pyrolysis of an expandable graphite-based intumescent flame retardant in polyethylene. Wang and Qu (2003) used FT-IR PAS to monitor both the photo and thermal degradation of ethylene–propylene–diene terpolymer. Nguyen (1989) analyzed a poly(vinyl formal) adhesive film as it thermally degraded.

6.3 Ageing of composites

FT-IR PAS has been used to study thermal ageing, photo-oxidation, and moisture degradation of composites. It has also been used to monitor the ambient-temperature ageing of uncured prepregs.

6.3.1 Thermal and photochemical ageing

One of the earliest applications of FT-IR PAS to the degradation of composites was by Murty and Yehl (1990). They studied a glass fibre/nylon 66 composite containing an undefined heat stabilizer as it degraded when heated in air at 150 °C for up to 5000 hours. Their spectra, presented in Fig. 6.2(a), show the growth of a carbonyl band at 1713 cm^{-1}. The difference spectrum in Fig. 6.2(b) shows this even more clearly. They demonstrated that the spectra could be correlated with the degradation in mechanical properties, as their plot of elongation vs. absorption at 1713 cm^{-1} in Fig. 6.3 shows. Unfortunately, they do not give an exact definition of what 'per cent retention of absorbance ratio' is. It is approximately equal to the absorbance at 1713 cm^{-1} divided by the absorbance at 1651 cm^{-1}, scaled so that this ratio in the spectrum of the fresh material is 100%. The 1651 cm^{-1} peak is the nylon Amide I band. They also proposed a reaction mechanism for the degradation based on a previous study of nylon thermal degradation (Do *et al.*, 1987). The process is a free radical chain scission from reaction with oxygen. The end product has the form —$(CH_2CH_2(C{=}O)NH(C{=}O)CH{=}CH)$—. The authors identify the peak at 1638 cm^{-1} in the difference spectrum (Fig. 6.2(b)) as arising from the C=C bond in this product.

Another FT-IR PAS study of thermal degradation was by Cole and Casella (1993). They studied the degradation of carbon fibre/poly(ether ether ketone) (PEEK) composites using FT-IR spectroscopy. They used diffuse reflectance, however, for the majority of their study because it provided them a better signal-to-noise ratio than PAS. Nevertheless, diffuse reflectance created a problem for them because they could not completely eliminate specular reflection from their measurements. The strongest spectrum peaks appeared inverted, and moderately strong peaks were distorted as well. They noted that the observed rate of change in peak heights during thermal degradation was higher for their diffuse reflectance measurements on the composite than for their earlier measurements on unfilled PEEK (Cole and Casella, 1992) using attenuated total reflectance (ATR) and transmission spectroscopy. They could not tell if this difference was real or a diffuse reflectance artefact, so they turned to PAS to resolve the question. They acquired spectra of both

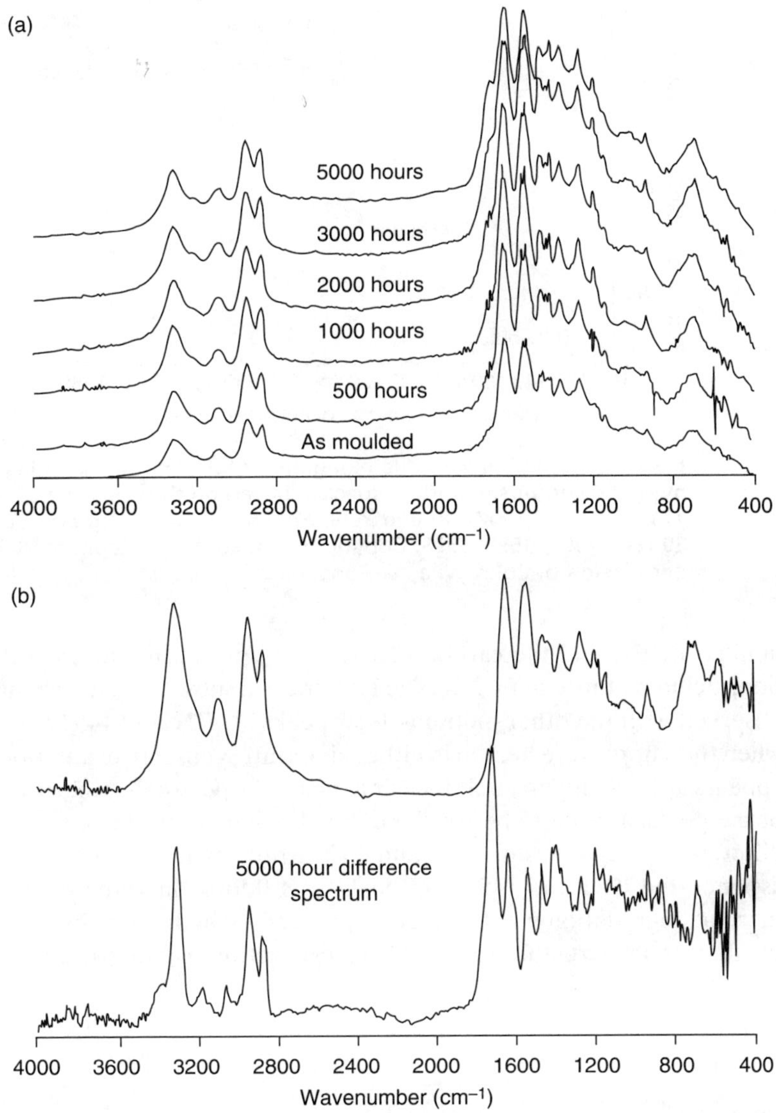

6.2 Spectra of glass fibre/nylon 66 composite thermally aged at 150 °C. (a) Spectra after ageing for the time periods indicated. (b) Spectrum of the difference between the 5000 hour and as-moulded spectra compared with the as-moulded spectrum. (From Murty and Yehl, *Polymer Engineering and Science*, **30** (24), 1990, 1595–1598. Copyright © 1990 Wiley. Reprinted with permission of John Wiley & Sons, Inc.)

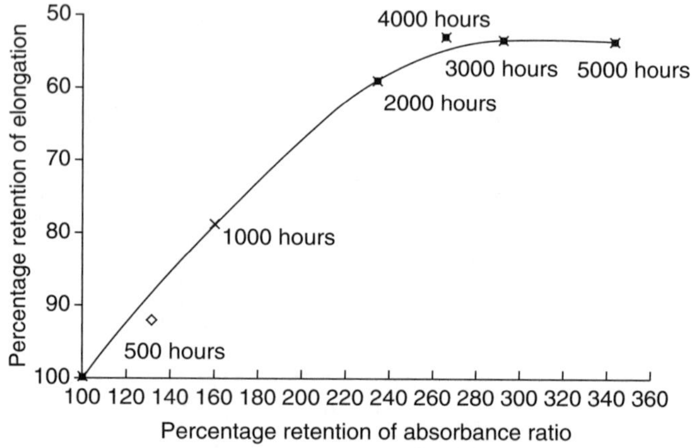

6.3 Correlation between the elongation of thermally aged glass fibre/nylon 66 composite and its spectra, based on the absorbance at 1713 cm^{-1}. (From Murty and Yehl, *Polymer Engineering and Science,* **30** (24), 1990, 1595–1598. Copyright © 1990 Wiley. Reprinted with permission of John Wiley & Sons, Inc.)

unfilled PEEK and the carbon/PEEK composite during thermal degradation, including those in Fig. 6.4. The PAS spectra show the same changes they observed with the other methods. New peaks at 1711 and 1452 cm^{-1} appear when the samples are heated in either air or nitrogen, and an additional peak appears at 1739 cm^{-1} in air. It is clear from these spectra that the growth rates of the peaks are the same for the unfilled polymer and the composite. The difference observed using ATR and diffuse reflectance was an artefact. They assigned the 1711 and 1452 cm^{-1} peaks to a fluorenone-type structure. The thermal degradation was believed to proceed by hydrogen abstraction, and the fluorenone structure resulted from cyclization of a diradical:

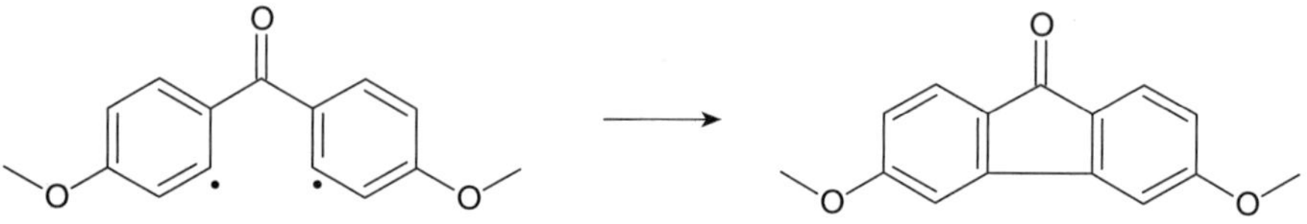

[6.6]

The additional carbonyl peak at 1739 cm^{-1} is assigned to an ester group formed by oxidation of the original carbonyl group. From measurements of the speeds at which the peaks grew at various temperatures, they calculated activation energies of 200 and 130 kJ/mol for the processes producing the 1711 cm^{-1} peak (in air) and the 1739 cm^{-1} peak, respectively.

Delor-Jestin *et al.* (2006) studied both thermal ageing at 100 °C and photochemical ageing of both epoxy resin and glass fibre/epoxy composite

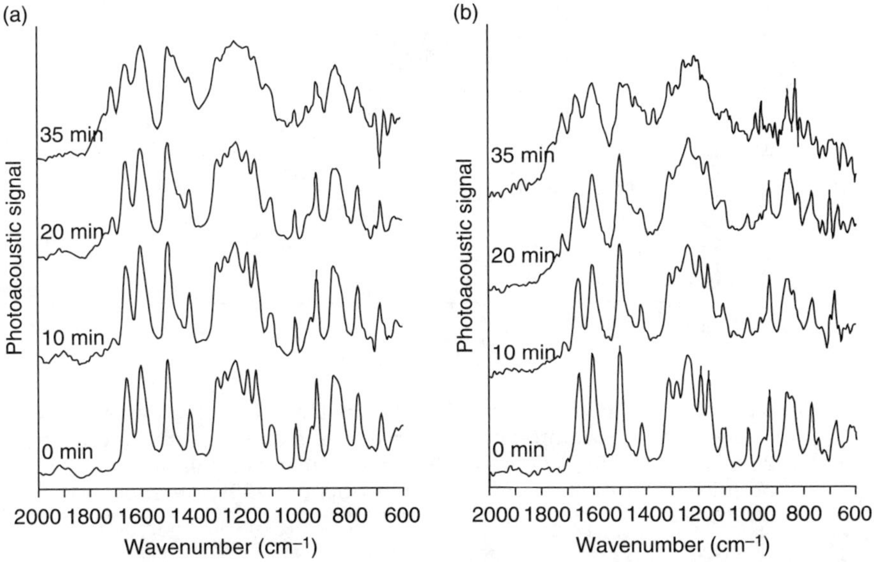

6.4 FT-IR PAS spectra of (a) Stabar K200 PEEK film and (b) NCS-1057 carbon fibre/PEEK laminate after being heated in air at 485 °C for the indicated times. (Reprinted from Cole and Casella (1993), Copyright 1993, with permission from Elsevier.)

using FT-IR PAS. They stated that PAS was the only technique they could use on filled epoxies. For the composite, they used the Dow epoxy system DER 331, which is based on diglycidyl ether of bisphenol A (DGEBA), with methyl nadic anhydride as the curing agent and 1-methyl imidazole as a crosslinking initiator. They studied the DGEBA-anhydride system both as unfilled polymer and as a composite, which they created by pultrusion. They also examined DER 331 with diethylene triamine as the curing agent, but only as the unfilled polymer. Photochemical degradation was done with a SEPAP 12-24 device, using borosilicate filtered ($\lambda > 300$ nm) mercury lamps and with the samples maintained at 60 °C. During photochemical ageing, spectra of both the DGEBA-anhydride polymer and its composite had peak growth at 3450 cm^{-1} from hydroxyl and at 1790, 1760, and 1710 cm^{-1} from carbonyls, and decreases in the 1740 cm^{-1} anhydride peak and the 1570 cm^{-1} aromatic peak. They found that the DGEBA-anhydride system degraded differently from, and was more stable than, the DGEBA-amine system, which resembled the uncrosslinked resin in its degradation. The uncrosslinked resin degrades by hydrogen abstraction along the polymer backbone. They assigned the increases at 1760 and 1710 cm^{-1} to phenyl alkylates and acids respectively. They did not identify other degradation products or specific mechanisms for the DGEBA-anhydride materials

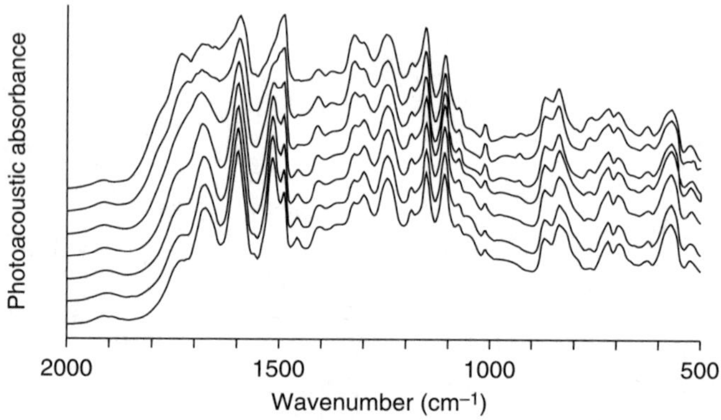

6.5 Spectra of carbon fibre/epoxy laminates after being heated for 4 hours at various temperatures. The temperatures are (top to bottom) 288, 260, 232, 204, 177, and 149 °C, and unheated. (Reprinted with permission from Jones *et al.* (2005) Copyright 2005, American Institute of Physics.)

but noted that the reduction in the 1740 cm^{-1} anhydride peak and the increased stability implied a pronounced influence by the anhydride. The thermal ageing results of Delor-Jestin *et al.* (2006) were analogous to the photochemical ageing. Increases were observed at 3520 cm^{-1} (hydroxyl) and at both 1780 and 1690 cm^{-1} (carbonyls) for the DGEBA-anhydride system. Bands at 1740 cm^{-1} (anhydride) and 1510 cm^{-1} (aromatic) shrank. The anhydride system had a longer induction period (900 hours) prior to the observable onset of degradation than the amine system did (400 hours).

The present authors with others also studied the thermal degradation of an epoxy composite by FT-IR PAS, but from a different perspective (Sweterlitsch *et al.*, 2004; Jones *et al.*, 2005). We investigated whether the spectral changes could be correlated with the losses in laminate strength caused by the thermal degradation. Samples of IM7 carbon fibre/977-3 epoxy composite 25-ply laminate were baked in air for 4 hours at 11 temperatures ranging from 149 to 288 °C, and then their FT-IR PAS spectra and their interlaminar shear strengths (ILSSs) were measured. The same measurements were made on a set of unbaked samples. The temperatures span a range from below the 177 °C service limit of the material to well above it. We acquired spectra of both intact blocks of the laminate panels and from small amounts of powder (0.64 ± 0.15 mg) sanded from the surface of the panels. Figure 6.5 shows some of the spectra. Numerous heat-induced changes are present including the growth of peaks at 1732, 1686, 1490, 1325, 1013, and 764 cm^{-1} and the shrinkage of peaks at 1458 and 1516 cm^{-1}. Small but definite changes are present even in the 149 and 177 °C spectra compared with the unbaked spectrum despite these being at or below the

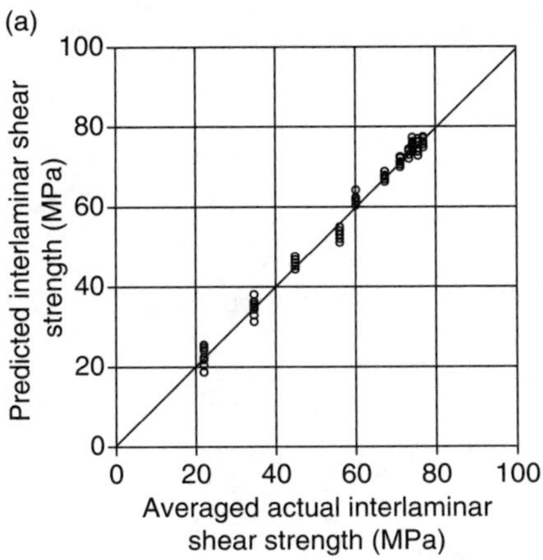

6.6 PLS cross-validation results for predicting ILSS of thermally aged carbon fibre/epoxy laminates from PAS spectra of the aged laminates. (a) Spectra from blocks of the laminate panels; (b) spectra from powder sanded from the panels. (Reproduced with permission of the Society for Applied Spectroscopy from Sweterlitsch *et al.* (2004); permission conveyed through Copyright Clearance Center, Inc.)

service limit, demonstrating the sensitivity of FT-IR PAS to composite degradation. We used partial least squares (PLS) chemometric analysis (Fuller *et al.*, 1988; Thomas and Haaland, 1990) to quantitatively correlate the spectra and ILSS measurements. In PLS, a training set of spectra is used to create a model quantitatively relating the spectra to the property of interest. The resulting model can then predict the property of interest from the spectra of unknown samples. We built a PLS model correlating the spectra of the intact blocks with ILSS measurements and another correlating the powder-sample spectra with ILSS, and then we used cross validations to test the accuracies of the PLS models. In a cross validation, one or a few samples are removed from the training set, a model is built based on the reduced set, and the removed samples are tested as unknowns. The removed members are then put back, other samples are removed, and another model is built and used to analyze the removed members. This process is repeated until all training set members have been removed and analyzed. We acquired spectra at each temperature from ten samples and measured ILSS values for ten samples. The average of the ten ILSS values was used in the PLS model building. Figure 6.6 plots the PLS-predicted ILSS values vs. the known ILSS values from the cross validations of the two training sets. Another measure of the accuracy of a PLS model is the standard error of cross validation (SECV), which is the root-mean-square error between the cross-validation predicted values and known values of the

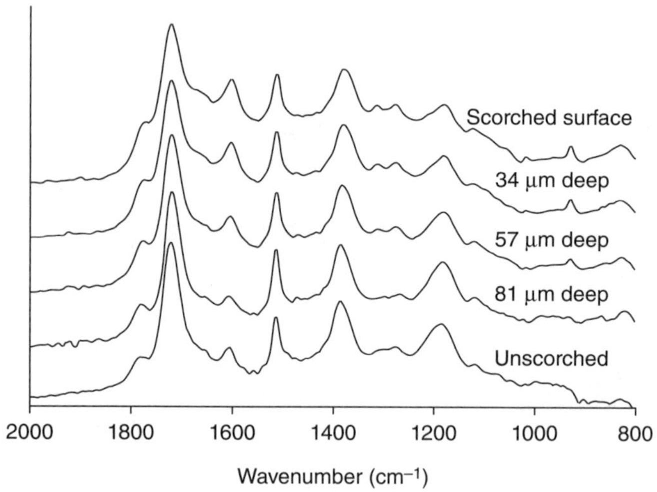

6.7 FT-IR PAS spectra of a scorched carbon fibre/BMI laminate panel taken at the indicated depths as the scorched material was removed. A spectrum of the unscorched laminate is shown at the bottom for comparison.

training set members. The SECVs are 1.60 and 2.02 MPa for the block and powder spectra models respectively, demonstrating that the models have good predictive capability.

The probe depth of FT-IR PAS can be altered by changing the FT-IR spectrometer scanning speed, but this adjustment is limited to a few tens of micrometres in the mid-infrared range by how far the infrared radiation can penetrate. An opaque layer, such as a carbon fibre mat, can limit the probe depth even more. Analysis over greater depths requires destructive sampling. Flat samples can be microlapped, in the same manner as gemstones are polished, to remove surface material in a series of steps as small as two or three micrometres. If FT-IR PAS spectra with short probe depths are taken at each step, a set of spectra showing the variation with depth within the material is built up. Each spectrum reflects the composition for a particular depth in the material. The present authors have used this approach to test FT-IR PAS as a method for examining the extent of heat damage from superficial scorching of a composite laminate panel. The surface of a panel of carbon fibre/BMI composite was slightly scorched by exposure to a stream of 617 °C air for 15 seconds. The panel was then microlapped in steps of two or three micrometres to a depth of 81 μm, with spectra being acquired at each step. A few of these spectra are shown in Fig. 6.7. The scorching did not produce a visible change in the surface, but the FT-IR PAS spectra clearly show changes. The bottom spectrum in the figure is from undamaged composite, and the top spectrum is from the scorched surface. The two most

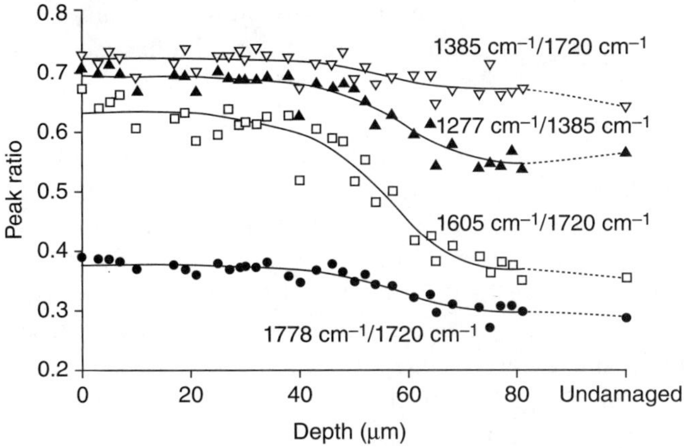

6.8 Spectrum height ratios for peaks at the indicated wavenumber positions as a function of depth removed from a sample of scorched carbon fibre/BMI composite.

obvious changes are the reduction of the large peak at 1180 cm^{-1} and growth of the previously small peak at 1605 cm^{-1}. The 1180 cm^{-1} peak is assigned to the C–N–C group in the succinimide ring (Wu *et al.*, 1998), so its reduction implies destruction of the imide group. The physical extent of the scorching becomes obvious when height ratios for various pairs of peaks in the spectra are plotted as a function of the depth of material removed, as in Fig. 6.8. The peak ratios are all nearly constant over roughly the first 35 μm, then they all change over the next 40 μm, and finally they level off at about 75 μm at values near those of the undamaged material, which are plotted along the right edge of the graph. Even without information on the chemistry of the material, it can be deduced that the scorching damage was approximately uniform in the uppermost 35 μm and then diminished over the next 40 μm. Material below 75 μm was undamaged. This is consistent with the appearance of the spectra in Fig. 6.7. The 34-μm-deep spectrum looks much like the scorched-surface spectrum, while the 81-μm-deep spectrum looks much like that of the undamaged material. The 57-μm-deep spectrum is intermediate between these two extremes.

6.3.2 Hydrolytic degradation

McDonald and Urban (1991) used FT-IR PAS to examine the matrix–fibre interface in glass/polyimide composites and the effects of moisture on the interface. They studied both Nextel AF10 fibre/PMR-15 and Nextel BF10 fibre/Ultem 1000 composites both with and without the fibres treated with a vinylbenzylaminosilane coupling agent. Each of these four combinations were examined both dry and after immersion in water at ambient temperature for 1 month. They acquired spectra of both the flat face and the fractured end of each panel. The fractured surface exposed the fibres and the matrix–fibre interface. They subtracted the spectrum of the glass fibres from the spectra of the flat and fractured surfaces to remove the spectral contributions from the glass. They then isolated the spectral effects of the interface by subtracting the flat-surface spectrum from the fractured-surface spectrum. They found no difference between the spectra of the flat and fractured surfaces for dry AF10/PMR-15 composite when no coupling agent was used. By contrast, their spectra of the dry, coupling-agent-treated AF10/PMR-15 composite showed a loss in intensity in the fractured-surface spectrum at 1730 cm^{-1}, which is the symmetric stretch for the polyimide carbonyl. There were also increases at 3475, 3325, 1655, and 1670 cm^{-1}. These are identified as O–H, hydrogen-bonded N–H, an amide carbonyl, and a carboxylic acid carbonyl respectively. Except for N–H, these spectral bands cannot arise from the coupling agent itself, so they indicate changes in the matrix at the interface. They indicate hydrolysis of the imide to both amide and acid functionalities as part of the fibre–matrix coupling.

Figures 6.9(a) and (b) show spectra of the AF10/PMR-15 without coupling agent treatment (after fibre spectrum subtraction) after the 1-month water immersion. The 1730 cm^{-1} peak in the flat-surface spectrum (Spectrum B) is somewhat weaker and broader than the same peak in the dry-composite spectrum (not shown), indicating some loss of imide to hydrolysis on the flat surface. Nevertheless, the fractured surface shows even more evidence of hydrolysis when the two spectra are subtracted. The broad increase in Spectrum C between 1650 and 1695 cm^{-1} represents the overlap of several hydrogen-bonded species. The difference spectrum shows larger increases at 3475 (hydroxyl), 3325 (hydrogen-bonded N–H), 1655 (amide carbonyl), and 1690 cm^{-1} (acid carbonyl) than the same spectrum for the dry composite with coupling agent. These again indicate hydrolysis of imide to amide and acid, implying that the hydrolysis is greater at the matrix–fibre interface than on the flat panel surface. Figures 6.9(c) and (d) show spectra of the coupling-agent-treated AF10/PMR-15 after immersion for 1 month in water. There is a strong decrease at 1730 cm^{-1}, and increases at 3475, 1780, 1690, and 1655 cm^{-1}. The large change at 1730 cm^{-1} suggests that the matrix–fibre interface has lost substantially more imide than the flat face of the sample. The increases at 3475, 1690, and 1655 cm^{-1} are the same changes observed in Fig. 6.9(a) and (b), but the increase at 1780 cm^{-1} is different. The 1780 cm^{-1} peak is the symmetric polyimide carbonyl stretch. This coupled with the large loss at 1730 cm^{-1}, the asymmetric polyimide carbonyl stretch, suggests a change in the environment around the polyimide carbonyl at the matrix–fibre interface. This change is probably from hydrogen bonding among carboxylic acids and amide hydrolysis products. Taken

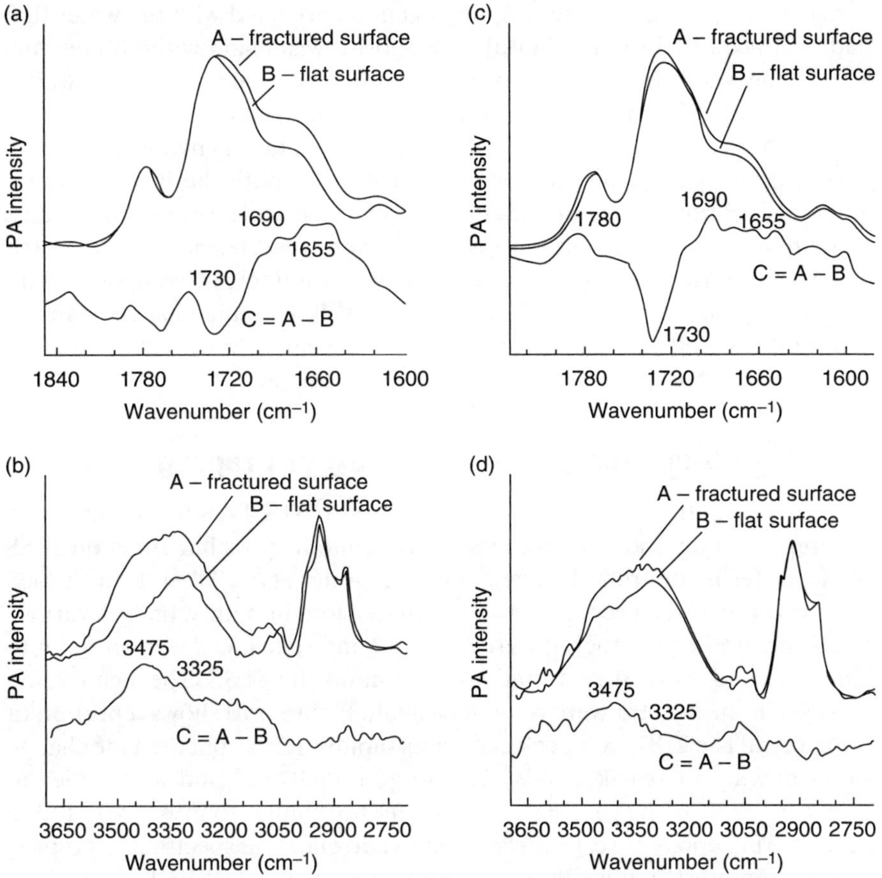

6.9 FT-IR PAS spectra of a Nextel AF10 glass fibre/PMR-15 polyimide composite plate after water immersion for 1 month. (a) and (b) Spectra of composite without coupling agent, and (c) and (d) spectra of composite with coupling agent. Spectra are of a flat side of the plate, of the fractured end exposing the matrix–fibre interface, and of the difference between these two. (Reprinted from McDonald and Urban (1991), Copyright 1991, with permission from Elsevier.)

together, these changes illustrate the hydrolytic instability of the interaction created by the coupling agent.

The spectra of the BF10/Ultem 1000 composites were similar to those of the AF10/PMR-15 composites, but the differences that were found were revealing. The spectra of dry BF10/Ultem 1000 composite without coupling agent did show hydrolysis-related differences between the flat and fractured surfaces, suggesting that free moisture had produced some hydrogen-bond coupling between the fibre and matrix. By contrast, the dry composite with coupling agent showed less difference between the flat and fractured surfaces, suggesting that the coupling agent had reacted with the water that had produced hydrolysis when the composite was made without coupling agent. The water-immersed composite without coupling agent showed a large drop at 1730 cm^{-1} in the difference spectrum but only small increases at the other relevant bands. McDonald and Urban (1991) hypothesized two chemical changes at the interface – hydrolysis of both the imide linkages and the neighbouring (unlinked) imides. This would release small fragments containing amide and acid groups, which were then leached away by the water. The difference spectrum for the water-immersed composite with coupling agent showed a positive-going but split (double-peaked) band at 1730 cm^{-1}, as well as increases at 1690 and 1655 cm^{-1}, showing that both the flat and fractured surfaces underwent hydrolysis.

6.4 Ambient temperature ageing of prepreg

The present authors and Yeow Ng have used FT-IR PAS to examine how prepreg sheet ages at ambient temperature and its possible effect on ILSS of laminates made from the aged material (Jones *et al.*, 2008). Toray T700S carbon fibre/epoxy prepreg was aged under low humidity, and at various times spectra of the prepreg were acquired and specimens were cured into laminate, and were then tested to determine the ILSS. The age-related changes in the spectra were relatively small. Figure 6.10 shows a portion of the spectra. The 2187 cm^{-1} peak decreases rapidly at first, but then the change slows down. Two weak bands at 1750 and 1800 cm^{-1} and a shoulder at 1670 cm^{-1} grow with age. These changes are not similar to those expected to occur during epoxy cure. PLS was used to correlate the spectra and prepreg age, and Fig. 6.11(a) shows the cross-validation results. The SECV is 1.47 days, which is less than the average spacing between successive measurements. The ILSS measurements did not show a definite trend relative to prepreg age. The shear strengths decreased by less than their uncertainty over the ageing period. PLS modelling successfully correlated the spectra with the ILSS measurements. Figure 6.11(b) shows the cross-validation between spectra and ILSS values. SECV is 1.4 MPa (0.21 ksi). There is substantially more scatter in the ILSS plot than in the prepreg-age plot, but the increased scatter is consistent with the accuracy of the ILSS measurements.

6.5 Acknowledgements

This manuscript has been authored by Iowa State University of Science and Technology under Contract No. DE-AC02-07CH11358 with the US Department of Energy. Some of the work reported here was supported by NASA under award No. NAG-1-02098 to the Center for Nondestructive Evaluation, Iowa State University.

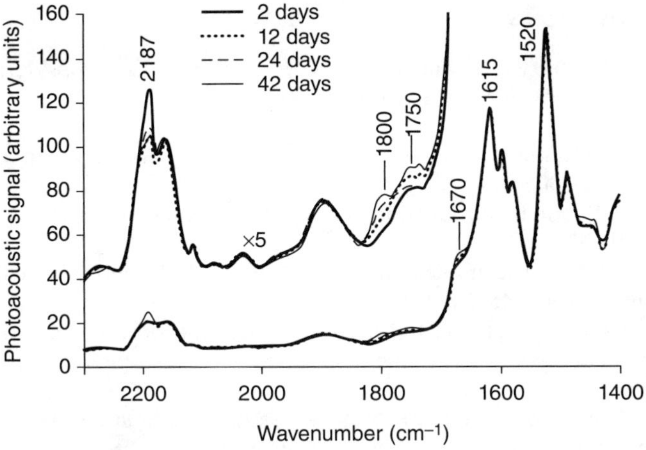

6.10 FT-IR PAS spectra of carbon fibre/epoxy prepregs after ageing at ambient temperature for the four indicated periods. Spectra have been scaled to the 1520 cm^{-1} peak. The upper four spectra are the lower four spectra enlarged by a factor of 5 ('×5').

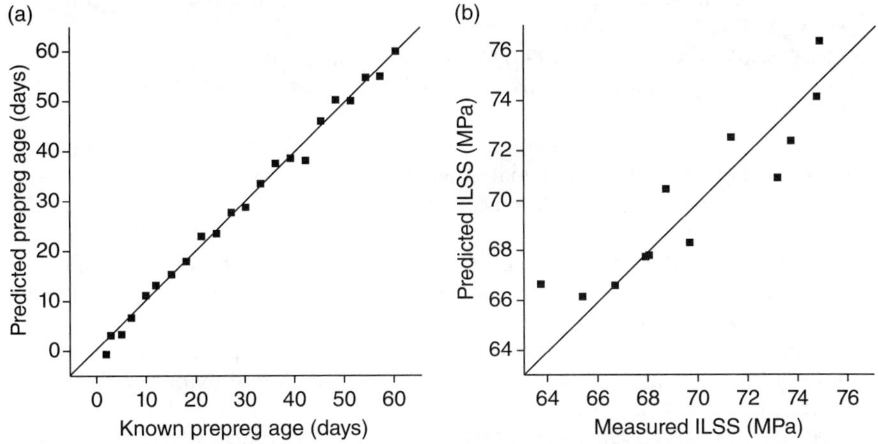

6.11 Cross validations of PLS models correlating FT-IR PAS spectra of aged prepreg with (a) prepreg age and (b) ILSS of laminates made from the aged prepreg. The straight lines are the ideal (i.e. predicted = actual).

6.6 References

AAMODT LC, MURPHY JC, and PARKER JG (1977), 'Size considerations in the design of cells for photoacoustic spectroscopy', *J Appl Phys*, **48** (3), 927–933.

BELL AG (1880), 'On the production and reproduction of sound by light', *Am J Sci*, series 3, **20** (118), 305–324.

BELL AG (1881), 'Upon the production of sound by radiant energy', *Phil Mag*, series 5, **11** (71), 510–528.

CARTER RO III (1992), 'The application of linear PA/FT-IR to polymer-related problems', *Appl Spectrosc*, **46** (2), 219–224.

CARTER RO III and MCCALLUM JB (1994), 'Photoacoustic infrared spectroscopy for evaluation of an ABS as an automotive interior material', *Polym Degrad Stab*, **45** (1), 1–10.

CARTER RO III, PAPUTA PECK MC, and BAUER DR (1989), 'The characterization of polymer surfaces by photoacoustic Fourier transform infrared spectroscopy', *Polym Degrad Stab*, **23** (2), 121–134.

COLE KC and CASELLA IG (1992), 'Fourier transform infrared spectroscopic study of thermal degradation in films of poly(etheretherketone)', *Thermochim Acta*, **211**, 209–228.

COLE KC and CASELLA IG (1993), 'Fourier transform infra-red spectroscopic study of thermal degradation in poly(ether ether ketone)-carbon composites', *Polymer*, **34** (4), 740–745.

DELOR-JESTIN F, DROUIN D, CHEVAL P-Y, and LACOSTE J (2006), 'Thermal and photochemical ageing of epoxy resin – Influence of curing agents', *Polym Degrad Stab*, **91** (6), 1247–1255.

DELPRAT P and GARDETTE J-L (1993), 'Analysis of photooxidation of polymer materials by photoacoustic Fourier transform infra-red spectroscopy', *Polymer*, **34** (5), 933–937.

DÉOM A, BOSCHER D, NOIROT L, ENGUEHARD F, and BALAGEAS D (1990), 'Imaging of the interface between fibers and matrix in the yarns of three-directional carbon-carbon composites by a photoacoustic method', *Mater Sci Eng B*, **5** (2), 135–141.

DITTMAR RM, CHAO JL, and PALMER RA (1991), 'Photoacoustic depth profiling of polymer laminates by step-scan Fourier transform infrared spectroscopy', *Appl Spectrosc*, **45** (7), 1104–1110.

DO CH, PEARCE EM, BULKIN BJ, and REIMSCHUESSEL HK (1987), 'FT-IR spectroscopic study on the thermal and thermal oxidative-degradation of nylons', *J Polym Sci A*, **25** (9), 2409–2424.

DRAPCHO DL, CURBELO R, JIANG EY, CROCOMBE RA, and MCCARTHY WJ (1997), 'Digital signal processing for step-scan Fourier transform infrared photoacoustic spectroscopy', *Appl Spectrosc*, **51** (4), 453–460.

DUBOIS M, CHOQUET M, MONCHALIN J-P, ENGUEHARD F, and BERTRAND L (1993), 'Absolute optical absorption spectra in graphite epoxy by Fourier transform infrared photoacoustic spectroscopy', *Opt Eng*, **32** (9), 2255–2260.

DUBOIS M, ENGUEHARD F, BERTRAND L, CHOQUET M, and MONCHALIN J-P (1994), 'Optical penetration depth determination in a graphite-epoxy laminate by photoacoustic F.T.I.R. spectroscopy', *J Physique IV*, **4** (C7), 377–380.

FISHMAN VA and BARD AJ (1981), 'Open-ended photoacoustic spectroscopy cell for thin-layer chromatography and other applications', *Anal Chem*, **53** (1), 102–105.

FULLER MP, RITTER GL, and DRAPER CS (1988), 'Partial least-squares quantitative analysis of infrared spectroscopic data. Part I: Algorithm implementation', *Appl Spectrosc*, **42** (2), 217–227.

GONON L, VASSEUR OJ, and GARDETTE J-L (1999), 'Depth profiling of photooxidized styrene-isoprene copolymers by photoacoustic and micro-Fourier transform infrared spectroscopy', *Appl Spectrosc*, **53** (2), 157–163.

GONON L, MALLEGOL J, COMMEREUC S, and VERNEY V (2001), 'Step-scan FTIR and photoacoustic detection to assess depth profile of photooxidized polymer', *Vib Spectrosc*, **26** (1), 43–49.

GRAHAM JA, GRIM WM III, and FATELEY WG (1985), 'Fourier transform infrared photoacoustic spectroscopy of condensed-phase samples', in Ferraro JR and Basile LJ, eds, *Fourier Transform Infrared Spectroscopy*, Vol. 4, Orlando, FL, Academic Press, pp. 345–392.

GRIFFITHS PR and DE HASETH JA (2007), *Fourier Transform Infrared Spectrometry*, 2nd edn., New York, Wiley-Interscience, Chapters 2 and 20.

HANDLEY J (2001), 'The sound of compounds', *Anal Chem*, **73** (23), 685A–688A.

JIANG EY, PALMER RA, and CHAO JL (1995), 'Development and applications of a photoacoustic phase theory for multilayer materials: The phase difference approach', *J Appl Phys*, **78** (1), 460–469.

JONES RW and MCCLELLAND JF (2001), 'Phase references and cell effects in photoacoustic spectroscopy', *Appl Spectrosc*, **55** (10), 1360–1367.

JONES RW and MCCLELLAND JF (2002), 'Quantitative depth profiling using saturation-equalized photoacoustic spectra', *Appl Spectrosc*, **56** (4), 409–418.

JONES RW, SWETERLITSCH JJ, WAGNER AJ, MCCLELLAND JF, HSU DK, POLIS DL, and SOVINSKI MF (2005), 'FT-IR photoacoustic spectroscopy applied to the curing and aging of composites', in Thompson DO and Chimenti DE, eds, *Review of Progress in Quantitative Nondestructive Evaluation*, Vol. 24B, Melville, NY, American Institute of Physics, pp. 1094–1099 (AIP Conference Proceedings Vol. 760).

JONES RW, NG Y, and MCCLELLAND JF (2008), 'Monitoring ambient-temperature aging of a carbon-fiber/epoxy composite prepreg with photoacoustic spectroscopy', *Composites A,* **39** (6), 965-971.

KARABUTOV AA, PELIVANOV IM, and PODYMOVA NB (2006), 'Laser optoacoustic method for nondestructive porosity evaluation of graphite–epoxy composites', in Arecchi T, Gurov IP, Yatagai T, Panchenko VY, and Golubev VS, eds, *International Conference on Lasers, Applications, and Technologies 2005: Laser Sensing, Imaging, and Information Technologies*, Bellingham, WA, SPIE, 61620I (SPIE Proceedings Vol. 6162).

KIM H and URBAN MW (2000), 'Molecular level chain scission mechanisms of epoxy and urethane polymeric films exposed to UV/H_2O. Multidimensional spectroscopic studies', *Langmuir*, **16** (12), 5382–5390.

MACRANDER AT, ED. (2003), *Proceedings of the 12th International Conference on Photoacoustic and Photothermal Phenomena*, Woodbury, NY, American Institute of Physics (*Rev Sci Instrum*, **74** (1, Pt 2).

MANDELIS A, PERALTA SB, and THOEN J (1991), 'Photoacoustic frequency-domain depth profiling of continuously inhomogeneous condensed phases – theory and simulations for the inverse problem', *J Appl Phys*, **70** (3), 1761–1770.

MANNING CJ, PALMER RA, and CHAO JL (1991), 'Step-scan Fourier-transform infrared spectrometer', *Rev Sci Instrum*, **62** (5), 1219–1229.

MCCLELLAND JF (1983), 'Photoacoustic spectroscopy', *Anal Chem*, **55** (1), 89A–105A.

MCCLELLAND JF, JONES RW, LUO S and SEAVERSON LM (1993), 'A practical guide to FT-IR photoacoustic spectroscopy', in Coleman PB, ed., *Practical Sampling Techniques for Infrared Analysis*, Boca Raton, FL, CRC Press, pp. 107–144.

MCCLELLAND JF, BAJIC SJ, JONES RW, and SEAVERSON LM (1998), 'Photoacoustic spectroscopy', in Mirabella FM, ed., *Modern Techniques in Applied Molecular Spectroscopy*, New York, Wiley, pp. 221–265.

MCCLELLAND JF, JONES RW, and BAJIC SJ (2002), 'Photoacoustic spectroscopy', in Chalmers JM and Griffiths PR, eds, *Handbook of Vibrational Spectroscopy*, Vol. 2, Chichester, UK, Wiley, pp. 1231–1251.

MCDONALD FA and WETSEL GC JR (1978), 'Generalized theory of the photoacoustic effect', *J Appl Phys*, **49** (4), 2313–2322.

MCDONALD WF and URBAN MW (1991), 'Chemical structure at the Nextel fibre/polyimide matrix interface detected by photoacoustic FTIR spectroscopy', *Composites*, **22** (4), 307–318.

MICHAELIAN KH (2003), *Photoacoustic Infrared Spectroscopy*, New York, Wiley-Interscience.

MICHAELIAN KH (2007), 'Linearization and signal recovery in photoacoustic infrared spectroscopy', *Rev Sci Instrum*, **78** (5), 051301.

MURATIKOV KL, GLAZOV AL, ROSE DN, and DUMAR JE (2002), 'Thermoelastic photoacoustic effect near tips of radial cracks in ceramics under external loading', *High Temp High Press*, **34** (5), 585–590.

MURTY EM and YEHL TW (1990), 'Adaptation of photoacoustic Fourier transform infrared spectroscopy for studying the thermal oxidation of nylon 66 at 150 °C correlated to mechanical properties', *Polym Eng Sci*, **30** (24), 1595–1598.

NGUYEN TT (1989), 'Thermal degradation of a poly(vinyl formal)-containing adhesive film on a polyethylene rubber surface. A photoacoustic Fourier transform infrared spectroscopic investigation', *J Appl Polym Sci*, **38** (4), 765–778.

NODA K, TSUJI M, TAKAHARA A, and KAJIYAMA T (2002), 'Aggregation structure and molecular motion of (glass-fiber/matrix nylon 66) interface in short glass-fiber reinforced nylon 66 composites', *Polymer*, **43** (14), 4055–4062.

PALMER RA and JIANG Y (1994), 'Photoacoustic molecular spectroscopy and chemical characterization', *J Physique IV*, **4** (C7), 337–344.

PAO Y-H, ED. (1977), *Optoacoustic Spectroscopy and Detection*, New York, Academic Press.

PARKER JG (1973), 'Optical absorption in glass: Investigation using an acoustic technique', *Appl Opt*, **12** (12), 2974–2977.

POWER JF and PRYSTAY MC (1995), 'Expectation minimum (EM): a new principle for the solution of ill-posed problems in photothermal science', *Appl Spectrosc*, **49** (6), 709–724.

QUINTANILLA L and PASTOR JM (1994), 'Structural analysis of polyamide-6,6 reinforced with glass fibre by the use of Fourier transform infra-red spectroscopy with photoacoustic detection and differential scanning calorimetry', *Polymer*, **35** (24), 5241–5246.

QUINTANILLA L, RODRÍGUEZ-CABELLO JC, JAWHARI T, and PASTOR JM (1994), 'Structural analysis of poly(ethylene terephthalate) reinforced with glass fibre: 1. A photo-acoustic Fourier transform infra-red study', *Polymer*, **35** (3), 514–518.

ROSENCWAIG A (1973a), 'Photoacoustic spectroscopy of solids', *Opt Commun*, **7** (4), 305–308.

ROSENCWAIG A (1973b), 'Photoacoustic spectroscopy of biological materials', *Science*, **181** (4100), 657–658.

ROSENCWAIG A (1975), 'Photoacoustic spectroscopy: a new tool for investigation of solids', *Anal Chem*, **47** (6), 592A–604A.

ROSENCWAIG A (1980), *Photoacoustics and Photoacoustic Spectroscopy*, New York, Wiley.

ROSENCWAIG A and GERSHO A (1976), 'Theory of the photoacoustic effect with solids', *J Appl Phys*, **47** (1), 64–69.

SCHMID T (2006), 'Photoacoustic spectroscopy for process analysis', *Anal Bioanal Chem*, **384** (5), 1071–1086.

SCUDIERI F and BERTOLOTTI M, EDS (1999), *Photoacoustic and Photothermal Phenomena: 10th International Conference*, Woodbury, NY, American Institute of Physics (AIP Conference Proceedings, Vol. 463).

SMITH MJ, MANNING CJ, PALMER RA, and CHAO JL (1988), 'Step scan interferometry in the mid-infrared with photothermal detection', *Appl Spectrosc*, **42** (4), 546–555.

SWETERLITSCH JJ, JONES RW, HSU DK, and MCCLELLAND JF (2004), 'Prediction of inter-laminar shear strength of a thermally aged carbon/epoxy composite material by Fourier transform infrared photoacoustic spectroscopy,' *Appl Spectrosc*, **58** (12), 1420–1423.

TAKAMOTO R, NAMBA R, MATSUOKA M, and SAWADA T (1992), 'Human in vivo percutaneous absorptiometry using the laser-photoacoustic method', *Anal Chem*, **64** (21), 2661–2663.

THOMAS EV and HAALAND DM (1990), 'Comparison of multivariate calibration methods for quantitative spectral analysis', *Anal Chem*, **62** (10), 1091–1099.

VARGAS H, ED. (2005), *Photoacoustic and Photothermal Phenomena*, Les Ulis, France, EDP Sciences (*J Physique IV*, **125**).

VIDRINE DW (1982), 'Photoacouistc Fourier transform infrared spectroscopy of solids and liquids', in Ferraro JR and Basile LJ, eds, *Fourier Transform Infrared Spectroscopy*, Vol. 3, Orlando, FL, Academic Press, pp. 125–148.

WANG W and QU B (2003), 'Photo- and thermo-oxidative degradation of photo-crosslinked ethylene–propylene–diene terpolymer', *Polym Degrad Stab*, **81** (3), 531–537.

WU W, WANG D, and YE C (1998), 'Preparation and characterization of bismaleimide-diamine prepolymers and their thermal-curing behavior', *J Appl Polym Sci*, **70** (12), 2471–2477.

XIE R and QU B (2001), 'Thermo-oxidative degradation behaviors of expandable graphite-based intumescent halogen-free flame retardant LLDPE blends', *Polym Degrad Stab*, **71** (3), 395–402.

7
Modeling physical ageing in polymer composites

H. HU, National Pingtung University of Science and Technology, Taiwan

7.1 Introduction

Physical ageing is a phenomenon displayed as changes in physical and mechanical properties of polymers over time. Above its glass transition temperature (T_g), a polymer can reach thermodynamic equilibrium instantaneously; below T_g, the polymer requires a very long time to achieve thermodynamic equilibrium. During this evolution time, the material properties (i.e. creep rate, stiffness, and strength) change continuously. This gradual process to establish the equilibrium state is known as physical ageing. Physical ageing is not evident over short time periods, but it continues indefinitely over the service life of the polymer or polymer composites. It is therefore essential to understand and be able to model physical ageing in order to assess the long-term durability of polymer composites.

Struik (1978) was the first to conduct a series of systematic creep tests on various polymers to characterize their ageing behaviors. Some of the remarkable conclusions drawn were: (a) physical ageing is a basic feature and is found in all polymers whether polymeric, monomeric, organic, or inorganic; (b) the mechanical properties of glassy polymers strongly depend on the ageing time; (c) in all glassy materials, physical ageing proceeds in a very similar way – during the ageing process, the time-dependent behavior of the glassy material is found to be independent of the specific chemical structure of the material; (d) physical ageing is thermoreversible – this means that when the temperature is above the T_g, physical ageing in a polymer can be erased, below the T_g, physical ageing in polymer starts and can persist for several years. Struik's conclusions are very important for the modeling of physical ageing in polymers and polymer composites.

In Struik's approach, the polymer is first heated to a temperature above the T_g for several minutes and then quenched to a test temperature below the T_g. This process is called rejuvenation. Since physical ageing is thermoreversible, it can be erased from the material by rejuvenation.

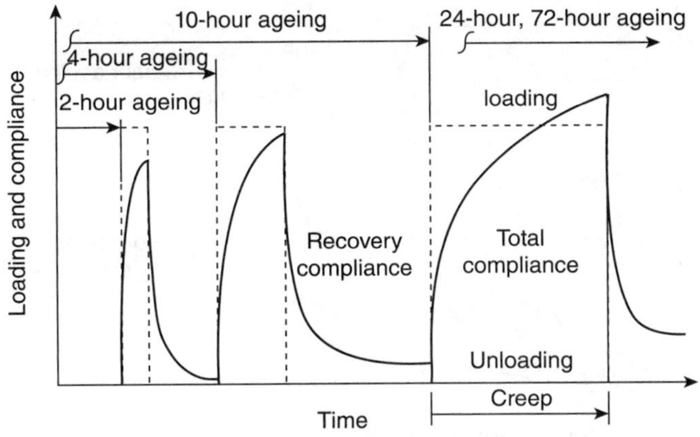

7.1 Momentary creep test process.

After rejuvenation, the temperature remains constant and physical ageing in the material initiates throughout the test. At the end of a certain ageing cycle, a momentary creep test is performed. Figure 7.1 shows the processes involved in a momentary creep test. Momentary creep tests are conducted on the material after 2, 4, 10, 24, and 72 hours of initial ageing cycles. The momentary creep test is defined for a time of less than one-tenth of the initial ageing time. (Struik, 1978). For instance, the creep test time should be less than 0.2 hours following a 2-hour initial ageing. During the momentary creep, the ageing effect on the individual creep compliances is not significant and is negligible so that the material properties can be regarded as constant. However, creep compliances obtained from tests with different initial ageing times are significantly different from one another. Consequently, physical ageing in polymer can be characterized by establishing ageing-related material constants for each momentary compliance curve. Struik's approach has been widely employed to characterize physical ageing in polymer composites. Many experimental works have demonstrated that the ageing behavior of polymer composites is very much like that of polymers.

7.2 Modeling physical ageing in short-term creep

7.2.1 Exponential model

The exponential model is typically employed to characterize physical ageing in polymers and polymer composites. For example, momentary compliances obtained from momentary creep tests can be singled out and plotted on a double logarithmic scale. Figure 7.2 shows the momentary

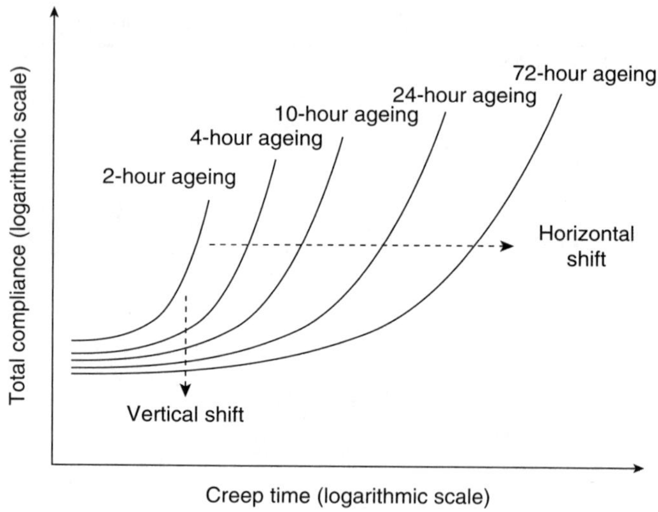

7.2 Exponential model for momentary compliance.

compliances with initial ageing cycles of 2, 4, 10, 24, and 72 hours. The individual momentary compliance can be fitted to an exponential function (Struik, 1978), i.e.

$$S = S_o e^{(t/\tau)^\beta} \qquad [7.1]$$

where S_o is initial compliance, t is creep time, τ is relaxation time, and β is a shape factor. It is assumed that the initial compliance and shape factor are constant. Ageing effects can be observed by a horizontal shift of the momentary compliance curves. All momentary compliance curves can be shifted and superposed to a reference curve by introducing a shift factor, i.e.

$$a = \left(\frac{t_{a,ref}}{t_a}\right)^\mu \qquad [7.2]$$

where t_a is ageing time, $t_{a,ref}$ is a reference ageing time associated with the reference curve of momentary compliance, and μ is the shift rate which can be obtained from the slope of the test data for shift factor and ageing time, i.e.

$$\mu = \frac{d\log(a)}{d\log(t_a)} \qquad [7.3]$$

Struik (1978) observed that shift rate in all cases is about unity over wide ranges of test temperatures below T_g. When the test temperature is close to T_g, the shift rate significantly decreases to about zero. Moreover, at very low temperatures, the ageing begins to cease and the shift rate decreases as well. In equation [7.1], relaxation time is considered as the only ageing-dependent parameter, and is given by

$$\tau(t_{\mathrm{a}}) = \tau(t_{\mathrm{a,ref}})/a \qquad [7.4]$$

Using shift factor and shift rate, all momentary compliances can be horizontally shifted and superposed to a reference curve, which enables us to fabricate any momentary compliance for any initial ageing time without a creep test.

Struik's model has been generally applied to characterize physical ageing in continuous fiber-reinforced polymer composites. Sullivan (1990) studied physical ageing in glass fiber/thermosetting composites and found that physical ageing significantly affects transverse compliances S_{22} and shear compliance S_{66}, but not S_{11}, S_{12} and S_{21}. In other words, there is no physical ageing in fiber properties and Poisson's ratio. Later, Sullivan and Blais (1993), and Hastie and Morris (1993) obtained similar ageing effects on graphite fiber/thermoset and thermoplastic composites. Brinson and Gates (1995), and Gates and Feldman (1995, 1996) also characterized physical ageing well in transverse, in-plane shear, and quasi-isotropic laminates by using [90], $[\pm 45]_s$, and $[0/\pm 45/90]_s$ specimens.

Nevertheless, in the horizontal shift of momentary compliance, most of the previous works were unable to achieve an ideal superposition and sometimes a vertical shift is inevitable, as shown in Fig. 7.2. This implies that initial compliance S_o is potentially ageing-dependent. Another challenge is to characterize physical ageing in composites with various fiber orientations. This may result in a large number of creep tests in order to obtain momentary compliances in every fiber orientation.

Hu and Sun (2000b) suggested that elastic and creep compliances can be separated from the total compliance by choosing the data point whose slope is abruptly changed. It was found that creep deformation is very small in comparison with elastic deformation during the initial loading process. Figure 7.3 shows the experimental data for ageing effects on composite AS4/3501-6 with 45° fiber orientation under a test temperature of 148 °C (Hu, 2007). Apparently, elastic compliance decreases to approach a steady-state value as ageing time increases. Momentary creep compliance rate also decreases as ageing time increases. Therefore, physical ageing in elastic and creep compliances can be individually characterized. The experimental data shown in Fig. 7.3(a) can be fitted into an exponential function as

$$S_x^{\mathrm{e}} = S_x^{\mathrm{o}}/(1 - \alpha \mathrm{e}^{-\gamma t}) \qquad [7.5]$$

where S_x^{o} is the steady-state elastic compliance in the loading direction; α and γ are constants.

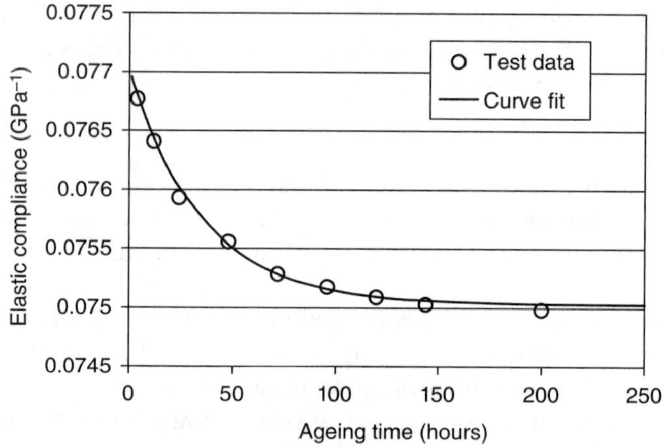

(a) Elastic compliance

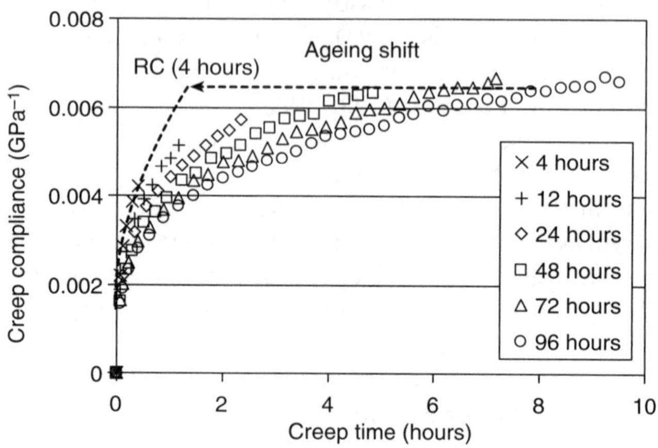

(b) Momentary creep compliances

7.3 Physical ageing in composite AS4/3501-6 (45° fiber orientation). RC, reference curve.

7.2.2 Power law model

Creep compliance, as shown in Fig. 7.3(b), can be obtained from the total creep compliance by subtracting elastic compliance. For each initial ageing time, the momentary creep compliance can be fitted into a power law of the form

$$S_x^c(t) = \left(\frac{t}{\tau}\right)^{\beta} \qquad [7.6]$$

where t is creep time, τ is relaxation time, and β is a shape factor. Basically, τ and β are constant for an individual momentary creep compliance curve.

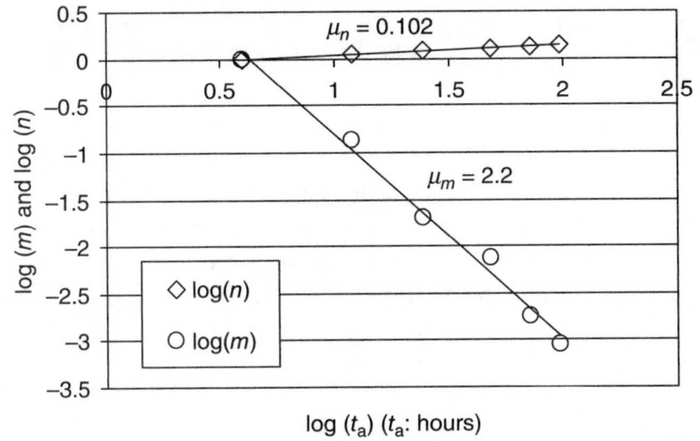

7.4 Shift factors of the momentary creep compliances in Fig. 7.3(b).

For instance, in Fig. 7.3(b), all momentary creep compliances can be shifted and superposed to a reference curve by introducing shift factors, m and n, for relaxation time and shape factor, respectively. The reference curve of momentary creep compliance is given by

$$S^{c}_{x,\mathrm{ref}}(t)=\left(\frac{t}{\tau_{\mathrm{ref}}}\right)^{\beta_{\mathrm{ref}}}=\left(\frac{t}{m\tau}\right)^{n\beta} \qquad [7.7]$$

where $n = \beta_{\mathrm{ref}}/\beta$ and $m = \tau_{\mathrm{ref}}/\tau$. For no loss of generality, the momentary creep compliance curve of 4-hour ageing is taken as a reference curve and the other five curves are shifted and superposed to the reference curve. Five shift factors of m and five shift factors of n are consequently obtained, and they are used to plot logarithmic shift factors versus logarithmic ageing times as shown in Fig. 7.4. In this double logarithmic plot, $\log(m)$ and $\log(n)$ appear to be linear functions of $\log(t_a)$. The slopes, μ_m and μ_n, are obtained from the linear fits, respectively, i.e.

$$\mu_m=-\frac{d\log(m)}{d\log(t_a)} \quad \text{and} \quad \mu_n=\frac{d\log(n)}{d\log(t_a)} \qquad [7.8]$$

where μ_m and μ_n are defined as shift rates. The shift factors m and n can be well fitted in a power law as

$$m=\left(\frac{t_{a,\mathrm{ref}}}{t_a}\right)^{\mu_m} \quad \text{and} \quad n=\left(\frac{t_a}{t_{a,\mathrm{ref}}}\right)^{\mu_n} \qquad [7.9]$$

where $t_{a,\mathrm{ref}}$ is reference ageing time corresponding to the reference curve. Consequently, the relaxation time and shape factor of a given ageing time t_a can be related to those of the reference ageing time $t_{a,\mathrm{ref}}$:

$$\tau(t_a) = \tau(t_{a,ref})\left(\frac{t_a}{t_{a,ref}}\right)^{\mu m} \quad \text{and} \quad \beta(t_a) = \beta(t_{a,ref})\left(\frac{t_{a,ref}}{t_a}\right)^{\mu n} \qquad [7.10]$$

Theoretically, for the reference curve we should have $m = n = 1$. However, from Fig. 7.4, it is apparent that the reference ageing time that gives the best fit is not the 4-hour ageing. Instead, an ageing time of 4.3 hours with $m = n = 1$ seems to serve better for curve fitting purposes. Using the reference curve, the relaxation time and shape factor for any given ageing time can be fabricated without another momentary creep test.

7.2.3 Effective creep compliance model

The complexity of stress state and creep strain in fiber-reinforced composites leads to the speculation of effective stress and effective creep strain. In general, constant load is applied to the off-axis coupon specimen of composites in creep test. Thus, deformation can be expressed in terms of creep compliance. If off-axis creep compliances can be transformed to effective creep compliances so that all cases can be simply expressed by a single master curve (Hu, 2006).

Parallel to the concept of potential function in plasticity, a potential function f for creep is proposed for a state of plane stress (Chung and Sun, 1993), i.e.

$$2f(\sigma_{ij}) = a_{11}\sigma_{11}^2 + a_{22}\sigma_{22}^2 + 2a_{12}\sigma_{11}\sigma_{22} + 2a_{66}\sigma_{12}^2 \qquad [7.11]$$

where σ_{ij} refer to the principal material directions and the coefficients a_{ij} describe the amount of anisotropy in the initial creep deformation. By using the flow rule, the creep strain increments can be expressed in terms of the creep potential

$$d\varepsilon_{ij}^c = \frac{\partial f}{\partial \sigma_{ij}} d\lambda \qquad [7.12]$$

where $d\lambda$ is a positive scale factor of proportionality. The increment of creep work is given by

$$dW^c = \sigma_{ij} d\varepsilon_{ij}^c = 2f d\lambda \qquad [7.13]$$

Effective stress is defined as

$$\bar{\sigma} = \sqrt{3f} \qquad [7.14]$$

Effective strain increment $d\bar{\varepsilon}^c$ is defined as

$$dW^c = \sigma_{ij} d\varepsilon_{ij}^c = \bar{\sigma} d\bar{\varepsilon}^c \qquad [7.15]$$

Substitution of equations [7.13] and [7.14] into equation [7.15] yields

$$d\lambda = \frac{3}{2}\left(\frac{d\bar{\varepsilon}}{d\bar{\sigma}}\right)\left(\frac{d\bar{\sigma}}{\bar{\sigma}}\right) \quad [7.16]$$

Recalling equations [7.11] and [7.12], we set $a_{22} = 1$ without loss of generality.

Creep strain increment can be expressed as

$$\begin{Bmatrix} d\varepsilon_{11}^{c} \\ d\varepsilon_{22}^{c} \\ d\gamma_{12}^{c} \end{Bmatrix} = \begin{bmatrix} a_{11} & a_{12} & 0 \\ a_{12} & 1 & 0 \\ 0 & 0 & 2a_{66} \end{bmatrix} \begin{Bmatrix} \sigma_{11} \\ \sigma_{22} \\ \sigma_{12} \end{Bmatrix} d\lambda \quad [7.17]$$

The coefficients a_{ij} can be determined by uniaxial creep tests of off-axis specimens.

Experiment shows that the creep deformation in the fiber direction is negligible (Hu, 2006). Thus a_{11} and a_{12} are determined to be zero. The creep potential in equation [7.11] is simplified to one parameter function, i.e.

$$2f(\sigma_{ij}) = \sigma_{22}^2 + 2a_{66}\sigma_{12}^2 \quad [7.18]$$

and equation [7.17] is simplified to

$$\begin{Bmatrix} d\varepsilon_{11}^{c} \\ d\varepsilon_{22}^{c} \\ d\gamma_{12}^{c} \end{Bmatrix} = \begin{Bmatrix} 0 \\ \sigma_{22} \\ 2a_{66}\sigma_{12} \end{Bmatrix} d\lambda \quad [7.19]$$

Coefficient a_{66} is the only unknown parameter to be determined. By using coordinate transformation, principal stresses are correlated to the uniaxial stress, i.e.

$$\begin{aligned} \sigma_{22} &= \sigma_x \sin^2(\theta) \\ \sigma_{12} &= -\sigma_x \sin(\theta)\cos(\theta) \end{aligned} \quad [7.20]$$

where θ is the fiber orientation related to the loading direction x. Substitution of equations [7.20] into [7.14] and [7.18] yields

$$\bar{\sigma} = h(\theta)\sigma_x \quad [7.21]$$

where $h(\theta) = \sqrt{3/2}\,(\sin^4\theta + 2a_{66}\sin^2\theta\cos^2\theta)^{1/2}$. Thus effective stress is related to the applied uniaxial stress σ_x. Substitution of equation [7.21] into [7.14] and [7.15] yields

$$d\bar{\varepsilon}^{c} = \frac{\sigma_{ij} d\varepsilon_{ij}^{c}}{\bar{\sigma}} = \frac{2}{3}\bar{\sigma}d\lambda = \frac{2}{3}h(\theta)\sigma_x d\lambda \quad [7.22]$$

From coordinate transformation law

$$d\varepsilon_x^{c} = d\varepsilon_{11}^{c}\cos^2\theta + d\varepsilon_{22}^{c}\sin^2\theta - \frac{1}{2}d\gamma_{12}^{c}\sin 2\theta \quad [7.23]$$

where $d\varepsilon_x^c$ is the creep strain increment, which can be measured in the loading direction. Substituting equations [7.19] and [7.20] into equation [7.23], we obtain

$$d\varepsilon_x^c = \frac{2}{3}h^2(\theta)\sigma_x d\lambda \quad [7.24]$$

Comparison of equations [7.22] and [7.24] leads to

$$d\bar{\varepsilon}^c = d\varepsilon_x^c / h(\theta) \quad [7.25]$$

For monotonic loading, equation [7.25] is integrable. Therefore, the effective creep strain is obtained, i.e.

$$\bar{\varepsilon}^c = \varepsilon_x^c / h(\theta) \quad [7.26]$$

The axial creep strain ε_x^c can be measured under uniaxial tension creep test. Combining equations [7.21] and [7.26], effective creep compliance can be expressed as

$$\bar{S}^c(t) = \frac{\bar{\varepsilon}^c(t)}{\bar{\sigma}} = \frac{\varepsilon_x^c(t)}{\sigma_x}\frac{1}{h^2(\theta)} = \frac{S_x^c(t)}{h^2(\theta)} \quad [7.27]$$

Therefore, in a given ageing time, momentary creep compliance $S_x^c(t)$ with respect to fiber orientation θ can be represented by effective creep compliance. Momentary creep compliance $S_x^c(t)$ can be obtained from uniaxial creep test and a_{66} is the only unknown parameter in $h(\theta)$ to be determined by experiment. In order to validate this model, off-axis coupon specimens with 90°, 45°, 30°, and 15° fiber orientations are tested isothermally at 104 °C after 5, 12, 24, 48, 72, and 96 hours initial ageing (Hu, 2006). Figure 7.5 shows the test data of creep compliances after 5 hours ageing. By choosing a_{66} = 0.95, creep compliances are transformed to effective creep compliances and

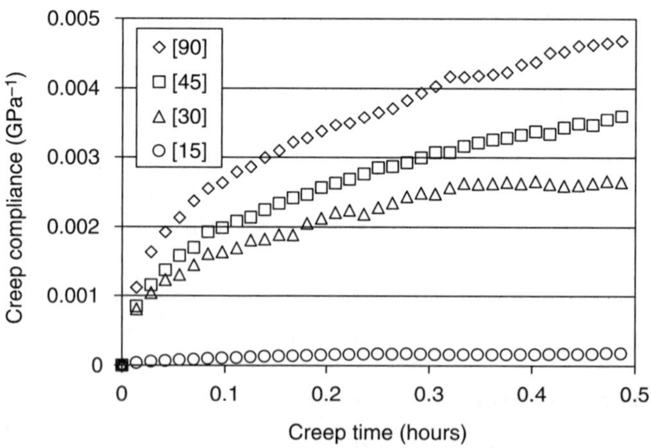

7.5 Momentary creep compliances of 5-hour ageing.

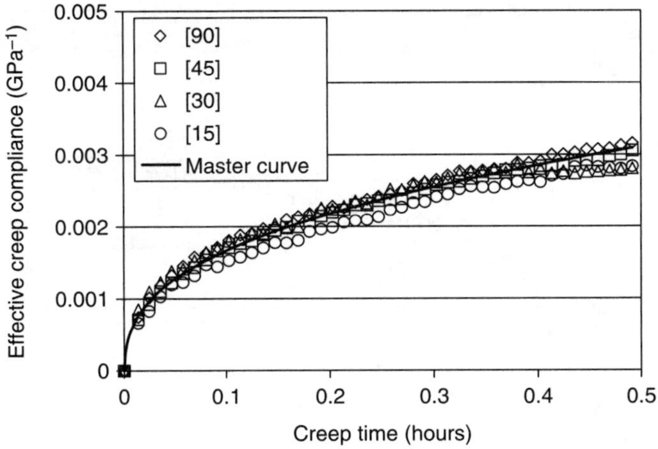

7.6 Effective creep compliances and master curve of 5-hour ageing.

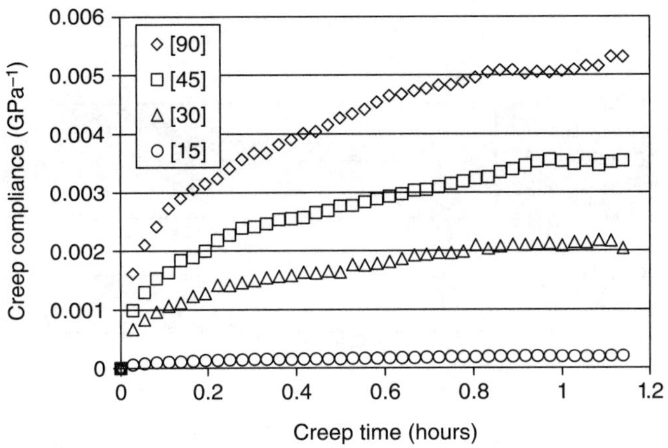

7.7 Momentary creep compliances of 12-hour ageing.

successfully collapsed into a single master curve as shown in Fig. 7.6. The test results obtained from different initial ageing times are shown in Figs 7.7 to 7.16. In all cases, effective creep compliances are successfully collapsed into a single master curve by using the same parameter $a_{66} = 0.95$. As a result, six momentary master curves for different initial ageing times, i.e. 5, 12, 24, 48, 72, and 96 hours, are obtained. Recalling the approach of ageing time shift, the individual momentary master curve can be fitted into a power law function. By introducing shift factors for relaxation time and shape factor, each momentary master curve can be shifted and superposed to a reference

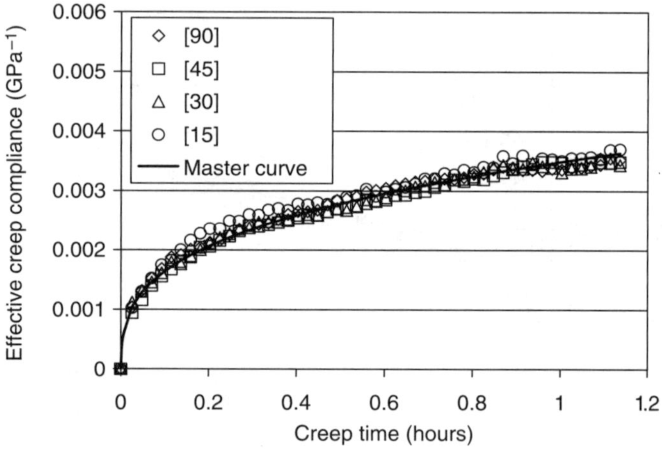

7.8 Effective creep compliances and master curve of 12-hour ageing.

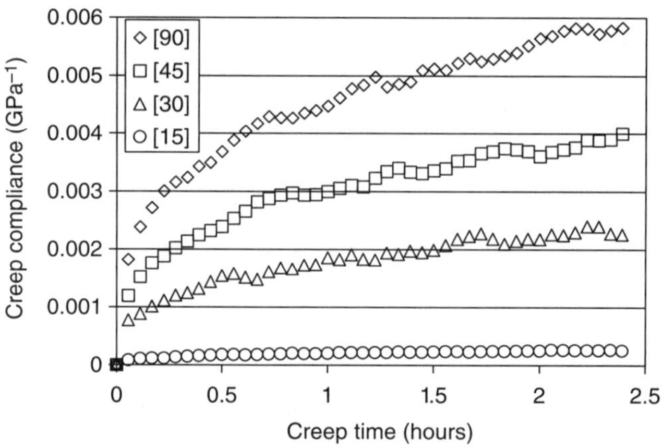

7.9 Momentary creep compliances of 24-hour ageing.

master curve. The shift rates can be obtained from the linear fits of logarithmic shift factor and logarithmic ageing time. Using the reference master curve, the relaxation time and shape factor of a master curve for any given ageing time can be fabricated without another momentary creep test.

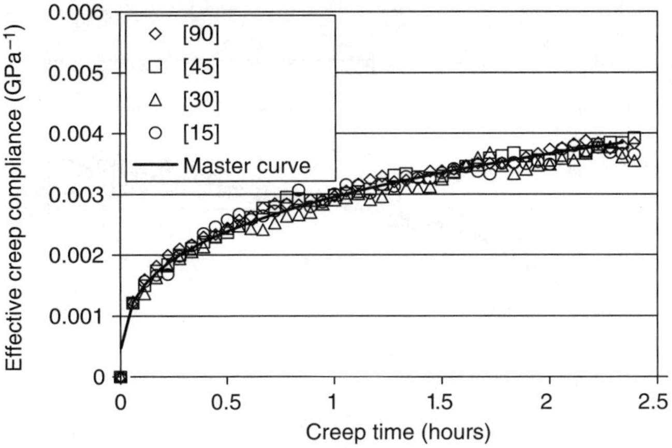

7.10 Effective creep compliances and master curve of 24-hour ageing.

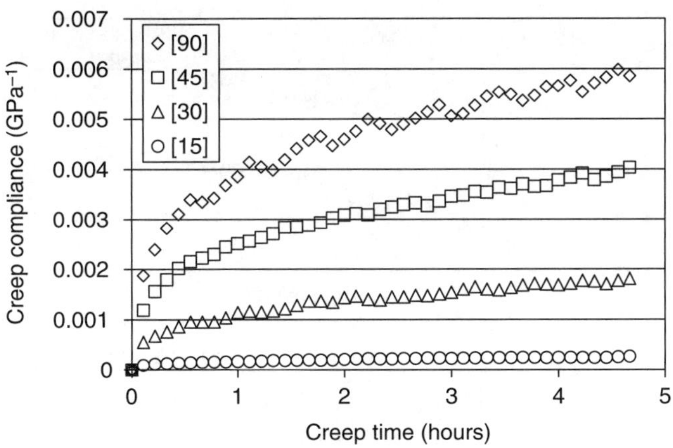

7.11 Momentary creep compliances of 48-hour ageing.

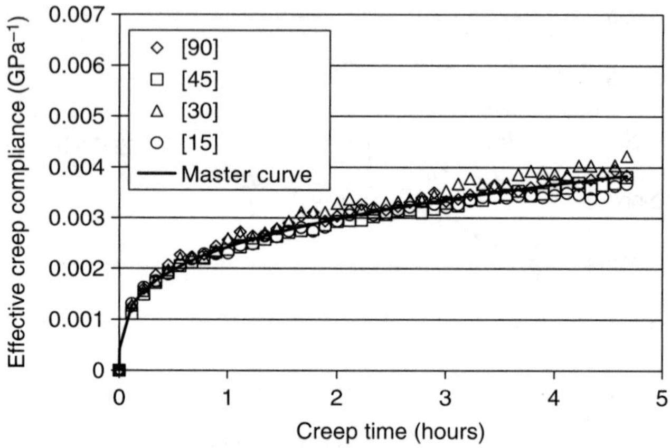

7.12 Effective creep compliances and master curve of 48-hour ageing.

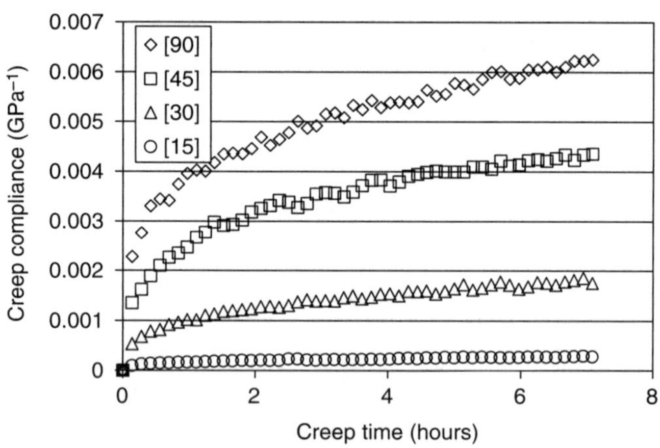

7.13 Momentary creep compliances of 72-hour ageing.

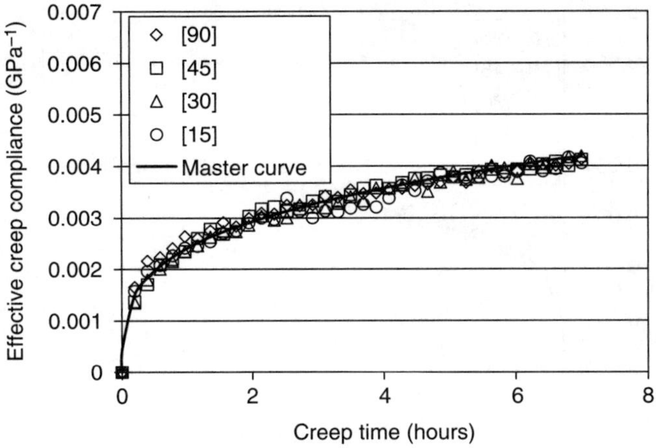

7.14 Effective creep compliances and master curve of 72-hour ageing.

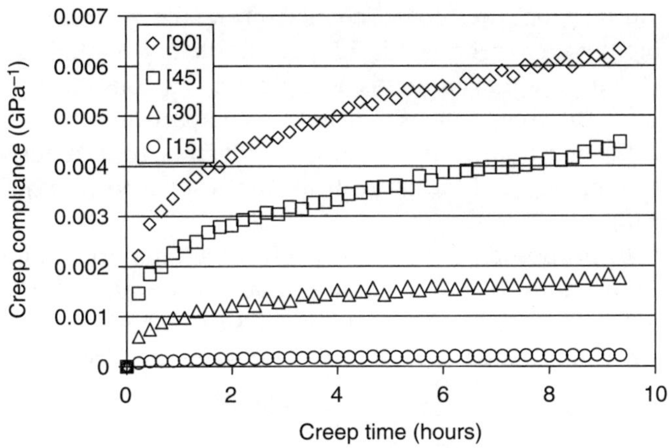

7.15 Momentary creep compliances of 96-hour ageing.

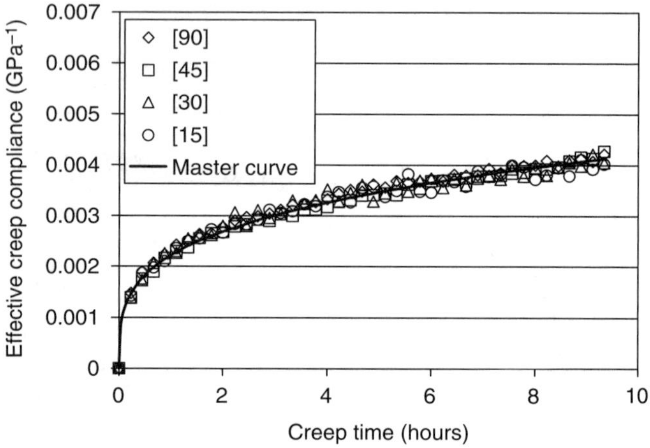

7.16 Effective creep compliances and master curve of 96-hour ageing.

7.3 Modeling physical ageing in long-term creep

7.3.1 Effective time model

It is noted that Boltzmann's superposition principle is based on the assumption that material properties do not change during the mechanical test. This is why the momentary creep test is adopted in the above studies since the ageing effect is not significant and is negligible during the short-time creep test. However, if the creep time is not short in comparison with the previous ageing time, the mechanical properties of polymers continuously change due to physical ageing and therefore Boltzman's superposition principle is no longer valid.

Struik (1978) proposed the concept of 'effective time' to account for the time-dependent nature of the relaxation time and developed a model for the prediction of long-term creep under physical ageing. In his model, a time-dependent shift factor, $a_o(t)$, for relaxation time is introduced to equation [7.1], i.e.

$$S = S_o e^{\left(\frac{t}{\tau_o/a_o(t)}\right)^{\beta_o}} \qquad [7.28]$$

and

$$a_o(t) = \left(\frac{t_a^o}{t_a^o + t}\right)^{\mu} \qquad [7.29]$$

where τ_o is relaxation time associated with previous ageing time, t_a^o, and β_o is shift rate. To compare with equation [7.2], shift factor is no longer a constant but is time-dependent. In the time interval between t and dt, all relaxation processes are therefore $1/a$ times slower than at $t = 0$. Thus the interval is equivalent to an effective time interval $d\lambda$ given by

$$d\lambda = \alpha_o(t)dt \qquad [7.30]$$

Based on this equation, the effective time λ can be defined as

$$\lambda = \int_0^t a_o(\xi)d\xi \qquad [7.31]$$

where ξ is an integration parameter on the time scale. Struik defined the effective time model in the form

$$S = S_o e^{(\lambda/\tau_o)^{\beta_o}} \qquad [7.32]$$

Equation [7.32] has been widely adopted to predict the long-term creep of polymer composites under physical ageing.

In the power law model, a master curve for a given initial ageing time t_a^o can be expressed as

$$\bar{S}^c(t) = \left(\frac{t}{\tau_o/m_o}\right)^{\beta o/no} \qquad [7.33]$$

where τ_o and β_o are associated with previous ageing time t_a^o. The shift factors m_o and n_o are ageing time-dependent, i.e.

$$m_o(t) = \left(\frac{t_a^o}{t_a^o + t}\right)^{\mu_m} \quad \text{and} \quad n_o(t) = \left(\frac{t_a^o + t}{t_a^o}\right)^{\mu_n} \qquad [7.34]$$

Equation [7.33] can be expressed in the form

$$\bar{S}^c(t) = \left(\frac{t}{\tau_o/G_o(t)}\right)^{\beta_o} \qquad [7.35]$$

where

$$G_o(t) = \left[m_o(t)\left(\frac{t}{\tau_o}\right)^{(1-n_o(t))}\right]^{(1/n_o(t))} \qquad [7.36]$$

is a time-dependent shift factor for relaxation time. An effective time interval is given by

$$d\lambda = G_o(t)dt \qquad [7.37]$$

Subsequently, the effective time λ is obtained from the integral

$$\lambda = \int_0^t G_o(\xi)d\xi \qquad [7.38]$$

The integration of equation [7.38] can be carried out by numerical methods, therefore the effective time model is obtained

$$\bar{S}^c(t) = \left(\frac{\lambda}{\tau_o}\right)^{\beta_o} \qquad [7.39]$$

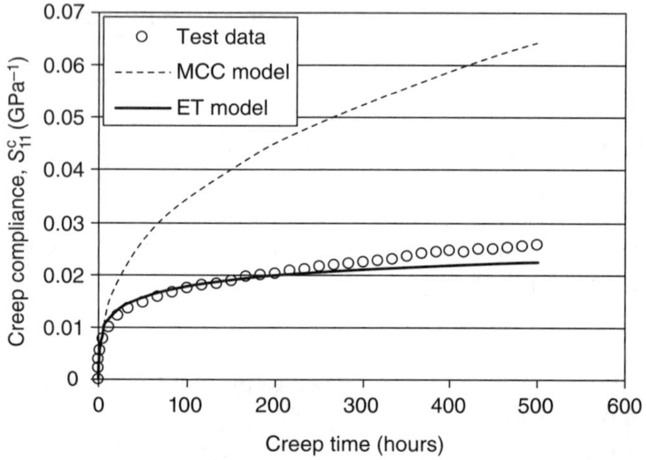

7.17 Long-term creep compliance of 45° off-axis specimen. MCC, momentary creep compliance; ET, effective time.

7.3.2 Long-term creep prediction

The predictions of long-term creep using the power law model have been validated by long-term creep tests (Hu, 2007). After rejuvenation and 5-hour initial ageing, the long-term creep test was conducted on 45° off-axis specimens. Figure 7.17 shows the test result and predictions of long-term creep. During the momentary creep, test data and the predictions of the effective time model and the momentary creep compliance model are almost overlapping. It is again demonstrated that the ageing effect is not significant and is negligible in momentary creep. When creep continues, the prediction of the momentary creep compliance model (dashed line) over-estimates the long-term creep. The effective time model (solid line) appears to provide relatively good predictions for long-term creep.

Another validation test was conducted on composite laminate $[90/(+45/-45)_3]_s$ (Hu, 2007). The coupon specimen was initially aged for 5 hours and then subjected to creep test at isothermal 107 °C. Creep time lasted for more than 500 hours. Figure 7.18 demonstrates that the effective time model provides a good prediction for long-term creep after isothermal ageing.

In the exponential model, physical ageing in composite laminate has been characterized by substituting total compliances for elastic compliances in classical laminate theory (Gates and Feldman, 1995, 1996). However, it is necessary to characterize physical ageing in the composite with various fiber orientations in the first place. Heavy experimental workloads are therefore inevitable. Hu (2007) suggested that creep tests can be conducted on the composite laminates directly to obtain the momentary creep compliances

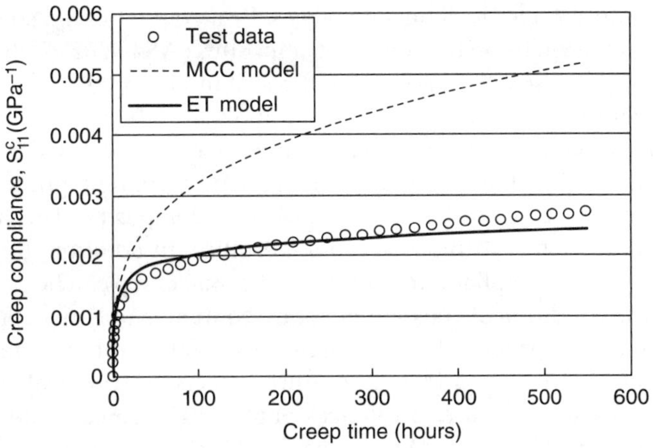

7.18 Long-term creep compliance of laminate $[90/(+45/-45)_3]_s$.

for different ageing times. The power law and effective time models are then employed to predict the long-term creep after physical ageing. This approach avoids the heavy test workload for the composite with various fiber orientations and provides relatively good predictions for long-term creep.

7.4 Temperature and moisture effects

In the course of this study, modeling of physical ageing has been investigated using the isothermal creep test. For cases of non-isothermal creep, an empirical time–temperature/ageing-time shift factor can be expressed (Sullivan, 1990) as

$$\log(a) = \log(\alpha t_a^{\mu}) + \log(a_T) \quad [7.40]$$

where α and μ are function of temperature; a_T is function of both temperature and ageing time. This time–temperature shift factor can be experimentally determined and applied to both exponential and power law models.

It has been recognized that the polymeric matrix utilized in reinforced fiber composites absorbs moisture from high-humidity environments. This moisture absorption causes significant changes in the physical and mechanical properties of polymeric composites (Shen and Springer, 1976; Vinson, 1977). In addition, Bueche (1962) found that mixing a polymer with a miscible liquid that contains more free volume than the pure polymer can lower its T_g. Later, Browning *et al.* (1977) also found a depression of T_g in moisture-absorbed 3501-5 neat resin and AS/3501-5 composites using the heat distortion temperature test. DeIasi and Whiteside (1978) and Springer (1982) obtained similar results in neat resins (3501-1, 3501-6, 5208, NMD2373, 3502, and 934) using thermomechanical analysis. Boil *et al.* (1985) also obtained the same conclusion by testing several composites (AS4/4502, AS4/3501-5A, AS4/3501-6, and AS4/2220-3) with the flexural modulus test.

Hu and Sun (2000a) indicated that the moisture effect is similar to the temperature effect on physical ageing in polymer composites. For instance, for different ageing times, increases in moisture content proportionally relax the elastic compliance of the polymer composite. This behavior resembles that seen with increasing temperature. In addition, the ageing-time shift of creep compliance curves in isothermal creep can be completely applied to the condition of constant moisture content in polymer composites. Hu and Sun (2003) proposed empirical moisture–temperature equivalences to interchange moisture and temperature effects on physical ageing in polymer composites. Such equivalences enable us to replace the complex tests for moisture effects on physical ageing by the tests at the equivalent temperature. Consequently, the modeling of physical ageing developed in isothermal conditions can be applied to model the physical ageing of polymer composites in constant moisture contents.

7.5 Conclusions

Physical ageing is a complex phenomenon that exists in all polymers and the polymer-based matrix of fiber-reinforced composites. Some models have been well developed and, as introduced in this chapter, have characterized ageing effects on stiffness of polymers and polymer composites; however, ageing effects on the strength and damage of polymer composites have not been well studied. Struik (1978) did note that that ageing effects on yielding, charpy impact strength, and environmental stress cracking are not as significant as those on creep compliances; however, whether these conclusions are valid for polymeric composites remains unanswered. In particular, at high stress levels, polymer composites exhibit significant rate-dependent nonlinear stress–strain behavior. Linear viscoelasticity is not able to describe the creep and relaxation behavior of the composite, and nonlinear constitutive models are needed.

7.6 References

BOIL, D. J., BASCOM, W. D. and MOTIEE, B. 'Moisture Absorption by Structural Epoxy-Matrix Carbon-fiber Composites,' *Composites Science and Technology*, **24**, 1985, 253–273.

BRINSON, L. C. and GATES, T. S. 'Effects of Physical Aging on Long Term Creep of Polymers and Polymer Matrix Composites,' *Journal of Solid Structures*, **32** (6/7), 1995, 827–846.

BROWNING, C. E., HUSMAN, G. E. and WHITNEY, J. M. 'Moisture Effects in Epoxy Matrix Composites,' in *Composite Materials: Testing and Design*, ASTM STP 617 (ed. J. G. Davis), ASTM, Philadelphia, PA, 1977, pp. 481–496.

BUECHE, F. *Physical Properties of Polymers*, Interscience Publishers, New York, 1962, pp. 112–118.

CHUNG, I. and SUN, C. T. 'Modelling Creep in Thermoplastic Composites,' *Journal of Composite Materials*, **27**, 1993, 1009–1029.

DEIASI, R. and WHITESIDE, J. B. 'Effect of Moisture on Epoxy Resins and Composite,' in *Advanced Composite Materials – Environmental Effects*, ASTM, STP 658 (ed. J. R. Vinson), ASTM, Philadelphia, PA, 1978, pp. 2–20.

GATES, T. S. and FELDMAN, M. 'Time-dependent Behaviour of a Graphite/Thermoplastic Composite and the Effect of Stress and Physical Aging,' *Journal of Composites Technology and Research*, **17** (1), 1995, 33–42.

GATES, T. S. and FELDMAN, M. 'Effects of physical aging at elevated temperatures on the viscoelastic creep of IM7/K3B,' in *Composites Materials: Testing and Design: Volume*, ASTM STP 1274 (ed. R. B. Deo), ASTM, Philadelphia, PA, 1996, pp. 7–36.

HASTIE, R. L. J. and MORRIS, D. H. 'The Effects of Physical Aging on Creep Response of a Thermoplastic Composites,' in *High Temperature and Environmental Effects on Polymeric Composites*, ASTM STP 1174 (eds C. E. Harris and T. S. Gates), ASTM, Philadelphia, PA, 1993, pp. 163–185.

HU, H. 'Master Curve of Creep in Off-axis Polymeric Composite Laminate,' *Journal of Mechanics*, **22** (3), 2006, 229–234.

HU, H. 'Physical Aging in Long Term Creep of Polymeric Composite Laminates,' *Journal of Mechanics*, **23** (3), 2007, 245–252.

HU, H. and SUN, C. T. 'Moisture Effect on Physical Aging in Polymeric Composites,' *Proceedings of The American Society for Composites*, 2000a, pp. 548–556.

HU, H. and SUN, C. T. 'The Characterization of Physical Aging in Polymeric Composites,' *Composites Science and Technology*, **60**, 2000b, 2693–2698.

HU, H. and SUN, C. T. 'The Equivalence of Moisture and Temperature in Physical Aging of Polymeric Composites,' *Journal of Composites Materials*, **37** (10), 2003, 913–928.

SHEN, C. H. and SPRINGER, G. S. 'Moisture Absorption and Desorption of Composite Materials,' *Journal of Composite Materials*, **10**, 1976, 2–20.

SPRINGER, G. S. 'Moisture Absorption in Fiber-Resin Composites,' in *Developments in Reinforced Plastics* (ed. R. Pritchard), Applied Science Publishers, New York, 1982, pp. 43–65.

STRUIK, L. C. E. *Physical Aging in Amorphous Polymers and Other Materials*, Elsevier Scientific Publishing Company, New York, 1978.

SULLIVAN, J. L. 'Creep and Physical Aging of Composites,' *Composites Science and Technology*, **39**, 1990, 207–232.

SULLIVAN, J. L., BLAIS, E. J. and HOUSTON, D. 'Physical Aging in the Creep Behaviour of Thermosetting and Thermoplastic Composites,' *Composites Science and Technology*, **47**, 1993, 389–403.

VINSON, J. R. (ed.) *Advanced Composite Materials – Environmental Effects*, ASTM STP 658, ASTM, Philadelphia, PA, 1978.

8
Ageing of silicon carbide composites

S. M. SKOLIANOS, Aristotle University of Thessaloniki, Greece

8.1 Introduction

In the metal matrix and ceramic matrix composite materials, silicon carbide (SiC), a hard ceramic, is the most commonly used reinforcement and exhibits the largest commercial volume by a significant margin over others such as Al_2O_3 and titanium carbide (TiC). This reinforcement is combined with a variety of commercial alloys, used as matrices, frequently age-hardenable, to produce the composite materials. While aluminium-based alloys are the most commonly used age-hardenable matrices, other alloys such as copper- and magnesium-based alloys are also of interest.

These kind of composite materials, and especially SiC particle-reinforced aluminum matrix composites, have a very attractive combination of properties and cost. As a result they have been widely accepted in various industries such as ground transportation (road and rail), aerospace, thermal management, recreational and military. In order to achieve the desired properties for these applications, the matrix of the composites is usually age treated after the production of the composite.

The presence of SiC reinforcement can have a great effect on the kinetics of precipitation reactions in the matrix. As a consequence, the properties of the composites can be significantly altered. Thus knowledge of the age-hardening behaviour of SiC-reinforced composites is of great interest both from scientific and technological points of view. This chapter aims to present an overview of the effect of SiC on the ageing mechanisms and kinetics of precipitation in SiC composites. The effect of microstructural features of the composites, of the volume fraction and size of SiC reinforcement on the ageing process is reviewed and the impact of the ageing treatment on the mechanical and corrosion properties is presented.

8.2 Silicon carbide composites

SiC is presently the most widely used reinforcement for metal matrix and ceramic matrix composites. SiC-reinforced composites are widely used in a variety of applications, much more so than any other composite material. This is due to the fact that SiC possesses such properties that, when combined with the corresponding properties of the matrices, gives rise to composite materials with very good mechanical properties and high performance.

SiC is a ceramic material with a diamond-like structure; it exhibits properties such as low density, high stiffness and hardness, good thermal conductivity and stability. These properties combined with its availability and relativity low production cost, make it a very attractive reinforcement material. Table 8.1 shows the different properties of SiC particles (SiC_p).[1] SiC can be produced in different forms such as particulates, fibers and whiskers. Particulate SiC can be readily produced in large quantities by milling while the production of fibers and whiskers is more complicated. Fibers are made by chemical vapor deposition (CVD) on substrates such as tungsten and carbon. Whiskers, on the other hand, are small single crystals and are usually encountered in two forms: a-SiC (wurtzite hexagonal) and b-SiC (zinc-blend cubic structure). They can be produced using several techniques; nowadays, however, the most widely used are the carbothermal reduction of silica and the vapour–liquid–solid (VLS) process. Among the different forms of SiC, particulates are extensively used because of their low cost and the relative ease with which particulate-reinforced composites can be produced. Whiskers, on the other hand, have little commercial application in general as a consequence of their potential health hazard and handling difficulties. However, SiC whiskers are presently the most widely used whisker reinforcements for ceramic matrix composites as well as metal matrix composites, particularly aluminum.

Since the mechanical, physical, electrical and thermal properties of the SiC reinforcement differ to a great extent from the corresponding properties of the metallic matrix, the properties of the metallic matrix composite (MMC) can be varied over a very broad range depending on several factors, including the size, morphology, distribution, and volume fraction of the SiC. In general, the addition of SiC to a metallic matrix enhances its wear

Table 8.1 Properties of SiC particulates

Physical	
Density (solid)	3.21 g/cm^3
Mechanical	
Knoop hardness	2480
Elastic modulus	430 GPa
Compressive strength	2800 MPa
Thermal	
Thermal conductivity	132 W/m K
Coefficient of thermal expansion	3.4×10^{-6}/K

behaviour, strength and stiffness, and decreases the coefficient of thermal expansion.

MMCs reinforced with SiC are usually manufactured by casting techniques (stir casting, melt infiltration) and powder metallurgy routes. Liquid-state processing has some advantages over powder techniques which makes it the preferred process in most commercial applications. Liquid metal can be handled more easily than powders, near-net-shape components can be produced and it is generally less expensive than powders.

8.3 Ageing kinetics

The presence of SiC reinforcement appears to accelerate the precipitation processes during artificial ageing in many MMC systems, as concluded by several investigations.[2–5] Such behaviour has been attributed to the increased dislocation density that appears near the interface between the matrix and the reinforcement during quench treatment, as shown in Fig. 8.1.[2–4,6–10] These zones of high-density dislocations are introduced by the differential thermal contraction between the two components of the material: the ceramic particles and the surrounding matrix. This differential thermal contraction is produced during solution heat treatment as a result of the notable, approximately one-tenth for aluminum alloys, difference in the coefficient of thermal expansion (CTE) between SiC reinforcement and the metallic matrix.[6] This thermal mismatch generates thermal stresses in the matrix, mainly in the vicinity of SiC reinforcements, during cooling from the solution heat-treated temperature. When the magnitude of these stresses locally

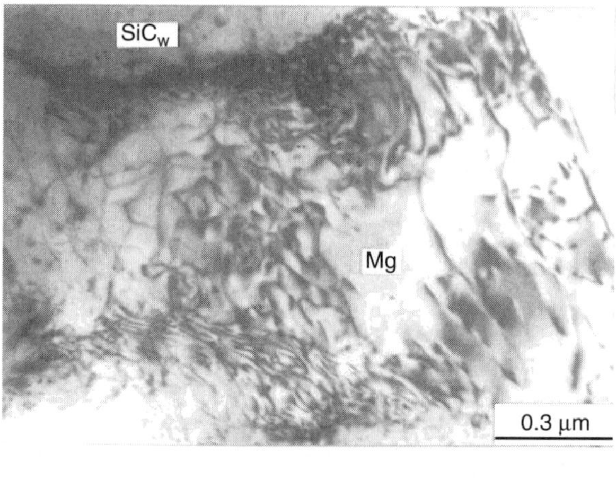

8.1 High-density dislocations in the matrix in solution-treated and quenched SiC_w/AZ91 magnesium matrix composite (transmission electron microscopy (TEM) micrograph).[3]

exceeds a certain limit, plastic deformation occurs leading to plastic relaxation and generation of dislocations. Moreover, electron microscopy studies indicate that high-density dislocations constitute preferential sites for the heterogeneous nucleation of new phases.[3,11]

Christman and Suresh[10] reported that in powder metallurgy aluminum alloy 2124 reinforced with 15 wt% SiC whisker (SiC_w) composites, the presence of increased dislocation density facilitates the nucleation of S' (Al_2CuMg). As a result the required time to achieve peak hardness is reduced. In addition to the increased dislocation density, it has also been reported[6–10] that the precipitation hardening behaviour, in aluminum alloy matrix composites, is affected by another interrelated factor: the reduction of the retained vacancy sites in composites (after solution heat treatment) *vis-à-vis* that of the unreinforced matrix. This is related to the increased dislocation density of the composites. Papazian,[9] using differential scanning calorimetry (DSC) to study the age-hardening precipitation reaction in 2124 and 2219 aluminum composites reinforced with SiC_w (0–20 vol.%) observed that the addition of SiC reinforcements leads to a decrease of Guinier–Preston (GP) zone formation, in all of the aluminum alloys studied. This reduction of GP zone formation was attributed to a reduction in the retained vacancies after solution heat treatment in the composites. It was also stated that dislocations acted as vacancy annihilation sites. The precipitation sequence remained the same but alteration of the ageing kinetics was observed. As described in Section 8.5 accelerated ageing occurs when SiC volume fraction exceeds a critical point.

Kiourtsidis *et al.*[12] investigated the ageing response of aluminum alloy 2024 reinforced with SiC_p (0, 8, 14, 19 and 24 vol.%) in three different conditions – under-ageing, peak-ageing and over-ageing – using X-ray diffraction and lattice spacing evaluations. While no alteration in precipitation sequence was observed, the ageing kinetics changed as a function of SiC_p content. It has also been found that the lattice spacing of the composites' matrices – after solution heat treatment – decreased as SiC_p content increased. This reduced lattice spacing creates a diffusion obstacle that has to be overcome in order for nucleation of different phases to occur during ageing and appeared to be the main factor controlling ageing kinetics in the over-aged (OA) region of all the composites.

Although the addition of SiC reinforcements accelerates the precipitation hardening, the ageing efficiency is lower in the composites compared with that in the unreinforced alloys.[3,6] This can be explained by the preferential interfacial precipitation which causes a depletion of the alloying elements in the matrix. As a result the amount of the continuous precipitates decreases, the spacing between the inter-platelets of the precipitates increases and the distribution of the continuous precipitates in the matrix becomes uneven.

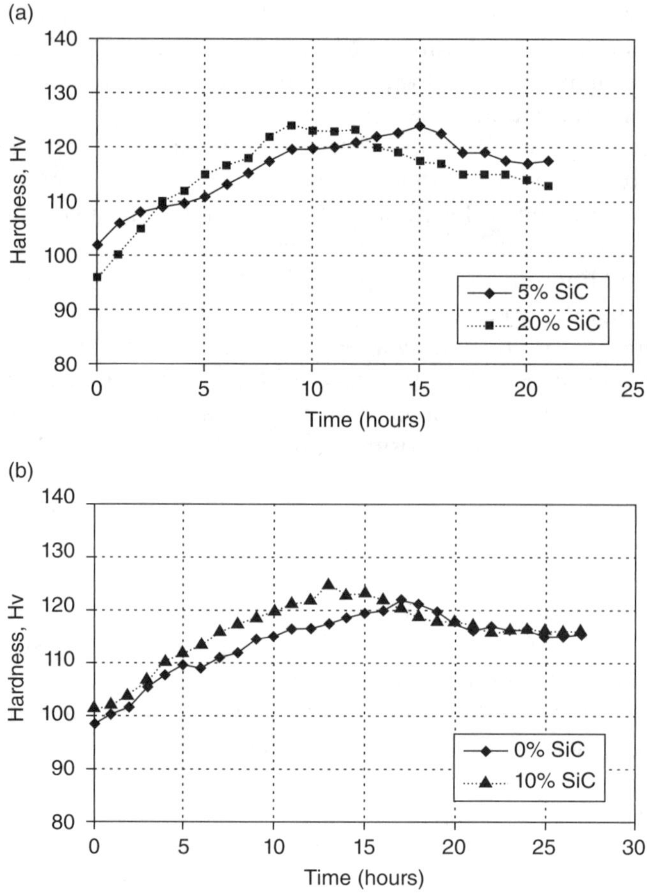

8.2 Hardness versus time of artificial ageing AA2024 (at 170 °C), solution treated at 500 °C for 2 h: (a) 5 and 20 wt% SiC; (b) 0 and 10 wt% SiC.[14]

Cold work can affect the ageing behaviour of SiC-reinforced composites in a similar way as SiC reinforcement.[13–15] This is due to the introduction of a large amount of dislocations into the matrix by severe cold deformation. As a result a higher precipitation rate is achieved which accelerates the ageing of the material (Figs 8.2 and 8.3). Dutta and Bourell[16] have also reported that cold work accelerates precipitation and reduces the incubation time of the precipitation in agreement with Bhargava *et al.*[17] In another investigation El-Baradie *et al.*[18] agree that cold work accelerates the ageing kinetics and alters the ageing precipitation sequence significantly, but they did not find an increase in the peak hardness values.

The ageing kinetics of SiC-reinforced composites varies with the ageing temperature.[15,19,20] As the ageing temperature increases, the time to peak

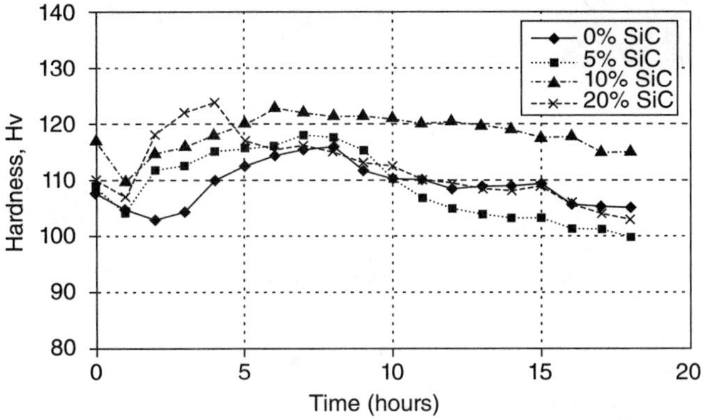

8.3 Hardness versus time of artificial ageing AA2024 (at 170 °C) after 10% reduction, solution treated at 500 °C for 2 h.[14]

age decreases and the difference in the peak ageing time between the unreinforced alloy and the composite, being minimal at low temperatures, increases (Fig. 8.4(a)). This can be attributed to the fact that, at low ageing temperatures, GP zones are nucleated mainly homogeneously with the support of the vacancies introduced by quenching from the solutionizing temperature. At higher temperatures, heterogeneous nucleation of the precipitates predominates and the formation of GP zones decreases. The appropriate path for heterogeneous nucleation of the precipitates is provided by the thermal stress-induced dislocations generated by the difference in the thermal expansion coefficients between the matrix and the SiC reinforcements. This means that the effect of excess dislocation density on the ageing kinetics seems to be more prominent at high than at low temperatures.

Similar results were also obtained for cold-worked composites.[15,19] Again the time to peak age decreases and the difference between the unreinforced alloy and the composite increases as ageing temperature increases (Fig. 8.4(b)). In addition, the cold-worked composites exhibit a significant lower peak ageing time than the unstrained ones (Figs 8.4(a) and (b)), due to the enhanced nucleation of the second phase at dislocations produced during prestraining.

8.4 Microstructural change

The different microstructure characteristics of the SiC-reinforced composites affect their ageing behavior to a great extent. Differences in the matrix alloy composition, matrix grain size, reactions between the matrix and the SiC reinforcement and inclusions are some of the features that can alter the ageing kinetics. These characteristics and the microstructure in general

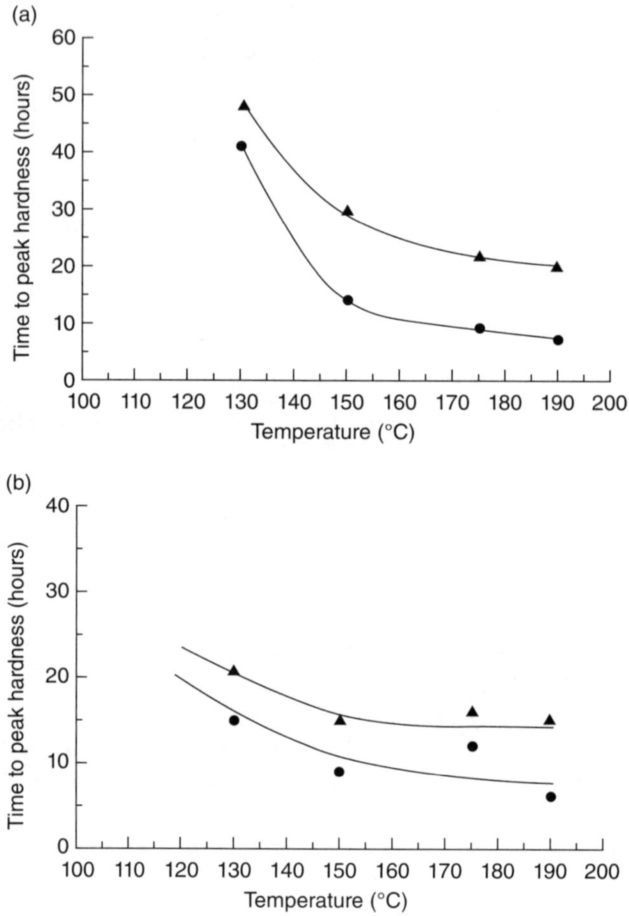

8.4 Time to peak age for the unreinforced AA8090 (▲) and the reinforced AA8090/SiC_p (●). (a) Samples without straining prior to ageing; (b) samples subjected to 2% strain prior to ageing.[15]

are strongly dependent on the production methods used to produce the composites. Processing temperature, contact time between the SiC reinforcements and the matrix at high temperatures, and rapid solidification are some examples of the manufacturing parameters that can influence the final microstructure of the SiC-reinforced composites.

8.4.1 Effect of grain size and porosity

One of the microstructure characteristics that affects the ageing kinetics is the matrix grain size. In general MMCs reinforced with SiC exhibit smaller matrix grain size than the monolithic alloys. In some liquid processing techniques, such as rheocasting, this can be attributed to the presence of broken dendrite arms as a result of stirring the melt in the semi-solid region.[21,22] These fragments can act as nucleation sites. In addition, the more uniform distribution of the SiC_p, observed in the composites produced by the rheocasting process, delays the grain growth.[23] At the same time SiC particulates can act as nucleation sites during the solidification of the melt.[24] Some other production routes, such as spray processes, are related to fast cooling rates that can give fine matrix grain size.[23,25–27] The finer matrix grain size results in an increase in grain boundary area which can affect the nucleation of the strengthening phases by reducing the activation barrier for the heterogeneous nucleation,[21,28] leading to accelerated ageing kinetics compared with other production techniques.

Similar results were obtained by minimizing the SiC particulates-associated porosity, as this can be achieved by the rheocasting process.[21,29] This kind of porosity is related to the physical properties of the matrix and the mode of solidification of the material. In conventional casting the high viscosity of the composite slurry cannot either wet the sharp corners of the reinforcement or infiltrate the interstitial voids created in clusters of SiC_p, leading to porosity. The misfit strains, introduced by the difference in coefficient of thermal expansion between SiC particulates and matrix, are reduced when porosity is present at the interface between the reinforcements and the matrix. This can be attributed to the ability of SiC particulates to adjust themselves in the pores during quenching. Reduction of misfit strains will reduce the average dislocation density and hence the heterogeneous nucleation sites around the particulates[30–32] thus decreasing the precipitation rate.

8.4.2 Effect of matrix composition

The ageing process of the SiC-reinforced composites can be altered by altering the matrix composition. Small differences in matrix composition can lead to the formation of different phases, modifying the precipitation sequence. In addition, when the composition of the matrix changes the physical properties of the metallic matrix can also be altered, resulting in a change in the ageing behaviour. In the case of Al–Cu/SiC composites produced by the rheocasting technique, a decrease in weight percent of copper in the Al-based matrix (from 4.5% to 1%), accelerates the ageing kinetics, for constant SiC particulates.[29,33] This was attributed to the change in physical properties, of the Al–Cu matrix. Decreasing the amount of Cu reduces the density and viscosity of metallic matrix, and increases the thermal conductivity of the matrix[34] as well as the difference in coefficient of thermal expansion between the Al–Cu matrix and the SiC reinforcement. These variations of physical properties lead to the entrapment of finer sized SiC particulates, to a more uniform distribution of SiC particulates, to a reduction in SiC cluster formation as well as in grain size, and to an increase in dislocation density. As a result the number of heterogeneous nucleation sites in the metallic matrix increases, thus accelerating the ageing kinetics.

The matrix composition can also change during the fabrication. As has been reported by numerous investigators[12,35–37] SiC can react with molten aluminum during processing of the composites forming Al_4C_3 and releasing Si into the melt:

$$4Al + 3SiC \rightarrow Al_4C_3 + 3Si$$

As a result the matrix Si content increases. This change of the matrix composition can result in an alteration of the precipitation sequence. As has been reported[12], the increase of matrix Si content in Al–Cu–Mg–Mn alloys decreases the Mg/Si ratio leading to the formation of $CuAl_2$ and Mg_2Si precipitate particles, and not Al_2CuMg as was anticipated.

Since the extent of the change in the matrix composition is a function of the contact time between the reinforcement SiC and the matrix at high temperatures, the appropriate choice of processing parameters in the liquid production methods appears to be of great significance. The formation of Al_4C_3 can also be prevented by adding Mg into the Al matrix. Mg can form $MgAl_2O_4$-spinel or MgO at the interface of oxidized SiC avoiding the reaction of the SiC reinforcement with the matrix.[38,39] At the same time Mg enhances the wetting of SiC, the interfacial bonding between the matrix and the reinforcement and the strength of the matrix.[40–42]

8.5 Effect of volume fraction and size of silicon carbide reinforcement

Several investigators[5–7,13,26,43–45] have reported that the ageing process of SiC composites can be affected by the size and content of the reinforcement. In general, it is accepted that the ageing kinetics is increased by decreasing the particle size and increasing the volume fraction of reinforcement which may be explained by a change in the type of precipitation.[26,13,46–47]

However, since the ageing behaviour depends on dislocation densities introduced by the differential thermal contraction between the matrix and the SiC_p, it seems that the combination of particle size and volume fraction of the reinforcement is the main factor, and not the size and content of the SiC independently.

Janowski and Pletka[7] examined the ageing response of a SiC particulate-reinforced powder metallurgy 201 aluminum alloy composite as a function of particle size and volume fraction. They reported that there is a critical ceramic volume fraction, which depends on the particle size and the alloy composition, below which the ageing behaviour is unaltered. This was attributed to the fact that a reinforcement volume fraction smaller than the critical one generates an insufficient amount of thermal expansion misfit dislocations to affect the age-hardening behaviour of the alloy 201. When SiC volume fraction exceeds this critical point, accelerated ageing occurs. Small particle sizes and high volume fractions were more effective in varying the ageing behaviour of composites. Additional SiC_p leads to loss of initial hardening due to suppression of GP zone formation, as no excess vacancies were present owing to thermal misfit dislocations.

Similar results were reported by Srivastava *et al.*[26] and Tekmen and Cocen[13,48] for the ageing behaviour of AA2014/SiC_p composites produced by spray co-deposition and Al–Si–Mg/SiC composites produced by compocasting techniques, respectively. In the first investigation, the ageing response of the composite materials was sensitive to the size of the reinforcement particulates and was enhanced with the reduction of the size. As far as the volume fraction of the reinforcement is concerned, both investigations draw the conclusion that acceleration in ageing kinetics occurs only if the SiC volume fraction level is above a certain amount, depending on the particle size.

Sannino and Rack[6] studied the effect of particle size on both the ageing sequence and the precipitation hardening behaviour of AA2009 composites reinforced with 20 vol.% SiC particulates. They reported that the normal precipitation sequence of the AA2009:

$$\text{Supersaturated } \alpha \text{ (fcc)} \rightarrow \text{GPI zones} \rightarrow \text{GPII} \rightarrow S' \rightarrow S$$

is affected by SiC particulate size at constant volume fraction. As shown in Fig. 8.5, the ageing sequence involves either two or three stages depending upon SiC particulate size.

During the first stage both the rate of hardness increase and the extent of the α (fcc) $\rightarrow$ GPI transformation decrease as particulate size decreases. This was attributed to the decrease in vacancy supersaturation owing to the interfacial matrix/SiC_p area increase, as the SiC particulate size decreases. These interface areas provide an additional number of vacancy sinks after solution heat treatment.

At longer ageing times (second stage), the transition of GPI zones to GPII zones occurs, for larger reinforcement size composites. For small reinforcement size composites, the first stage is missing and the development of GPII zones occurs directly. In general, decreased particle size is associated with gradual substitution of GPI by GPII zones as the main strengthening factor. The third stage of age hardening is common in all SiC-reinforced composites and consists of heterogeneous nucleation of S'/S and GPII $\rightarrow$ S'/S transformation.[7]

The particle size does not seem to affect the hardening rate during peak hardness and over-ageing but it does affect the time required for peak-

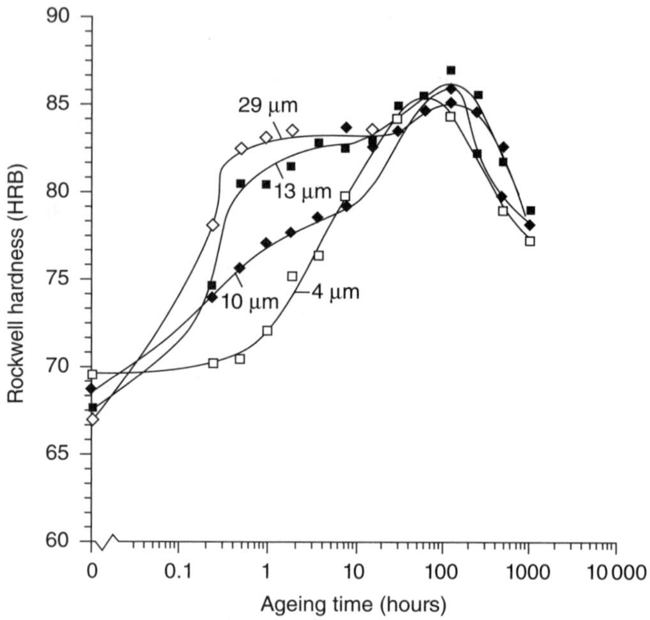

8.5 Ageing response of AA2009/20 vol.% SiC_p solution treated at 495 °C for 1 h, water quenched and aged at 150 °C.[6]

ageing (Fig. 8.5). The time required to achieve peak hardness in the composite with the small SiC reinforcement size (4 μm), is half of the corresponding time in the other composites.

Gupta *et al.*[43] investigated the influence of SiC particulates size (34.4 μm and 8 μm) on the ageing response of Al–4.5% Cu metallic matrix and correlated this to the microstructural characteristics. The results of ageing studies revealed that the composite with the small SiC reinforcement size exhibited an accelerated ageing kinetics compared with the other composite and with the unreinforced alloy, in agreement with the work of Sannino and Rack.[6] This was attributed to several interrelated factors associated with the decrease of the particulate size, such as the decrease of grain size and of the particulate porosity, a more uniform distribution of SiC reinforcement and an increased number of heterogeneous nucleation sites, as described in Section 8.4.

In the case of SiC_w-reinforced 6061 aluminum alloy composites fabricated via squeeze casting and subsequent hot extrusion, it was reported[5] that a correlation exists between the length of the SiC reinforcements and the ageing rate. The longer the average length of the whiskers, the more marked was the acceleration of the precipitation hardening processes, the former depending on mechanical working undergone during the production process.

8.6 Changes in properties

Although the properties of SiC composites are influenced by several factors – such as reaction products formed at the reinforcement/matrix interface; reinforcement clustering; size and volume fraction of SiC reinforcement; residual stresses induced in the matrix due to differences in the coefficient of thermal expansion between the reinforcement and the matrix – only the effect of the ageing treatment on the properties is considered in this chapter.

8.6.1 Tensile properties

The effect of ageing on tensile properties in SiC-reinforced composites should be approached with caution since both the addition of SiC and the precipitation heat treatment strengthen the composites. In other words, in some cases ageing may not be the main factor in strengthening. Instead, the presence of SiC reinforcements and more specifically their size, shape and distribution are the fundamental issues for structural applications. As a consequence, some results on tensile properties appear to be contradictory.

It has been reported that ageing of SiC_p-reinforced composites produced by the stir casting technique increases the yield and ultimate tensile strength, and the elastic modulus of the material, whereas the ductility decreases.[49,50] The effect of the reinforcement size on the tensile properties was studied by Doel and Bowen.[45] They found that reinforcement with 5 and 13 μm SiC_p increases the yield stress and tensile strength of AA7075 in all ageing conditions, while reinforcement with 60 μm particles reduced the yield stress of AA7075 in the under-aged and peak-aged conditions, and increased it only in the over-aged condition.

The effect of ageing treatment on the mechanical properties of the hybrid composites reinforced with different contents of both SiC_w and SiC_p was investigated by Ko and Yoo.[51] The results led to the suggestion that the addition of both SiC_w and SiC_p into AA2124 matrix alloy increased the strength and the work-hardening rate. In addition, the reinforcement with SiC_w was more effective in increasing the ultimate tensile strength and work-hardening rate of the hybrid composites than the reinforcement with just SiC_p. The SiC_w:SiC_p = 1:1 composite exhibited a lower work-hardening rate in the over-aged condition than in the under-aged owing to the precipitation of coarse Al_2CuMg phase near the matrix/reinforcement interfaces which lowered the load transfer from the matrix to the reinforcement. However, the single composite reinforced with only SiC_w showed a higher ultimate tensile strength and work-hardening rate than the hybrid composites and the single composite reinforced with only SiC_p.

8.6.2 Wear

The effect of thermal ageing on the wear resistance of SiC-reinforced composites has been studied by several investigators.[52,53] The results showed that the abrasive wear resistance may be controllably altered by thermal ageing. As the ageing temperature rises from the under-aged to peak-aged condition the wear resistance of the composite improves (Fig. 8.6). This can be attributed to the different type of precipitates present (coherent and semi-coherent in the under-aged condition and incoherent in the peak-aged) and their resistance to plastic deformation. In the over-aged condition the wear resistance of the composites is reduced due to the coarsening of matrix precipitates. However, it has also been reported that in some age-hardenable composites containing SiC particulates,[54,55] the abrasive wear resistance was higher in the over-aged than in the peak-aged condition. Two possible mechanisms were proposed. One is the improvement of the interfacial bonding between the SiC and Al alloy matrix due to over-aged precipitates formed on the SiC_p/Al alloy interfaces. The other is the dimensional change of the matrix during ageing, which induced a compression force on the SiC_p reinforcements which made them more difficult to remove by abrasion. An important parameter that affects the wear behaviour of aged SiC composites is the volume fraction of the reinforcement phase. It has been reported that the influence of peak-ageing on the wear decreases as SiC_p content increases and that in high-SiC_p volume fraction composites over-ageing no longer plays an important role.[56]

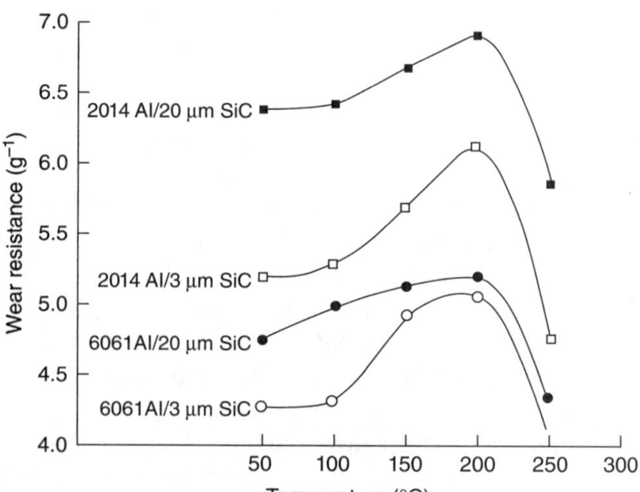

8.6 Effect of ageing temperature on the abrasive wear resistance of composites.[52]

8.6.3 Corrosion

Several corrosion studies have indicated that age hardening may alter the corrosion behaviour of MMCs reinforced with SiC fibers or particles. This alteration mainly concerns the dominant corrosion mechanism and the point at which it takes place. It seems that there is general agreement that the main corrosion mechanism of the artificially aged Al matrix composites reinforced with SiC is pitting, as shown in Fig. 8.7. According to these studies, the cathodic precipitation sites are related to the protective passivation film breakdown points and are usually located in the matrix or in the matrix/SiC interface as these are preferential positions for precipitation.

Kiourtsidis and Skolianos[57] reported that pitting of the AA2024/SiC_p composites occurs immediately after the immersion in solution in every ageing state (under-ageing, peak-ageing, over-ageing) and have related it to the effect of SiC_p on the ageing kinetics. The formation of different precipitate phases (GPI, GPII, S', θ', β') during ageing alters the corrosion characteristics of the matrix alloy, due to their corrosion behaviour (anodic or cathodic). Pitting corrosion potentials (E_{pit}) of SiC-reinforced Al alloy matrix composites and the unreinforced alloys were found to be very similar and were not affected significantly by the SiC_p content in a given ageing condition.[57,58] Similar results were presented by Trzaskoma *et al.*[59] The pitting corrosion potentials (E_{pit}) of the T6 Al alloy 5056, 6061/SiC_w composites were identical to those of the monolithic alloy while the E_{pit} of the 2024/SiC_w composite was found to be 100 mV more electronegative than the plain alloy AA2024. In addition, the pits had different morphologies, being more numerous and smaller in size and depth in composites. This

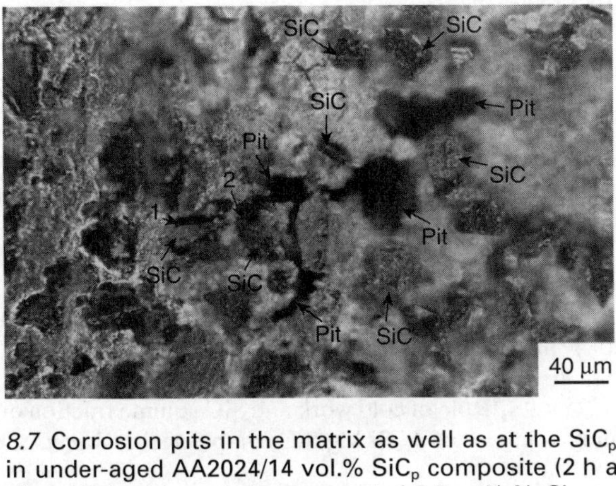

8.7 Corrosion pits in the matrix as well as at the SiC_p/matrix interface in under-aged AA2024/14 vol.% SiC_p composite (2 h aged at 177 °C) after 40 days immersion in aerated 3.5 wt% NaCl aqueous solution.[57]

could lead to the conclusion that SiC_w acts as a barrier to the development of the pits. It has also been observed[58,60,61] that pits occurred preferentially both at the matrix areas adjacent to the matrix/SiC interface and at the matrix/precipitate interfaces.

McIntyre *et al.*[62] observed that pitting corrosion shifted to the electronegative direction for AA2124/SiC_w composites with ageing time. This was attributed to the formation of $CuAl_2$ and CuAlMg precipitation phases which led to the successive reduction of Cu concentration in solid solution of the Al alloy matrix. The initiation of corrosion pits in AA2124 and AA6061 matrix composites reinforced with SiC, was also related to some precipitation phases serving as local cathodes.[63,64]

8.7 References

1 D. B. MIRACLE and S. DONALDSON, *ASM Handbook, Volume 21: Composites*, ASM International, Materials Park, OH (2001), p. 51.

2 E. HUNT, P. D PITCHER, and P. J GREGSON, 'Precipitation reactions in 8090 SiC particulate reinforced MMC', *Scripta Metall. Mater.* **24** (1990) 937–941.

3 M. Y. ZHENG, K. WU, S. KAMADO, and Y. KOJIMA, 'Aging behavior of squeeze cast SiCw/AZ91 magnesium matrix composite', *Mater. Sci. Eng. A* **348** (2003) 67–75.

4 D. B. MIRACLE, 'Metal matrix composites – From science to technological significance', *Comp. Sci. Technol.* **65** (2005) 2526–2540.

5 C. BADINI, F. MARINO, and A. TOMASI, 'Kinetics of precipitation hardening in SiC whisker-reinforced 6061 aluminium alloy', *J. Mater. Sci.* **26** (1991) 6279–6287.

6 A. P. SANNINO and H. J. RACK, 'Effect of reinforcement size on age hardening of PM 2009 Al-SiC 20 vol % particulate composites', *J. Mater. Sci.* **30** (1995) 4316–4322.

7 G. M. JANOWSKI and B. J. PLETKA, 'The effect of particle size and volume fraction on behavior of a liquid-phase sintered SiC/aluminum composite', *Metall. Mater. Trans. A* **26A** (1995) 3027–3035.

8 C. T. KIM, J. K. LEE, and M. R. PLICHTA, 'Plastic relaxation of thermoelastic stress in aluminum/ceramic composites', *Metall. Mater. Trans. A* **21A** (3) (1990) 673–682.

9 J. M. PAPAZIAN, 'Effects of SiC whiskers and particles on precipitation in aluminum matrix composite', *Metall. Trans. A* **19A** (1998) 2945–2953.

10 T. CHRISTMAN and S. SURESH, 'Microstructural development in an aluminum alloy-SiC whisker composite', *Acta Metall.* **36** (1988) 1691–1704.

11 R. J. ARSENAULT and R. M. FISHER, 'Microstructure of fiber and particulate SiC in 6061 Al composites', *Scripta Metall.* **17** (1) (1983) 67–71.

12 G. E. KIOURTSIDIS, S. M. SKOLIANOS, and G. A. LITSARDAKIS, 'Aging response of aluminium alloy 2024/silicon carbide particles (SiC_p) composites', *Mater. Sci. Eng. A* **382** (2004) 351–361.

13 C. TEKMEN and U. COCEN, 'Role of cold work and SiC volume fraction on accelerated age hardening behavior of Al Si Mg/SiC composites', *J. Mater. Sci. Lett.* **22** (2003) 1247–1249.

14 N. E. BEKHEET, R. M. GADELRAB, M. F. SALAH, and A. N. ABD EL-AZIM, 'The effects of aging on the hardness and fatigue behavior of 2024 Al alloy SiC composites', *Mater. Design* **23** (2002) 153–159.

15 R. U. VAIDYA, Z. R. XU, X. LI, K. K. CHAWLA, and A. K. ZUREK, 'Ageing response and mechanical properties of a SiC_p/Al-Li (8090) composite', *J. Mater. Sci.* **29** (1994) 2944–2950.

16 I. DUTTA and D. L. BOURELL, 'Influence of dislocation density and distribution on the aging behavior of 6061 Al-SiC_w composites', *Acta Metall. Mater.* **38** (11) (1990) 2041–2049.

17 N. R. M. R. BHARGAVA, I. SAMAJDAR, S. RANGANATHAN, and M. K. SURAPPA, 'Role of cold work and SiC reinforcements on the β'/β precipitation in Al-10 pct Mg alloy', *Metall. Mater. Trans. A* **29** (11) (1998) 2835–2842.

18 Z. M. EL-BARADIE, O. A. EL-SHAHAT, and A. N. ABD EL-AZIM, 'Accelerated aging processes in SiC-7020 aluminium composite', *J. Mater. Process. Technol.* **79** (1998) 1–8.

19 M. M. SHARMA, M. F. AMATEAU, and T. J. EDEN, 'Aging response of Al–Zn–Mg–Cu spray formed alloys and their metal matrix composites', *Mater. Sci. Eng. A* **424** (2006) 87–96.

20 K. K. CHAWLA, A. H. ESMAEILI, A. K. DATYE, and A. K. VASUDEVAN, 'Effect of homogeneous/heterogeneous precipitation on aging behavior of SiC_p/Al 2014 composite', *Scripta Metall. Mater.* **25** (1991) 1315–1319.

21 M. GUPTA, L. LU, and S. E. ANG, 'Effect of microstructural features on the ageing behaviour of Al–Cu/SiC metal matrix composites processed using casting and rheocasting routes', *J. Mater. Sci.* **32** (1997) 1261–1267.

22 F. A. GIROT, L. ALBINGRE, J. M. QUENISSET, and R. NASLAIN, 'Rheocasting Al matrix composites', *J. Metals* **39** (11) (1987) 18–21.

23 M. GUPTA, F. A. MOHAMED, E. J. LAVERNIA, and T. S. SRIVATSAN, 'Microstructural evolution and mechanical properties of SiC/Al_2O_3 particulate-reinforced spray-deposited metal-matrix composites', *J. Mater. Sci.* **28** (1993) 2245–2259.

24 A. N. ABDEL-AZIM, Z. M. EL-BARADIE, and M. F. SALAH, Precipitation hardening of SiC particle-reinforced Al-Mg-Si alloy composites, in *6th International Conference on Mining, Petroleum and Metallurgy*, Cairo University, Egypt (1999), vol. 3, pp. 413–428.

25 M. GUPTA, J. JUAREZ-ISLAS, W. E. FRAZIER, F. A. MOHAMED, and E. J. LAVERNIA, 'Microstructure, excess solid solubility, and elevated-temperature mechanical behavior of spray-atomized and codeposited Al-Ti-SiC_p', *Metall. Trans. B* **23B** (1992) 719–736.

26 V. C. SRIVASTAVA, A. SCHNEIDER, V. UHLENWINKEL, and K. BAUCKHAGE, 'Spray processing of 2014-Al + SiC_p composites and their property evaluation', *Mater. Sci. Eng. A* **412** (2005) 19–26.

27 M. GUPTA, C. LANE, and E. J. LAVERNIA, 'Microstructure and properties of spray atomized and deposited Al-7Si/SiC_p metal matrix composites', *Scripta Metall. Mater.* **26** (1992) 825–830.

28 P. G. SHEWMON, *Transformation in metals*, McGraw Hill Book Company, New York (1969), pp. 156–309.

29 C. W. WONG, M. GUPTA, and L. LU, 'Effect of matrix constitution on microstructure and mechanical properties of rheocast metal matrix composites', *Mater. Manuf. Proc.* **13** (1) (1998) 27–52.

30 R. J. ARSENAULT and N. SHI, 'Dislocation generation due to differences between the coefficients of thermal expansion', *Mater. Sci. Eng.* **81** (1986) 175–187.
31 M. TAYA, K. E. LULAY, and D. J. LLOYD, 'Strengthening of a particulate metal matrix composite by quenching', *Acta Metall. Mater.* **39** (1) (1991) 73–87.
32 M. GUPTA and M. K. SURRAPA, 'Effect of increase in heterogeneous nucleation sites on the aging behavior of 6061/SiC metal matrix composites', *Mater. Res. Bull.* **30** (8) (1995) 1023–1030.
33 C. W. WONG, M. GUPTA, and L. LU, 'Effect of variation in physical properties of the metallic matrix on the microstructural characteristics and the ageing behaviour of Al-Cu/SiC metal matrix composites', *J. Mater. Sci.* **34** (1999) 1681–1689.
34 K. R. VANHORN, Ed., *Aluminum: Vol. I. Properties, Physical Metallurgy and Phase Diagrams*, American Society for Metals, Ohio (1968), pp. 163–192.
35 M. SUERY and L. SALVO, 'Matrix-reinforcement interactions during fabrication and thermal treatment of cast Al-matrix composites', *Metall. New Mater.* **111** (1995) 119–137.
36 U. COCEN and K. ONEL, 'The production of Al-Si alloy-SiC_p composites via compocasting: Some microstructural aspects', *Mater. Sci. Eng.* **A221** (1996) 187–191.
37 D. J. LLOYD, 'Particle reinforced aluminium and magnesium matrix composites', *Int. Met. Rev.* **39** (1994) 1–23.
38 L. SALVO, G. L'ESPERANCE, M. SUERY, and J. G. LEGOUX, 'Interfacial reactions and age hardening in Al–Mg–Si metal matrix composites reinforced with SiC particles', *Mater. Sci. Eng.* **A177** (1994) 173–183.
39 J. C. LEE, J-P. AHN, J-H. SHIM, Z. SHI, and H-I. LEE, 'Control of the interface in SiC/Al composites', *Scripta Mater.* **41** (8) (1999) 895–900.
40 D-S. HAN, H. JONES, and H. V. ATKINSON, 'The wettability of silicon-carbide by liquid aluminum: the effect of free silicon in the carbide and of magnesium, silicon and copper alloy additions to the aluminum', *J. Mater. Sci.* **28** (10) (1993) 2654–2658.
41 K. B. LEE and H. KWON, 'Interfacial reactions in SiC_p/Al composite fabricated by pressureless infiltration', *Scripta Mater.* **36** (8) (1997) 847–852.
42 M. I. PECH-CANUL, R. N. KATZ, and M. M. MAKHLOUF, 'The role of silicon in wetting and pressureless infiltration of SiCp preforms by aluminum alloys', *J. Mater. Sci.* **35** (9) (2000) 2167–2173.
43 M. GUPTA, M. O. LAI, M. S. BOON, and N. S. HERNG, 'Regarding the SiC particulates size associated microstructural characteristics on the aging behavior of Al-4.5% Cu metallic matrix', *Mater. Res. Bull.* **33** (2) (1998) 199–209.
44 Y. FLOM and R. J. ARSENAULT, 'Effect of particle size on fracture toughness of SiC/Al composite material', *Acta Metall.* **37** (9) (1989) 2413–2423.
45 T. J. A. DOEL and P. BOWEN, 'Tensile properties of particulate-reinforced metal matrix composites', *Composites Part A* **27A** (1996) 655–665.
46 L. SALVO, M. SUERY, D. TOWLE, and C. M. FRIEND, 'Age-hardening behaviour of liquid-processed 6061 alloy reinforced with particulates and short fibres', *Composites Part A* **27A** (1996) 1201–1210.
47 L. SALVO and M. SUERY, 'Effect of reinforcement on age hardening of cast 6061 Al-SiC and 6061 Al-Al_2O_3 particulate composites', *Mater. Sci. Eng. A* **A177** (1–2) (1994) 19–28.

48 C. TEKMEN and U. COCEN, 'The effect of Si and Mg on age hardening behavior of Al–SiC_p composites', *J. Comp. Mater.* **37** (2003) 1791–1800.

49 S. SKOLIANOS, 'Mechanical behavior of cast SiC_p-reinforced Al-4.5%Cu-1.5%Mg alloy', *Mater. Sci. Eng.* **A210** (1996) 76–82.

50 Z. M. EL-BARADIE, 'Structure and properties of magnesium–zinc composite alloys thermomechanically treated', *Mater. Lett.* **57** (2003) 3269–3275.

51 B-C. KO and Y-C. YOO, 'The effect of aging treatment on the microstructure and mechanical properties of AA2124 hybrid composites reinforced with both SiC whiskers and SiC particles', *Comp. Sci. Technol.* **59** (1999) 775–779.

52 W. Q. SONG, P. KRAUKLIS, A. P. MOURITZ, and S. BANDYOPADHYAY, 'The effect of thermal ageing on the abrasive wear behaviour of age-hardening 2014 Al/SiC and 6061 Al/SiC composites', *Wear* **185** (1995) 125–130.

53 A. WANG and H. J. RACK, *Metal and Ceramic Matrix Composites: Processing Modeling and Mechanical Behavior*. The Mineral, Metals and Materials Society Warrendale, PA (1990), pp. 487–498.

54 S.-J. LIN and K.-S. LIU, 'Effect of aging on abrasion rate in Al-Zn-Mg-SiC composite', *Wear*, **121** (1) (1988) 1–14.

55 H.-L. LEE, W.-H. LU, and S. L.-I. CHAN, 'Effect of aging on the sliding abrasive wear of P/M 2014 and 6061 Al alloy-SiC particle composites', *Mater. Lett.* **15** (1992) 49–52.

56 G. E. KIOURTSIDIS and S. M. SKOLIANOS, 'Wear behavior of artificially aged AA2024/40 μm SiC_p composites in comparison with conventionally wear resistant ferrous materials', *Wear* **253** (2002) 946–956.

57 G. E. KIOURTSIDIS and S. M. SKOLIANOS, 'Pitting corrosion of artificially aged T6 AA2024/SiC_p composites in 3.5 wt.% NaCl aqueous solution', *Corros. Sci.* **49** (2007) 2711–2725.

58 B. DIKICI, M. GAVGALI, and, C. TEKMEN, 'Corrosion behavior of an artificially aged (T6) Al–Si–Mg-based metal matrix composite', *J. Comp. Mater.* **40** (2006) 1259–1269.

59 P. P. TRZASKOMA, 'Corrosion behavior of SiC/Al metal matrix composites', *J. Electrochem. Soc.* **130** (9) (1983) 1804–1809.

60 P. P. TRZASKOMA, 'The effects of silicon carbide whiskers on the initiation and propagation of pits on silicon carbide/aluminum metal matrix composites', in Tenth International Congress on Metallic Corrosion, Madras, India, November 1987, Trans. Tech. Publications, Zurich, Switzerland (1989), vol. V, pp. 231–239.

61 P. P. TRZASKOMA, 'Pit morphology of aluminum alloy and silicon carbide/aluminum alloy metal matrix composites', *Corrosion* **46** (5) (1990) 402–409.

62 J. F. MCINTYRE, P. K. CONRAD, and S. L. COLLEDGE, 'Technical note: the effect of heat treatment on the pitting behavior of SiC_w/AA2124', *Corrosion* **46** (11) (1990) 902–905.

63 S. L. COLEMAN, V. D. SCOTT, and B. MCENANEY, 'Corrosion behaviour of aluminium-based metal matrix composites', *J. Mater. Sci.* **29** (1994) 2826–2834.

64 J. E. CASTLE, L. SUN, and H. YAN, 'Use of scanning Auger microscopy to locate cathodic centres in SiC_p/6061 Al MMC and to determine the current density at which they operate', *Corros. Sci.* **36** (6) (1994) 1093–1110.

9 Modelling accelerated ageing in polymer composites

G. MENSITIERI, CR-INSTM – University of Naples Federico II, Italy; and M. IANNONE, Alenia Aeronautica s.p.a., Italy

9.1 Introduction

A design life of 10–50 years is required for important areas of application of reinforced polymers (polymer matrix composites, PMCs); these applications include the automotive and aeronautical industries, bridge structures, water and waste systems, and offshore exploration and oil production. There is, hence, a strong need for accelerated life-time characterization methodologies, which can predict the evolution of stiffness and strength of PMC materials in order to assure the integrity and safety of structural components.

The design of composite parts for advanced applications, where a lack of prior service experience is present, depends upon predictive models to characterize long-term behaviour. Prediction of real-time, long-term behaviour of composites with a high degree of confidence has to rely upon costly and time-consuming long-term (real-time) testing to fully assess the durability of materials. Accelerated ageing methodologies may significantly reduce the expense and time involved by narrowing or screening the field of acceptable candidate materials that will go into long-term qualification tests. Moreover, accelerated testing may be fruitfully exploited to evaluate residual service life of existing structures and to supply indications for product improvements.

However, the setting up of meaningful accelerated testing methodologies is a difficult task in view of complex, time-dependent phenomena related to materials ageing, especially when multiple ageing processes and degradation mechanisms are involved.[1] In fact, it is only from the understanding of ageing effects on a given material that it can be determined whether that ageing process can actually be accelerated and how. To this aim, composite material behaviour should be fully characterized and theoretically modelled to properly develop methods finalized to the prediction of material performance under various conditions of load, temperature, and environment (e.g. radiation, oxygen, moisture and fluid exposure). Other chapters of this book discuss the various ageing mechanisms in depth. In this chapter, these mechanisms are reviewed briefly, mainly to introduce in a self-consistent way the approaches and equations that are relevant for the discussion of the main topic addressed in the chapter, i.e. the issue of accelerated ageing methodologies for PMCs.

PMCs consist of high-strength and/or high-modulus fibres, embedded in a high-temperature-resistant polymeric matrix. Matrix polymeric materials can be divided into two main categories: thermosets and thermoplastics. Thermosets, mainly high-performance ones, are relatively brittle, resulting in a low fracture toughness, reduced damage tolerance, and lower transverse tensile, interlaminar tensile, and in-plane shear failure strains; under mechanical loading or thermal cycling they could display micro-cracking. Thermoplastics, on the other hand, are characterized by high toughness, high damage tolerance, good mechanical properties, and excellent shelf-life; however, they display a lower chemical resistance and creep deformation is increased. Moreover, thermoplastics are frequently semi-crystalline, the crystalline domains being responsible for good mechanical properties but also determining phase boundaries which represent stress discontinuities.

In view of the time-dependent properties of polymeric materials, the analytical formulations useful for predictive models generally rely upon time-based superposition principles.[2] In fact, first approaches[3,4] used elevated temperatures to accelerate ageing phenomena and the prediction of long-term mechanical performances from short-term tests was based on these principles. In view of the viscoelastic behaviour of PMC, creep testing was frequently used to generate material constants, adopting largely empirical accelerated test schemes. There is a renewed interest in accelerated ageing, but based on the harmonization of individual test methods combined in an integrated scheme to provide an accurate method for understanding the different contributions of various degradation mechanisms to the durability of PMC materials. In this endeavour, the modelling of the chemical and physical phenomena, and of their dependence on accelerating factors, plays a major role.

Ageing processes, which we need to accelerate, derive from long-term exposure of advanced PMC materials to the operative environment and eventually result in irreversible changes in the original properties of the material. Primary ageing mechanisms may be categorized as chemical, physical, and mechanical. The possible interaction between these three mechanisms is related to material characteristics and ageing environment.

In the context of this chapter there are few important definitions[2] to be given.

- *Environmental degradation factor*: specific use-environment conditions, i.e. temperature, moisture, mechanical load, etc.
- *Critical degradation mechanism*: the mechanism that occurs due to the action of specific environmental degradation factors. It includes

chemical, physical, and mechanical processes occurring when the environmental degradation factors are inside the boundaries of the use-environment. It results in a significant loss in important bulk physical properties.

- *Accelerated ageing*: defined as the process or processes required to accelerate a specific critical degradation mechanism or mechanisms relative to a baseline ageing condition; thereby resulting in the material reaching the same aged end-state as a real-time-aged material, but in less time. It is obvious that an ageing mechanism can be properly accelerated only on the basis of the mechanistic understanding of how ageing affects a given material system. The matter is rather complex, and is additionally so in view of the fact that different ageing mechanisms may act synergistically. The need to predict performances in a broad range of service conditions requires the development of empirical techniques coupled with analytical methods. All the tests undertaken should provide insight into how a material behaves and establish input for the development of analysis methods to predict material performance under various conditions of load, temperature, and environment. Validation of accelerated testing methods takes place through a comparison of the changes of mechanical properties, damage mechanisms, and physical parameters (e.g. weight loss, changes in glass transition temperature, T_g, or fracture toughness) during accelerated testing with those from real-time testing. Schematically, the accelerated ageing methodology can, in general, be summarized as follows:
- individuation of proper environmental conditions and variables;
- individuation, understanding, and characterization of critical ageing mechanisms and their possible interactions;
- modelling of ageing behaviour in the conditions of interest;
- design of accelerated testing on the basis of the modelling;
- validation of accelerated testing methodology by comparison with selected real-time ageing tests.

9.2 Definition of environmental conditions and important variables

One or more environmental degradation factors are active during ageing of PMCs, their action being operated through a molecular mechanism that acts on the chains or the network: these modifications, in turn, determine observable changes at the level of macroscopic bulk mechanical properties. Consequently, a well-designed accelerating protocol should accelerate the action of degradation factors without departing from the underlying molecular mechanisms. It is worth noting that the ageing processes may bring about both reversible and irreversible (damage accumulation) effects on the PMCs. When evaluating the effects of accelerated ageing protocols, this issue should be taken into account.

Continuing damage accumulation is induced and driven by several material characteristics and degradation factors: type of polymer matrix and polymer morphology, fibre volume fraction, fibre architecture, service temperature and temperature cycling, relative humidity, oxidative attack, solvent infusion, internal moisture concentration, air pressure, radiation, mechanical loads (and combined cyclic loads) and related applied stress level, and degree of damage and ageing time. The coupling process linking the growth of various damage modes and the external environmental drivers will certainly prove complex.

For example, typical ground ambient exposure in aeronautical applications consists of temperatures ranging from about 60 to –55 °C, moisture conditions ranging from extreme humidity to desert conditions, and irradiation (especially ultraviolet). During the flight, the most severe conditions are encountered with temperatures that, depending on the speed, can reach a value as high as 150 °C (supersonic speed). Significant stresses also occur in supersonic aircrafts because of thermal variations and thermal spikes. Differences in temperatures throughout the structure cause different parts of the structure to expand by different amounts, giving rise to thermal stresses which are added to other imposed stresses.

Therefore, the accelerated testing must replicate the changes occurring in real-life conditions for long-term applications, reproducing the degradation mechanisms. The complete knowledge that is required of all degradation processes, of their interaction and of their kinetics, is quite difficult to obtain and a simpler approach can be adopted based on determination of primary degradation mechanisms that are easy to measure.

In order to evaluate the state of a material exposed to long-term ageing, it is also important to use indicator properties (e.g. weight, T_g, damage state – such as crack density) that are easy to measure and can be correlated in a reproducible manner to the modification of bulk properties relevant for the material application.

Finally, the availability of mechanical properties data from both accelerated ageing and real-time ageing experiments is extremely important to determine the accelerating factor for each degradation mechanism.

9.3 Degradation mechanisms and processes

The general degradation mechanisms that should be included for designing accelerated ageing protocols for materials and structures are listed below.

1 *Time-dependent mechanical behaviour* of polymers and resultant damage accumulation (involving viscoelastic–viscoplastic creep).

2 *Mechanical degradation mechanisms*, which are irreversible processes related to high stress levels and/or fatigue. These degradation mechanisms include matrix cracking, delamination, interface degradation, fibre breaks, and inelastic deformation, and thus have a direct effect on engineering properties such as stiffness and strength. In some cases, mechanical degradation mechanisms dominate only after chemical or physical ageing mechanisms have altered the polymer properties.
3 *Physical ageing*, which occurs for polymers at temperatures below the T_g. In this condition the material evolves towards the thermodynamic equilibrium. This evolution is characterized by changes in free volume, enthalpy, and entropy of the polymer and produces measurable changes in the mechanical properties.
4 *Hygrothermal effects*, which are related to the uptake of water molecules into the material. Sorbed moisture promotes plasticization of the polymer matrix and, eventually, hydrolytic degradation affecting instantaneous as well as long-time mechanical properties. These effects are enhanced by concurrent application of stress.
5 *Chemical degradation mechanisms*, including thermo-oxidative, thermal, and hydrolytic ageing. At typical PMC operating temperatures, cross-linking and oxidation are the dominant chemical ageing mechanisms. Frequently, such ageing results in an increase in cross-linking density that can severely affect the mechanical properties (e.g. increase of elastic modulus and decrease of toughness) by densification and increasing the T_g.
6 *Synergistic effects* are determined by the concurrent action of several of the above-mentioned physical–chemical phenomena. They determine consequences that can be more severe than those promoted by simple addition of the action of the single mechanisms.

The design of composite parts is based on the properties of the fibres, the fibre–matrix interface, and the polymer matrix. However, during long-term ageing, most of the design properties (e.g. composite stiffness, strength, and fatigue life) may change, mainly in relation to changes in the mechanical properties of the polymer matrix. Thermal oxidation, physical ageing, viscoelastic behaviour, hygrothermal effects, matrix cracking, and microstructural changes need to be further understood, as well as the synergistic accelerations in damage accumulation deriving from combinations of these phenomena. In fact, the combined effect of these factors on the failure mechanism should be accounted for in any reliable modelling process.

9.3.1 Time-dependent mechanical behaviour

Polymers, both in the rubbery and in the glassy state, are characterized by viscoelasticity, i.e. a time-dependent mechanical behaviour. The characteristic times (retardation or relaxation times) of this constitutive behaviour are much longer in the glassy state and decrease at temperatures near and above T_g. Typical evidences of viscoelasticity are *creep* (a time-dependent deformation, consisting of three stages, under a constant load) or *stress-relaxation* (time-dependent stress evolution under a constant deformation). In composite materials this issue is important when considering the stress transfer between viscoelastic matrix and elastic reinforcement or between laminae with different fibre directions. This effect has to be included in a long-term durability model, mainly when considering matrix-dominated properties. With the aim of setting up accelerated testing methodologies, long-term viscoelastic behaviour can be accelerated by adopting the time–temperature superposition principle, which is based on the concept that the behaviour for long time periods at low temperatures corresponds to that for short time periods at higher temperatures. In a polymer, higher temperatures relate to easier rearrangements of the polymer chain backbones and thus to higher compliance (lower modulus). The time–temperature superposition principle indicates that the compliance curves for creep experiments or modulus curves for stress-relaxation experiments at different temperatures are related to one another by a simple (temperature-dependent) shift on the log time scale (expressed by means of a shift factor, a_T). This result is based on the physical model that the relaxation times for a material, which represent the ease of motion of different segments of the polymer chain and are related to the available *free volume*, are all scaled by temperature in an identical manner, with the scaling factor decreasing as temperature increases.

This matter is further complicated if we consider that polymers below their T_g exhibit a time-dependent rearrangement of their non-equilibrium structure (i.e. a physical ageing phenomenon that will be illustrated later). The mutual interaction between viscoelasticity and physical ageing (in fact physical ageing affects the viscoelastic properties) imposes the introduction of time–ageing time–temperature shift factors. Moreover, other time-shifting effects on material behaviour may be determined, in addition, by state of stress, moisture, and other environmental factors.

Time-dependent evolution of PMCs concerns not only the stiffness properties but also the strength characteristics. In fact, the durability analysis of such composite systems has to be based on strength criteria with time-dependent coefficients or on a non-linear viscoelastic–viscoplastic damage analysis. In general, it has been observed that the creep and recovery behaviour of a polymeric material can be described by the viscoelastic constitutive law only when the material has been subjected to mechanical preconditioning. For un-preconditioned specimens, the recovery behaviour following creep can not be adequately described by viscoelasticity alone, but a combined viscoelastic–viscoplastic model is necessary.

As indicated previously, the dependency of mechanical behaviour on time is increased in the proximity of T_g. In fact, the service temperature, although it does not attain high values, might approach or even exceed the actual T_g of the polymer matrix due to moisture plasticization which lowers the T_g of the polymer as measured in the dry state. This decrease in T_g can be quite relevant (e.g. in epoxies it can be of the order of 30 °C per percentage % b-w. of absorbed moisture, for the first few weight percent of moisture uptake) determining a decrease in yield strength and, generally, an increase in fracture toughness as well as the magnitude of the viscoelastic response.

9.3.2 Mechanical degradation

Irreversible processes such as matrix cracking, delamination, fibre-bundle rupture, local compressive instability, interfacial failure, and inelastic deformation (plastic strain) are relevant examples of mechanical degradation. Although any one of these damage modes may be viewed as sub-critical, the combined effects of several damage modes may be dangerous. The most critical damage mechanism, operating in high-temperature polymeric composites, is the formation of transverse ply cracks and in-plane micro-cracks in matrix polymers of multi-axial composites. Matrix cracking can result from initial laminate processing, mechanical static and fatigue loading, residual stresses resulting from hygrothermal exposure, thermal cycling, and combined effects of mechanical and environmental cycles. Residual stress resulting from differences in thermal expansion of the lamina in longitudinal (parallel to fibre orientation) and transverse (perpendicular to fibre orientation) directions is the primary factor that causes thermally induced transverse matrix cracking. Hygrothermal cycling (combined moisture and temperature cycling) of composite laminates can produce transverse matrix cracks that initiate in surface plies and progress deeply in the laminate with accumulating cycles. Composite strength, stiffness, and thermal properties as well as failure modes can be affected by transverse matrix cracking which can also promote higher uptake of moisture deeper in the laminate.

Moreover, since high-temperature thermosets commonly incorporate discrete toughening phases to improve impact resistance, problems related to phase separation of the toughening agent, which proceeds from the surface to the interior, should be accounted for, since it could not only determine deleterious effects on matrix toughness, but could also lead to matrix cracking.

9.3.3 Physical ageing

Sub-T_g ageing of polymer-based composites results in thermoreversible physical phenomena, known as physical ageing, that should be addressed by any durability analysis.[5] In fact, polymers below T_g are in a non-equilibrium state and tend to evolve toward the equilibrium state, displaying changes with time in volume (slow continuous decrease in volume), enthalpy, and entropy at an ever-decreasing rate. This process induces an increase in stiffness and brittleness, increasing the likelihood of more rapid progression of various damage states. As a consequence, mechanical and viscoelastic behaviours of glassy polymers and their composites depend on their thermal history and correlated decrease in the free volume and, in turn, segmental mobility. In fact, polymers display an evolution of typical viscoelastic relaxation times[6] with the ageing time that, as illustrated later, can be interpreted by introducing properly defined time–ageing time–temperature shift factors. This complicates the analysis of a loaded PMC since, under a mechanical load, physical ageing accompanies the time-dependent viscoelastic behaviour that is typical of polymers both above and below T_g.

9.3.4 Hygrothermal effects

Low molecular weight substances, like water, can be absorbed and diffuse in the polymeric matrices of composites.[7] The equilibrium sorbed amount and the mass transport kinetics depend heavily upon the chemical structure and morphology (e.g. degree of crystallinity) of a polymer. In the case of a composite, these properties are also related to the arrangement and amount of the impermeable fibres and to the nature of fibre–matrix interfacial behaviour. Mass transport mechanisms can be rather complex, in view of the viscoelastic nature of polymeric materials.[7,8] However, when the sorbed amount is vanishingly small, the time evolution of penetrant concentration profiles can be determined by using a simple approach based on Fick's law,[9] as will be illustrated in Section 9.7.

Combined exposure to heat and moisture affects reinforced plastics in a variety of ways. The long-term effect of moisture is reflected, predominantly, in changes in matrix-dominated mechanical properties (i.e. shear modulus, compressive strength, and transverse tensile strength, although degradation in strength is more than that for modulus) that are closely coupled to T_g. In fact, high-temperature and -humidity exposure effectively plasticizes the matrix, accelerating the sub-T_g polymer relaxation process towards thermodynamic equilibrium.[1] Moreover, the hygrothermal swelling causes a change in the residual stresses within the composite that could lead to micro-crack formation. These micro-cracks, in turn, provide fast diffusion paths (for both liquid water and water vapour) and then alter the moisture absorption characteristics of the laminate. Other relevant phenomena are fibre–matrix debonding and eventual coalescence of micro-cracks and debonded regions to form macro-cracks.

Short-term hygrothermal effects on composites include matrix cracking, microvoid generation, outer-ply delamination, or surface blistering during rapid heating, mainly under hygrothermal cycling conditions, even with no applied loads. In fact, concentration gradients, from high moisture levels at the surface to near dryness at the midplies, along with subsequent desorption in the near-surface plies and redistribution of moisture, can lead to significant residual stresses, or outer-ply delamination, or blistering during rapid heating.

9.3.5 Thermo-oxidative degradation

Another important degradation mechanism of composite matrix polymers is probably due to thermal instability and to the accelerating effects of oxidative attack, involving chain scission, cross-linking and thermo-oxidative reactions.[10,11] Thermal degradation of PMCs, in an inert atmosphere, is exclusively a thermolysis phenomenon while, in air, it is dominated by oxidation. Thermolysis is the result of breaking of covalent bonds in the polymer network and, in general, thermal stability of high-performance polymers in an inert atmosphere is very good. Oxidation is usually characterized in terms of weight loss and is largely a surface phenomenon. In fact, chemical changes and related weight changes are basically limited to the polymeric matrix and occur from the surface inwards,[12,13] penetrating deeply only at high temperatures and for long exposure times. Hence, even small weight changes could be a symptom of significant surface changes. Oxidation reactions in the interior rely on mass transport processes, including diffusion of oxygen inwards and subsequent diffusion of reaction products outwards. Resin fracture toughness also plays a large role in thermo-oxidative resistance. The presence of the toughener significantly reduces the weight loss rate during ageing, probably through better resistance to crack formation (due to resin degradation and shrinkage) which could provide pathways for oxygen diffusion. Another issue that is of particular importance when dealing with accelerated ageing, is that the pattern of oxidation reactions and related induced chemical changes, depend on temperature determining changes in mechanical properties which can differ significantly as a function of sample thickness, surface protection, and ageing temperature. As an example, for low-cure-temperature epoxies,[12] at lower temperatures the oxidative degradation is very selective for chemical groups representing defects in the network structure while at higher temperatures, a combination of further cure reactions and generalized oxidative degradation changes makes for a highly complex ageing pattern.

Reaction patterns obviously change significantly with the nature of the polymer matrix; for example, the different chemical behaviours of epoxy and epoxy-cyanate matrices[14] have different consequences on mechanical and viscoelastic behaviour. Relevant problems are also related to the presence of unreacted and defect groups (e.g. loss of unreacted hardener can promote micro-cracking).

The examination of the thermo-oxidative ageing resistance of polymer composites should focus not only on the characterization of the degradation of mechanical properties, but also on failure mechanisms.[15] By identifying the degradation mechanisms that will be the most significant for the desired application, an appropriate test program may be designed to incorporate tests concentrating on material properties that reflect these mechanisms.

9.4 Modelling time-dependent mechanical behaviour

Polymers exhibit viscoelastic behaviour which, depending on the deformation and rate of deformation, can be linear or non-linear.[16] In order to predict the long-term behaviour of polymer-based composites, it is, then, important to analyse relevant viscoelastic phenomenology, for example creep behaviour. Since these phenomena, at operative temperatures, are frequently rather slow, they should be accelerated to keep the experimental characterization time in a reasonable range. One of the analytical approaches generally used to accelerate testing consists of utilizing the principle of time–temperature superposition, which can be adopted if the material is thermorheologically simple, i.e. if a change in temperature promotes the same shift of all the various relaxation times involved in the viscoelastic behaviour. Accelerated testing methods based on superposition have been used with creep and stress-relaxation phenomena as well as for damage or fatigue life properties; wherein testing over reasonable (short) time scales is performed at multiple elevated temperatures and results are extrapolated to lower use temperatures and longer times.[2] In order to implement such accelerated procedures, modelling of the underlying degradation phenomena should be available. For thermorheologically simple materials, a simple horizontal shift can be applied to the time scale to construct a time-dependent creep compliance curve at a certain temperature from the values at another temperature. The shift consists of correcting the time scale by means of a time–temperature shift factor, a_T, which is temperature dependent. The underlying physics is based on the concept of free volume and its effect on macromolecular mobility and, in turn, on the value of relaxation times. The effect of other environmental factors such as humidity, can also be accounted for by similar arguments, introducing a horizontal shift factor which reflects, for example, the plasticizing effect of absorbed moisture.

This time–temperature horizontal shift approach works in the framework of linear viscoelasticity. However, polymers can display a non-linear viscoelastic behaviour and, in many cases, cannot be assumed to be thermorheologically simple. As a consequence, in order to accelerate experiments properly, other horizontal shift factors for the time axis have to be introduced and also a vertical (compliance, modulus axis) shift is often necessary.[17–19] In fact, for the case of creep compliance, a stress–time superposition is introduced, beside the temperature–time superposition, defining a stress–time shift factor, a_σ, that is generally a function of both temperature and stress. To complicate the matter, glassy polymers are characterized by the already-mentioned physical ageing phenomenon that consists of a structural rearrangement that promotes a gradual change in free volume towards the equilibrium state. This free volume change, in turn, influences the viscoelastic behaviour, slowing down time-dependent phenomena. In addition, this effect has a temperature-dependent rate which can be accounted for by introducing a time–ageing time shift factor, a_δ, which depends on the extent of the departure from equilibrium, the temperature, and the stress state. The physics behind the introduction of all these horizontal shifts can be lumped into a single concept: the *effective* (or *material* or *reduced*) time, which is a reference time scale obtained by correcting real time by means of all the introduced horizontal shift factors.

9.4.1 Linear viscoelastic polymers. Stress-based superposition: the Boltzmann principle

For linear viscoelastic materials, the strain response to a combined loading of a complex stress history can be evaluated by simple superposition, based on the assumption that the modulus ($E(t)$) and the compliance ($S(t)$) are not functions of stress. This concept is the basis of the Boltzmann superposition principle which is used to create a hereditary integral form for a viscoelastic constitutive law. As an example, the total elongational strain as a function of time for a material subjected to any load is

$$\varepsilon(t) = \sigma_0 \cdot S(t) \cdot H(t) + \int_0^t S(t-t') \cdot \frac{d\sigma(t')}{dt'} \cdot dt' \qquad [9.1]$$

where σ_0 is the initial stress applied at time zero and $H(t)$ is the Heaviside unit function. An analogous expression can be obtained for the stress:

$$\sigma(t) = \varepsilon_0 \cdot E(t) \cdot H(t) + \int_0^t E(t-t') \cdot \frac{d\varepsilon(t')}{dt'} \cdot dt' \qquad [9.2]$$

where ε_0 is the initial stress applied at time zero.

9.4.2 Time–temperature superposition for a thermorheologically simple material

In a polymer, higher temperatures relate to easier rearrangements of the polymer chain backbones and thus to higher compliance (lower modulus). The time–temperature superposition principle indicates that the compliance curves for creep experiments (see Fig. 9.1) and modulus curves for stress-relaxation experiments (or other relevant viscoelastic properties) at different temperatures are related to one another by a simple (temperature-dependent) shift on the log time scale. This result implies that, as already

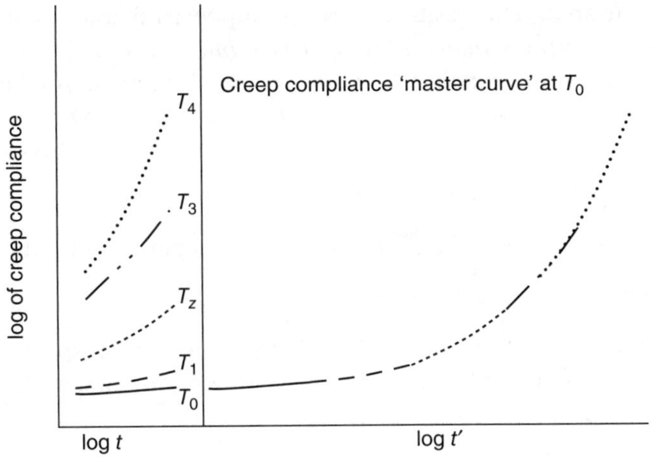

9.1 Example of construction of creep compliance master curve for a themorheologically simple polymer using the time–temperature superposition principle.

discussed, the relaxation times for a material, which represent the ease of motion of different segments of the polymer chain, are all scaled by temperature in an identical manner. The basic idea is that the higher the temperature, the shorter the relaxation time, and, consequently, relaxation time can be evaluated at a reference temperature by using a simple shift factor, a_T.

Above T_g the method of reduced variables can be used to arrive at the WLF (Williams–Landel–Ferry) equation[20]

$$\log a_T = \frac{-c_1 \cdot (T - T_0)}{c_2 + (T - T_0)} \qquad [9.3]$$

while, below T_g, the exponential expression

$$a_T = \exp\left[\frac{\Delta H}{R} \cdot \left(\frac{1}{T} - \frac{1}{T_0}\right)\right] \qquad [9.4]$$

should be used. Frequently, in both expressions, T_g is used as reference temperature.

9.4.3 Effective or material time concept

For viscoelastic materials, the effect of many factors on material response can be expressed as a simple time shift, analogous to what we have seen for the case of temperature. The effects of ageing, moisture, and stress levels can all be captured by use of the shift factor concept and associated method of reduced variables. This result give rise to a powerful concept, called the *effective* (or, sometimes, *material* or *reduced*) *time*.

The *effective time*, ξ, is defined by noting that all relaxation mechanisms in *real* time increment $\mathrm{d}t$ at temperature T are a_T times slower/faster than those in a time increment $\mathrm{d}\xi$ at T_R (reference temperature). So that

$$\mathrm{d}\xi = a_T \cdot \mathrm{d}t \rightarrow \xi(t) = \int_0^t a_T(\zeta)\cdot \mathrm{d}\zeta \qquad [9.5]$$

where ζ is a dummy variable. Consequently, if temperature is constant

$$E_T(t) = E_{T_R}(\xi(t)) = E_{T_R}(a_T \cdot t) \qquad [9.6]$$

where $E_T(t)$ and $E_{T_R}(t)$ are the stress-relaxation moduli, respectively at T and T_R. Thus, the constitutive law for linear viscoelastic materials at T can be written in the *real* time domain as

$$\sigma(t) = \int_{0^-}^t E_{T_R}(\xi(t) - \xi(t'))\cdot \frac{\mathrm{d}\varepsilon(t')}{\mathrm{d}t'}\cdot \mathrm{d}t' \qquad [9.7]$$

where $\sigma(t)$ is the Cauchy stress at time t. Equation [9.7] in the *effective* time domain becomes

$$\sigma(\xi) = \int_{0^-}^{\xi} E_{T_R}(\xi - \xi')\cdot \frac{\mathrm{d}\varepsilon(\xi')}{\mathrm{d}\xi'}\cdot \mathrm{d}\xi' \qquad [9.8]$$

The effective time concept is used in viscoelasticity to perform calculations of material response in conditions of variable temperature, moisture, and/or damage by using different forms of equation [9.5] where the product of several shift factors appears (e.g. a_T for time–temperature shift, a_{RH} for time–moisture shift). As will be discussed later (see Section 9.4.4), the concept of *effective* time has been generalized to also include the effects on relaxation times of stress and physical ageing, which introduces non-linearity. In fact, in the case of non-linear viscoelasticity, the effective time concept incorporates the effect of stress and ageing by defining other shift factors: a_σ for the time–stress shift and a_δ for the time–ageing time shift. Moreover, as already mentioned, non-linear viscoelastic materials also require a vertical shift in order to obtain reliable 'master curves'.

9.4.4 Non-linear viscoelastic polymers

The most comprehensive attempt to include the relevant features of non-linear viscoelasticity and temperature dependence was developed by Schapery.[21,22] He derived, on the grounds of irreversible thermodynamics, starting from Clausius–Duhem inequality, the following expression for uni-axial deformation:

$$\varepsilon(t) = g_0(t)\cdot\sigma(t)\cdot S_0 + g_1(t)\cdot\int_0^t \Delta S(\xi(t)-\xi(\zeta))\cdot\frac{\mathrm{d}}{\mathrm{d}\zeta}[g_2(\zeta)\cdot\sigma(\zeta)]\cdot \mathrm{d}\zeta \qquad [9.9]$$

where S_0 is the instantaneous compliance and $\Delta S(\xi(t))$ is a transient creep compliance function. In general, g_0, g_1, and g_2 are material properties that are non-linearizing functions of stress level, temperature, and moisture concentration. The factor g_0 defines stress and temperature effects on the instantaneous elastic compliance and is a measure of state-dependent reduction (or increase) of stiffness. Transient compliance factor g_1 has a similar meaning, operating on the creep compliance component. The factor g_2 accounts for the influence of loading rate on creep. $\xi(t)$ represents a reduced time scale parameter defined by

$$\xi(t) = \int_0^t a_{\sigma_T}(\zeta)\cdot \mathrm{d}\zeta \qquad [9.10]$$

where $a_{\sigma_T}(\zeta)$ is a material property and represents the time–stress shift factor that, in general, is a function of stress level, temperature, and moisture concentration.

After a choice for the transient compliance (e.g. simple power law, generalized power law, Cole–Cole function, Kolrausch function) based on equation [9.9], the construction by curve fitting of the experimental results from creep or creep-recovery tests results in a numerical or analytical expression for g_i and a_{σ_T}.

9.4.5 Viscoelastic–viscoplastic behaviour

In general, it has been observed that the creep and recovery behaviour of a polymeric material can be described by viscoelastic constitutive law only when the material has been subjected to mechanical preconditioning. For un-preconditioned specimens, the recovery behaviour following creep can not be adequately described by viscoelasticity alone, but a combined viscoelastic–viscoplastic model is necessary. Zapas and Crissman[23] employed a functional and account for stress history-dependent, non-linear viscoplastic strain. The combined Schapery–Zapas–Crissman model for uniaxial loading, as employed by Pasricha *et al.*,[24] is given by

$$\begin{aligned}\varepsilon(t) = {} & g_0(t)\cdot\sigma(t)\cdot S_0 + g_1(t)\cdot\int_0^t \Delta S(\xi(t)-\xi(\zeta))\cdot\frac{\mathrm{d}}{\mathrm{d}\zeta}[g_2(\zeta)\cdot\sigma(\zeta)]\cdot \mathrm{d}\zeta \\ & + \varphi\cdot\left[\int_0^t g_3(\zeta)\cdot(\sigma(\zeta))\cdot \mathrm{d}\zeta\right] + \alpha\cdot\Delta T\end{aligned} \qquad [9.11]$$

The third term on the right-hand side of equation [9.11], is the Zapas–Crissman functional and it accounts for the load history-dependent viscoplastic strain, while the fourth term on the right-hand side is the thermal strain.

As an example, if we consider a specific expression for the viscoplastic component of the type[25]

$$\varepsilon_{vp}(t) = C\cdot\sigma_0^{Nm}\cdot t^m = \varepsilon_{vp}(\sigma_0)\cdot t^m \qquad [9.12]$$

we obtain for the analysis of creep strain (ε_C) at stress level σ_0 during the time period $[0, t_1]$:

$$\varepsilon_C = g_0 \cdot S_0 \cdot \sigma_0 + g_1 \cdot g_2 \cdot \sigma_0 \cdot \frac{S_\infty - S_0}{1+\left(\frac{\tau_0 \cdot a_\sigma}{t}\right)} + C \cdot \sigma_0^{Nm} \cdot t^m \qquad [9.13]$$

Finally, it is important to note that the shortening of the life-time and the acceleration of the creep behaviour after a first loading followed by a static loading may be explained by the damage introduced by the first dynamic loading. In order to analyse this behaviour a damage component, ε_d – based on a viscoplastic model for which the yield stress is a decreasing function of the number of initial cycles, of time, of environmental changes, and based on an *effective* time to account for ageing – can be added to the viscoelastic–viscoplastic model:

$$\varepsilon = \varepsilon_e + \varepsilon_{ve} + \varepsilon_{vp} + \varepsilon_d \qquad [9.14]$$

where subscripts e, ve, vp and d stand for elastic, viscoelastic, viscoplastic, and damage, respectively.

9.4.6 Application to composite materials

Modelling of the long-time mechanical behaviour of composites is the basis for developing reliable acceleration procedures of the time-dependent responses of a PMC. The behaviour in the fibre direction is generally assumed to be purely elastic or elasto-plastic while in other directions the viscoelastic–viscoplastic model, similar to that adopted to describe the response of the neat polymer, should be applied. Time–temperature and time–ageing time superposition principles can still be applied and properly coupled with classical laminated plate theory to provide a framework for the analysis of the long-time mechanical response of laminated composite materials. Each unidirectional fibre-reinforced lamina is geometrically and physically anisotropic, changing from elastic to viscoelastic. For a single lamina under a plane stress state the only time-dependent compliance terms are the transverse, S_{22}, and shear, S_{66}, ones. As a consequence the total laminate compliance will also be time-dependent.

As for polymers, PMCs also exhibit non-linear viscoelastic behaviour in tension as well as in shear[5] and among many non-linear integral constitutive equations, the already introduced Schapery's law (equation [9.9]) is a good candidate to be used since the non-linear parameters are relatively simple to characterize and numerical implementation of the governing equations is straightforward and robust.[26]

Time-dependent evolution of PMCs concerns not only the stiffness properties but also the strength characteristics.[25] The durability analysis of such composite systems can be based on strength criteria with time-dependent coefficients or on a damage analysis (based on a combined non-linear viscoelastic–viscoplastic damage model). As for neat polymers, an important time-dependent behaviour is related to ageing, while viscoelastic behaviour plays a major role in the interaction level of elastic and viscoelastic parts, regulating stress transfer from the viscoelastic to the elastic regions, i.e. matrix and reinforcement (if the interface conditions are perfect) or between laminae with different fibre directions. In real composites the interface conditions are not perfect and these local, time-dependent effects are smaller.[22] The long-term behaviour of composite materials may be affected by physical and chemical ageing (change in molecular weight, oxidation, change in density of reticulation). Environmental effects have to be integrated into the constitutive equations of viscoelasticity.

In order to illustrate a concrete example of the modelling of long-time mechanical response of a composite material, the approach proposed by Guedes *et al.*[26] is presented here; this introduces a method based on the concept of accelerating factors in viscoelastic and viscoplastic models. The accelerating factors considered are temperature and stress, accounted for by introducing the already-mentioned time–temperature and time–stress shift factors. With a similar approach, the effect of other accelerating factors could be included, such as moisture, physical ageing, and mechanical degradation. A complete picture of the long-term durability of composites should also consider the coupled effects of all these non-linear mechanisms.

For viscoelastic behaviour, the non-linear constitutive equation introduced by Schapery has been used.[22] In the framework of a 'macro-mechanic' approach, the constitutive models are defined at the 'ply level'. The plane of a single ply is defined by the 1–2 coordinate system where the 1 and 2 axes are parallel and perpendicular to the fibres, respectively. Consider a ply subjected to a plane stress state $\{\sigma_{11}, \sigma_{22}, \tau_{12}\}$ that may change with time, at a fixed temperature. The total strains induced are

$$\begin{Bmatrix} \varepsilon_{11}(t) \\ \varepsilon_{22}(t) \\ \gamma_{12}(t) \end{Bmatrix} = \begin{bmatrix} S_{11} & S_{12} & 0 \\ S_{21} & S_{22}(t) & 0 \\ 0 & 0 & S_{66}(t) \end{bmatrix} \cdot \begin{Bmatrix} \sigma_{11}(t) \\ \sigma_{22}(t) \\ \tau_{12}(t) \end{Bmatrix} + \begin{Bmatrix} 0 \\ \varepsilon_{22}(t) \\ \gamma_{12}(t) \end{Bmatrix}_{vp} \qquad [9.15]$$

where subscript vp denotes the viscoplastic component of the strain.

Here only the matrix-dominated compliance terms (i.e. $S_{22}(t)$ and $S_{66}(t)$) are time-dependent. The total strains associated with the matrix-dominated compliances S_{22} and S_{66} are modelled using the modified Schapery theory to include the viscoplastic behaviour, first presented by Tuttle:[27]

$$\begin{Bmatrix} \varepsilon_{11}(t) \\ \varepsilon_{22}(t) \\ \gamma_{12}(t) \end{Bmatrix} = \begin{bmatrix} S_{11} & S_{12} & 0 \\ S_{21} & g_{0,22}S_{0,22} & 0 \\ 0 & 0 & g_{0,66}S_{0,66} \end{bmatrix} \cdot \begin{Bmatrix} \sigma_{11}(t) \\ \sigma_{22}(t) \\ \tau_{12}(t) \end{Bmatrix}$$
$$+ \begin{Bmatrix} 0 \\ g_{1,22} \cdot \int_0^t \Delta S_{22}(\psi - \psi') \cdot \dfrac{\mathrm{d}(g_{2,22} \cdot \sigma_{22}(\tau))}{\mathrm{d}\tau} \cdot \mathrm{d}\tau \\ g_{1,66} \cdot \int_0^t \Delta S_{66}(\varphi - \varphi') \cdot \dfrac{\mathrm{d}(g_{2,66} \cdot \tau_{12}(\tau))}{\mathrm{d}\tau} \cdot \mathrm{d}\tau \end{Bmatrix}$$
$$+ \begin{Bmatrix} 0 \\ \left(C_{22} \cdot \int_0^t \sigma_{22}^{N_{22}}(\tau) \cdot \mathrm{d}\tau\right)^{n_{22}} \\ \left(C_{66} \cdot \int_0^t \tau_{12}^{N_{66}}(\tau) \cdot \mathrm{d}\tau\right)^{n_{66}} \end{Bmatrix} \qquad [9.16]$$

where C_{22}, C_{66}, N_{22}, N_{66}, n_{22}, and n_{66} are stress-independent but temperature-dependent material properties. The reduced times ψ, ψ', φ, and φ', are given by

$$\psi = \int_0^t \frac{\mathrm{d}\tau'}{a_{\sigma,22}}, \quad \psi' = \int_0^\tau \frac{\mathrm{d}\tau'}{a_{\sigma,22}}$$
$$\varphi = \int_0^t \frac{\mathrm{d}\tau'}{a_{\sigma,66}}, \quad \varphi' = \int_0^\tau \frac{\mathrm{d}\tau'}{a_{\sigma,66}} \qquad [9.17]$$

In order to eliminate the Volterra-type integrals, the transverse and shear compliances can be expressed using Prony series.

Once time-dependent stress–strain relationships for unidirectional plies are known, the response of a multidirectional composite can be determined by using a modified laminate plate theory subjected to in-plane loads {**N**} and out-of-plane loads {**M**}.

9.5 Modelling mechanical degradation

Composite structural parts are subjected to both cyclic mechanical loading and temperature variations. Due to the greatly anisotropic thermo-mechanical behaviour of each unidirectional ply of a PMC, these two types of loading induce cyclic stresses in each layer. The mismatch of thermal expansion coefficients of fibres and matrix, as well as the difference of ply orientation in the lay-up, are such that local stresses appear, which can take part in the degradation of the laminate. When these thermal variations are cyclic, they induce, at the ply level, cyclic stress variations that can be compared, at this scale, to a fatigue phenomenon leading to transverse matrix cracking, fibre–matrix debonding, and delamination.

9.6 Modelling physical ageing

In traversing the glass transition through a decrease in temperature, a polymeric material is brought into an out-of-equilibrium state and is expected to subsequently recover into equilibrium with time. Structural recovery behaviours similar to those promoted by temperature jumps follow relative humidity jumps (humidity decrease), which induce transition from a plasticized rubbery state to a glassy polymer–water mixture containing a reduced amount of water. The molecular rate processes, responsible for structural recovery below T_g, depend both on temperature and on microscopic order or structure of the system, which is defined by a set of values of ordering parameters. The normalized difference between actual volume and the volume that the system would display if it were at equilibrium at that temperature and pressure (referred to in the following as δ), is a measure of the departure of the structure from equilibrium and is an example of a possible *order* parameter. Actually, δ is considered to be the sum of several contributions, δ_i, each of which can be assumed to be an *order* parameter constituting the internal state variables for the out-of-equilibrium glassy polymer.

Struik[28] originally proposed a method to model physical ageing through the use of a momentary (i.e the test duration is much less than the time the sample was aged prior to loading) creep master curve obtained from a series of short-term creep experiments performed at various ageing times. The momentary creep master curve was then used in conjunction with the *effective time* theory to predict long-term creep in a polymer in the presence of physical ageing.[29,30] Struik demonstrated that the creep properties of glassy polymers are profoundly influenced by the physical ageing process that can persist for a long period of time and can influence toughness, yielding, and non-linear creep.

The alternative TNM–KAHR (Tool, Narayanaswamy, Moynihan–Kovacs, Aklonis, Hutchinson, Ramos) model[31–37] is known to provide a good representation of the structural recovery behaviour of glassy materials in thermal histories. The kinetics of structural recovery in the glassy state depend on both the temperature and the instantaneous structure of the glass. In the following we present, in some detail, this model and subsequent modifications to account for a wide range of phenomena related to physical ageing.

As anticipated, it is possible to attribute to each ordering parameter, δ_i, a fractional contribution to δ:

$$\delta = \sum \delta_i \qquad [9.18]$$

The time evolution of each δ_i can be assumed to be expressed as

$$-\frac{d\delta_i}{dt} = \Delta\alpha_i \cdot \frac{dT}{dt} + \frac{\delta_i}{\tau_i} \qquad [9.19]$$

where $\Delta\alpha_i = \alpha_{il} - \alpha_{ig}$, i.e. the difference between the thermal expansion coefficient of the equilibrium liquid and the non-equilibrium glass, evaluated at the glass transition. The model results in a distribution of retardation mechanisms, each controlling a definite fraction δ_i of the total departure δ and each involving a distinct relaxation time τ_i associated with the rate of change of the i-th *order* parameter and vice versa. As a consequence, a spectrum of retardation times can be defined and each characteristic relaxation time, τ_i, evolves during structural recovery and can be expressed as a function of temperature, T, and departure from equilibrium, δ (i.e. $\tau_i(T, \delta)$):

$$\tau_i(T,\delta) = \tau_{i,\mathrm{REF}} \cdot \exp[-\theta_\mathrm{T} \cdot (T - T_\mathrm{REF})] \cdot \exp[-(1-x) \cdot \theta_\mathrm{T} \cdot \delta/\Delta\alpha] \qquad [9.20]$$
$$= \tau_{i,\mathrm{REF}} \cdot a_\mathrm{T} \cdot a_{\delta_\mathrm{T}}$$

where a_T and a_{δ_T} have the usual meaning of (time–temperature and time–ageing time) shift factors. Here, x is a non-linearity parameter, θ_T is a form of activation energy, T_REF is the reference temperature, $\tau_{i,\mathrm{REF}}$ is the retardation time at the reference temperature and $\Delta\alpha$ is the difference between the coefficient of thermal expansion in the liquid state and in the glassy state. The shift factor a_T incorporates the temperature dependence of τ_i at equilibrium or at constant δ, while the shift factor a_{δ_T} represents the structure-dependent adjustment of time scale at constant temperature. Knowing the retardation times, it is possible to integrate the differential equations for δ_i obtaining their time evolution and, consequently, the time evolution of δ. In particular, the material response in terms of structural evolution as a consequence of an arbitrary temperature history, can be obtained by Boltzmann superposition of responses to differential temperature steps. The final expression for δ reflects the non-exponential nature of relaxation behaviour of glass forming systems: it can be represented either through a spectrum of relaxation times (sum of exponentials) or in terms of a single correlation time (τ_R) using the stretched exponential Kohlrausch–Williams–Watts (KWW) formalism, as introduced by Moynihan *et al.*:[35,36]

$$\delta(z) = -\Delta\alpha \cdot \int_0^z \exp\left[-\left(\frac{z-z'}{\tau_\mathrm{R}}\right)^\beta\right] \frac{\mathrm{d}T}{\mathrm{d}z'} \cdot \mathrm{d}z' \qquad [9.21]$$

where $\beta \leq 1$. $\beta = 1$ means that the system displays a single retardation time. The response of the polymer is expressed in terms of an *effective* time scale, z, which is the time interval measured in the reference state:

$$z = \int_0^t \frac{\mathrm{d}\zeta}{a_\mathrm{T} \cdot a_\delta} \qquad [9.22]$$

The expression for structural recovery has been extended[38] to account for the combined effect of pressure and temperature changes on the structural recovery of glass forming materials. In order to also account for the effect of pressure, equation [9.21] takes the following form:

$$\delta(z)=\int_0^z\left\{-\Delta\alpha\cdot\exp\left[-\left(\frac{z-z'}{\tau_R}\right)^{\beta}\right]\cdot\frac{dT}{dz'}+\Delta k\cdot\exp\left[-\left(\frac{z-z'}{\tau_R}\right)^{\beta}\right]\cdot\frac{dP}{dz'}\right\}\cdot dz' \tag{9.23}$$

where Δk is the difference of compressibility, evaluated at the glass transition, and now z takes the form:

$$z=\int_0^t\frac{d\zeta}{a_T\cdot a_\delta\cdot a_P} \tag{9.24}$$

Here a_T takes the same form as above, a_P can take several forms, while for a_δ the following expression has been proposed:

$$a_\delta=\exp\left[-(1-x)\cdot\delta\cdot\left(-\frac{\theta_T}{\Delta\alpha(P)}+\frac{\theta_P}{\Delta k(P)}\right)\right] \tag{9.25}$$

This model has been further extended to account for the effect of a non-isotropic state of stress in order to solve problems where structural recovery occurs during loading. This extension[39] is based on substituting –⅓ × (*trace of stress tensor*) for the pressure in the previous expressions.

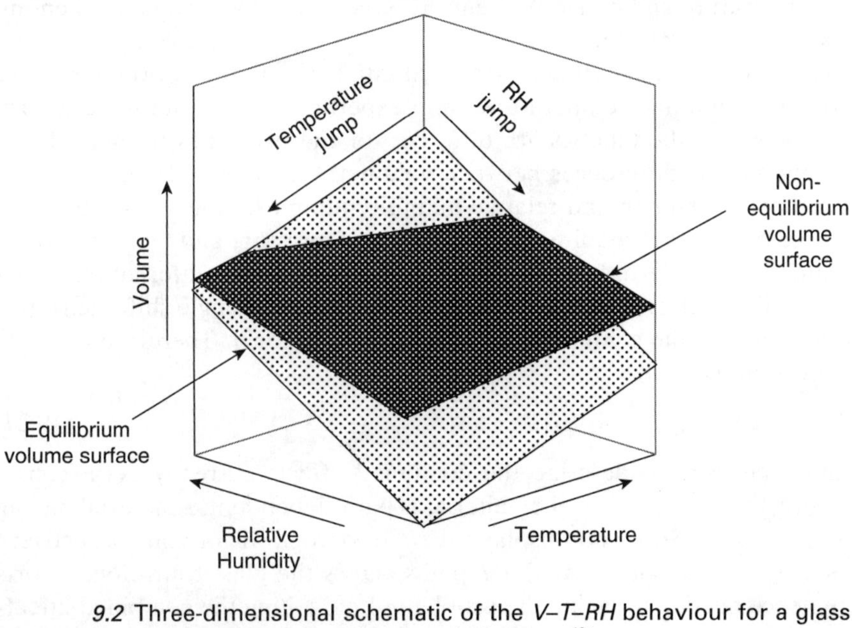

9.2 Three-dimensional schematic of the *V–T–RH* behaviour for a glass forming material (after Zheng and McKenna[40]).

The same theoretical framework has been utilized to interpret the effect of relative humidity (RH) on structural recovery.[6,40] Plasticizing molecules (as is the case of water molecules absorbed in polymer matrices) act similarly to temperature in altering the viscoelastic response (i.e. structural recovery and physical ageing) of the material at a given temperature, simply because the distance from the T_g is altered. A three-dimensional *V–T–RH* plot reported in Fig. 9.2 illustrates the effect on T_g of both temperature and relative humidity.

Moisture, like temperature, merely shifts the retardation spectrum on the time axis without changing its shape. At a constant relative humidity (isopiestic) one traverses the glass transition as temperature is reduced. Similarly, at each T (isotherm), one traverses the glass transition as relative humidity decreases. In fact the change of relative humidity has a similar effect to temperature changes on structural recovery and physical ageing.

The TNM–KAHR model can be extended to isothermal relative humidity histories by simply replacing T, T_R, θ_T and $\Delta\alpha_T$ with RH, RH_R, θ_{RH} and $\Delta\alpha_{RH}$. As with the effect of temperature, a separation between structural and relative humidity effects is possible, obtaining:

$$\tau_i(RH, \delta) = \tau_{i,\mathrm{REF}} \cdot a_{\mathrm{RH}} \cdot a_{\delta_{\mathrm{RH}}} \qquad [9.26]$$

Hence, a time–temperature–moisture superposition principle (TTMSP) can be applied to evaluate the combined effects on relaxation times determined by temperature and moisture themselves as well as by the ageing phenomena associated with them.

However, although there is a qualitative similarity between relative humidity jump and temperature jump experiments, quantitative comparison show that the kinetics are different for the two types of jump. There are anomalous differences between volumetric and viscoelastic responses in temperature-jump and relative humidity-jump experiments to the same final condition that require that the effects of moisture and temperature be treated differently. In fact the dependence of a_{δ_T} on δ is different from that of $a_{\delta_{RH}}$. This difference makes it impossible to have a single shift factor (a_δ) to account for the combined effect of δ_{RH} and δ_T on τ_i. Therefore a simple expression like

$$\tau_i = \tau_{i,\mathrm{REF}} \cdot a_{\mathrm{RH}} \cdot a_{\mathrm{T}} \cdot a_\delta \qquad [9.27]$$

cannot be used to describe the combined effects correctly. As a consequence, the behaviour of the material is path-dependent: a material having the same total δ does not display the same viscoelastic or volume recovery response for two different *RH*/*T* paths across the glass transition. Efforts are hence needed to tackle the problem of modelling the combined effects of temperature jump and relative humidity jump.

9.6.1 Application to composites

As an example of application to composites, we present here the prediction of long-term creep which accounts for the effect of physical ageing, consisting of an analysis performed with off-axis composite specimens by means of momentary tensile creep tests.[41] The experimental data for the elastic compliance of all off-axis specimens were fitted into an exponential function:

$$S_x^e = \frac{S_x^0}{(1-\alpha\cdot\exp(-\gamma\cdot t))} \quad [9.28]$$

S_{11}^e, being dominated by the fibres, is basically independent of time. Moreover, there is no physical ageing effect on Poisson's ratio, υ_{12}. Thus, only shear compliance, S_{66}^e, and transverse compliance, S_{22}^e, are affected by physical ageing. The effect of physical ageing on elastic properties may be different from that on creep compliance (creep compliance can be obtained from the total creep compliance by subtracting the elastic compliance). By defining[41] a creep potential function f as

$$2\cdot f(\sigma_{ij}) = \sigma_{22}^2 + 2\cdot a_{66}\cdot\sigma_{12}^2 \quad [9.29]$$

where a_{66} is the only unknown parameter, it is possible, at each ageing time, to represent the axial creep compliance for an off-axis specimen under uniaxial tension by an *effective* creep compliance expressed in the following way:

$$\bar{S}^c(t) = \frac{\bar{\varepsilon}^c(t)}{\bar{\sigma}} = \frac{\varepsilon_x^c(t)}{\sigma_x}\cdot\frac{1}{h^2(\theta)} = \frac{S_x^c(t)}{h^2(\theta)} \quad [9.30]$$

where ε_x^c is the axial creep strain along x, $\bar{\varepsilon}^c(t)$ is the effective creep strain defined as

$$\bar{\varepsilon}^c = \frac{\varepsilon_x^c}{h(\theta)} \quad [9.31]$$

σ_x is the applied uniaxial stress, $\bar{\sigma}$ is the effective stress defined as

$$\bar{\sigma} = h(\theta)\cdot\sigma_x \quad [9.32]$$

and

$$h(\theta) = \sqrt{\frac{3}{2}}\cdot(\sin^4\theta + 2\cdot a_{66}\cdot\sin^2\theta\cdot\cos^2\theta)^{1/2} \quad [9.33]$$

with θ the fibre orientation relative to the loading direction.

For each ageing time, the momentary effective creep compliance curve can be fitted into a power law of the form

$$\bar{S}^{c}(t)=\left(\frac{t}{\tau}\right)^{\beta} \qquad [9.34]$$

where τ and β are functions of the ageing time.

By introducing shift factors, m and n, for relaxation time and shape factor respectively, the curve for effective creep compliance can be shifted to a reference master curve:

$$\bar{S}^{c}_{ref}(t)=\left(\frac{t}{\tau_{ref}}\right)^{\beta_{ref}}=\left(\frac{t}{m\cdot\tau}\right)^{n\cdot\beta} \qquad [9.35]$$

where

$$m=\left(\frac{t_{a,ref}}{t_{a}}\right)^{\mu_m}; \qquad n=\left(\frac{t_{a}}{t_{a,ref}}\right)^{\mu_n} \qquad [9.36]$$

Thus the relaxation time and shape factor are functions of the ageing time, t_a. Their values at a given ageing time can be related to those at a reference ageing time, $t_{a,ref}$:

$$\tau(t_a)=\tau(t_{a,ref})\cdot\left(\frac{t_{a}}{t_{a,ref}}\right)^{\mu_m}; \qquad \beta(t_a)=\beta(t_{a,ref})\cdot\left(\frac{t_{a,ref}}{t_{a}}\right)^{\mu_n} \qquad [9.37]$$

In contrast to the KAHR-type approach, here the effect of ageing on τ is expressed in terms of ageing time (t_a) instead of δ and also β is considered to be a function of ageing time.

The treatment illustrated above describes the time–ageing time shift effect as applied to creep tests of short duration compared with the elapsed ageing time (momentary tests). If the loading time of the creep test is not short compared with the elapsed ageing time, the ageing occurring as the creep test progresses must be accounted for. As a consequence, the expression for the effective creep compliance is

$$\bar{S}^{c}(t)=\left(\frac{\xi}{\tau_0}\right)^{\beta_0} \qquad [9.38]$$

where τ_0 and β_0 are referred to a reference ageing time, t_a^0, while ξ is the *effective* or *material* time.

It is a relatively simple matter to extend this approach to account properly for non-isothermal histories by implementing a temperature shift factor in a similar way as for the KAHR approach (defining also the appropriate effective or material time which accounts for temperature effects).

9.7 Modelling hygrothermal effects

Low molecular weight substances, such as water, absorb and diffuse into the polymeric matrices of composites, involving chemical and physical aspects and affecting several properties of the polymeric matrices of composites and, in particular, mechanical properties. In the analysis of viscoelastic deformation presented above, we have referred to the effect of mechanical loading, also alluding to the possible influence of moisture content on material response, implicitly assuming a uniform moisture distribution inside the material. In actual service conditions, the mechanical loads act in combination with the environmental factors, such as moisture, temperature, and radiation. This coupled action gives rise to a complex phenomenon at the molecular level inside the material, which induces changes in its macroscopic viscoelastic properties.[26] The understanding of these phenomena and the modelling of time–property evolution is a crucial task for the long-term durability analysis of a polymer-based composite material.

Post-cure reactions, hydrolysis, and leaching of low molecular weight species are some of the reported chemical effects of moisture. The physical effects of moisture are the well-known plasticization and swelling of the polymer matrix, which are only partially reversible. The underlying mechanism is that of a time- and material state-dependent reorganization of the internal structure of the material which accompanies absorption/desorption cycles, probably consisting of a rearrangement of the molecular network with associated changes in the amount of the available free volume.

These phenomena, when coupled with exposure to high temperatures and temperature cycling, are the origin, in PMCs, of micro-structural damage such as fibre debonding and matrix cracking. In fact, the hygrothermal swelling develops a significant stress at the interface between the matrix and un-swollen fibres and causes a change in the residual stresses within the composite, leading to possible micro-crack formation. In fact, humidity induces cracking at temperatures that do not produce noticeable cracking and debonding in a dry environment. These micro-cracks, in turn, provide fast diffusion paths and then alter the moisture absorption characteristics of the laminate.

However, even when the hygrothermal conditions are such that the only physical mechanisms are active, the modelling of moisture effects on the mechanical behaviour is a difficult task and remains, in many respects, an unresolved problem. Since one of the most relevant physical effects due to the presence of absorbed moisture is the depression of T_g and related changes in mechanical properties, much effort has been devoted to the modelling of the compositional dependence of T_g. Such modelling can supply important information to establish the effects of moisture on time–temperature shift factors, by scaling the temperature axis on the basis of T_g change. Free-volume theories have been used to this aim with varying degrees of success. A classical equation obtained on the grounds of free volume theory is as follows:[16,42–44]

$$T_g = \frac{v_1 \cdot \alpha_{f1} \cdot T_{g1} \cdot v_2 + \alpha_{f2} + T_{g2} + k_v' \cdot v_1 \cdot v_2}{v_1 \cdot \alpha_1 + v_2 \cdot \alpha_2} \quad [9.39]$$

where T_g is the resulting glass transition temperature of the polymer–penetrant mixture, v_1 and v_2 are, respectively, the penetrant (water) and polymer volume fractions, T_{g1} and T_{g2} are, respectively, the glass transition temperatures of penetrant and polymer, α_{f1} and α_{f2} are the expansion coefficients, and k'_v is an interaction parameter. Similar expressions were also obtained by Kelley and Bueche.[45]

In general, the physical/chemical action of absorbed water is not uniform inside a composite part, due to the presence of a three-dimensional profile of water concentration, inside the material. In fact, a finite time is needed for water to penetrate inside the material to attain the concentration throughout dictated by the thermodynamic equilibrium with external humidity conditions. Moreover, boundary conditions outside the material can change with time. As a consequence, a concentration profile is established inside the material that changes in space and time. To simplify the matter, we may assume that the characteristic time for transport of water molecules into the composite material is small compared with the characteristic times of other time-dependent processes occurring in the polymer. Hence we may, in the first instance, assume that there is a uniform moisture concentration inside the material and, as illustrated previously, deal with the influence of temperature and moisture on the viscoelastic response of polymeric matrix composites, by using the already-mentioned combined time–temperature–moisture superposition principle (TTMSP). It is worth remembering here that it is only if the effects of temperature and moisture can be uncoupled that the superposition principle can be put in the form of a factorizable time–temperature–moisture shift factor a_{TH}, expressed as $a_{TH} = a_T(T) \cdot a_H(H)$. Frequently, this is not the case, and the time–temperature–moisture shift factor takes a more involved form.

However, the assumption of uniform water concentration is often inadequate to perform a reliable long-term durability analysis of a PMC. In fact, we need to know the time-dependent evolution of the spatial profile of the water concentration and, in turn, the non-uniform distribution of the values of shift factors inside the material. In order to determine the evolution of the profiles of sorbed water, it is necessary to know the mass transport mechanism of water molecules into the material. In fact, mass transport mechanism in polymers can be rather complex, in view of the viscoelastic nature of polymeric materials. However, when the sorbed amount is vanishingly small, the time evolution of penetrant concentration profiles can be determined simply by solving the mass balance equation which, for a one-dimensional geometry case, takes the form

$$\frac{\partial C}{\partial t} = -\frac{\partial J}{\partial x} \qquad [9.40]$$

where the constitutive expression for mass flux, J, is Fick's law:

$$J = -D \cdot \frac{\partial C}{\partial x} \qquad [9.41]$$

where C is the penetrant concentration and D is the polymer–penetrant mutual diffusivity, which, in general, depends upon C and temperature. Equation [9.40] has to be associated to boundary and initial conditions appropriated for the specific physical situation. In imposing boundary conditions, it is generally assumed that in the material, at the interface with the external environment, sorption equilibrium is instantaneously attained. Hence, information about water sorption thermodynamics is also needed. Three-dimensional problems concerning sample shapes where one dimension is much smaller than the others can still be solved by keeping the one-dimensional formalism, but accounting, in the expression for D, for corrections due to edge effects;[5] otherwise, the proper three-dimensional form of mass balance and of constitutive expression for mass flux have to be used. The previous equations hold in the case of an isotropic medium, but Fick's law can be extended to model three-dimensional diffusion in anisotropic media, by introducing a diffusion tensor.

In the case of inhomogeneous materials, as is the case of composite materials, the value of D relative to the neat polymer must be corrected to account for the diffusive path tortuosity and flux area decrease induced by the presence of the fibres, to obtain the value for the composite, D_C, which is an effective anisotropic property that is dependent on diffusion direction. Fibre amount, size, aspect ratio, and orientation all affect the value of D and can be properly accounted for in the expression for D_C. The mass balance and flux constitutive law still retain the form of expressions [9.40] and [9.41], but in laminates with laminae with different orientation the composite material should be considered as a multilayer material with a different diffusivity in each layer.

Solutions for equation [9.40] under a wide range of initial and boundary conditions are available in the excellent, classic book by Crank.[9] One of the distinctive features of ideal Fickian behaviour is that, in a sorption experiment, mass uptake increases linearly with the square root of time for values of $M_t/M_\infty \leq 0.5$ (where M_t is the total amount of penetrant absorbed at time t, while M_∞ is the total amount sorbed at sorption equilibrium). Moreover, sorption kinetics for samples of different thickness collapse on a single curve when (for rectangular one-dimensional geometry) the normalized abscissa $\sqrt{t}/l$ is used (where l is the sample thickness). This feature allows time–thickness scaling enabling long-term prediction of sorption in thick samples from short-term measurement in thin samples. This scaling is obviously applicable only for cases in which the simple Fick's law holds true. If the mass transport mechanism does not change with temperature, the diffusion phenomenon can be accelerated by increasing the temperature. In fact, diffusivity can be assumed to depend on temperature according to an Arrhenius relationship:

$$D = D_0 \cdot \left[\exp\left(\frac{-E_{\text{act}}}{R \cdot T} \right) \right] \quad [9.42]$$

where R is the universal gas constant and E_{act} is the activation energy of the diffusion rate controlling step. It is to be noted, however, that in order to accelerate these processes, it is also necessary to know how water solubility (upon which boundary conditions depend) depends upon temperature. In general the following Vant'Hoff expression is used to describe the dependence of solubility (*Sol*) on temperature:

$$Sol = Sol_0 \cdot \left[\exp\left(\frac{-\Delta H}{R \cdot T} \right) \right] \quad [9.43]$$

where ΔH is the enthalpy of water solubilization.

In fact, mass transport is often coupled with the development of a stress state in the material due to concentration profiles (with related local volume increase) and different absorption capabilities of different constituents in heterogeneous materials, as is the case of composites. These effects can become quite relevant in the case of significant absorption amounts and of significant dependence of polymer properties on the concentration of penetrant molecules. The experimental evidence of such anomalous behaviour is part of a wide scientific literature (see references 7 and 8, to mention some contributions from our group). Several approaches have been proposed to model these phenomena ranging from stress-dependent diffusivities,[9] to approaches based on lumping the effect of the stress state in a flux-relaxation expression for mass flux (rate-type constitutive equation for diffusive flux),[46] to the explicit coupling of mass and momentum balances, accounting for the dependence of penetrant chemical potential on the stress state and of viscoelastic mechanical response on concentration of sorbed penetrant.[47] As a consequence, the simplified view based on Fick's constitutive law (equation [9.41]) reported above frequently cannot be applied and, therefore, accelerating methodologies based on the assumption of validity of such constitutive equations have to be considered with extreme caution. In fact, transport mechanisms could range from purely diffusive to relaxation-controlled (see, for instance, reference 7) as temperature is changed. In view of the larger activation energies typical of polymer relaxation phenomena as compared with those associated with diffusive jumps, relaxational effects tend to dominate at lower temperatures. The matter is further complicated in the presence of strong temperature and stress gradients, which are known to engender non-Fickian driving forces.

Some of the approaches reported above, which account for anomalous effects, involve rather complex constitutive expressions which can imply different levels of difficulty in modelling, related to analytical or numerical issues, or to the excessive number of parameters to be determined in order to perform a reliable prediction of behaviour. In fact, deviations from Fick's law have been treated with simpler approaches based on *two-stage* models,[48,49] or *non-Fickian* models.[50,51] These are based on the idea that anomalies in mass transport (non-Fickian mechanisms) arise if the rate of viscoelastic relaxation in a resin is comparable with the rate of moisture diffusion.

In another approach, the concept of free volume was adapted by Roy *et al.*[52] to unify the effects of polymer viscoelasticity, volumetric change due to strain, and penetrant concentration on the diffusion process within a polymer. The resulting non-linear diffusion coefficient is given by

$$D = \frac{D_0}{T_0} \cdot T \cdot \exp\left\{ \frac{B}{f_0} \cdot \left[\frac{3 \cdot (\alpha \cdot \Delta T + \gamma \cdot C^N) + \delta \cdot \left(\varepsilon_{kk} - \frac{1}{3} \cdot M_0 \cdot \sigma_{kk} \right)}{f_0 + 3 \cdot (\alpha \cdot \Delta T + \gamma \cdot C^N) + \delta \cdot \left(\varepsilon_{kk} - \frac{1}{3} \cdot M_0 \cdot \sigma_{kk} \right)} \right] \right\} \quad [9.44]$$

where T_0 is the reference temperature, T is the current temperature, B is a material constant, f_0 is the reference free volume fraction, α is the linear coefficient of thermal expansion, N is the exponent for the moisture concentration term, ΔT is the difference T-T_0, γ is the linear coefficient for the change of free volume due to dilatational strain, ε_{kk} is the dilatational strain, σ_{kk} is the dilatational stress, C is the moisture concentration at a material point, M_0 is the instantaneous bulk compliance of the polymer and δ is a factor whose value is typically equal to 1.

Various types of coupled non-linear Fickian diffusion processes were numerically simulated using the free-volume approach used in equation [9.44], as well as non-Fickian transport. The non-Fickian transport was modelled as a stress-induced mass flux that typically occurs in the presence of non-uniform stress fields normally present in complex structures.

9.7.1 Application to composites

A fibre-reinforced composite can be envisaged as a three-phase system: the matrix, the fibre–matrix interface region, and the fibres (generally assumed to be water impermeable). Although moisture diffusion in a PMC is mostly dominated by the polymer matrix, the presence of the inter-phase region is also important since, as the fibre diameter and the inter-fibre distances are both very small, a large amount of interface is present. The fibre–matrix interface and fibre architecture may affect moisture diffusion in the composite in a complex manner. In fact, fibres, in view of their rigidity, may restrict matrix segmental motions near the interface when strong adhesion is present, constraining the segmental mobility of macromolecules which is involved in the elementary diffusive steps. As a consequence of water sorption and associated swelling of the matrix, stresses can develop in the polymer phase in the proximity of fibres, affecting, in turn, the mass transport mechanism. Moreover, the diffusion properties of the inter-phase may be different from those of the bulk matrix and point-dependent diffusivities should be used. Finally, absorbed moisture can damage the interface over time and poor wetting of the fibres by the matrix often results in open space at the interface, in which case capillary effects could cause fast diffusion along the fibre direction.

In the following, we give a brief description of a possible simple approach to model water transport in a PMC using a Fick's constitutive expression for mass flux or accounting, in the simplest way possible, for mass transport anomalies. For an ideal composite, where no effect on equilibrium moisture uptake is present, the moisture uptake can be normalized to resin weight gain according to

$$M_{\mathrm{M}} = M_{\mathrm{c}}/w_{\mathrm{r}} \qquad [9.45]$$

where M_{M} and M_{c} are the moisture uptake of the matrix and the composite respectively, and w_{r} is the resin weight fraction in the dry specimen.

In the following we consider the case of a unidirectional composite. As mentioned before, the *effective* diffusivity in a composite, considered as an equivalent homogeneous medium, is anisotropic and depends upon the direction of the mass flow vector. In view of the impermeability of fibres to water, the longitudinal (fibre direction) diffusivity, $D_{\parallel}$, is given by:[53]

$$D_{\parallel} = D_{\mathrm{M}} \qquad [9.46]$$

For diffusivity perpendicular to the direction of fibres, $D_{\perp}$, several expressions have been proposed. Shen and Springer[54] proposed an expression for transverse diffusivity for the case of a composite with impermeable fibres arranged in a square array. This expression, after the correction by Kondo and Taki,[55] reads:

$$D_{\perp} = D_{\mathrm{M}} \cdot \frac{1 - 2\cdot\sqrt{v_{\mathrm{f}}/\pi}}{1 - v_{\mathrm{f}}} \qquad [9.47]$$

where v_{f} is the fibre volume fraction. This model, due to its simplified nature, does not agree with experimental data and underestimate transverse diffusivity. A good estimate of transverse diffusivity, for low volume concentration of impermeable fibres, can be derived from the expression originally proposed by Lord Rayleigh[56] for the electrical conductivity of a medium with embedded cylinders arranged in a rectangular array. In the case of non-conducting cylinders:

$$D_{\perp} = D_{M} \cdot \frac{1 - v_{f} = 0.3058 \cdot v_{f}^{4}}{(1 + v_{f} - 0.3058 \cdot v_{f}^{4}) \cdot (1 - v_{f})} \qquad [9.48]$$

By analogy between diffusivity and shear stiffness, Shirrel and Halpin[57] proposed the following expression for the case of unidirectional composites with impermeable fibres which, after the correction by Kondo and Taki, reads:

$$D_{\parallel} = D_{M} \qquad D_{\perp} = \frac{D_{M}}{1 + v_{f}} \qquad [9.49]$$

Kondo and Taki[55] indicated that a random array of fibres results in a considerably lower transverse diffusivity than a regular array. They introduced a specific parameter, β, to characterize the degree of randomness ($\beta = 0$ being equivalent to a hexagonal array). The higher the value of β, the higher the randomness of packing and the lower the transverse diffusivity.

In order to determine the time and spatial evolution of water concentration profile in a unidirectional composite part, the three-dimensional form of differential mass balance with a Fickian constitutive equation (which holds if the material behaves *ideally*) has to be used with proper diffusivity in each direction (as calculated using one of the above-reported expressions for diffusivity in different directions). If the sample geometry and the boundary conditions determine a one-dimensional problem, edge effects can be neglected and diffusion occurs only along the fibre direction or across the fibre direction, a one-dimensional form of the differential mass balance can be considered, using in the constitutive expression for the mass flux $D_{\parallel}$ or $D_{\perp}$ respectively. In the case where edge effects cannot be neglected, the simplified one-dimensional form of the mass balance can still be retained for short diffusion times, but using an apparent diffusion coefficient defined as (for a rectangular specimen)

$$\sqrt{D_{\parallel,a}} = \sqrt{D_{\parallel}} + \left(\frac{h}{l} + \frac{h}{n}\right) \cdot \sqrt{D_{\perp}}; \qquad \sqrt{D_{\perp,a}} = \sqrt{D_{\perp}} \cdot \left(1 + \frac{h}{n}\right) + \frac{h}{l}\sqrt{D_{\parallel}} \qquad [9.50]$$

where h is the thickness and l and n are the other two dimensions.

However, as already mentioned, composite materials frequently display mass transport anomalies. One of the most common 'anomalous' sorption behaviours observed in a PMC is the attainment of an apparent sorption equilibrium by Fickian diffusion followed by a long-term slow increase in weight gain. This two-stage sorption can be successfully described by two-step diffusion models: the first and second stage of sorption are assumed to be, respectively, diffusion- and relaxation-controlled. The gradual increase in weight gain in the second stage is related to moisture-enhanced structural relaxation:[53]

$$M_t = M_{\infty 0} \cdot \left(1 + k \cdot \sqrt{t}\right) \cdot \left\{1 - \mathrm{epx}\left[-7.3 \cdot \left(\frac{D_x \cdot t}{h^2}\right)^{0.75}\right]\right\} \qquad [9.51]$$

where M_t is the amount of moisture absorbed at time t, $M_{\infty 0}$ is the moisture amount absorbed at equilibrium related to the diffusive controlled stage, D_x is the diffusivity in the x direction and the term $(1 + k \cdot \sqrt{t}\,)$, which characterizes the relaxation-controlled second stage, contains the parameter k that is related to the rate of relaxation.

This kind of behaviour is similar to that displayed by the neat resin, confirming that the behaviour is generally dominated by the polymer matrix. However, quantitative differences are evident when the behaviour of the neat resin is compared with that of the composite, after normalization to account for the presence of the impermeable fibres. This supports the idea that the presence of fibres does influence the structural rearrangement in the second stage, slowing down the relaxation rate due to constraining effects of the rigid fibres. In fact, transport in the first stage involves only small-scale segmental motions which are isolated and local. Larger-scale cooperative motions are possibly enhanced by the plasticizing action of absorbed water and are driven by the swelling stresses. The network gradually relaxes, accommodating additional water. As a consequence, in the second stage, moisture absorption is controlled by the rate of network relaxation. The presence of fibres, which is unlikely to affect the isolated local motion characteristics of the first stage, creates instead a constraint for relatively long-ranged segmental motions involved in relaxation phenomena.

9.8 Modelling chemical ageing

In this context, we will consider only thermo-oxidative degradation phenomena. In general, the two sequential processes that lead to chemical degradation of a polymer due to oxidation are: (a) oxygen diffusion and (b) chemical reaction.[5,58] Modelling of the phenomena is based on a system of differential mass balance equations which incorporate chemical reactions and mass fluxes. As an example, for the case of epoxy matrices, it is possible to simulate the weight loss and the density increase of an epoxy matrix induced by oxidation with a kinetic model of radical chain oxidation coupled with the equation of oxygen diffusion.[59] The reaction can be modelled[59] with a closed-loop radicalic oxidation mechanistic scheme. From this scheme a set of differential equations can be derived (balance of species) which describe the concentration evolution of reactive species. All reactions can be assumed to be irreversible. The oxygen balance is a diffusion-reaction equation and represents the balance between oxygen consumed in chemical reactions, oxygen generated in chemical reactions, and oxygen that is brought in by diffusion. Oxygen diffusivity is assumed to change with diffusion direction due to the anisotropy of the composite. Diffusion transport also has to be taken into account for the balance equations of volatile products. The weight change as a function of time can be calculated by evaluating the weight increase due to grafting of oxygen on the backbone of the polymer chain and the weight decrease due to elimination of water and other volatile compounds. A kinetic model of radical chain oxidation coupled with the equation of oxygen diffusion predicts the concentration profile of oxidation products, the weight loss, and the shrinkage profile in a thick part of neat resin.[60]

Other researchers have used[61] the so-called 'unreacted core' model to characterize and predict thermo-oxidative degradation in a composite laminate. The composite weight loss due to oxidation can be expressed in a simple way with an expression that lumps all the phenomenology in a power law function of time:

$$q_c = D_E \cdot t^n \qquad [9.52]$$

where D_E and n are material constants. For composite degradation controlled by oxygen diffusion, $n = 0.5$, while for degradation processes controlled by chemical reactions, $n = 1$.

In composite materials, the degradation is faster in the 'through the thickness fibre-end surface' than in the compressed surface and much faster than in the surface cut parallel to the fibres. The oxidation mechanism is limited to superficial layers and from a kinetic point of view, in general, it is diffusion-controlled in the parallel direction to fibres whereas it is controlled by the chemical reaction kinetics in the transverse direction.[59] The presence of carbon fibres does not promote a specific effect on the weight loss induced by oxidation at the ageing temperature under consideration.

Another important factor to be accounted for when evaluating long-term performances of an organic matrix is the combination of the effects of thermomechanical cyclic loading and long-term ageing due to temperature and environment.[44] Samples subjected to thermal cycles develop thermal stresses stemming from the mismatch of thermal expansion coefficients of plies with different orientations. When cyclic temperature variations occur in the presence of an oxidative environment, other damaging phenomena appear, due to matrix oxidation. Studies carried out on epoxy resins[60] emphasize the superficial character of oxidation, which results in the creation of a low-thickness oxidized layer on the free edges of the samples subjected to ageing. Oxidation of an epoxy-resin involves a weight loss and a density increase and, consequently, some shrinkage of the skin layer. The oxidation mechanism strongly depends on temperature, specimen geometry, anisotropy, and matrix–fibre debonding.

There are few published works dealing with the long-term behaviour of composites subjected to high temperature under an oxidative atmosphere.

Favre *et al.*[62] highlighted an interaction between matrix micro-cracking of the laminate layers due to the cyclic thermal stresses and oxidation already observed under isothermal ageing. Examples of relevant damages related to these processes are listed below.

1 *The permanent deformation of the matrices due to their shrinkage* that is accompanied, in the presence of an oxidative environment, by matrix–fibre debonding and crack initiation consequent to the high strain gradients present in the matrix areas close to the high-stiffness fibres. Matrix shrinkage does not occur when the atmosphere is pure nitrogen.
2 *The transverse matrix cracking* that is a consequence of thermal cycling. The transverse cracking development is much faster and more important in an oxidative atmosphere than in a neutral one. In an oxidative environment there is also some evidence of delamination between central and external layers.
3 *The weight loss* that is related to thermo-oxidation of the matrix. At the highest operative temperatures (180 °C) mass loss occurs in a nitrogen environment due to thermolysis of the matrix. In oxidizing atmospheres, weight loss is higher due to oxidation.

9.9 Methodology for accelerated testing based on the modelling approach

Modelling of durability in long-term operative conditions and the consequent design of accelerated testing methodologies, rely upon the knowledge of material properties and their relationship with environmental conditions (temperature, pressure, loads, relative humidity, etc.) and upon fundamental damage mechanisms.[1] To this aim, the threshold values of factors that measure acceptable level of performances have to be established and the possible synergistic effects of different degradation mechanisms should be known, although this is, at the moment, a difficult task.

The scope of accelerated methods for the characterization of ageing response is to speed up the accumulation of damage or deformation, potentially leading to failure, by establishing relationships between time of testing at the selected accelerating conditions and real-time or operative cycles. Accelerated testing also aims to determine the material microstructure and damage at the end-of-life.

For a PMC, the main degradation processes to be accelerated are related to mechanical fatigue (for interlaminar and out-of-plane loading), impact damage, and humidity or other fluid exposure. Possible ways of accelerating consist of increasing exposure temperatures and loads, pre-damaging samples, shortening hold times on cyclic exposures, or increasing the concentration of a degradative chemical or compound. It is important to avoid regimes where other degradation mechanisms become active and to consider that accelerated methods become much more complex when multiple mechanisms and synergistic effects are involved because the relationship between accelerated and service conditions is not usually the same for different mechanisms. For this reason it is very difficult to test multiple accelerated conditions directly and simultaneously. The best approach is, probably, to subject samples incrementally to accelerated conditions designed to advance single-failure mechanisms. An approach based on a fundamental understanding of material response, of degradation processes, and models and simulations based on validated accelerated test methods will lead to increased confidence on ageing conditions.

In summary, the accelerated testing protocol must replicate those changes occurring in long-term, real applications (on the grounds of a mechanistic understanding of the fundamental degradation processes), evaluate post-test indicator properties, and determine the accelerating factors for each mechanism as well as the dominant environmental degradation factors and the critical degradation mehanism. The key steps in performing accelerated ageing characterization are as follows:[2] (a) identification of material class; (b) establishment of the critical degradation mechanism (thermo-oxidative, matrix cracking, etc.); (c) selection of an environmental degradation factor; (d) conduction of accelerated ageing tests under conditions (temperature, concentration, loads, sample geometry, etc.) and using procedures dictated by the theoretical modelling of the degradation phenomena and of the dependence of these mechanisms on accelerating factors, performing the tests for experimental times that are equivalent to actual exposure times in the operative life; (e) post-ageing evaluation of indicators (weight change, physical changes, mechanical properties, crack density) that measure variations in performances (key properties) of the PMC.

9.10 Accelerated long-time mechanical behaviour

In this section we present some recently developed methodologies for the prediction of long-time mechanical behaviour and of ultimate properties of a PMC, based on the concepts illustrated in the previous sections. In fact, we present two applications, one illustrating a methodology to accelerate time-dependent viscoelastic behaviour based on the concept of *effective* time (Section 9.10.1) and the other, based on a similar approach, illustrating a methodology to accelerate time-dependent failure of a PMC (Section 9.10.2).

9.10.1 Long-time viscoelastic behaviour of polymer matrix composites

We start illustrating the case of predicting long-time viscoelastic behaviour, based on the assumption that the behaviour of each lamina is linear viscoelastic and thermorheologically simple.[63] At temperatures elevated, but still below T_g, the PMCs exhibit significant time dependence as a consequence of several factors: (a) the overall viscoelastic nature of a polymer close to T_g; (b) changes in the chemical make-up of the polymer (chemical ageing); (c) evolution towards the equilibrium state (physical ageing); (d) moisture absorption (hygrothermal effects); (e) evolving material defects due to loading or environment (damage).

Several complete and general treatments of viscoelastic composite laminate response have been performed, but only a few solutions have been extended to predict the effects of ongoing physical ageing. The effect of ageing, in the framework of composite materials that can be described with linear viscoelastic and thermorheologically simple models, can be included by modifying the *effective* (or *reduced*, or *material*) time function in the analysis. Since the effects of physical ageing in composite laminates have been observed experimentally to vary in the shear and transverse lamina directions, the effective time function needs to be to be directionally dependent for the most general solution.

For the composite solution methods using non-linear material models, inclusion of physical ageing effects is more problematic; two of the issues that must be addressed are the degree to which ageing has already been incorporated into determining the non-linear response of the material and what effect large loads have upon the state of physical ageing of the polymer matrix.

If the lamina is assumed to be thermorheologically simple, the analysis is reduced to linear viscoelasticity, when the problem is mapped to the appropriate effective time domain. This decouples the ageing and the mechanical response, since the ageing shift factors can be determined by the temperature history alone, without regard to the loading applied. A system of convolution integral equations govern the lamina stress–strain relationship. This result can also be adjusted to account for hygrothermal strain effects. Finally, classical thin laminate theory employs the resulting equations to create an algebraic relationship between the laminate force-moment vector and the laminate displacement-curvature vector.

The momentary response for thermorheologically simple materials can be described, in the case of a creep experiment, as

$$S(t)_{te,T} = S_{\mathrm{ref}}(a_{\mathrm{te}} \cdot a_{\mathrm{T}} \cdot t)_{te\ ref,T_{\mathrm{ref}}} \qquad [9.53]$$

where $t_{e\,ref}$ and T_{ref} are the isothermal ageing time and temperature at which the reference curve was defined, t_e and T are the ageing time and temperature at which the short-term test is taking place, and a_{te} and a_T are the horizontal shift factors due to ageing and temperature effects. The above expression describes mechanical response functions (compliance) immediately after a step loading. The momentary nature of the response, S, requires that the time range considered is short enough so that the state of ageing in the material (and, consequently, a_{te} and a_T) remains approximately constant throughout the load. In the case of isothermal ageing:

$$a_{te}(t_e, T) = \left(\frac{t_{e\,ref}}{t_e}\right)^{\mu(T)} \qquad [9.54]$$

where μ is the shift rate and the ageing time t_e is the time elapsed since the quench from $T > T_g$ to $T < T_g$. For non-isothermal ageing, a_{te} can no longer be described using a simple formula in terms of the time since the quench. One approach consists of encapsulating the effects of non-isothermal physical ageing in the calculation of the effective time, ξ, as will be detailed later. Such an approach can be generalized to include, at the same time, the concurrent effects of physical ageing, chemical ageing, and temperature (provided that the material response satisfies the short-time relationship):

$$S(t)_{te,\Omega,T} = S_{ref}(a_{te} \cdot a_\Omega \cdot a_T \cdot t)_{te\,ref,\Omega ref,T ref} \qquad [9.55]$$

here Ω defines, for example, the state of chemical ageing in the material and a_Ω has to be included in the corresponding effective time functions. It is important to note, however, that ageing and temperature effects tend to be strongly coupled when they occur together and, generally, cannot be considered independently of one another (e.g. $a_\Omega = f(t_e,\Omega,T)$).

Effective (or *material* or *reduced*) *time theory* can be employed to account for ongoing ageing in long-term loading cycles. The combined momentary shift factor, a, is used (in analogy to a_T, see equation [9.5]) to create an alternative time scale to determine the response, by mapping the time since load initiation t into effective (or material) time $\xi(t)$:

$$\mathrm{d}\xi = a(t)\cdot \mathrm{d}t \rightarrow \xi(t) = \int_0^t a(\zeta)\cdot \mathrm{d}\zeta \qquad [9.56]$$

The following expression is an example of how the response to an arbitrarily varying stress can be calculated:

$$\sigma(t) \xrightarrow{\text{map to } \tilde{\sigma}(\xi)} \tilde{\varepsilon}(\xi) = \int_0^\xi S_{ref}(\xi - \zeta)\cdot \frac{\mathrm{d}\tilde{\sigma}}{\mathrm{d}\zeta}\cdot \mathrm{d}\zeta \xrightarrow{\text{evaluate } \tilde{\varepsilon}(\xi) \text{ and map to } \varepsilon(t)} \varepsilon(t)$$

[9.57]

or, alternatively

$$\varepsilon(t)=\int_0^t S_{\text{ref}}[\xi(t)-\xi(\zeta)]\cdot\frac{\mathrm{d}\sigma}{\mathrm{d}\zeta}\cdot\mathrm{d}\zeta \tag{9.58}$$

The linear viscoelasticity model used in the following requires that the material response can be both scaled and superposed in the effective time domain. For this condition to be generally satisfied, as already mentioned, the state of ageing in the material must be decoupled from mechanical loading (i.e. the state of ageing depends only on thermal history and is not affected by the applied mechanical loads). In this case, the effective time can be determined prior to the mechanical response solution by an appropriate method, so that at all desired solution times, t_i, the associated effective time $\xi(t_i)$ is known. Struik has demonstrated that, for some cases, such as high stress loadings, the state of ageing is affected by the mechanical loading.[28] If the stress–ageing relationship can be satisfactorily modelled using effective time methods (i.e. via a modified a_{te} or a stress shift factor, a_σ), the model described in the following will also be applicable (although the effective time will need to be calculated during each solution step to account for the changing stress state).

The in-plane lamina response is governed by a matrix of four reference curve functions (compliance $\underline{\underline{S}}$, modulus $\underline{\underline{Q}}$).[63] For an ageing lamina, these functions can be characterized by a series of physical ageing tests at various ageing times and temperatures. To determine the lamina strain, ε, in terms of the stress, σ, and of the compliance, $\underline{\underline{S}}$, in the most general case, an understanding of four ageing compliance responses must be known:

S_{11}: fibre direction compliance;
S_{22}: transverse direction compliance;
S_{66}: shear compliance;
S_{12}: fibre-transverse coupling compliance;

where S_{11} and S_{12} exhibit little time dependence. Only S_{22} and S_{66} display a time dependence, even though the associated effective times are different. If we simplify the matter by assuming that the effective times associated to all extension (non-shear) components are the same (i.e. the effective time $\xi_2(t)$) and that the shear behaviour is governed by the effective time $\xi_6(t)$, we have:

$$\begin{Bmatrix}\sigma_1(t)\\ \sigma_2(t)\end{Bmatrix}\xrightarrow{\xi_2}\int_0^{\xi_2}\begin{bmatrix}S_{11}(\xi_2-\zeta) & S_{12}(\xi_2-\zeta)\\ S_{12}(\xi_2-\zeta) & S_{22}(\xi_2-\zeta)\end{bmatrix}\cdot\frac{\mathrm{d}}{\mathrm{d}\zeta}\begin{Bmatrix}\tilde{\sigma}_1(\zeta)\\ \tilde{\sigma}_2(\zeta)\end{Bmatrix}\cdot\mathrm{d}\zeta$$
$$=\begin{Bmatrix}\tilde{\varepsilon}_1(\xi_2)\\ \tilde{\varepsilon}_2(\xi_2)\end{Bmatrix}\xrightarrow{t}\begin{Bmatrix}\varepsilon_1(t)\\ \varepsilon_2(t)\end{Bmatrix} \tag{9.59}$$

$$\sigma_6(t)\xrightarrow{\xi_2}\int_0^{\xi_6}S_{66}(\xi_6-\zeta)\cdot\frac{\mathrm{d}\tilde{\sigma}_6}{\mathrm{d}\zeta}\cdot\mathrm{d}\zeta=\tilde{\varepsilon}_6(\xi_6)\xrightarrow{t}\varepsilon_6(t) \tag{9.60}$$

The same effective time parameters can be associated to the corresponding modulus functions. As a consequence the lamina stress vector can be expressed in terms of lamina strain vector and lamina modulus $\underline{\underline{Q}}$ as follows:

$$\begin{Bmatrix} \varepsilon_1(t) \\ \varepsilon_2(t) \end{Bmatrix} \xrightarrow{\xi_2} \int_0^{\xi_2} \begin{bmatrix} Q_{11}(\xi_2-\zeta) & Q_{12}(\xi_2-\zeta) \\ Q_{12}(\xi_2-\zeta) & Q_{22}(\xi_2-\zeta) \end{bmatrix} \cdot \frac{\mathrm{d}}{\mathrm{d}\zeta} \begin{Bmatrix} \tilde{\varepsilon}_1(\zeta) \\ \tilde{\varepsilon}_2(\zeta) \end{Bmatrix} \cdot \mathrm{d}\zeta$$
$$= \begin{Bmatrix} \tilde{\sigma}_1(\xi_2) \\ \tilde{\sigma}_2(\xi_2) \end{Bmatrix} \xrightarrow{t} \begin{Bmatrix} \sigma_1(t) \\ \sigma_2(t) \end{Bmatrix} \quad [9.61]$$

$$\varepsilon_6(t) \xrightarrow{\xi_2} \int_0^{\xi_6} (\xi_6-\zeta) \cdot \frac{\mathrm{d}\tilde{\varepsilon}_6}{\mathrm{d}\zeta} \cdot \mathrm{d}\zeta = \tilde{\sigma}_6(\xi_6) \xrightarrow{t} \sigma_6(t) \quad [9.62]$$

Note that the lamina strain in these expressions is that due to load effects alone; other hygrothermal strains required to account for thermal expansion, moisture effects, etc., can be added to these expressions.

Once the behaviour of a lamina has been characterized, it is possible to determine the response of the overall laminate by applying lamination theory. A matrix equation is obtained that can be used to solve for strain and curvature vectors. Once the strain and the curvature vectors are known, the ply-level mechanical strains as well as the stress state can be determined. Thus, all the information about the response of the material is available at the current step and the procedure can be repeated for the next step.

9.10.2 Time-dependent failure criteria for polymer matrix composites

Theoretical life-time models for viscoelastic materials proposed during the last 40 years have focused on a molecular scale and are based on rate theory for breakage of molecular scale bonds, or on a continuum approach based on energy criteria, or on the fracture mechanics theory.[64]

Creep rupture

The Reiner–Weissenberg (R–W) criterion[65] is a classical example of time-dependent failure energy criterion applied to viscoelastic materials. This type of criteria establishes that failure takes place when the stored energy reaches a limit value. The energy limit value is considered a material constant. In accordance with the R–W criterion, it has been found that, for elevated temperatures or very long times, a constant value for the critical fracture energy is considered suitable. Basically, these energy-based criteria are defined as a function of the total energy is the free stored energy. According to this type of approach, it is possible to obtain the life-time under constant load for each criterion, as a function of the applied load, σ_0, and the strength under instantaneous conditions, σ_R.

Let us suppose that the unidirectional strain response of a linear visco-elastic material, under arbitrary stress $\sigma(t)$, is given by

$$\varepsilon(t) = S_0 \cdot \sigma(t) + S_1 \cdot \int_0^t \left(\frac{t-\tau}{\tau_0} \right)^n \cdot \frac{\partial \sigma(\tau)}{\partial \tau} \cdot d\tau \qquad [9.63]$$

The life-time, t_f, under constant load (creep) for the R–W criterion, as a function of applied stress, σ_0, and strength under instantaneous conditions, σ_R, is given as

$$\left(\frac{t_f}{\tau_0} \right) = \left(\frac{1}{2-2^n} \right)^{1/n} \cdot \left(\frac{S_0}{S_1} \right)^{1/n} \cdot \left(\frac{1}{\gamma} - 1 \right)^{1/n} \qquad [9.64]$$

where $\gamma = \sigma_0^2/\sigma_R^2$.

The time-dependent failure criteria can also be deduced using fracture mechanics principles. Originally developed for elastic materials, the fracture mechanics analysis was more recently extended to viscoelastic media to predict time-dependent growth of flaws or cracks. Schapery developed a theory of crack growth and used it to predict the crack speed and failure time for an elastomer under uniaxial and biaxial stress states.[66–68] For a centrally cracked viscoelastic plate, with a creep response give by equation [8.63] under constant load, Schapery deduced a simple relation between stress and failure time:

$$\left(\frac{t_f}{\tau_0} \right) = (B \cdot \sigma_0)^{-2 \cdot (1+(1/n))} \qquad [9.65]$$

where n is the exponent of the creep compliance power law and B a parameter that depends on the geometry and properties of the material.

Recently Christensen[69] developed a kinetic crack formulation to predict the creep rupture life-time of polymers. The life-time was found from the time needed for an initial crack to grow to sufficiently large size as to cause instantaneous further propagation. The method assured quasi-static conditions and only applies to the critical crack failure. The polymeric material was taken to be in the glassy elastic state, as would be normal in most applications:

$$1 - \left(\frac{\sigma(t)}{\sigma_R} \right)^{1/m} = \int_0^{t_f/(\alpha \cdot \tau_0)} (\sigma(\tau)/\sigma_R)^{((1/m)+1)} d\tau \qquad [9.66]$$

Static versus creep strength

The previous time-dependent failure criteria enable us to predict static strength (constant stress/constant strain rate tests) curves from creep strength or vice versa. If we consider a constant stress rate (CSR) loading ($\sigma(t) = R \cdot t$), the strain response and strain rate of a linear viscoelastic material are given by

$$\varepsilon(t) = \left[S_0 + \frac{S_1}{n+1} \cdot \left(\frac{t}{\tau_0} \right)^n \right] \cdot R \cdot t \qquad [9.67]$$

$$\dot{\varepsilon}(t) = \left[S_0 + \frac{S_1}{n+1} \cdot \left(\frac{t}{\tau_0} \right)^n \right] \cdot R \qquad [9.68]$$

In general $S_0 \gg S_1$, and CSR tests produce similar results to constant strain rate tests except for long times or very low rates, i.e.

$$\dot{\varepsilon} \approx S_0 \cdot R \qquad [9.69]$$

It is hence very easy to deduce the life-time under CSR loading from the previous energy-based and fracture mechanics principles:

R–W criterion $$\left(\frac{t_f}{\tau_0} \right) = \left(\frac{(n+1) \cdot (n+2)}{(n+2) + (1 + 2^{n+1})} \right)^{1/n} \cdot \left(\frac{S_0}{S_1} \right)^{1/n} \cdot \left(\frac{1}{\gamma} - 1 \right)^{1/n} \qquad [9.70]$$

Christensen criterion $$\left(\frac{t_f}{\tau_0} \right) = \frac{\alpha}{\sqrt{\gamma}} \cdot \left(\frac{1}{\sqrt{\gamma^{1/m}}} - 1 \right) \cdot \left(\frac{1}{m} + 2 \right) \qquad [9.71]$$

where $\gamma = (R \cdot t_f / \sigma_R)^2$.

Schapery criterion $$\left(\frac{t_f}{\tau_0} \right) = \left[2 \cdot \left(n + \frac{1}{n} \right) + 1 \right] \cdot (B \cdot \sigma_0)^{2 \cdot (1 + (1/n))} \qquad [9.72]$$

with $\sigma_0 = R \cdot t_f$.

These relationships can be obtained also using alternative damage propagation laws.[64] Linear cumulative damage (LCD) integral law has been used to predict life-time for PMCs under constant stress rate from creep life-time, by establishing a relationship between the master curve for *creep strength* and that for the *static strength* (constant stress rate or strain rate).[70,71] In addition, the concept of the strength evolution integral (SEI) has been used to calculate the static strength based on creep–rupture curves.

All the presented criteria predict the same curve shape for *stress–time to failure* curves under creep and CSR loading, i.e. both curves differ only by a constant parameter.

Fatigue

Existing accelerated testing methodologies for metals, which are based on the assumption that fatigue life depends on cycles, but not on time, cannot be simply applied to PMCs, since these methodologies are not intended for viscoelastic materials exhibiting strong time and temperature dependencies.

Miyano *et al.*[70–75] proposed an accelerated testing methodology for the long-term durability (flexural fatigue) of polymer composites based on the time–temperature superposition principle (TTSP) to be held for the viscoelasticity of polymer matrix. As already mentioned, the time–temperature superposition methodology uses elevated temperature to accelerate the mechanical degradations, which occur under loads over long periods of time at lower temperatures. In fact, this principle, originally developed for non-destructive material properties, has recently been shown to be also applicable to failure properties of composite materials. Using this principle, Miyano *et al.* developed a methodology that actually encompasses prediction of several types of long-term life of polymer-based composite materials under various temperature and loading conditions, such as CSR, creep, and fatigue loadings. By this method, the long-term fatigue strength at any time, temperature, and number of cycles to failure, can be predicted using the master curve of fatigue strength obtained based on an accelerated testing methodology. This methodology rests on the following hypotheses.[73]

A The same failure process occurs for the accelerated loading history under elevated temperatures as compared with the low-temperature failure process. Moreover, the failure mechanisms under CSR, creep, and fatigue loading are identical.
B The same TTSP applies for all three types of strengths (i.e. CSR, creep, and fatigue).
C The life under complex loading can be estimated by summing up the damages for individual load steps, i.e. LCD law holds for monotonic loading.
D There is a linear dependence of fatigue strength on stress ratio.

Hypothesis A allows us to relate to relate the three different loading types (i.e. CSR, creep, and fatigue) as follows. Here, a creep loading is considered as a fatigue loading with a stress ratio $R = \sigma_{min}/\sigma_{max} = 1$, and a CSR loading is considered to be equivalent to a half-cycle of fatigue loading with $R = 0$.

The CSR test data for different temperatures are shifted along the logarithmic scale of time to form a smooth curve (*master curve*) at a suitably chosen reference temperature. Values of shift factor as determined from viscoelastic tests performed on the matrix polymer can be used for this purpose (on the basis of hypothesis B the same values of shift factors can be used for the three types of strength). Once obtained, the master curve can then be used to predict the strength under any combination of temperature and time to failure.

The creep strength master curve (i.e. fatigue strength master curve for $R = 1$) can now be constructed starting from the CSR strength master curve, on the basis of hypothesis C (it has to be noted, however, that this hypothesis was proved to be inappropriate to estimate the life under complex loading by summing up the damages for individual load steps). Let $t_S(\sigma)$ and $t_C(\sigma)$ be the CSR and creep failure time for the stress σ. Suppose that the material experiences a non-decreasing stress history $\sigma(t)$ for $0 < t < t^*$, where t^* is the failure time under this stress history. The LCD law states that

$$\int_0^{t^*} \frac{\mathrm{d}t}{t_c[\sigma(t)]} = 1 \qquad [9.73]$$

When $\sigma(t)$ is a constant stress (σ_0), this formula predicts $t^* = t_C(\sigma_0)$. The experimentally determined creep failure master curve obtained by shifting (using the shift factors determined from stress-relaxation flexural tests or CSR tests) the creep strength versus time to failure curves measured at several temperatures, agrees well with that predicted using the procedure based on equation [9.73].

On the other hand, the master curve of fatigue strength at $R = 0$ is constructed from the CSR master curve and the curves relating flexural fatigue strength and number of cycles to failure (S–N curves) at a single frequency and various temperatures, according to hypotheses A and B. Starting from fatigue failure tests at a single frequency for various temperatures at $R = 0$ (reporting flexural fatigue strength versus number of cycles to failure), plots can be obtained reporting fatigue strength at $R = 0$ at reference temperature versus the reduced time to failure t_f', for a fixed frequency, defined as

$$f' = f \cdot a_{T_0}(T); \qquad t_f' = \frac{t_f}{a_{T_0}(T)} = \frac{N_f}{f'} \qquad [9.74]$$

where, again, the shift factor is the same as determined from flexural viscoelastics tests. It is then possible to construct the master curve for fatigue strength at $R = 0$ versus reduced time to failure for fixed N_f (number of cycles to failure) by connecting the points of the same N_f on the curves at fixed frequency. These fatigue master curves can be used to predict the fatigue life under any combination of temperature, time to failure, and cycles to failure, but are obviously applicable only to the case $R = 0$.

At this point of the methodology, the creep strength master curve is available at any temperature, which may be interpreted as the fatigue strength at $R = 1$, as well as the master curves of fatigue strength at $R = 0$. The fatigue strength at an arbitrary value of R can be predicted from the master curve of creep strength, based on hypothesis D, which assumes the linear dependence of fatigue strength upon the stress ratio (R). The proposed simple expression formula is deduced which can be used to predict fatigue master curves for various values of R:

$$\sigma_f(t_f; f, R, T) = \sigma_{f:1}(t_f; f, T) \cdot R + \sigma_{f:0}(t_f; f, T) \cdot (1 - R) \qquad [9.75]$$

where $\sigma_{f:0}$ and $\sigma_{f:1}$ represent, respectively, the fatigue strength at a certain fatigue failure time, frequency, and temperature, and the creep failure

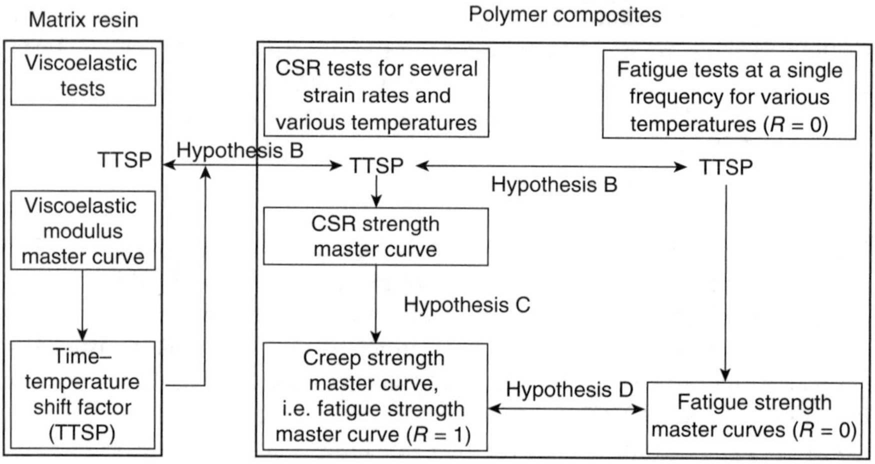

9.3 Scheme of the proposed accelerated testing methodology (after Miyano *et al.*[73]).

strength at a creep failure time equal to the fatigue failure time ($t_C = t_f$), at the same temperature and at an arbitrary frequency. The whole procedure is outlined schematically in Fig. 9.3.

An alternative formulation of this approach has been proposed,[64] which is based on the SEI concept to predict fatigue strength for arbitrary load ratios. If we assume that static fatigue ($R = 1$) and fatigue ($R = 0$) effects on strength degradation have linear dependence upon R, then the remaining strength can be calculated by the following expression:

$$Fr = 1 - R \cdot \int_0^{\tau_1} (1 - Fa) \cdot j_1 \cdot \tau^{j_1 - 1} \mathrm{d}\tau - (1 - R) \cdot \int_0^{\tau_2} (1 - Fa) \cdot j_2 \cdot \tau^{j_2 - 1} \mathrm{d}\tau \qquad [9.76]$$

where $z_1 = t/\tau_1$; $z_2 = t/\tau_2$; τ_1 and τ_2 are characteristic times associated, respectively, with static fatigue ($R = 1$) and fatigue ($R = 0$) processes, j_1 and j_2 are material parameters, Fa is the normalized failure function that applies to a specific controlling failure mode. The failure criterion is given by $Fr = Fa$. This accelerated testing methodology has been experimentally verified both for flexural strength of glass fibre-reinforced polymer (GFRP)[73] and carbon fibre-reinforced polymer (CFRP)[75] laminates.

The illustrated approach can be extended to find similar relationships using – in place of temperature – moisture or other agents.[60] A methodology with time–temperature–moisture superposition would enable the prediction of the life under any temperature and moisture condition, allowing also the use of moisture and other agents for the acceleration of the tests.[76] To this aim, preliminary work performed by Miyano and co-workers, based on an experimental characterization of three-point bending CSR strength or three water absorption conditions, has demonstrated that, for CFRP laminates, the time–temperature–water absorption superposition principle (TTWSP) holds for both the creep compliance and the flexural CSR strength.

Flexural creep compliance and flexural CSR strength master curves have been obtained. In the construction of creep compliance master curves, horizontal time–temperature shift factors and vertical temperature shift factors are used. As a result, the horizontal shift factors are scarcely affected by water absorption, while the vertical shift factors change with water absorption and return perfectly, after drying, to those for the dry specimen. This points to a coupling between moisture and temperature effects. Moreover, it is found that the master curve for creep compliance for dry and wet specimens can be superimposed smoothly, indicating that the TTWSP is applicable for creep compliance. Furthermore, it is demonstrated that the TTWSP holds also for the construction of master curves for flexural CSR strength. It is clear from the master curves that the degradation rate of CSR strength of these CFRP laminates is determined only by increasing of time, temperature, and water absorption, and is independent of the type and weave of carbon fibres.

The next step, and the authors' stated intention, is the extension of the applicability of TTSP and TTWSP to the fatigue strength of CFRP laminates, paving the way for the prediction of long-term strength for CFRP laminates under arbitrary temperature and water absorption conditions using master curves.

Effect of combined states of loading and degradation factors

Most of the following discussion rests upon the results reported in the paper by Reifsnider *et al.*[77] Defining strength in fibrous composite materials cannot generally be done by simply identifying a single stress level that causes failure.[77] Owing to anisotropy and non-homogeneity, the stress state in the material is always rather complex. Moreover, in addition, the proper material strength at local level is anisotropic and spatially non-uniform. Specific stress values must be selected and compared with the correct corresponding strength values to construct a proper strength concept at the global level. To this aim, for a lamina, *failure functions* are commonly defined: the more 'natural' choice is to compare the stresses in the composite directions with their respective strengths. There are failure functions, like the Tsai–Hill function, that are able to account for complex combinations of stresses. In the case of a laminate, the evaluation of the strength for each of the laminae is the same as described above for a single lamina, but the calculation has to account for the constraint effect of the neighbouring laminae. In fact, each ply has to accommodate its deformation to be consistent with the strain of conforming plies. For the laminate, each ply may fail, in various ways, and the failure of the laminate can be said to be defined by 'last ply failure', in some sense.

The most typical approach to the definition of the strength of laminates is to address the failure of each ply with increasing load level or time. In that sense, one can define a 'first ply failure' and subsequent ply failure as the internal stress state changes as a result of internal relaxation. The 'ply discount' method is used which assumes that stress in the matrix of those plies is redistributed, with the effect of increasing various stress components in the other plies.

If failure is defined by fracture, then the last ply to fail can be called a *critical element*, in the sense that global failure is defined by the local failure of that ply. Then, one might consider the failure function in that critical element as a canonical parameter for the definition of strength, or remaining strength in the process of progressive failure of the plies in the laminate. This is, in fact, the fundamental foundation of the *critical element theory*. The concept of a local failure function in the critical element changing with loading history, or more precisely, with changes in internal stress state or material state is the foundation of the remaining strength philosophy. The application of this concept should account for the non-uniform stress state and strength as a point function.

The previous concepts and methods can be extended to the most general case of the calculation of remaining strength and life of composite materials and structures under mechanical, thermal, and environmental applied conditions that produce combinations of fatigue, creep, and stress rupture (time-dependent failure). The first step in the philosophy is to carefully identify the failure mode that is induced by the applied conditions, using experimental methods. Then, using the precepts described above, a failure function form is selected to describe the final failure event. All of the processes that cause changes in the stress state or material state in that critical element are then characterized by rates as a function of the applied conditions and generalized time. Failure modes are primarily a function of the stress state in the critical element, under the thermal conditions expected in service. It is important to note that failure modes can change as a result of applied environments such as temperature, chemical agents, and time or cycles. The failure mode must be determined for the conditions to be modelled, preferably by experimental characterization. The initial stress state and the material state of the material are greatly altered by the history of loading. It is important to track the changes in the stress state and material state in the critical element as a function of the duration and history of the applied conditions, as well as the properties of the constituents and the rates of the degradation processes that act to change the properties. Those elements have to be combined in a self-consistent way, such that the interactions and collective effects of fatigue, creep, stress rupture, and other phenomena are retained.

The concept of *continuity*, Ψ, is now introduced; this is defined in the usual way with a value of 1 when the material is undamaged and 0 when the material is completely damaged. This continuity parameter represents the normalized probability of survival of the material and can be referred as the 'state of the material'. The probability of survival of the material as a function of the occupied energy level of the damage states, e, takes the form

$$\Psi = A^{*} \cdot \exp\left[-\frac{e}{e_{\mathrm{av}}}\right] \qquad [9.77]$$

If we make the postulate that the occupied energy is proportional to the total time over which energy is supplied to the system, at a steady rate, we have

$$\Psi = A^{*} \cdot \exp[-\eta \cdot \tau^{j}] \qquad [9.78]$$

where $\tau = t/\bar{\tau}$, in which t is a time variable, j is a material parameter, and $\bar{\tau}$ is a characteristic time associated with the process. From equation [9.75] one obtains the rate equation for the change in material state due to damage accumulation as a function of generalized time:

$$\frac{\delta\Psi}{\delta\tau} = -\eta \cdot \Psi \cdot j \cdot \tau^{j-1} \qquad [9.79]$$

Another postulate is introduced, i.e. that the continuity of the material can be set equal to the quantity (1 – Fa), where Fa is the *failure function* that applies to a specific controlling failure mode and, in general, is a function of local stress components divided by the corresponding material strength components. On the basis of this approach it is possible to obtain an expression for the 'remaining strength' of the material in the form

$$Fr = 1 - \int_{0}^{\tau_1}\left(1 - Fa\left[\frac{\sigma_{ij}(\tau)}{X_{ij}(\tau)}\right] j \cdot \tau^{j-1}\mathrm{d}\tau\right) \qquad [9.80]$$

which has to be evaluated at the local level, in the critical element. The inputs for this equation must be characterized by the fact that they can be measured using straightforward engineering tests (i.e. fatigue, creep, stress rupture) and that such measurements produce independent input data (e.g. creep should be carefully measured in such a way to avoid cracking and fatigue). These independent characterizations are recombined in equation [9.80] by their collective effect on the arguments in Fa. The effects are combined, affecting in a piecewise linear manner the numerator (local stress components) or denominator (material strength components) of the argument of the Fa function. It should be noted that the coupling of the effects is however accounted for in this procedure. If creep reduces the local stress which is driving crack initiation, for example, the reduced stress level

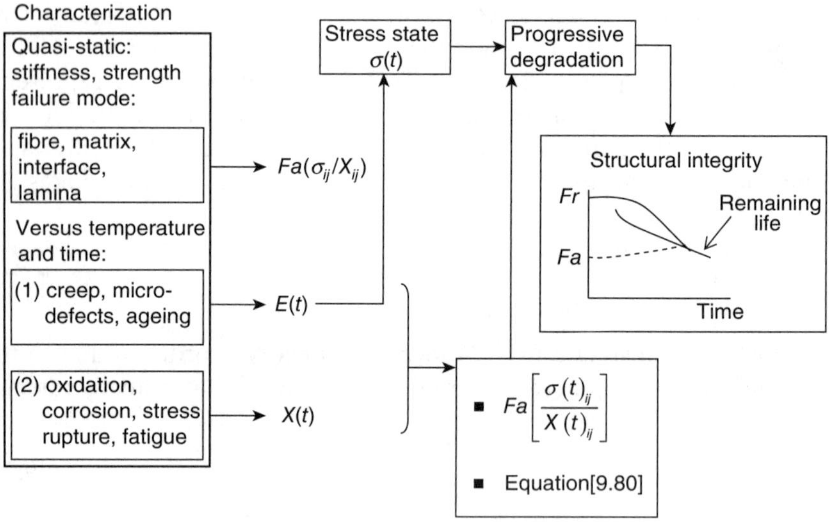

9.4 Schematic diagram of the critical element application methodology (after Reifsnider *et al.*[77]).

is used in the calculation of the next incremental matrix cracking rate. The incremental evaluation of the integral in equation [9.80] brings all coupled effects together by updating the independent variables that enter the rate equations for all degradation processes with each incremental evaluation of the integral. Figure 9.4 shows a schematic outline of the engineering methodology associated with the method. Numerous predictions of remaining strength and life for numerous combinations of fatigue, creep, and stress rupture effects have been made and checked against experimental data.

In conclusion, *life-time* models for viscoelastic materials – i.e. energy-based criteria and fracture mechanics principles extended to viscoelastic media – can be applied to predict the life-time of composite materials under special cases of constant load (creep rupture), CSR to failure, and fatigue strength. In particular, energy-based failure criteria show good prediction capabilities, remarkable potential to extrapolate experimental creep–rupture data with a high degree of confidence.

9.11 Accelerated mechanical degradation

Initiation of damage or the growth of pre-existing damage due to the application of load or deformation are evidence of mechanical degradation mechanisms that result from concurrent physical or chemical ageing that changes the constituent mechanical properties. The dominant mechanisms include matrix cracking, delamination, fibre failures, and inelastic deformation and these will be outlined below.

1 *Transverse matrix cracking* or in-plane micro-cracking can be thermally, hygrothermally, or mechanically induced. Test acceleration is gained by rapid cycling, increased temperature range, or increased ply thickness obtaining a damage progression plot. The results of this empirical approach are difficult to apply to conditions different from those tested. These tests have proven to be useful in screening or ranking materials, but have not enabled predictive capabilities.

   Mechanistic modelling is necessary to allow results to be applied over a broad range of conditions. Modelling is needed for two steps: (a) modelling of the development of matrix cracking and (b) modelling of the effect of matrix cracking on mechanical properties. Models that could be used to predict the onset and development of matrix cracking based on constituent properties must account for the dependence on processing conditions, residual stresses, and thermal cycle range, and the effects of moisture content and distributions on the stresses that lead to cracking. Regarding step (a), examples of mechanistic modelling to predict the development of matrix cracking caused by both mechanical and thermal stresses are available.[78] Regarding step (b), there are a number of analytical methods to determine stiffness reduction in a composite laminate because of matrix cracking,[79] including a shear-lag model, a self-consistent scheme, a strain energy method, a complementary strain energy method, and an internal state variable approach.
2 *Delamination* in PMC laminates occurs when through-the-thickness strain energy release rate exceeds interlaminar fracture toughness. Methods of accelerating this mechanism involve increasing the rate at which the mechanical and thermal spectra are applied. Physical changes (such as void formation) and mechanical properties (such as the change in stiffness and strength) are indicators of this type of mechanical degradation.
3 *Fibre failure* occurs when mechanical loads exceed fibre tensile or compressive strength or from shear-induced failure during out-of-plane loads; it is often increased by environmental degradation factors which promote a decrease in fibre strength. Increase of the fatigue loads and thermal cycle frequency are used to accelerate this mechanism. Indicators are physical changes (such as failure surfaces) and change of mechanical properties (such as stiffness and strength).
4 *Inelastic deformation* is associated with plasticity, creep, or stress relaxation and mainly affects matrix-dominated properties, such as shear and transverse stiffness. Acceleration can be obtained by imposing higher

stresses and temperatures. Indicators are physical changes and changes in mechanical properties (such as reduction of stiffness and strength, non-proportional stress–strain behaviour, and violation of the Boltzmann superposition principle).

9.12 Accelerated physical ageing

The concept of ageing-induced shift factors has been thoroughly discussed in previous sections, where we have illustrated how quantitative relationships can be obtained among the short-term, time-dependent experimental data, ageing-free creep master curves, and the long-term behaviour. The basic assumption is that the time-based response can be accelerated by increasing temperature, humidity, and applied stresses. These acceleration procedures can be modelled using the predictive models illustrated in Section 9.6, for which shift factors and other parameters form the basis of the materials-related inputs.

The indicators used to characterize the physical ageing of polymer composites are the mechanical properties, time-dependent tests (creep and stress relaxation), and static properties (tension and compression strength and stiffness of notched and un-notched specimens). It is worth noting that, for the physical ageing mechanism, loading a material to a non-linear stress level will apparently 'de-age' or reverse-age the material.

9.13 Accelerated hygrothermal degradation

As already discussed, the combined actions of moisture diffusion and thermal exposure can lead to several possible damage modes. Moisture distributions and the effect of hygrothermal cycling on moisture gradients can be characterized. Work needs to be undertaken to allow prediction of residual stress and mechanical property effects from moisture concentration gradients and diffusion histories and to elucidate the synergistic effects of matrix cracking and moisture absorption. To this aim, the more sophisticated models (see Section 9.7) based on the coupling of mass and momentum balances should be applied to design accelerated ageing protocols on a physically sound basis.

If, to simplify, we assume that the water concentration profile is independent of the local state of stress and each lamina is assumed to be linear viscoelastic and thermorheologically simple, we can easily extend the results[63] illustrated in the first part of Section 9.4.6 to include the effect of hygrothermal strain on the long-time mechanical behaviour of PMCs. For a free-standing lamina, the total strain can be determined by simply adding the hygrothermal and mechanical strains without affecting the mechanical response:[63]

$$\begin{Bmatrix} \varepsilon_{11} \\ \varepsilon_{22} \\ \varepsilon_{66} \end{Bmatrix} = \begin{Bmatrix} \varphi_{11} \\ \varphi_{22} \\ \varphi_{66} \end{Bmatrix} - \begin{Bmatrix} e_{11} \\ e_{22} \\ 0 \end{Bmatrix} \quad [9.81]$$

where φ_{ii} are the total strains in the lamina, while e_{11} and e_{22} are the hygrothermal strains in the fibre and transverse directions respectively. Obviously, in this case, the concept of *effective time* used in this approach should account for the contribution of the time–humidity shift factor. As reported in Section 9.7, this physical effect mainly consists of plasticization phenomena, which determine a depression of T_g that is also amenable to a quantitative theoretical evaluation. Again, once the behaviour of a lamina has been characterized, it is possible to determine the response of the overall laminate by applying lamination theory.

Besides the physical effect on long-time mechanical behaviour, hygrothermal degradation also consists of other relevant changes in the chemical–physical structure of the material related to possible hydrolytic reactions. Accelerated ageing protocols should obviously include these important processes, and preferably based on mechanistic modelling.

Ageing by exposure to low molecular weight compounds different from water (e.g. solvents) is also of importance. Damage may occur due to chemical changes in the polymer and loss of polymer material, and can lead to matrix cracks or delamination. Indicators for tracking solvent degradation are: weight increase or decrease, physical changes (e.g. decrease of surface quality, crack density, delamination) and changes of mechanical properties (such as stiffness loss).

9.14 Accelerated thermal degradation and oxidation

The characterization of thermal degradation, including degradation kinetics, of a polymeric composite, under both oxidizing and inert conditions, should be based on the determination of the dominant degradation mechanism/s and on the changes in the dominant mechanism/s, and should account for the anisotropy of composite degradation.[61,80] Moreover, the synergistic effects of time, pressure, and atmosphere on composite degradation should be established.[81,82] This characterization is frequently based on weight-loss measurements as a function of time and temperature, although, as discussed later, the correlation of weight loss with mechanical performance is often unsatisfactory. The speed of weight loss (k) versus time has been modelled by using an Arrhenius' law to calculate an estimate of material life-time:[14]

$$k = A \cdot \exp(-E_{act}/R \cdot T) \quad [9.82]$$

where k represents the ratio between the weight loss and the starting weight of the sample, E_{act} is the activation energy of the phenomenon and A is a pre-exponential factor. However, since degradation pathways are very complex and temperature- and humidity-dependent, the modelling of real-life degradation by thermally accelerated ageing and Arrhenius extrapolations of the results are unlikely to produce very accurate predictions.[13] In fact, care should be applied in establishing time–temperature correlations since the chemistry of degradation of the composite matrix at the higher ageing temperatures can be very different from that seen at the lower temperatures.[12] In particular, testing too near T_g gives degradation rate results that are non-linear with respect to temperatures below T_g, thereby making estimates of useful life-time difficult at best.

With particular reference to thermo-oxidative ageing, there are two ways to accelerate the process:[59] increasing temperature and increasing pressure of oxygen. It has been suggested by Tsotsis and Lee[15] that more realistic accelerated ageing results are obtained by the use of high-pressure air ageing at lower temperatures. In fact, increasing oxygen pressure (thus accelerating the rate of degradation due to oxidation) seems better, since, as already mentioned, a higher test temperature possibly involves other chemical reactions than those observed at moderate temperatures. By comparing such results with real-time ambient pressure results, a scaling law for calculating actual useful life-time might be developed.

Indicators of progression of degradation as determined by thermo-oxidative conditions, elevated temperatures in the absence of oxygen, and cryogenic temperatures are: (a) weight change; (b) physical changes (colour, surface texture, crack density); (c) T_g (as a way of indicating that chain extension or network cross-linking has occurred); (d) mechanical properties (with particular reference to those properties such as fracture toughness and plasticity that are sensitive indicators of short-term ageing owing to their dependence on matrix-dominated behaviour). As an example, in the aeronautical field it is very common to calculate material life-time by using a weight loss that does not exceed 5 wt%. For higher weight loss, the repercussions on residual mechanical properties become too extended. A rather linear correlation between loss in properties (interlaminar shear strength, ILSS) and weight loss has been found in several cases[14] and the slope of the property loss – mass loss curve has been reported to change significantly with temperature due to different behaviours involved at different temperatures (e.g. volatiles evolution, cross-linking, and structural rearrangement of the matrix). With the mentioned limitations, on the basis of this type of analysis it is possible to anticipate evolution of properties in the long term. Nevertheless, for extrapolations, it is again necessary to deepen the study of mechanisms of thermal degradation to separate the different mechanisms that could intervene at temperatures other than those studied.

Accelerated methods should reproduce the failure mechanisms that occur during ageing, instead of relying on changes such as weight loss or decreases in unidirectional tensile strength that occur only after severe matrix or interface degradation.[15] A more useful approach is to view the matrix structure as a series of very different individual groups, some of which are much more susceptible to degradation than others. In fact, it should be considered that thermo-oxidative mechanisms of composite ageing are very surface-selective leading to molecular stiffening, shrinking, and microcracking,[13] whereas many of the commonly used physical testing methods are not. As a consequence, changes in weight and other indicators may not correlate with many of the important mechanical properties, whose changes, often measured in accelerated ageing tests, can lead to very errant predictions, depending on the type of test used. For example, double cantilever beam (DCB) testing measures the strength of the centre of the sample. Moreover, since these composites are often associated with other elements as adhesives and honeycombs in sandwich structures, it would also be necessary to indentify possible synergistic mechanisms of degradation. In fact, analytical models capable of predicting the long-term changes due to chemical ageing are rare, a promising modelling technique being computer simulation of molecular dynamics.

Finally, it is worth noting that chemical degradation makes analytical approaches based on the principles of viscoelasticity and ageing-based superposition quite complicated since the ageing shift rate is a function not only of the temperature, but also of ageing time itself.

9.15 Validation of acceleration procedure by comparison with real-time data

Validation of a newly designed accelerated ageing procedure should be based on real-time testing, within the limits of the time requirements discussed in the introduction to this chapter. It should take place through a comparison of mechanical properties, damage mechanisms, and physical parameters (e.g. weight loss, changes in glass transition or fracture toughness) from accelerated testing with those from real-time testing.

The requirements for real-time testing are that material should be exposed to loads and environmental degradation factors that reproduce the critical aspects of the service environment. The real-time testing provides the baseline data against which all accelerated work must be validated. It should include individual and combined effects of specific environmental degradation factors on the degradation process. From real-time data, it can be determined which degradation factors and degradation modes should be tracked closely in the accelerated test studies. Once an understanding of the material behaviour is obtained, models can be developed to predict material performance under different exposure and loading conditions. These models can be used to simulate the effects of different ageing profiles to help formulate an accelerated ageing scenario.

9.16 Future trends

Some of the goals still to be reached are: reduction of the time required to develop and validate short-time tests; development of reliable non-destructive *in situ* methods to monitor the state of the material; tackling the problems related to the impossibility of accelerating some degradation mechanisms; further investigation and understanding of synergy among different degradation mechanisms; development of standards for experimental data collection; integration of laboratory and field data.

Further development of the capability to perform reliable accelerated ageing procedures in order to predict the long-term durability of resins and fibre-reinforced plastics, should rely upon the ever-increasing sophistication and accuracy of comprehensive analytical models for degradation processes. These models ideally should be further developed to include all the complex phenomena that come into play, including their coupling, taking into account viscoelastic–viscoplastic behaviour, physical and chemical ageing, the effect of sorbed humidity and hygrothermal factors, and tackling the complex stress analysis problems involved – including interactions between non-linear material constitutive behaviour and environmental effects.

A promising field of development is that of molecular dynamics (mainly with reference to coarse-graining procedures that guarantee predictive capability on the mesoscopic scale) which could provide useful information to enable the building of equations describing the constitutive behaviour of materials.

9.17 References

1 NATIONAL MATERIALS ADVISORY BOARD, 'Accelerated aging of materials and structures', Publication NMAB-479, National academy press, Washington DC, 1996.
2 GATES TS, '*On the use of accelerated test methods for characterization of advanced composite materials*', NASA/TP-2003–212407, May 2003.
3 BRINSON HF, MORRIS DH, GRIFFITH WI, DILLARD DA, 'The viscoelastic response of a graphite/epoxy laminate', in *Composite Structures* I/H Marshal, Ed., Applied Science Publishers, London, 1981, pp. 285–300.
4 BRINSON HF, GRIFFITH WL, MORRIS DH, 'Creep rupture of polymer–matrix composites. The accelerated characterization of graphite/epoxy viscoelastic moduli and strengths and creep-rupture predictions', *Experimental Mechanics*, 1981, **21** (9), 329–336.
5 ROY S, 'Computer models for predicting durability', in '*Reinforced Plastics Durability*, G Pritchard, Ed., CRC Press, Boca Raton and Woodhead Publishing Limited, Cambridge, UK, Chapter 12, 1998.

6 ZHENG Y, PRIESTLEY RD, MCKENNA GB, 'Physical aging of an epoxy subsequent to relative humidity jumps through the glass concentration', *Joural of Polymer Science: Part B: Polymer Physics*, 2004, **42**, 2107–2121.
7 MENSITIERI G, DEL NOBILE MA, APICELLA A, NICOLAIS L, 'Moisture–matrix interactions in polymer based composite materials', *Revue de l'Institut Français du Pétrole*, 1995, **50** (4), 1.
8 DEL NOBILE MA, MENSITIERI G, NETTI PA, NICOLAIS L, 'Anomalous diffusion in poly-ether–ether–ketone (PEEK)', *Chemical Engineering and Science*, 1994, **49** (5), 633.
9 CRANK J, *The Mathematics of Diffusion*, Clarendon Press, Oxford, 1975.
10 BOWLES KJ, 'Transverse flexural tests as a tool for assessing damage to PMR-15 composites from isothermal aging in air at elevated temperatures', *SAMPE Quarterly*, 1993, **24** (2), 49–53.
11 HIPP RC, MALLOW A, MCLELLAN PS, RENIERI MD, 'Thermal aging screening of composite materials', paper presented at the 4th NASA/DOD Advanced Composite Technology Conference, Salt Lake City, Utah, June 7–14, 1993.
12 DAO B, HODGKIN J, KRSTNA J, MARDEL J, TIAN W, 'Accelerated aging versus realistic aging in aerospace composite materials. I. The chemistry of thermal aging in a low-temperature-cure epoxy composite', *Journal of Applied Polymer Science*, 2006, **102** (5), 4291–4303.
13 DAO B, HODGKIN J, KRSTNA J, MARDEL J, TIAN W, 'Accelerated aging versus realistic aging in aerospace composite materials. II. Chemistry of thermal aging in a structural composite', *Journal of Applied Polymer Science*, 2006, **102** (4), 3221–3232.
14 MORTAIGNE B, RÉGNIER N, 'Study of epoxy and epoxy–cyanate networks thermal degradation to predict materials lifetime in use conditions', *Journal of Applied Polymer Science*, 2000, **77**, 3142–3153.
15 TSOTSIS TK, LEE SM, 'Long-term thermo-oxidative aging in composite materials: failure mechanisms', *Composites Science and Technology*, 1998, **58**, 355–368.
16 FERRY JD, *Viscoelastic Properties of Polymers*, 3rd edition, John Wiley & Sons, New York, 1980.
17 HADLEY DW, WARD IM, 'Anisotropic and nonlinear viscoelastic behavior in solid polymers', *Rep. Prog. Phys.*, 1975, **38**, 1143–1215.
18 LAI J, BAKKER A, 'Analysis of the non-linear creep of high-density polyethylene', *Polymer*, 1995, **36** (1), 93–99.
19 RAND JL, HENDERSON JK, GRANT DA, 'Nonlinear behavior of linear low-density polyethylene', *Polymer Engineering Science*, 1996, **36** (8), 1058–1064.
20 WILLIAMS ML, LANDEL RF, FERRY JD, 'The temperature dependence of relaxation mechanisms in amorphous polymers and other glass-forming liquids', *J. Am. Chem. Soc.*, 1955, **77**, 3701.
21 SCHAPERY RA, 'A theory of nonlinear thermoviscoelasticity based on irreversible thermodynamics', in *Proceedings of the 5th US National Congress on Applied Mechanics*, ASME 1966, pp. 511–530.
22 SCHAPERY RA, 'Characterization of nonlinear viscoelastic materials', *Polymer Engineering and Science*, 1969, **9** (4), 295–310.
23 ZAPAS LJ, CRISSMAN JM, 'Creep and recovery behaviour of ultra-high molecular weight polyethylene in the region of small uniaxial deformations', *Polymer*, 1984, **25**, 57–62.

24 PASRICHA A, TUTTLE M, EMERY A, 'Time-dependent response of IM7/5260 composites subjected to cyclic thermo-mechanical loading', *Composites Science and Technology*, 1996, **56**, 55–62.
25 CARDON AH, QIN Y, VAN VOSSOLE C, 'Durability analysis of polymer matrix composites for structural application', *Computers and Structures*, 2000, **76**, 35–41.
26 GUEDES RM, MORAIS JJL, TORRES MARQUES A, CARDON AH, 'Prediction of long-term behaviour of composite materials', *Computers and Structures*, 2000, **76**, 183–194.
27 TUTTLE ME, PASRICHA A, EMERY AF, 'Time dependent behaviour of IM7/5260 composites subjected to cyclic loads and temperatures', in *Mechanics of Composite Materials: Non-linear Effects*, AMD vol. 159, American Society of Mechanical Engineers, New Yock, 1993, pp. 343–357.
28 STRUIK LCE, *Physical Aging in Amorphous Polymers and Other Materials*, Elsevier Scientific Publishing Co., Amsterdam, 1978.
29 SULLIVAN JL, 'Creep and physical aging of composites', *Composites Science and Technology*, 1990, **39**, 207–232.
30 BRINSON LC, GATES TS, 'Effects of physical aging on long term creep of polymers and polymer matrix composites', *International Journal of Solids and Structures*, 1995, **32** (6/7), 827–846.
31 TOOL AQ, EICHLIN CG, 'Variations caused in the heating curves of glass by heat treatment', *Journal of the American Ceramic Society*, 1932, **14**, 276–308.
32 TOOL AQ, 'Relation between inelastic deformability and thermal expansion of glass in its annealing range', *Journal of the American Ceramic Society*, 1946, **29**, 240–253.
33 TOOL AQ, 'Effect of heat treatment on the density and constitution of high silica glasses of the borosilicate type', *Journal of the American Ceramic Society*, 1948, **31**, 177–186.
34 NARAYANASWAMY OS, 'A model of structural relaxation in glass', *Journal of the American Ceramic Society*, 1971, **54**, 491–498.
35 MOYNIHAN CT, EASTEAL AJ, TRAN DC, WILDER JA, DONOVAN EP, 'Heat capacity and structural relaxation of mixed alkali glasses', *Journal of the American Ceramic Society*, 1976, **59**, 137–140.
36 DEBOLT MA, EASTEAL AJ, MACEDO PB, MOYNIHAN CT, 'Analysis of structural relaxation in glass using rate heating data', *Journal of the American Ceramic Society*, 1976, **59**, 16–21.
37 KOVACS AJ, AKLONIS JJ, HUTCHINSON JM, RAMOS AR, 'Isobaric volume and enthalpy recovery of glasses. II. A transparent multiparameter theory', *Joural of Polymer Science, Polymer Physics*, 1979, **17**, 1097–1162.
38 RAMOS AR, KOVACS AJ, O'REILLY JM, TRIBONE JJ, GREENER J, 'Effect of combined pressure and temperature changes on structural recovery of glass-forming materials I. Extension of the KAHR model', *Journal of Polymer Science, Part B Polymer Physics*, 1988, **26**, 501–513.
39 GRASSIA L, D'AMORE A, 'Constitutive law describing the phenomenology of subyield mechanically stimulated glasses', *Physical Review E*, 2006, **74**, 021504.
40 ZHENG Y, MCKENNA GB, 'Structural recovery in a model epoxy: comparison of responses after temperature and relative humidity jumps', *Macromolecules*, 2003, **36**, 2387–2396.

41 HU H, SUN CT, 'The characterization of physical aging in polymeric composite materials', *Composites Science and Technology*, 2000, **60**, 2693–2698.
42 JENCKEL E, HEUSCH R, 'Lowering the freezing temperature of organic glasses with solvents', *Kolloid-Zeitschrift*, 1953, **130**, 89–105.
43 GARFIELD LJ, PETRIE SE, 'Viscosity and glass-transition behaviour of polymer-diluent systems', *Journal of Physical Chemistry*, 1964, **68**, 1750–1754.
44 BRAUN G, KOVACS AJ, 'Variations in the glass transition temperature of binary systems of statistical distribution', in *Proceedings of the Conference on Physics of Non-Crystalline Solids*, JA Prins, Ed., North Holland, Amsterdam, 1965, p. 303.
45 KELLEY FN, BUECHE F, 'Viscosity and glass-temperature relations for polymer-diluent systems', *Journal of Polymer Science*, 1961, **50**, 549–556.
46 CAMERA RODA C, SARTI GC, 'Mass transport with relaxation in polymers', *AIChE Journal*, 1990, **36**, 851–860.
47 LUSTIG SR, CARUTHERS JM, PEPPAS NA, 'Continuum thermodynamics and transport theory for polymer-fluid mixtures', *Chemical Engineering Science*, 1992, **47**, 3037–3057.
48 LO SY, HAHN HT, CHAIO TT, 'Swelling of Kevlar 49/epoxy and S2-glass/epoxy composites', in *Progress in Science and Engineering of Composites*, ICCM IV, Tokyo, 1982, pp. 987–1000.
49 GURTIN ME, YATOMI C, 'On a model for two phase diffusion in composite materials', *Journal of Composite Materials*, 1979, **13**, 126–130.
50 SHIRRELL CD, *'Diffusion of water vapour in graphite/epoxy composites'*, ASTM STP 658, 1978, pp. 21–42.
51 WHITNEY JM, BROWNING CE, *'Some anomalies associated with moisture diffusion in epoxy matrix composite materials'*, ASTM STP 658, 1978, pp. 43–60.
52 ROY S, LEFEBVRE DR, DILLARD DA, REDDY NJ, 'A model for the diffusion of moisture in adhesive joints. Part III: numerical simulations', *Journal of Adhesion*, 1989, **27**, 41–62.
53 BAO LR, YEE AF, 'Moisture diffusion and hygrothermal aging in bismaleimide matrix carbon fibre composites – part I: uni-weave composites', *Composites Science and Technology*, 2002, **62**, 2099–2110.
54 SHEN CH, SPRINGER GS, 'Moisture absorption and desorption of composite materials', *Journal of Composite Materials*, 1976, **10**, 2–20.
55 KONDO K, TAKI T, Moisture diffusivity of unidirectional composites, *Environmental Effects in Composite Materials*, GS Springer, Ed., Technomic Publishing Lancaster, PA 1984, Chapter 24, pp. 288–298.
56 RAYLEIGH L, 'On the influence of obstacles arranged in rectangular order upon the properties of a medium', *Philosophical. Magazine*, 1892, **34**, 481–502.
57 SHIRRELL CD, HALPIN J, 'Moisture absorption and desorption in epoxy composite laminates', in *Composite Materials: Testing and Design*, 4th Conference, ASTM STP 617, 1977, pp. 514–528.
58 PARVATAREDDY H, WANG JZ, DILLARD DA, WARD TC, 'Environmental aging of high-performance polymeric composites: effects on durability', *Composites Science and Technology*, 1995, **53**, 399–409.
59 BELLENGER V, DECELLE J, HUET N, 'Aging of a carbon epoxy composite for aeronautic applications', *Composites Part B: Engineering*, 2005, **36**, 189–194.
60 DECELLE J, HUET N, BELLENGER V, 'Oxidation induced shrinkage for thermally aged epoxy networks', *Polymer Degradation and Stability*, 2003, **81**, 239–248.

61 NAM JD, SEFERIS JC, 'Anisotropic thermo-oxidative stability of carbon fibre reinforced polymeric composites', *SAMPE Quarterly*, 1992, October, 10–18.
62 FAVRE JP, LEVADOUX H, OCHIN T, CINQUIN J, 'Viellissement des composites à matrice organbique aux temperatures moyennes: Un premier bilan', in D Baptiste, A Vautrin, Eds, *10èmes Journées Nationales sur les Composites* JNC10, Paris, France, AMAC, 1996, pp. 205–214.
63 BRADSHAW RD, BRINSON LC, 'Mechanical response of linear viscoelastic composite laminates incorporating non-isothermal physical aging effects', *Composites Science and Technology*, 1999, **59**, 1411–1417.
64 GUEDES RM, 'Durability of polymer matrix composites: Viscoelastic effect on static and fatigue loading', *Composites Science and Technology*, 2007, **67**, 2574–2583.
65 REINER M, WEISSENBERG K, 'A thermodynamic theory of the strength of the materials', *Rheology Leaflet*, 1939, **19** (1), 12–20.
66 SCHAPERY RA, 'Theory of crack initiation and growth in viscoelastic media. I. Theoretical development', *International Journal of Fracture*, 1975, **11** (1), 141–159.
67 SCHAPERY RA, 'A theory of crack initiation and growth in viscoelastic media. II. Approximate methods of analysis', *International Journal of Fracture*, 1975, **11** (3), 369–388.
68 SCHAPERY RA, 'A theory of crack initiation and growth in viscoelastic media', *International Journal of Fracture*, 1975, **11** (4), 549–562.
69 CHRISTENSEN RM, 'An evaluation of linear cumulative damage (Miner's law) using kinetic crack growth theory', *Mechanics of Time-Dependent Materials*, 2002, **6** (4), 363–377.
70 MIYANO Y, NAKADA M, MCMURRAY MK, MUKI R, 'Prediction of flexural fatigue strength of CRFP composites under arbitrary frequency, stress ratio and temperature', *Journal of Composite Materials*, 1997, **31** (6), 619–638.
71 MIYANO Y, NAKADA M, KUDOH H, MUKI R, 'Prediction of tensile fatigue life for unidirectional CFRP', *Journal of Composites Materials*, 2000, **34** (7), 538–550.
72 MIYANO Y, MCMURRAY MK, ENYAMA J, NAKADA M, 'Loading rate and temperature dependence on flexural fatigue behavior of a satin woven CFRP laminate', *Journal of Composite Materials*, 1994, **28** (13), 1250–1260.
73 MIYANO Y, NAKADA, SEKINE M, 'Accelerated testing for long-term durability of GFRP laminates for marine use', *Composites Part B: Engineering*, 2004, **35**, 497–502.
74 MIYANO Y, NAKADA, SEKINE M, 'Accelerated testing for long-term durability of FRP laminates for marine use', *Journal of Composites Materials*, 2005, **39** (1), 5–20.
75 MIYANO Y, NAKADA M, NISHIGAKI K, 'Prediction of long-term fatigue life of quasi-isotropic CFRP laminates for aircraft use', *International Journal of Fatigue*, 2006, **28**, 1217–1225.
76 MIYANO Y, NAKADA M, ICHIMURA J, HAYAKAWA E, 'Accelerated testing for long-term strength of innovative CFRP laminates for marine use', *Composites Part B: Engineering*, 2008, **39**, 5–12.
77 REIFSNIDER K, CASE S, DUTHOIT J, 'The mechanics of composite strength evolution', *Composites Science and Techology*, 2000, **60**, 2539–2546.

78 NAIRN JA, HU S, 'The formation and effect of outer-ply microcracks in cross-ply laminates: a variational approach', *Engineering Fracture Mechanics*, 1992, **41**, 203.

79 ALLEN DH, LEE JW, 'Matrix cracking in laminated composites under monotonic and cyclic loadings' in *Microcracking-Induced Damage in Composites*, AMD vol. 111, MD vol. 22, GJ Dvorak and DC Lagoudas Eds, American Society of Mechanical Engineers, New Yock, 1990, pp. 65–75.

80 SALIN IM, SEFERIS JC, LOECHELT CL, ROTHSCHILDS R, 'Time-temperature equivalence in thermogravimetry for BMI composites', *SAMPE Quarterly*, 1992, **24** (1), 54.

81 PRIDE RA, STEIN BA, SCHMIDT FW, 'Mechanical properties of polyimide-resin/glass fibre laminates for various time, temperature and pressure exposure', in *Proceedings of the 23rd Annual Technical Conference of the SPI Composites Institute*, Washington DC, SPI Composites Institute, 1968 pp. 1–8.

82 LARSON FR, MILLER J, 'A time-temperature relationship for rupture and creep stresses', *Transactions of ASME*, 1952, **74**, 765–771.

Part II

Ageing of composites in transport applications

10

Ageing of composites in the rail industry

K. B. SHIN, HANBAT National University, Korea

10.1 Introduction

10.1.1 The use of composites in the rail industry

A reduction in the mass of rail vehicles could lead to weight savings in the traction system, suspension, brakes and other subsystems. A reduced total weight of rail vehicles means less wear on the rails, wheels and bearings, which would then require less maintenance. Lightweight design and a significant reduction in the production costs are the driving forces for the introduction of new material systems in railway applications. Therefore, the use of composites in rail vehicles has increased quite substantially in recent years as designers have come to appreciate the benefits afforded by such systems. The major reasons for the use of composites in rail vehicles could be outlined as follows:[1]

- a specific design approach when using composite materials;
- high design flexibility for particular three-dimensional profiles;
- use of composites leads to lighter and cheaper products as well as high technical performances;
- excellent durability and dimensional stability, etc.

The composites typically used in the rail vehicle consist of low-cost grades of thermoplastics or thermoset polymers reinforced with E-glass fibers. Composites containing high-modulus fibers, such as carbon fiber, and higher-performance resins such as epoxies, are used where the higher cost can be justified to meet special product requirements. Materials such as polymer foams, balsa and honeycomb are used for the production of lightweight, stiff paneling which has for many years found widespread application in the rail industry. Balsa is used as the core material in all sections other than around the headlight box, where it is replaced by polyurethane to simplify manufacturing. Aluminum honeycomb is the ideal material for buffers, fenders and driver protection in the rail industry.

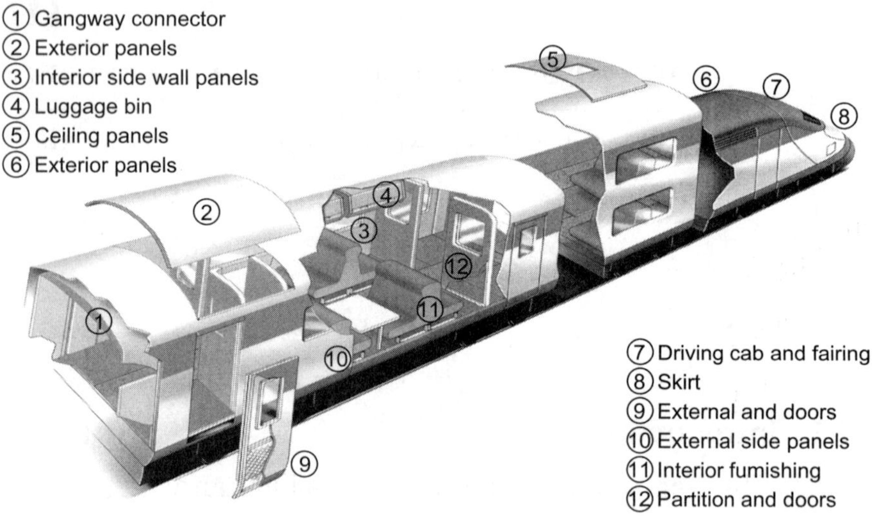

10.1 Detailed structure of rail vehicle made of composites (Hexcel composites).

Rail vehicles are usually divided into major modules such as cab, roof, side wall, end wall and underframe. Figure 10.1 shows the parts of the train commonly made in composites. Europe and Japan are leading the world in the application of composites to passenger trains. One of the most exciting developments in the field of bodyshell construction was unveiled by the Swiss company Schindler Wagon in 1995.[2] Their prototype three-car tilting train features a bodyshell fabricated entirely from composites using a pioneering automated production process. The Japanese Railway Technical Research Institute developed and tested hybrid aluminium–carbon fiber-reinforced polymer (CFRP) structures in 1993. In order to keep production costs down, an automated pultrusion process was used to produce the CFRP panels.

Recently (2006), the Korea Railroad Research Institute (KRRI) has developed the Korean tilting train, as shown in Fig. 10.2. The carbody of the Korean tilting train has been developed using a hybrid design concept combined with composite structures for the bodyshell and a stainless steel structure for the underframe to match the challenging demands with respect to cost-efficient, lightweight design for railway carriage structures. Table 10.1 lists representative trains made of composites.

10.1.2 The importance of research into ageing of composites for the rail industry

The use of composites in the rail industry is becoming widespread and their use for primary load-carrying structures is becoming usual as the advantages

(a)

(b)

10.2 Korean tilting train made of composites (KRRI and Hankuk Fiber).

of composite structures are enhanced. However, in service or when used as stock, composite structures used on ground transportation applications such as rail vehicles will be exposed to the external environment during long-term missions. Rail vehicles will be exposed to a severe external environment for over 30 years. Up to a certain period of exposure, the composite may retain its strength and stiffness above its allowable limits. However, as time passes by, the strength and stiffness may become so low that the material cannot sustain the imposed loads to the structure or maintain the prescribed allowable deflections. The mechanical properties of composite structures when exposed to environmental influences, such as temperature, moisture, ultraviolet light, ozone, salt, chemicals in liquid solutions and gaseous mixtures, may be degraded with time. The reduction of the stiffness

Table 10.1 Representative trains made of composites

Applied part	Composite used	Representative train
Driver cabs	Moulded epoxy prepreg component	Intercity 125 (UK), ETR 500 (Italy), ICN (Switzerland), C20 (Sweden)
Internal fittings	Fabricated with panel materials or sandwich molded construction	Electrostar train (UK), KTX (Korea), Amtrak Surf Liner (USA)
Bodyshells	Fabricated with panel materials or sandwich molded construction	Schindler Waggon (Switzerland), TTX (Korea), Amtrak NEC (USA), Tren Urbano (Puerto Rico), APM Otis (Japan)
Bogies	Specialist components from carbon/glass prepreg	Intercity coach (German), FRP bogie prototype (Japan)

and strength of composite structures will possibly result in a decrease in performance. Therefore, it is a very important issue to predict and evaluate the structural responses of composite structures when exposed to external environments.

10.1.3 A brief history of environmental ageing research on composites

Although much research exists on the effects of ageing of composites, new materials and configurations are continually being developed and need to be studied because understanding the effects of composite ageing is difficult. In particular, less work has been performed on composites for rail vehicle applications.

Baker[3] evaluated the long-term behavior of four composites subjected to environmental conditions for up to 10 years at five different locations in North America. Six specimens of each composite were tested after exposure for 1, 3, 5, 7 and 10 years. A statistically based procedure was applied to determine the mechanical properties of the composites. Shen and Springer[4] studied the effect of moisture and temperature on the tensile strength of composite materials. They measured the degradation of the ultimate tensile strengths of T300/1034 graphite/epoxy composites with material temperatures ranging from 200 to 422 K and moisture contents from 0 (dry) to 1.5% (fully saturated). All measurements were performed using 0°, 90° and π/45 laminates. They showed that changes in temperature in the range 200–380 K appeared to have negligible effects on the ultimate tensile strength of 0° and π/45 laminates, regardless of the moisture content of the material. However, for 90° laminates the moisture content and the temperature of the materials significantly affected the ultimate tensile strength. Larsson[5] evaluated the influence of degradation by ultraviolet light exposure on the mechanical properties of unidirectional Kevlar-49 epoxy laminates of varying thickness. He showed that an exposure for 1000 hours with a xenon burner was supposed to correspond to 3–4 years of outdoor sun exposure in Florida, USA. Starlinger[6] suggested a new design concept in the field of transportation industries to save production costs and to reduce the weight of railway carriages. He conducted structural analyses of a complex hybrid design concept to verify the structural integrity of the components. However, he did not consider the effects of degradation of composite structures due to exposure to external environments. Shin *et al.*[7] evaluated the effects of degradation induced by external environment factors on the T300/AD6005 graphite/epoxy composite materials being considered for use on the Korean tilting train carriage structures. T300/AD6005 graphite/epoxy composite specimens were exposed to natural environments for 5 years and accelerated environmental conditions – including ultraviolet radiation, temperature and moisture – for 2000 hours. In addition, in order to achieve the same effect on the material in a shorter time period, the acceleration factor, a, was introduced. Shin and Hahn[8] studied the structural integrity of the composite carriage structures of the Korean tilting train and the effects of degradation of the properties of T300/AD6005 graphite/epoxy composite materials in the external environment. They showed that the structural integrity of composite railway carriage structures is affected by the degradation of composite materials during its mission life in external environments.

10.1.4 Objectives of the chapter

This chapter describes the ageing test methods and evaluation procedures for predicting the degradation of mechanical properties of composite materials as a result of external environmental agents during long-term missions of rail vehicles. In addition, consideration is given to (a) the correlation of the test results with the data acquired from real-life conditions and (b) the necessity to generate data at a reasonable speed for design requirements. The relationships between natural and accelerated ageing tests are investigated in order to assess the accuracy of predicting the long-term performance of composites using degradation equations.

This chapter also describes the evaluation and verification of railway carriage structural integrity induced by the aged composites through comprehensive examples from field applications of the Korean tilting train in design stage.

10.2 The major environmental ageing factors and their effects on composites for rail vehicle applications

Railroad structures made of polymeric composites may be exposed to severe environmental conditions for over 30 years. Although composite structures may retain their properties at levels above the inherent design allowances for a certain exposure period, with increased ageing time the properties of the composites may become so low that they cannot satisfy the design requirements. Temperature, moisture, ultraviolet light, ozone, chemicals in liquid solutions, and gaseous mixtures are some of the environmental conditions that are of significance. Therefore, it is important to understand the response of composites to environmental exposure. Figure 10.3 shows the major environmental factors and their effects on composites.

10.2.1 Moisture

Most polymers absorb water and, consequently, most composites will absorb moisture, despite the fact that there may not be favorable paths such as voids or debonds along which the water might travel.[9] Generally, the changes in matrix properties induced by moisture uptake will degrade the composite properties. However, in aligned composites the longitudinal tensile strength may be unaffected while the compression strength and interlaminar properties may fall significantly.

10.2.2 Thermal cycling

Thermal cycling can induce microcracking of the resin matrix and degradation of the mechanical properties of a composite as a result of

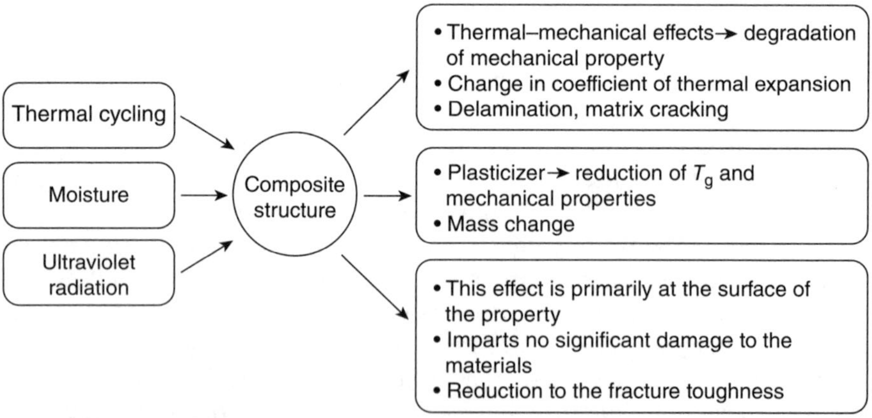

10.3 Environmental factors and their effects on composites. T_g, glass transition temperature.

thermal expansion mismatch. As the thermal cycling progresses the residual stresses are effectively being cycled and this will lead to damage such as fiber fracture, matrix cracking and delamination, with a consequent degradation in mechanical properties.

10.2.3 Ultraviolet radiation

Ultraviolet radiation leads to matrix loss at the surface of composites and changes in optical properties such as discoloration. Ultraviolet radiation at the 330 nm wavelength has sufficient energy to break many bonds found in the polymer materials.

10.2.4 Salt water

If the rail vehicle runs near or along the seashore, the effects of salt should be considered. Exposure tests on glass fiber-reinforced polymer (GFRP), aramid fiber-reinforced polymer (AFRP) and CFRP specimens of 65% fiber weight content (submerged in sea water and under unstressed modes) showed generally that after 2.75 years, the tensile retention ratios of the GFRP and the AFRP specimens were about 65 and 50% of their initial value, respectively.[10]

10.2.5 Protective coatings

When an environmentally resistant composite material cannot be utilized, protection of the material through the use of coatings is necessary. A variety of coatings have been developed for protecting composites from various environments. Pigmented coatings and polyurethanes have been used to prevent ultraviolet radiation and the weathering erosion of salt water.[11]

10.3 Environmental test methods and evaluation procedures for ageing of composites

Understanding the effects of environmental factors on a composite system in the rail industry is the most significant problem in its ground applications. Therefore, evaluation of the degradation of material properties as a result of environmental agents is required to ensure safe operating during long-term rail vehicle missions. In addition, consideration must also be given to (a) the correlation of the test results with the data acquired from real-life conditions and (b) the necessity to generate data at a reasonable speed for design requirements. An understanding of the relationships between natural and accelerated ageing tests is essential.

10.4 Outdoor rack for the natural ageing test.

10.3.1 Natural and accelerated ageing tests

In order to predict the long-term performance of composite materials exposed to outdoor environmental conditions using short-term exposure tests, it is necessary to conduct evaluations of mechanical properties through natural and accelerated ageing tests.

Natural ageing tests

The exposure rack for natural ageing tests on composite materials is constructed according to ASTM D1435.[12] Test specimens should be mounted on wood panels and exposed at a 45° elevation to simulate the real-life environment, for about 5–10 years. Figure 10.4 shows a typical outdoor rack for a natural ageing test.[7]

Accelerated ageing tests

In order to determine the performance and the long-term behavior of a composite in a particular environment, the composite is exposed to the environment in a precisely controlled manner. Specialized equipment is required to ensure that the exposure condition is maintained for the required duration of the test. In many cases, combinations of exposure are used to assess the synergistic effects of exposure types. Accelerated ageing test equipment should be used to control the combined environmental factors of temperature, moisture and ultraviolet light. In addition, the radiation source should be arranged in the center of the chamber and the emitted radiation should be similar to that of natural sun exposure.

(a) (b)

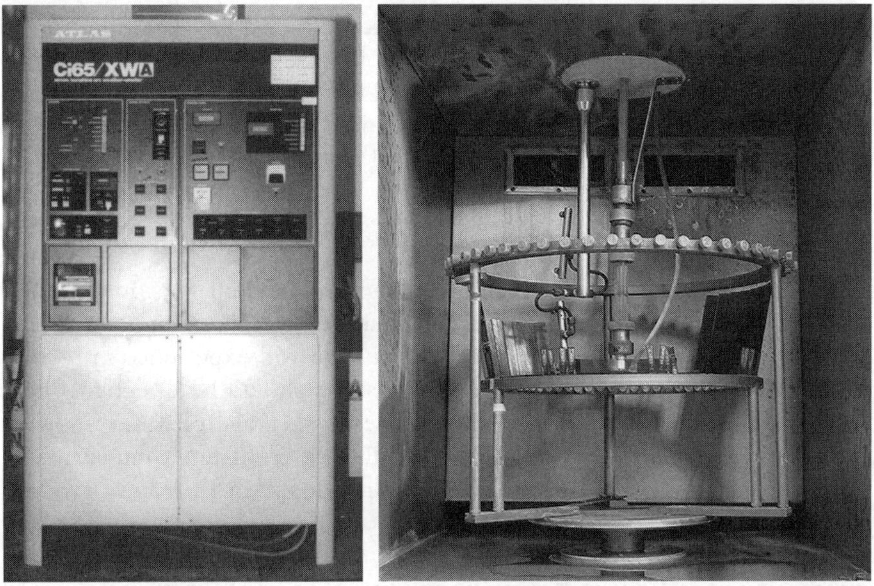

Front view Specimen rack and holder

10.5 Overview of accelerated ageing test equipment.

The accelerated ageing test is generally performed in an Atlas Weather-Ometer using a Sunshine xenon arc lamp as shown in Fig. 10.5.[13] At specific time intervals, specimens are removed from the instrument and tested to determine the weathering effects versus time on various mechanical properties. Ageing conditions should be selected according to ASTM G26,[14] and the program number in the Weather-Ometer should be set to match the operating environments of the composite structures.

10.3.2 Evaluation of the degration of composite properties through natural and accelerated ageing tests

In order to evaluate the degradation of the mechanical and physical properties of composites under the operating environments of rail vehicles, T300/AD6005 graphite/epoxy composite specimens were selected as an example and exposed to natural environments for 5 years and to accelerated environmental conditions – including ultraviolet radiation, thermal cycling and moisture – for 2000 hours. T300/AD6005 graphite/epoxy composites are cured by the filament winding method, and are typical of the commercial composite materials either being considered for or being used in land transportation structures such as rail vehicles in Korea.[15]

Material preparation

After cutting and shaping to predetermined dimensions according to the ASTM standard, specimens should be labeled and their dimensions measured. The specimens used by Shin and Koo[15] were unidirectional. Each subsequent group of specimens should be stored under natural environments and accelerated ageing environments until testing.

The evaluation of ageing of composites through accelerated ageing tests

Firstly, engineers have to select the ageing environment and the program number of the Weather-Ometer accelerated ageing equipment according to the operating environment that the rail vehicle will experience in its lifetime. Generally, the railway carriage structures should be effectively protected against corrosion and should not permit leakage of water, snow or dust when operated at design speed under external environments. The environmental operating conditions for rail vehicles in Korea are given in Table 10.2.

Program 5 in the Weather-Ometer is the best choice for simulating weathering conditions in Korea (from 33 to 43° north) for the accelerated ageing test.[16] A description of the program is given in Table 10.3. The

Table 10.2 Typical environmental conditions for rail vehicles operating in Korea

Ambient temperature	−35 °C~50 °C
Humidity	5~95%
Maximum rain quantity	120 mm/hour (414 mm/day)
Maximum snow quantity	125 mm/hour (296 mm/day)
Wind (10 m above sea level)	Continuous wind: 45 m/s Sudden wind: 50 m/s
Level	1000 m above sea level

Table 10.3 Description of program 5 of the Weather-Ometer

Accelerated conditions	60 min D and SS; 40 min L; 20 min L and SS; 60 min L
Black panel temperature	Light cycle, 60 °C; Dark cycle, 10 °C
Relative humidity	85%
Irradiance level	0.37 W/m^2
Light source	6500 W, water-cooled xenon arc

D, dark; L, light; SS, specimen spray.

program can be set up to test specimens periodically at 0 (baseline), 500, 1000, 1500 and 2000 hours or more. Changes in the mechanical properties of composites after exposure to accelerated ageing environments are measured according to appropriate ASTM standards: (a) tensile properties (ASTM D3039); (b) compressive properties (ASTM D3410); (c) shear properties (ASTM D5379); (d) flexural properties (ASTM D790); (e) interlaminar shear strength (ASTM D2344). The number of specimens per testing is usually six. In order to attach a strain gage to the surface of the specimen, surface treatment of the specimen will be needed. However, this might result in the removal of the exposed surface of the specimen and cause an inaccurate measurement. Therefore, in order to solve this problem, it is recommended that a non-contacting extensometer is used for the measurement of strain on the specimen.

The measured values (averaged)[7] for mechanical properties after exposure to accelerated ageing environments are given in Tables 10.4 and 10.5. As a general rule, the stiffness and strength values of the composites after ageing were lower than those of unexposed specimens, and decreased as the ageing time increased. As shown in Table 10.4, it was found that longitudinal tensile stiffness (LTS) and transverse compressive stiffness (T CS) showed a rapid decrease with increasing ageing time, and the changes in longitudinal compressive stiffness (L CS) and shear stiffness in the 1–2 plane (S_{12}) were moderate. Unusually, the value of transverse tensile stiffness (T TS) was increased with increase in ageing time. This behavior is a direct result of post-curing of the epoxy resin due to exposure to temperature and xenon arc.

Unlike the stiffness behavior, most of the strength properties showed a significant drop, as seen in Table 10.5, except for the shear strength in the

Table 10.4 Variations in stiffness (GPa) for T300/AD6005 graphite/epoxy composite

T300/ AD6005	AT 0 hours	AT 500 hours	L (%)	AT 1000 hours	L (%)	AT 1500 hours	L (%)	AT 2000 hours	L (%)
L TS	122.97	118.34	−3.77	112.11	−8.63	101.39	−17.55	101.45	−17.50
T TS	8.32	7.93	−4.69	8.01	−3.73	8.77	5.41	8.99	8.05
L CS	109.78	105.05	−4.31	103.27	−5.93	102.22	−6.89	105.07	−4.29
T CS	9.90	8.00	−19.19	7.81	−21.11	9.41	−4.95	9.51	−3.94
S_{12}	5.23	4.98	−4.78	5.04	−3.63	4.99	−4.59	4.78	−8.60
L FS	119.85	109.91	−8.29	110.38	−7.90	115.93	−3.27	113.71	−5.12
T FS	8.13	7.36	−9.47	5.64	−30.63	5.53	−31.98	6.64	−18.33

L, longitudinal; T, transverse; TS, tensile stiffness; CS, compressive stiffness; S_{12}, shear stiffness in the 12 plane; FS, flexural stiffness; AT, ageing time; *L* (%), loss of property.

Table 10.5 Variations in strength (MPa) for T300/AD6005 graphite/epoxy composite

T300/ AD6005	AT 0 hours	AT 500 hours	L (%)	AT 1000 hours	L (%)	AT 1500 hours	L (%)	AT 2000 hours	L (%)
L TR	1752.70	1632.97	−6.83	1625.97	−7.23	1558.57	−11.07	1527.56	−12.85
T TR	27.43	21.17	−22.82	19.77	−27.93	24.97	−8.97	19.82	−27.74
L CR	1150.12	1087.94	−5.41	1015.48	−11.71	997.80	−13.24	940.91	−18.19
T CR	155.01	135.09	−12.85	133.99	−13.56	134.65	−13.13	153.76	−0.81
S_{12}	70.95	71.63	0.96	71.01	0.08	72.81	2.62	71.45	0.70
L FR	1222.95	1067.77	−12.69	1068.95	−12.59	1133.75	−7.29	1117.36	−8.63
T FR	40.82	34.62	−15.19	34.01	−16.68	32.73	−19.82	30.40	−25.53
ILSS	56.01	47.04	−16.02	46.48	−17.01	46.12	−17.66	45.65	−18.50

L, longitudinal; T, transverse; TR, tensile strength; CR, compressive strength; $S_{12,}$ shear strength in the 12 plane; FR, flexural strength; ILSS, interlaminar shear strength; AT, ageing time; L (%), loss of property.

1–2 plane (S_{12}) which remained relatively constant. The property that showed the largest change was the transverse tensile strength (T TR). The loss was approximately 28% after 2000 hours of exposure.

Tables 10.4 and 10.5 show that the matrix-dominated mechanical properties, such as transverse flexural strength (T FR) and transverse tensile strength (T TR), showed a sharp reduction with the increase in ageing time. The severe decrease in the matrix-dominated mechanical properties was a direct result of the removal of the matrix due to thermal cracking. The removal of the matrix at the surface of composite specimens with increasing exposure time is shown in Fig. 10.6. The removal of the matrix can lead to the degradation of the mechanical properties of a composite and a reduction in the lifetime in composite structures.[17]

The evaluation of ageing of composites through natural ageing tests

The location for natural ageing tests should be chosen to simulate and represent the environmental conditions in which the rail vehicle will be operating. Therefore, composite specimens were exposed at a 45° elevation facing due south for 5 years in Daejon, Korea, where weathering conditions are average due to its location in the center of the country. The tests that were conducted on the composites were longitudinal and transverse flexural strength and stiffness, and shear in the 1–2 plane. There were six specimens per test. The values measured for mechanical properties after exposure to natural environments are given in Table 10.6 (averaged values). As shown in Table 10.6, the mechanical properties of T300/AD6005 graphite/epoxy composites decreased considerably after exposure to the natural environment for 5 years.

(a)

Baseline (0 hours ageing)

(b)

Accelerated ageing for 500 hours

(c)

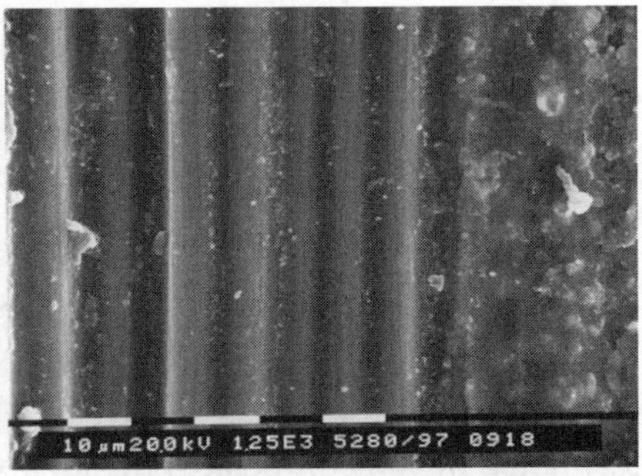

Accelerated ageing for 1000 hours

(d)

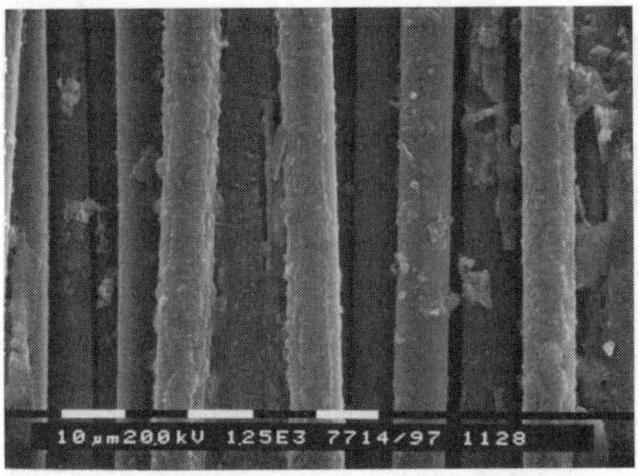

Accelerated ageing for 1500 hours

(e)

Accelerated ageing for 2000 hours

10.6 Surface morphology of aged T300/AD6005 graphite/epoxy composite specimens using scanning electron microscopy.

Table 10.6 Variations in stiffness and strength for T300/AD6005 graphite/epoxy composite after exposure to the natural environmental for 5 years

	0 years (baseline)	After 5 years	Loss (%)
Longitudinal flexural strength	1222.95 MPa	1109.58 MPa	–9.27
Longitudinal flexural stiffness	119.85 GPa	92.18 GPa	–23.09
Transverse flexural strength	40.82 MPa	33.59 MPa	–17.71
Transverse flexural stiffness	8.13 GPa	4.84 GPa	–40.47
Shear stiffness	5.23 GPa	4.04 GPa	–22.75

10.3.3 The degradation equation for stiffness and strength in the aged composites

It is important to know the equivalent time of natural ageing that gives the same conditioning as accelerated ageing in order to predict long-term performance of the composites through short-term exposure test. The empirical approach to the prediction of changes in strength and stiffness of the composites can be expressed by equations [10.1] showing an exponential relationship between the extent of degradation and the ageing time.[18]

$$\begin{aligned} M_i &= A_i \exp[-B_i t]; \qquad i = 1, 2, \ldots, 7 \\ T_j &= C_j \exp[-D_j t], \qquad j = 1, 2, \ldots, 8 \end{aligned} \qquad [10.1]$$

where M_i and T_j are the properties under consideration (stiffness and strength), and t is the natural or accelerated ageing time defined as a function of temperature, ultraviolet radiation and moisture. A_i, B_i, C_j and D_j are exposure constants.

The parameters A_i and C_j are defined as unexposed baseline values of stiffness and strength, and the parameters B_i and D_j are defined as loss rates or extents of degradation. That is, if B_i and D_j have large values, strength or stiffness drops severely with an increase in ageing time. The strength and stiffness loss are expressed by

$$\begin{aligned} \text{Loss}(\%)_{\text{stiffness}} &= \{1 - \exp[-B_i t]\} \times 100 \\ \text{Loss}(\%)_{\text{strength}} &= \{1\text{-}\exp[-D_j t]\} \times 100 \end{aligned} \qquad [10.2]$$

The parameters A_i, B_i, C_j and D_j for the strength and stiffness of the T300/AD6005 graphite/epoxy composite after exposure are given in Table 10.7.

Table 10.7 Exposure constants A_i, B_i, C_j and D_j for T300/AD6005 graphite/epoxy composite

	$M_i = A_i \exp(-B_i t_a)$: Stiffness (GPa)			$T_j = C_j \exp(-D_j t_a)$: Strength (MPa)		
	i	A_i	B_i	j	C_j	D_j
L TP	1	122.97	1.048×10^{-4}	1	1752.70	7.487×10^{-5}
T TP	2	8.32	-2.292×10^{-5}	2	27.43	1.664×10^{-4}
L CP	3	109.78	3.705×10^{-5}	3	1150.12	1.023×10^{-4}
T CP	4	9.90	9.916×10^{-5}	4	155.01	5.892×10^{-5}
S_{12}	5	5.23	4.159×10^{-5}	5	70.95	-7.797×10^{-6}
L FP	6	119.85	3.742×10^{-5}	6	1222.95	6.622×10^{-5}
T FP	7	8.13	1.865×10^{-4}	7	40.82	1.581×10^{-4}
ILSS	—	—	—	8	56.01	1.299×10^{-4}

L, longitudinal; T, transverse; TP, tensile property; CP, compressive property; S_{12}, shear property in the 12 plane; FP, flexural property; ILSS, interlaminar shear strength; t_a, accelerated ageing time (hours).

10.3.4 Relationship between natural and accelerated ageing time

In order to perform the exposure test in a reasonable time period, an acceleration of real time is needed. The primary difficulty in this acceleration of the testing time is in verifying that the data obtained from the accelerated test is that which would be obtained from the real-time test. Therefore, in order to achieve the same effect on the material in a shorter time period, it is important to know the equivalent time of natural ageing to give rise to the same conditioning in the accelerated ageing. In order to determine the relationship between natural and accelerated tests, the acceleration factor, a, is introduced.

In order to determine the relationship between natural and accelerated ageing time, we assumed that the rates of changes appeared to be different under two sets of conditions and that the relationship between them was linear because of relatively short exposure times.[19] On the basis of the experimental data available, the relationship between the two kinds of exposure can be expressed as

$$t_n = a t_a \qquad [10.3]$$

where t_a is time (in hours) under the accelerated exposure to achieve a given incremental change in the property being measured, t_n is the time in years of natural exposure to accomplish the same incremental change (hours), and a is the acceleration factor (constants specific to the weathering apparatus, location, and types of material and material property involved).

For example, the acceleration factor, a, of longitudinal flexural strength may be obtained as follows. As shown in Tables 10.6 and 10.7, the longitudinal flexural strength reduced by 1109.58 MPa after exposure to the natural environment for 5 years. The loss rate is 6.622×10^{-5}. From equations [10.1], we can obtain the equivalent time of accelerated natural ageing to give rise to the same conditioning of the natural ageing for 5 years as follows:

$$\begin{aligned} T_j &= C_j \exp[-D_j t_a], \qquad j = 6 \\ 1109.58 &= 1222.95 \exp[-6.622 \times 10^{-5} - 5t_a] \\ \therefore t_a &\cong 1469 \text{ hours} \end{aligned} \tag{10.4}$$

In the case of longitudinal flexural strength, an exposure of 1469 hours in the Weather-Ometer using a Sunshine xenon arc lamp is shown to correspond to 5 years of outdoor sun exposure in Daejon, Korea.

The acceleration factor, a, can be obtained from equation [10.3]:

$$a = \frac{t_n}{t_a} = \frac{43\,800 \text{ hours (5 years)}}{1469 \text{ hours}} \cong 30 \tag{10.5}$$

In order to achieve the same effect on the degradation of longitudinal flexural strength from environmental ageing tests, it is shown that the accelerated ageing test achieves a 30-fold reduction in the time required compared with the natural ageing test.

The acceleration factors for other properties of the T300/AD6005 graphite/epoxy composite are reported in Table 10.8.

10.3.5 Analytical model for prediction of failure time of the aged composites

In order to predict the reduced failure time (or lifetime) of composite structures induced by environmental factors during long-term missions of rail vehicles, a modified failure criterion as a function of time, t, could be used. It is assumed that failure occurs when a simple to use quadratic formula for stress is satisfied. For example, the failure criteria adopted in the present work are the Hill and Tsai-Wu failure criteria.

From equations [10.1], five fundamental strengths can be expressed as a function of ageing time, t.

Table 10.8 Correlation of accelerated ageing hours to natural exposure hours

	Longitudinal flexural strength	Longitudinal flexural stiffness	Transverse flexural strength	Transverse flexural stiffness	Shear stiffness
Acceleration factor, *a*	30	5.5	31	14	6.4

$$T_1 = X^T = C_1 e^{-D_1 t}, \qquad T_2 = X^c = C_2 e^{-D_2 t}$$
$$T_3 = Y^T = C_3 e^{-D_3 t}, \qquad T_4 = Y^c = C_4 e^{-D_4 t} \qquad [10.6]$$
$$T_5 = S_{12} = C_5 e^{-D_5 t}$$

Where X^T and Y^T are longitudinal and transverse tensile strengths, X^c and Y^c are longitudinal and transverse compressive strength, S_{12} is shear strength in the 1–2 plane. C_j ($j = 1, 2, 3, 4, 5$) are the baseline values and D_j ($j = 1, 2, 3, 4, 5$) are the loss rates due to ageing time.

Hill failure criterion

This theory is a generalization of the Von Mises–Hencky maximum distortional energy theory to include anisotropic materials, and is recommended or used by many authors for quick design checks.[20] For transversely isotropic lamina in the plane stress state, the Hill failure criterion is given by

$$F(\sigma_1, \sigma_2, \sigma_6)_{\text{Hill}} = \frac{\sigma_1^2}{X^2} + \frac{\sigma_2^2}{Y^2} - \frac{\sigma_1\sigma_2}{X^2} + \frac{\sigma_6^2}{S^2} = 1 \qquad [10.7]$$

where σ_1, σ_2 and σ_6 are normal and shear stresses in the direction of the principal material axes.

From equations [10.6], strengths (X and Y) are expressed as

$$X = X' e^{-x't}, \qquad Y = Y' e^{-y't} \qquad [10.8]$$

where: X' is C_1 ($\sigma_1 > 0$) or C_2 ($\sigma_1 < 0$); x' is D_1 ($\sigma_1 > 0$) or D_2 ($\sigma_1 < 0$); Y' is C_3 ($\sigma_2 > 0$) or C_4 ($\sigma_2 < 0$); and y' is D_3 ($\sigma_2 > 0$) or D_4 ($\sigma_2 < 0$). Then, the modified Hill failure criterion can be derived from equations [10.7] and [10.8].

$$F(\sigma_1, \sigma_2, \sigma_6, t)_{\text{mod ified\_Hill}} = \frac{e^{2x't}}{X'^2}(\sigma_1^2 - \sigma_1\sigma_2) + \frac{e^{2y't}}{Y'^2}\sigma_2^2 + \frac{e^{2D_5 t}}{C_5^3}\sigma_6^2 = 1 \qquad [10.9]$$

Tasi-Wu failure criterion

Tsai-Wu has proposed a tensor polynomial failure criterion and it is considered to be the general theory of strength for anisotropic materials.[21] For the case of plane stress, the Tsai-Wu failure criterion is reduced to

$$\begin{aligned} F(\sigma_1, \sigma_2, \sigma_6)_{(\text{Tsai-Wu})} &= F_1\sigma_1 + F_2\sigma_2 + F_6\sigma_6 + F_{11}\sigma_1^2 + F_{22}\sigma_2^2 + F_{66}\sigma_6^2 + 2F_{12}\sigma_1\sigma_2 \\ &= \left(\frac{1}{X^T} - \frac{1}{X^c}\right)\sigma_1 + \left(\frac{1}{Y^T} - \frac{1}{Y^c}\right)\sigma_2 + \left(\frac{1}{X^T X^c}\right)\sigma_1^2 \\ &\quad + \left(\frac{1}{Y^t Y^c}\right)\sigma_2^2 + \frac{\sigma_6^2}{S^2} - \sqrt{\frac{1}{X^T X^c Y^t Y^c}}\,\sigma_1\sigma_2 \\ &= 1 \end{aligned} \qquad [10.10]$$

where F_{12} is assumed to be $-½\sqrt{F_{11}F_{22}}$.

From equations [10.6] and [10.10], the modified Tsai-Wu failure criterion can be expressed as

$$F(\sigma_1, \sigma_2, \sigma_6, t)_{\text{mod ified\_Tsai\_Wu}} = \left(\frac{1}{C_1 e^{-D_1 t}} - \frac{1}{C_2 e^{-D_2 t}}\right)\sigma_1 + \left(\frac{1}{C_3 e^{-D_3 t}} - \frac{1}{C_4 e^{-D_4 t}}\right)\sigma_2 + \frac{\sigma_1^2}{C_1 C_2 e^{-(D_1+D_2)t}} + \frac{\sigma_2^2}{C_3 C_4 e^{-(D_3+D_4)t}} + \frac{\sigma_6^2}{C_5^2 e^{-2D_5 t}} - \sqrt{\frac{1}{C_1 C_2 C_3 C_4 e^{-(D_1+D_2+D_3+D_4)t}}}\,\sigma_1\sigma_2 = 1 \quad [10.11]$$

The exposure constants C_j and D_j for equations [10.9] and [10.11] can be obtained from Table 10.7. If the values of stress components are given or known, then the failure ageing time can be obtained using the modified failure criterion.

10.4 Case study: evaluation of the effect of increased composite ageing on the structural integrity of the bodyshell of the Korean tilting train

When designing a new vehicle, there are a number of requirements that need to be met before the design can be certified. In conditions of considerable load it is primarily the stiffness and strength of the structure that has to be proven. In addition, significant plastic deformations or local failures are not allowed. For these reasons, the yield stress in metal structures and the failure limit in composite structures have to be checked as design criteria.

The Korean tilting train (TTX) was designed for speeds of up to 200 km/h and was developed using a hybrid design concept combined with laminated fiber-reinforced composites for the drive cab, roof, side walls and end walls, and metal structures for the underframe to match the challenging demands – with respect to cost-efficient, lightweight design – of railway carriage structures, as shown in Fig. 10.2. In the TTX project, the railway carriage structures had to comply with the specified requirements shown in Table 10.9.

Composite usage has increased dramatically in railroad applications – for example in the magnetic levitation train, the light rail vehicle and the tilting train – owing to the advantages of light weight, specific strength and stiffness, dimensional stability and tailorability of properties such as coefficient of

Table 10.9 Design requirements for TTX railway carriage structures

	Parts	Limit values
Stress (MPa)	Metal structure	<Yield stress
	Composite structure	Failure criterion[1]
Deflection (mm)	Underframe	<15.97[2]

[1] Direct laminate approach was recommended.
[2] Distance between bogie bolsters/1000.

thermal expansion. Although composite structures may retain their properties above inherent design allowables for a certain exposure period, the properties of the composites may become so low that they cannot satisfy the design requirements as ageing time increases. The environmental effects on these properties may compromise a structure and must be considered during the design process.

The structural integrity of composite railway carriage structures is usually proven in two steps: (a) finite element (FE) analysis and (b) static load testing. The FE analysis is used to verify the design requirements of the hybrid railway carriage structures at the preliminary and detailed design stages prior to manufacture. Before running tests on the tracks, the static load test is performed using full-scale testing according to the specified standard.

In this case study, FE analyses have been conducted to check the design requirements of TTX composite railway carriage structures against static load cases. In addition, TTX hybrid railway carriage structures made of composite laminates, which can be influenced significantly by accelerated environmental factors, have been analyzed and evaluated with respect to the above-mentioned degradation properties of composite materials.

Usually in testing, the maximum loading conditions are subjected to the motorized car. Therefore, the motorized car was selected to evaluate the structural integrity of TTX composite railway carriage structures. In structural analyses, it was assumed that joints such as adhesive layers and rivets were perfectly bonded. The loads to be considered are vertical load induced by the weight of passengers and equipment multiplied by safety coefficients, and twist. In the case of the TTX motorized car, the distributed vertical load of 6.53×10^3 MPa (case 1) is uniformly applied to the upper surface of underframe as shown in Fig. 10.7(a), and the twisting moment of 39.2 kN m (case 2) is achieved by a concentrated load of 26.3 kN applied to both the lower corners of the front end wall, as shown in Fig. 10.7(b).

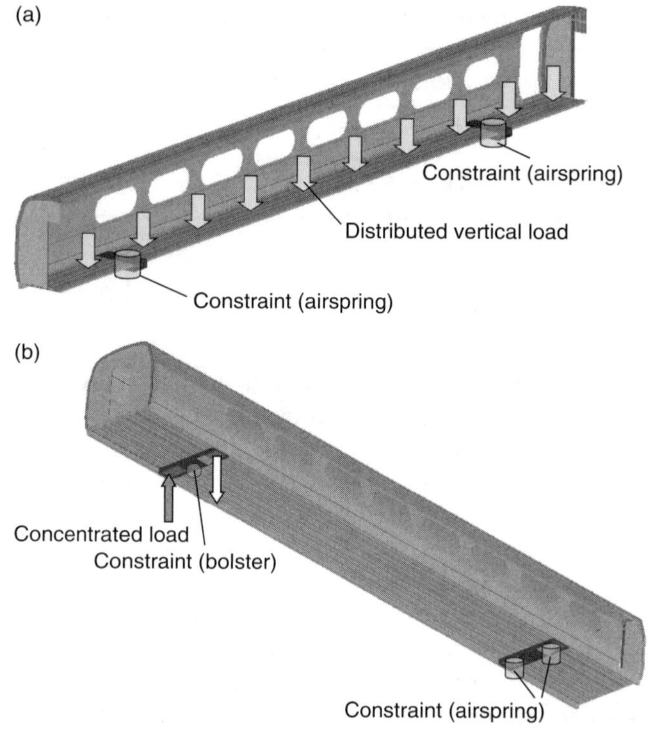

10.7 Boundary and loading conditions for railway carriage structures.

10.4.1 Structural performance before exposure to environmental factors

The bodyshell structures of the TTX have been assumed to be manufactured from T300/AD6005 graphite/epoxy quasi-isotropic laminates with the stacking sequence $[0_n/\pm45_n/90_n]_s$. The ply number, n, is determined by the thickness of the composite structures; and the minimum value of n is 3 and the ply thickness is 0.125 mm. The mechanical properties used in the structural analysis are shown in Table 10.10. These properties are the values for the baseline. Table 10.11 shows the analysis results of the TTX railway carriage structures before the influences of environmental factors on composite materials.

In order to check the design requirements of composite structures, the failure of a bodyshell made of composite laminates was evaluated using direct laminate approaches such as the Puppo–Evensen criterion[22] and the Guess–Gerstle criterion.[23] These criteria are recommended or used for quick checks during the design process[24] of railway carriage structures made of composite laminates. Figure 10.8 shows that all the stress components of

Table 10.10 Baseline mechanical properties of railway carriage structures

Parts	Materials	E_{11} (GPa)	E_{22} (GPa)	G_{12} (GPa)	ν_{12}
Bodyshell structure	T300/AD6005 graphite/epoxy laminate	48.06	48.06	18.53	0.297

E_{11}, longitudinal modulus; E_{22}, transverse modulus; G_{12}, 12 direction shear modulus; ν_{12}, 12 direction Poisson's ratio.

Table 10.11 Results of analysis of the unexposed motorized car

	Loading conditions	Parts	Present values	State
Maximum stress (MPa)	Distributed vertical load (case 1)	Composite structure	64.68	Safe
	Twisting moment (case 2)	Composite structure	66.30	Safe
Maximum deflection (mm)	Distributed vertical load (case 1)	Underframe	15.76	Safe

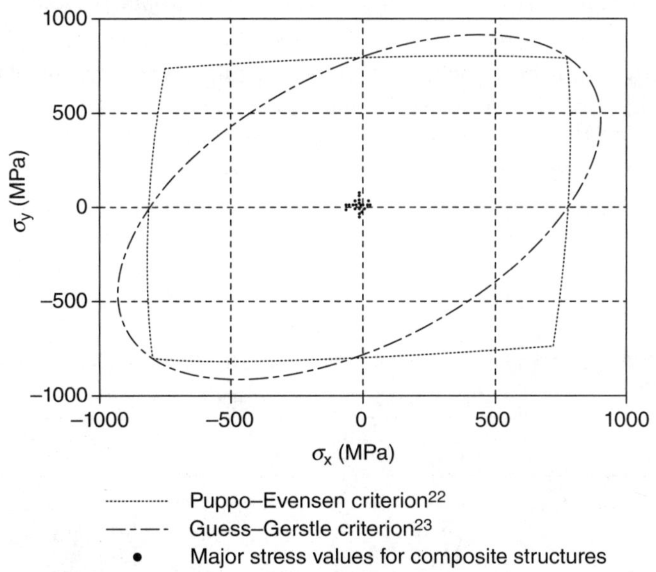

10.8 Failure surface for composite structures using the direct laminate approach.

the composite structures were inside the failure surface. Here, σ_x and σ_y are the stresses of composite structures in the direction of the general coordinate system x, y. The structural analyses of the hybrid railway carriage structures proved the structural integrity of the new design for the load condition specified. The stiffness and strength of the structure are sufficient to fulfill all design requirements.

10.4.2 Structural performances after exposure to environmental factors

The changes in the strength and stiffness of composite materials after exposure to external environments can lead to unexpected structural responses that correspond to the difference between the predicted structural behaviors and actual structural behaviors. Therefore, evaluation of the degradation of composite structures by external environmental factors is necessary to obtain viable hybrid railway carriage designs.

The degradation properties of composite laminates with increasing ageing time are shown in Table 10.12. These values were calculated using unidirectional properties of graphite/epoxy specimens (Table 10.4) obtained by accelerated ageing tests. Table 10.13 shows the results of analysis of the TTX hybrid railway carriage structures, including the degraded properties of composite materials influenced by environmental factors. The results show that all the stress components of hybrid railway carriage structures are inside the safety regions, independent of the degraded properties of the composite laminate. That is to say, the strength of the hybrid railway carriage structures has been sufficient to fulfill design requirements. However, the deflections of the underframe increased with increasing ageing time, and these values exceeded the limit of the design requirements between baseline (0 hour of exposure) and 500 hours of exposure.

The deflections along the central line of the underframe, with increasing ageing time, are shown in Fig. 10.9. These results show that the stiffness of

Table 10.12 Variations of mechanical properties for $[0°_n/\pm45°_n/90°_n]_s$ graphite/epoxy laminate

Ageing time (hours)	E_{11} (GPa) = E_{22} (GPa)	G_{12} (GPa)	ν_{12}
0 (baseline)	48.06	18.53	0.297
500	46.18	17.80	0.297
1000	44.16	17.05	0.295
1500	40.74	15.72	0.296
2000	40.65	15.64	0.299

Table 10.13 Effect of increasing ageing time on analysis results for the motorized car

	Loading conditions	Parts	AT 0 hours	AT 500 hours	AT 1000 hours	AT 1500 hours	AT 2000 hours
Maximum stress (MPa)	Distributed vertical load (case 1)	Composite structure (side walls)	64.48	62.79	60.69	57.41	57.46
	Twisting moment (case 2)	Composite structure (side walls)	66.30	65.82	65.21	64.28	64.38
Maximum eflection (mm)	Distributed vertical load (case 1)	Metal underframe	15.76	16.21 (fail)	16.48 (fail)	17.01 (fail)	17.02 (fail)

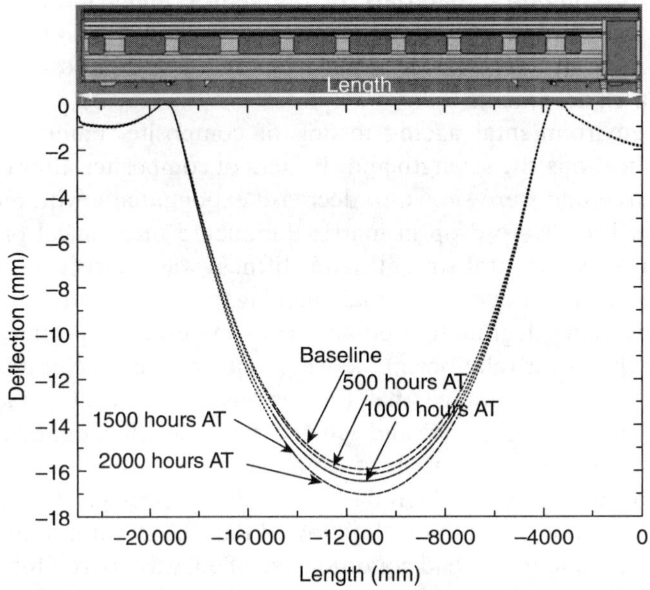

10.9 Comparisons of deflections along the central line of the underframe with increasing ageing time.

the hybrid railway carriage structures, when exposed to environmental influences, degrades with increasing ageing time, and the design requirements can not be guaranteed. Therefore, an understanding of the effects of environmental factors on the degradation of a composite system is the most important design factor in the control of the structural integrity of composite railway carriage structures. In addition, in order to fulfill design requirements, the prediction of the structural behavior of hybrid railway carriage structures following degradation of composite materials may provide some information for the modification of design parameters such as changes in thickness, or alteration of the composite system being considered, or the introduction of coating techniques for composite structures to minimize the influences of environmental factors. In fact, TTX have modified the structural design parameters, for example the insertion of a reinforced beam frame and changes in the jointing method between the bodyshell and the underframe, based on the results of the structural degradation due to the aged composites.

10.5 Conclusions

This chapter has discussed the use and ageing effects of composites in the rail industry and has described the ageing test methods and evaluation procedure for composite materials for rail vehicle applications. The TXX has been considered in order to help understand the degradation effects in railway carriage structures; effects that are induced by composite ageing in the external environment.

For the environmental ageing testing of composite materials for rail vehicle applications, the strength and stiffness of composites after exposure were measured and were shown to decrease exponentially with increasing ageing time. The severe drop in matrix-dominated mechanical properties such as transverse flexural strength and stiffness was a direct result of the removal of the matrix due to thermal cracking. The acceleration factor was evaluated by using degradation equations of material properties of composites and the linear relationship between natural and accelerated ageing time. This has the advantage that the proposed acceleration factor can predict long-term performance and achieve the same effect on the degradation of composite materials as short-term exposure testing.

For the evaluation of the degradation of railway carriage structures due to the ageing composites, the results showed that the structural integrity of the new design, under the load condition specified, was proven for baseline conditions. After exposure to accelerated ageing conditions, it was found that all stress components of the TTX railway carriage structures were inside the safety regions, independent of the effects of degradation of composite materials. However, the deflections of the underframe increased with increasing ageing time, and these values exceeded the limit of the design requirements. Therefore, although the design requirements of hybrid railway carriage structures made of composite materials are satisfied at the initial design stages, the structural integrity of the hybrid railway carriage and the effects of degradation of the composite system must be evaluated to guarantee safety of the train journey on the railway lines.

In summary, environmental effects on the composites used in the rail industry may compromise the integrity of a structure and must be considered during the design process.

10.6 References

1 *Composite car body for Korean Tilting Train Express Project*, KRRI Technical Report, 2004.

2 HERMANN S and ANDREAS S, 'Cost-effective manufacturing process for large foam-cored sandwich structures', *Composites*, 1996 **14** 12–21.

3 BAKER DJ, *Ten-year ground exposure of composite materials used on the bell model 206l helicopter flight service program*, NASA TP 3468, 1994.

4 SHEN CH and SPRINGER GS, 'Effects of moisture and temperature on the tensile strength of composite materials', *Journal of Composite Materials*, 1977 **11**(2) 2–16.

5 LARSSON F, 'The effect of ultraviolet light on mechanical properties of kevlar 49 composites', *Journal of Reinforced Plastics and Composites*, 1986 **5**(1) 19–22.

6 STARLINGER A, 'Transportation industry applications of ABAQUS at ALCAN', *ABAQUS User's Conference Proceedings*, Maastricht, 2001.

7 SHIN KB, KIM CG and HONG CS, 'Correlation of accelerated aging test to natural aging test on graphite-epoxy composite materials', *Journal of Reinforced Plastics and Composites*, 2003 **22**(9) 849–866.

8 SHIN KB and HAHN SH, 'Evaluation of the structural integrity of hybrid railway carriage structures including the ageing effects of composite materials', *Composite Structures*, 2005 **68**(2) 129–137.

9 KERR JR and HASKINS JF, 'Effects of 50 000 h of thermal aging on graphite/epoxy and graphite/polyimide composites, *AIAA Journal*, 1982 **22**(1) 96–102.

10 HOLLAWAY LC and HEAD PR, *Advanced Polymer Composites and Polymers in the Civil Infrastructure*, Oxford, Elsevier, 2001.

11 PETERS ST, *Handbook of Composites*, California, Chapman & Hall, 1998.

12 ASTM D1435, *Standard Practice for Outdoor Weathering of Plastics*, 1995.

13 SHIN KB, HAHN SH, YOON SH, KIM CG and HONG CS, 'Life prediction of graphite/epoxy composites under short-term exposure test', *SAMPE 2003 International Conference*, Long Beach, 2003.

14 ASTM G26, *Standard Practice for Operating Light-Exposure Apparatus (Xenon-Arc Type) With and Without Water for Exposure of Nonmetallic Materials*, 1996.

15 SHIN KB and KOO DH, 'A study of the evaluation of the failure for carbody structures made of laminated fiber-reinforce composite materials using total laminate approach', *The Korean Society for Composite Materials*, 2004 **17**(1) 18–28.

16 HONG SH, SHIN KB, JUNG B, HWANG TK, KIM JS, KIM CG and HONG CS, 'The evaluation of long-term performance of composite material by environmental effects using accelerated aging test', *The Korean Society for Composite Materials*, 1998 **11**(5) 1–13.

17 SHIN KB, KIM CG, HONG CS and LEE HH, 'Prediction of failure thermal cycles in graphite/epoxy composite materials under simulated low earth orbit environments', *Composites Part B: Engineering* 2000 **31**(3) 223–235.

18 SHIN KB, KIM CG, HONG CS and LEE HH, 'Thermal distortion analysis of orbiting solar array including degradation effects of composite materials', *Composites Part B: Engineering* 2000 **32**(4) 271–285.
19 HOWARD JB and GILROY HM, 'Natural and artificial weathering of polyethylene plastics', *Polymer Engineering and Science*, 1969 **9**(4) 286–294.
20 NAHAS MN, 'Survey of failure and post-failure theories of laminated fiber-reinforced composites', *Journal of Composites Technology and Research*, 1988 **8**(4) 138–153.
21 OWEN MJ and RICE DJ, 'Biaxial strength behavior of glass fabric-reinforced polyester resins', *Composites*, 1981 **12**(1) 13–25.
22 PUPPO AH and EVENSEN HA, 'Strength of anisotropic materials under combined stresses', *AIAA Journal,* 1972 **10**(4) 468–474.
23 GUESS TR and GERSTLE FP, 'Deformation and fracture of resin matrix composites in combined stress states', *Journal of Composite Materials*, 1977 **11**(1) 146–163.
24 SHIN KB and HAHN SH, 'A study on the evaluation of the failure for carbody structures made of laminated fiber-reinforced composite materials', *The 2003 Autumn Conference and Annual Meeting of the Korean Society for Railway*, Ansan, 2003.

11

Ageing of composites in the rotorcraft industry

K. DRAGAN, Polish Air Force Institute of Technology, Poland

11.1 Introduction to composite structures applied in the rotorcraft industry using the example of PZL

This chapter discusses the manufacturing and structure design of main rotor blades (MRBs); it will cover design schemes, materials used and applications. The main helicopter, as well as composite-elements, manufacturer in Poland is PZL Swidnik S.A. Examples of helicopters manufactured by PZL include W-3 and SW-4 (http://www.pzl.swidnik.pl/). According to data received from the manufacturer (PZL Swidnik S.A., unpublished data, July 2007), the following polymer matrix composite structures are used:

- composite structures with glass and carbon tapes, roving tows, fabrics such as reinforcement materials, mainly based on epoxy resin;
- composite sandwich structures with core materials such as Nomex®, glass–epoxy honeycomb and rigid foam plastics based on polyurethane or polimethacrylimide, and with epoxy–glass/carbon woven skin and epoxy adhesive films and adhesive pastes.

The composite materials for the structural elements of PZL helicopters – such as MRBs and tail rotor blades (TRBs) – consist of structures such as shells (glass fabrics with epoxy resin) and carrying straps (glass roving reinforced with epoxy resin). The non-critical elements of rotor blades are the trailing parts (sandwich structures using Nomex or epoxy–glass honeycomb or foam core).

The construction of the MRB of an SW-4 helicopter is presented in Fig. 11.1. Composite structures used in the construction of non-critical elements can be divided into two groups:

- elements working under static loads;
- elements without static and fatigue strength requirements.

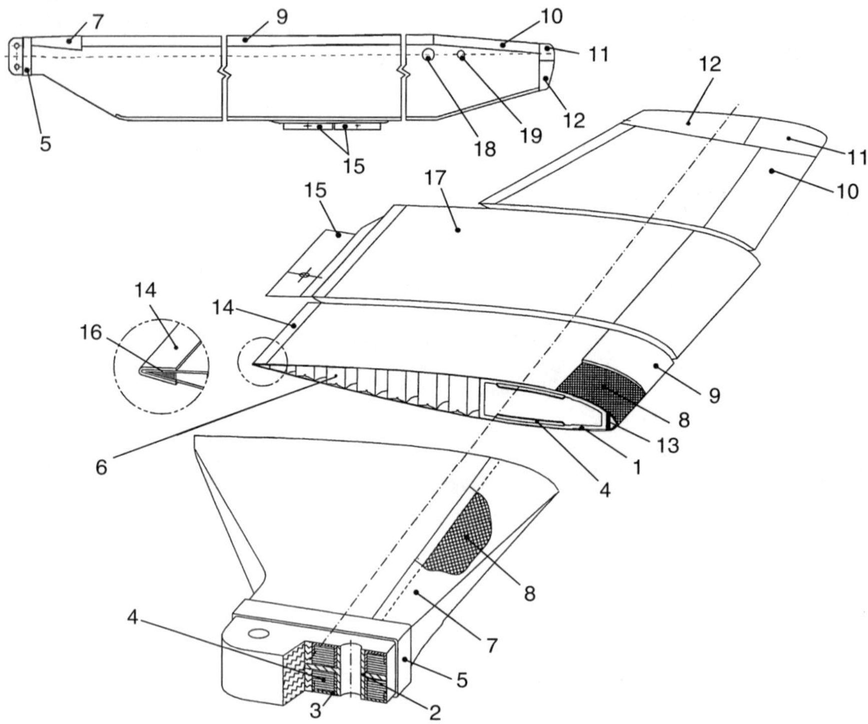

1. Spar (glass roving, glass fabrics/epoxy resin)
2. Blade grip
3. Link
4. Spar carrying strap (glass roving/epoxy resin)
5. Strap (glass fabric/epoxy resin)
6. Honeycomb core (Nomex: aramid fibre/phenolic resin)
7. Strap (glass fabric/epoxy resin)
8. Lightning system shielding (Cu–mesh/phenolic adhesive film)
9. Polyurethane tape
10. Leading edge grip (stainless steel)
11. Fore fairing (stainless steel)
12. Aft fairing (aluminium alloy)
13. Anti-flutter weight (lead)
14. Trailing edge strap (glass fabric/epoxy resin)
15. Trimming tab (aluminium alloy)
16. Trailing edge strap (glass roving/epoxy resin)
17. Aft section skin (glass fabric/epoxy resin)
18. Balance weight seat
19. Blade mooring seat

11.1 Construction of composite MRB of the SW-4 helicopter.

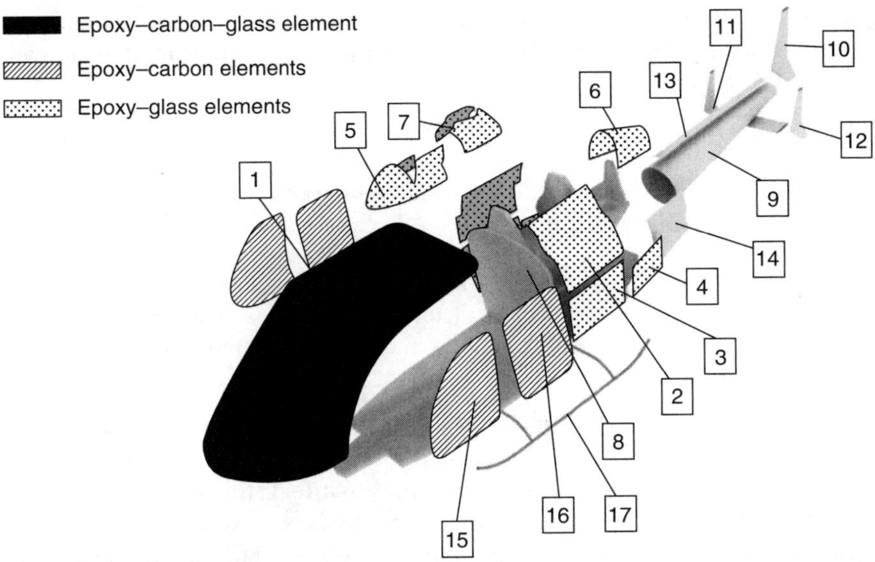

11.2 Composite share in the SW-4 helicopter structure.

Examples of the first group include the following composite elements of helicopters: the horizontal stabilizer of W-3, the cabin structure of SW-4, doors and some elements that are not part of the primary structure of the fuselage. The skin of the horizontal stabilizer of W-3 is made from a glass–epoxy solid structure (tube spar) and from glass honeycomb (trailing part). Figure 11.2 presents the structure of the cabin of an SW-4 helicopter. Composite elements in the cabin structure are made from:

- tapes, roving tows, woven hybrid composites with the use of glass and carbon reinforcement and epoxy resin;
- sandwich structures with the use of carbon and glass fibre skin and Nomex honeycomb.

Examples of elements that do not have static and fatigue strength requirements are cowlings, instrument panels, cowlings for instrument panels, linings, elements of the luggage compartment. The materials used for the helicopter elements are carbon fibres and glass fibres (as reinforcement) and epoxy matrix.

11.2 Potential damage that can occur in a composite main rotor blade

Information about damage, based on original equipment manufacturer (OEM) data as well as on maintenance data from users, will be provided in this chapter. Damage can occur during manufacturing as well as during the service life of MRBs. During the manufacturing of composite elements, and also during the maintenance of such elements, different defects can occur.[1] According to Birt and Smith:[1]

> One of the most serious is voids in the matrix, which can be further classified as:
>
> - delaminations – these are large planar voids occurring at the interfaces between the plies;
> - discrete voids;
> - porosity – this can be described as a large number of microvoids, each of which is too small to be of structural significance. It is usually produced during the curing cycle from entrapped air, moisture or volatile products.

Porosity has been a recognized problem for composites for a number of years. According to Birt and Smith, porosity influences the mechanical properties of composites and as a consequence affects the durability of such materials. Depending on the manufacturing process, as well as on manufacturing techniques, there are a number of defects that can occur in the composite structure: (PZL Swidnik S.A., unpublished data, July 2007 and reference 2):

- incompletely cured matrix;
- incorrect fibre volume fraction;
- ply misalignment;
- wavy fibres;
- ply cracking;
- delaminations;
- resin cracks;
- bonding defects;
- voids and porosity;
- foreign object damage (inclusions);
- resin rich/poor areas;
- bridging;
- fibre misalignments.

Different types of damage can occur during manufacturing and also during the service life of composites. Figure 11.3 shows the total percentage share of failure effects for metal and composite MRBs on the basis of data received from the users.[3] Ageing issues and manufacturer faults, as well as in-service damage, are the most significant failure effects influencing the MRBs during service life. Figure 11.4 shows the percentage share of failure modes for the MRBs of helicopters. Most of the in-service damage is connected with skin–honeycomb disbonds which lead to skin separation and composite ply cracking, as well as moisture (water ingress) in the honeycomb structure.

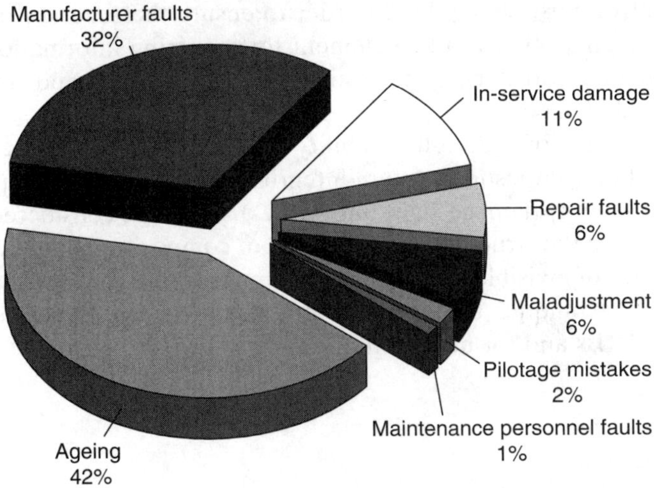

11.3 Percentage share of failure effects for the MRBs of helicopters.

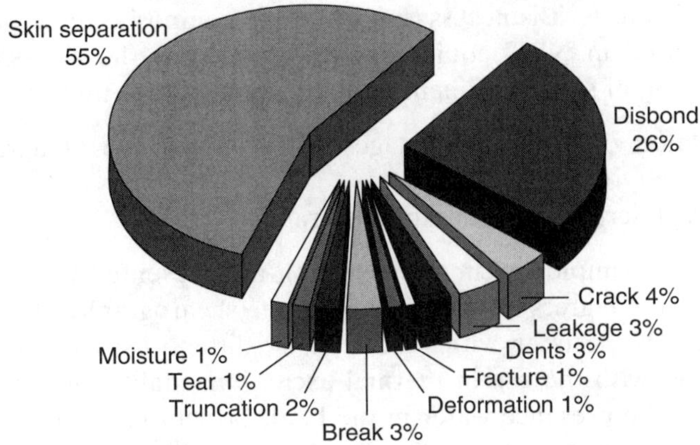

11.4 Percentage share of failure modes for the MRBs of helicopters.

Increased composite use in aerospace constructions has increased the importance of non-destructive testing (NDT) methods capable of identifying flaws in composites.[4] There is a necessity for composite inspection to detect manufacturing as well as in-service faults from the point of view of durability of composites.[5] The main reasons for the necessity of composite inspection are:

- human factors;
- improper manufacturing parameters (technology);
- improper materials, etc.

Quality control must be applied in order to ensure the accuracy of manufactured elements. NDT is a key element for gathering information about structure quality and it helps to achieve higher accuracy and quality of manufactured elements.

From the point of view of durability of composite materials, several factors affecting composite structure integrity (e.g. load cycles, temperature changes, low-energy damage, lightning strikes) should be considered. These factors influence the structural performance of composites and may lead to the occurrence of invisible damage in the composite structure. Based on the above, NDT techniques are necessary for the in-service maintenance of composite MRBs and for detection of damage such as:

- disbonds;
- delaminations;
- water ingress;
- low-energy impact damage.

In order to describe the likelihood of detection of the above-mentioned damage types, tests that use reference standards should be applied. In order to perform such tests, it is necessary to develop composite reference standards to be used in NDT equipment calibration for the assessment of damage.[6] Testing of such specimens enables, among other things:

- the detection of different damages types by the use of a range of technique;
- sensitivity description (signal to noise ratio).

The results of a sample specimen inspection are presented below. Figure 11.5 shows water ingress detection using laser shearography (for a glass fibre honeycomb specimen with water cells of different sizes and shapes). This technique, with the use of thermal excitation, enables an assessment to be made of the presence water in the blade structure. Figure 11.6 is an example of a carbon fibre plate with teflon inserts of different sizes between plies which create foreign object inclusions. A step sample made from glass fibre with teflon inserts and removable metal shims to create disbonds and delaminations was also considered.[7]

A range of techniques can be applied for specimen testing:

- ultrasonic testing (disbonds, delaminations, foreign object inclusion);
- mechanical impedance analysis (disbonds);
- shearography (disbonds, water ingress).

Results obtained during specimen testing gave information on which technique should be used in relation to the following factors: inspected material, thickness range, object geometry, required damage detection, the best sensitivity. A wide range of different techniques provides many possibilities for

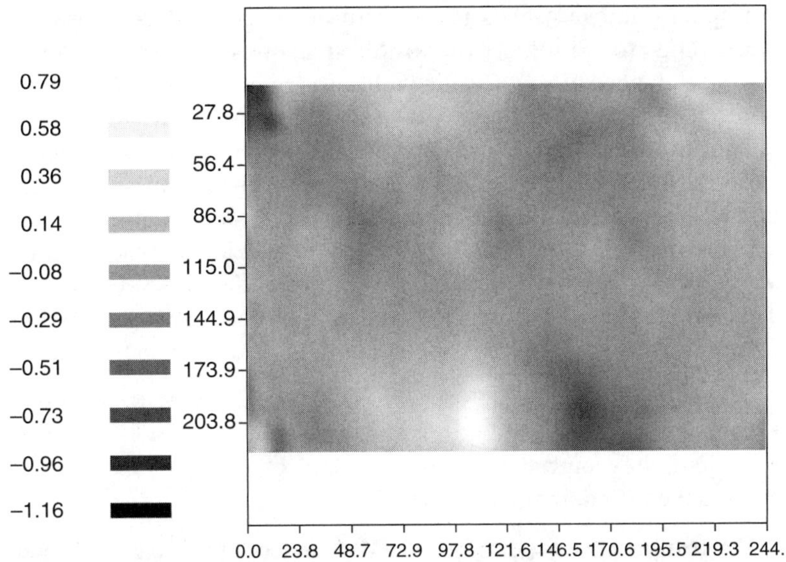

11.5 Water ingress indication with the use of NDT – shearography.

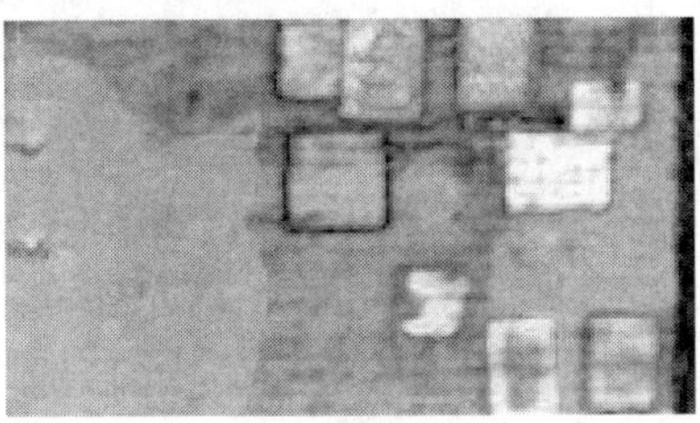

11.6 Carbon fibre specimen with artificial defects.

applying various damage and subsequent detection of damage scenarios. On the basis of this experiment the best techniques were selected for inspection of composite MRBs.

11.3 Low-energy impact damage and durability in a W-3 main rotor blade

Depending on the in-service conditions, the following events can occur: hail strikes and impact damage from bodies such as stones affecting the structural integrity of the MRB. On the basis of these events, low-energy impact tests were performed to assess the durability of composite MRBs to impact damage using advanced non-destructive inspection (NDI) techniques.

Impact damage affects the structural integrity of composite elements.[8] According to Smith, 'Composite structures can suffer quite severe impact damage without a noticeable surface indentation – known as barely-visible impact damage (BVID) – and this makes large-area defect detection a necessity for critical structural components'.[2] Such accidents, especially those that create non-visible damage, are called low-energy impact damages.[9] These damages are not visible to the human eye but can create damages affecting the structural integrity of elements. Very often, the results of low-energy impact damage are: delaminations, disbonds and multiple cracks. Regarding more complex structures such as composite rotor blades, defects that occur can be divided into the following groups:

- skin damage (delaminations, cracks);
- skin–honeycomb damage (disbonds);
- honeycomb damage (core crush).

For test purposes, the MRB of a W-3 helicopter was used. A special stand for the test was designed and the so-called 'blunt impactors' for the drop test were prepared. The term 'blunt impactors' indicates that these impactors do not affect the composite surface visually. The main aims of the test were:

- to deliver information about the influence of dropping a mass onto the blade structure (drop energy from 5 to 20 J);
- to determine whether any failure modes will occur during the blade test;
- to find appropriate, fast and reliable NDT techniques for the evaluation of drop test results;
- to describe the influence of energy levels on the damage size.

The reason for performing this test was to evaluate the behaviour of composite helicopter blades used in W-3 helicopters in situations that may occur during the service life. The next important issue was to deliver information on whether the maintenance techniques used are capable of finding damages that occur as a result of low-energy impact. The last stage was to evaluate which NDT technique was capable of finding such damages.

At the beginning of the test, calibration standards were prepared to determine the sensitivity and accuracy of the techniques selected. One of the specimens was a step sample, and also blade structures with mechanically induced disbonds and delaminations. Tests were applied to these specimens to detect disbonds and delaminations. For the tests, the following techniques were used: mechanical impedance analysis (MIA), pitch–catch, shearography and D-Sight.[10]

11.3.1 Test run

The first stage of the test involved NDT of prepared W-3 MRB samples in order to detect any manufacturing damages. The next step was to describe the region of interest where impactors could be dropped. For the selected areas of MRBs and referred to the skin and honeycomb interface, the drop test was applied.[9] After that stage, NDT was carried out and data were evaluated. The next step was again a drop test followed by NDT. The important issue was the fact that impactors were dropped on the same area as previously. The aim of this test was to analyse structure behaviour under repeated loading conditions and to evaluate the extent of the damage.

11.3.2 Test results

The techniques used for the inspection were based on a PC interface which made it possible to collect data in the form of pictures and C-scans. This method of data presentation makes it possible to compare data between consecutive test stages. Figure 11.7 presents the results of NDT on the composite skin of a W-3 MRB. Areas marked with polygons show disturbance in the signal and indicate skin–honeycomb disbonds. As shown, differences in the size of the disbond area are related to the different drop energies used. Figure 11.8 shows the impact damage area and its relation to drop energy. Furthermore, results of two stages of the test are shown. The first is connected with the first drop of the impactors and denoted by MIA-

- MIA-1, 2 – first drop, second drop results;
- P–C-1, 2 (pitch–catch) – first drop, second drop results.

Figure 11.8 shows that an increase in drop energy is associated with an increase in damage area. Moreover, the second stage of the test proved that the impact damage area almost doubles in size.

The damage size described differs depending on the NDT techniques used; this is a result of the different sensitivities of the techniques used. In

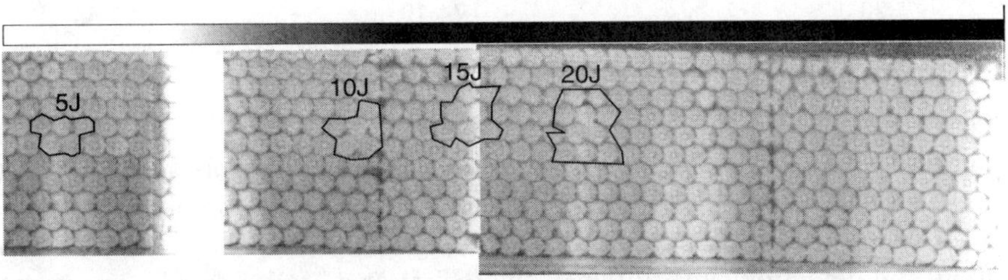

11.7 NDT results for the composite skin of the W-3 helicopter MRB.

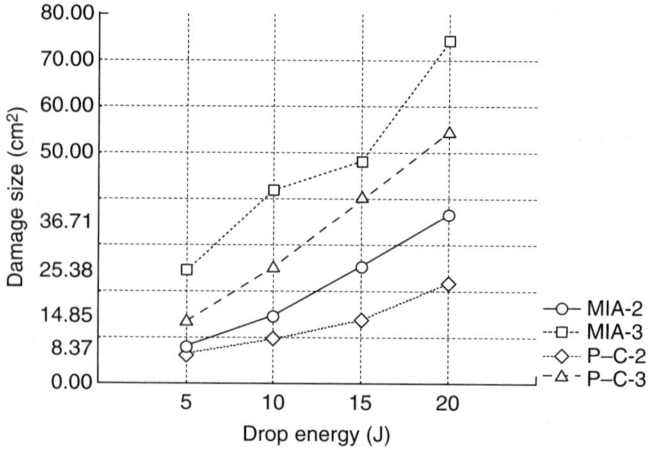

11.8 Damage size in relation to the drop energy.

Table 11.1 Detection capabilities of the selected techniques

Drop energy	MIA	Pitch–catch	Shearography	D-Sight
5 J	+	+	+	–
10 J	+	+	+	+
15 J	+	+	+	+
20 J	+	+	+	+

+, damage detected; –, damage not detected.

Table 11.2 Damage size (cm^2)

Drop energy	MIA	Pitch–catch
5 J	24.54	13.68
10 J	41.82	25.12
15 J	48.14	39.89
20 J	74.10	54.44

addition, qualitative techniques were also applied, such as shearography and D-Sight. These techniques do not provide information about damage size but only about the existence of damage. Table 11.1 gives information about the detection capabilities of the techniques used. Table 11.2 presents the damage sizes determined from the NDT.

11.3.3 Conclusions

All the techniques applied detected damage (disbonds), with the exception of D-Sight for the smallest damage size (13–24 cm^2). Generally, D-Sight is a good option for fast disbond detection. In the work of Heida,[10] it was concluded that D-Sight can reliably detect significant impact damage, with a damage area equal to or larger than 5 cm^2, within a field of view of about 0.25. However, one very important consideration is that surface curvature in the direction of observation leads to variations in the intensity of reflected light. On the other hand, the signature of surface defects is strongly reduced as a result of environmental light. On the inspected blade, the smallest damage may not be visible because of the geometry and also because of the presence of reflections from the structural elements of the blade.

It should be mentioned that the manufacturer does not permit further blade maintenance if the skin–honeycomb disbond area is greater than 80 cm^2. Data presented in Fig. 11.8 as well as in Table 11.2 show that the damage size for a drop energy of 20 J is very close to the acceptable limit for further blade maintenance.

An ultrasonic pulse–echo technique was also applied to determine the possibility of detection of skin–honeycomb disbonds, as well as the possibility of delamination detection. Figure 11.8 shows the results of an ultrasonic pulse–echo C-scan. For this technique, disbond detection was not as good as that of other techniques due to adhesive in the honeycomb structure which affects the ultrasonic signal. Using this technique it proved possible to detect small delaminations occurring in the drop area. These damages were not detected using any of the other techniques.

Further work is now being conducted with the aim of describing the effects of low-energy impacts on the composite blade structure in the critical areas, i.e. (a) the stress concentration area of the blade and (b) the spar structure. Problems to overcome include geometry changes and construction complexity, which may influence signal interpretation. For these purposes, special samples with artificial defects are prepared. At this stage, calibration work has been carried out and trials for the detection of defects are in preparation. Further results will be delivered in the near future.

11.4 Influence of moisture and temperature

The influence of environmental hazards will be discussed in this section, based on OEM information and tests performed in the manufacturer's facility. Length of service life (airworthiness limitation), expressed in years or months, is mainly affected by moisture (which the composite material may absorb). The influence of other factors such as ultraviolet radiation is negligible over the service life as a result of the protection provided by coatings. In addition, manufacturers' data showed that the environmental resistance of composites exposed to the ultraviolet light did not decrease over a 50-year period. After that period, fatigue strength degradation is approximately 20%, i.e. total fatigue strength equals 80% of the initial value. Laboratory tests carried out by PZL (ageing according to the ASTM D 5229 standard) proved that the composite material reaches total moisture saturation in a stable manner. This process continues over the service life in a similar way, the only difference being the rate of moisture absorption. On the basis of moisture diffusion in the composite material, it is possible to determine such parameters as: diffusion constant and effective equivalent moisture content (M_m). On the basis of the M_m determined, it is possible to calculate the time required for the ageing material to reach a moisture content equal to 99.9% M_m. Calculations performed by the manufacturer proved that for the main composite rotor blades these periods were much greater than in the real service life of these elements. It should be mentioned that during the calculation, the following factors were not taken into consideration:

- the influence of the existence of protective coatings;
- the material behaviour in real conditions – i.e. the material absorbing moisture and then drying according to external (weather) conditions.

The calculations assumed that moisture and temperature affect material in a constant (unchanged) way. The most significant factor is that the material reaches a maximum moisture level in a stable way and longer ageing process do not change moisture content or affect material strength.

Because composite material durability parameters decrease in higher temperatures, static strength tests were carried out at temperatures equal to the maximum service life temperature, taking the effect of the sun into consideration. During the tests it was concluded that the degree of strength degradation determined makes it possible to determine an additional static strength safety coefficient. It has been proven that the strength degradation coefficient under high temperature conditions decreases to as low as 0.75.

The next issue was to determine strength degradation coefficients during the ageing process in climate chambers. Results show that ageing of materials in conditioning chambers (temperature 80 °C; moisture equal to 85%), with the use of high and low temperatures, affects the static strength of composite materials. These coefficients express the ratio of the strength of the aged specimens (tested in the temperature range –45 °C to 80 °C) to the strength of specimens tested in the environment temperature; the coefficients have a value of about 0.5 depending on the type of structure and the kind of load. A comparison of painted and non-painted elements subjected to natural ageing (in the air) shows much lower strength decreases than those subjected to an accelerated ageing process.

During natural ageing, the decrease in strength of elements was 40%, in reality, these decreases are lower than 40%. Erosion occurring on the external surface of unpainted specimens was the direct reason for this. During service life, composite materials are protected by the painted coatings that protect the material from erosion as well as from moisture (continued a moisture barrier primer). Moreover, it was proven that composite degradation during accelerated ageing was greater than that in environmental ageing.

Tests made on the laboratory samples were confirmed by testing blades that had been aged in natural conditions. During tests the following measurements were made: weight, mass balance and stiffness, additional static tests and assessment of appearance. No significant changes in the static strength or in the mass and stiffness were found.

One of the effects of moisture absorption was a decrease in the matrix glass transition temperature; this temperature is very important in relation to composite materials. Above this temperature, rapid change in physical properties can occur that influence mechanical properties, in particular properties that are strongly dependent on the matrix – for example, its interlaminar shear strength. In this respect, it is essential to determine the influence of the ageing process on the glass transition temperature. Glass transition temperatures were determined for the composite materials that were submitted to ageing processes. The glass transition temperature was verified by results of material strength tests at elevated temperatures. These analyses proved that glass transition temperatures exceed maximum helicopter operating temperatures. In this respect, the materials described were recognized as safe for use in these helicopters.

11.5 New techniques for testing composite structures

Techniques such as structural health monitoring (SHM) will be discussed in this section, using information available in the published literature on modern structural integrity assessment. Applying NDT to composites is necessary not only because of their applications in the rotorcraft industry but also their applications in other aircraft structures. The continuous development of the extensive use of composites in aerospace applications creates great demand for advanced NDT techniques as well as for health monitoring techniques. The main advantages of the use of advanced NDT techniques is the possibility of obtaining information about the structural integrity of composites without affecting structure properties and also the collection of data for structure monitoring. The main disadvantages are the necessity to have the aircraft on the ground, the human staff requirements and the time required to inspect all relevant locations.[11] The use of network sensors, distributed in the structure, for periodical or even continuous monitoring could affect the time required for inspections. SHM enables NDT to be deployed with greater access to more complex structures. According to Roach, 'The core of SHM is the development of self-sufficient systems, that use built-in distributed sensors/actuators not only to detect structural failures but to monitor the effects of structural usage'.[11] The use of sensor networks could be achieved using three approaches, as listed below:[11]

- *in situ* sensors;
- sensor networks with *in situ* data acquisition;
- sensors networks with real-time data transmission to a remote site.

In the first group, sensors are permanently installed in the structure. For diagnostic purposes aircraft must be on the ground. All necessary items such as power supply and data acquisition electronics must be delivered to the aircraft. All sensors are connected to the data acquisition system, and data analysis and detection procedures are performed on-site. In this system, manual or automated data collection are substituted by distributed networks.

In the second group, distributed networks are equipped with the electronics and memory to record and to store data. 'Those items are equipped with programmable circuits having the power for automated data logging in flight or on the ground'.[11] Data must be gathered by technical staff while the aircraft is on the ground. The third group is quite similar to the second group; the only difference is the use of communication system applications. This system enables wireless transmission of the data collected. For this particular application, software could be developed to send data to maintenance personnel.

There is much work dedicated to the possibility of applying SHM to aircraft, as well as to composites. It appears that the use of *in situ* sensors for the SHM of aircrafts to will be a aviable option in the near future.[11]

11.6 References

1 E A BIRT, R A SMITH, A review of NDE methods for porosity measurement in fibre-reinforced polymer composites, *Insight,* **46**(11), November 2004, 681–686.

2 R A SMITH, An introduction to the ultrasonic inspection of composites, in *Non Destructive Evaluation of Composite Materials*, Course, Farnborough, November 2004.

3 K DRAGAN, S KLIMASZEWSKI, In-service NDI of aging helicopters main rotor blades used in Polish Armed Forces, in 9th *Joint FAA/DoD/NASA Aging Aircraft Conference*, Atlanta, Georgia March 2006.

4 D ROACH, Enhanced inspection methods to characterize bonded joints: moving beyond flaw detection to quantify adhesive strength, in *Air Transport Association Nondestructive Testing Forum*, Fort Worth, Texas, October 2006.
5 COMPOSITE QUALIFICATION CRITERIA, in *Proceedings of the 51st Annual Forum of the American Helicopter Society*, Fort Worth, Texas, May 1995.
6 D ROACH, *Development of composite honeycomb and solid laminate reference standards to aid aircraft inspections*, Sandia Report, SAND99-05405, Albuquerque, New Mexico.
7 J GIESKE, D ROACH, P WALKINGTON, *Ultrasonic inspection technique for composite doubler/aluminium skin bond integrity for aircraft*, Sandia Report, SAND98-0311C, Albuquerque, New Mexico.
8 K DRAGAN, S KLIMASZEWSKI, Low energy impact damage detection in the composite sandwich elements, in *35th Polish National Conference on Non Destructive Testing*, Szczyrk, Poland, October 2006.
9 K DRAGAN, S KLYSZ, J LISIECKI, Detection of damages from the low energy impact damage in the composite structures, in *Congress of Polish Mechanics*, Warszawa, Poland, August 2007.
10 JH HEIDA, D-Sight technique for rapid impact damage detection on composite aircraft structures, *NDT.net*, **4**(6), June 1999.
11 DH ROACH, 'Smart' aircraft structures: a future necessity. Health monitoring of aircraft structures using distributed in situ sensor system, *High-Performance Composites*, January 2007.

12
Ageing of composites in marine vessels

P. DAVIES and D. CHOQUEUSE, IFREMER Brest Centre, France

12.1 The use of composites in marine vessels

This chapter presents an overview of the ageing of composite marine vessels, defined here as craft navigating either on or below the water. Long fibre reinforced polymer matrix composites have been used in marine vessels for over 50 years now,[1–5] so there is extensive in-service experience to draw on when studying ageing effects. This is fortunate, as many laboratory evaluations of composite durability in the published literature are based on short, extremely severe tests of little relevance to the final application. In this chapter the use of composites in marine vessels will be described briefly first, as this provides an overview of the wealth of experience available. The materials employed will then be presented; these are mainly E-glass reinforced polyester and epoxy but some carbon fibre composites and a range of sandwich materials are also used. It is important to note that protective outer coatings or gel-coats are nearly always applied, and these must be considered when ageing is discussed. A short description of the marine environment follows – this is described in more detail in Chapter 18 – then a summary of published work is given. Three case studies are presented, which allow the various parameters that influence marine ageing to be discussed. Finally the relevance of accelerated test methods to the ageing of marine vessels is examined.

12.1.1 Leisure craft

Pleasure boats, defined as vessels built for recreational purposes, were one of the first applications of composites. There is some discussion over when the first fibreglass boat was built, some reports indicating that a phenolic composite prototype existed as early as 1937,[5] but commercial composite boat production dates from the 1950s. It is now a big industry, employing some 35 000 people in the UK. The leisure marine industry is very strong in Europe at present, with growth rates around 7% per year.[6–8] The

Bénéteau group in France, world leader in sailing yacht production, produces over 3500 sailboats and 3000 motorboats every year.[9] Over 90% of small boats are made from composites and the industry has evolved from small boatyards to large production line factories operated by international groups. The average lifetime of a composite pleasure boat is 30 years. This clearly indicates the excellent ageing performance of these materials.

12.1.2 Workboats

The heading 'Workboats' includes both civil and military vessels with specific functions, including fast ferries, fishing craft and anti-mine warfare vessels. The first composite fishing trawlers were built in South Africa in 1960,[10] and one boatyard in Cape Town built over 40 large vessels (around 25 m long) in 4 years. Composite fishing boats were also being built in Japan in the early 1960s to replace wooden boats. Scott Bader documents from the 1970s show various composite fishing vessels up to 28 m long.[11] The majority of ships' lifeboats are also made in glass reinforced plastic (GRP). A major landmark in marine composite development was their development for military craft. Feasibility studies into GRP warship construction started in the UK in 1965. HMS *Wilton*, 47 m long, was a prototype built in 1972 and the subsequent *Sandown* series were built to replace minesweepers with aluminium-framed wooden hulls.[12] Smith describes the different design philosophies for mine countermeasure vessels (MCMVs),[2] including unstiffened monolithic (Italian *Lerici* class), stiffened single-skin (UK *Hunt* and *Sandown* class and the France/Netherlands/Belgium *Tripartite* series) and poly(vinyl chloride) (PVC) core sandwich (Swedish *Landsort* and Australian *Bay* classes). Mouritz *et al.* provided a more recent overview of composite MCMVs, in a summary of composites for naval applications in 2001.[13] They noted that since the early 1980s over 200 all-composite vessels of this type have been built. Studies to establish the state of these vessels after 20 years or more in service have provided valuable information on the ageing of composites but few data have been published in the open literature. This will be discussed in more detail below. Composites have also found many applications in fast ferries, particularly in the Nordic countries where sandwich construction has been widespread.[14–16] The high-speed light craft classification rules from DNV (Det Norske Veritas) were developed for the design of such vessels, the first tentative set of rules dates from 1972.[17]

12.1.3 Racing vessels

Racing vessels are considered as a separate class here, as they are generally prototype structures. The manufacturing methods are closer to those of aerospace structures, and prepreg is widely used. Monohull and multihull designs have been developed with the largest trimarans such as *Geronimo* and *Groupama 3* being over 30 m long, over 20 m wide and carrying masts 40 m high.[18] The drive towards high performance and low weight provides a proving ground for new materials, particularly carbon fibre composites and sandwich structures, which can then be transferred to the leisure boat industry. Ageing is not a major concern for designers as the racing lifetime is quite short, often less than 5 years, but the use of very thin facings, often less than 1 mm, with little or no protection represents a severe test.

12.1.4 Underwater vessels

The low weight of composite materials makes them particularly attractive for underwater vessels, for which lighter hulls allow heavier pay-loads. As a result, large composite structures have been used on military submarines for many years, one of the earliest reported being a fairwater on the USS *Halfbeak* which entered service in 1954.[19] Published details of these applications are rare but some examples can be found[13,20,21] which present very large components such as sonar domes, external deck areas and fairings.

12.2 Marine composites

12.2.1 Laminates

The traditional marine composite is a polyester reinforced by woven glass fabric and/or chopped strand mat layers. Figure 12.1 shows a dry Rovimat

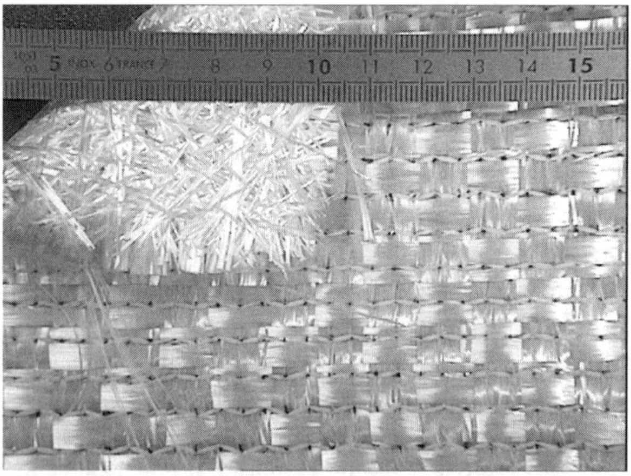

12.1 Woven roving/chopped strand mat cloth (Rovimat 500/300).

cloth, consisting of a 300 g/m^2 chopped strand mat layer stitched to a 500 g/m^2 woven roving layer which is a very common reinforcement. Other reinforcements such as stitched multiaxial cloths are more expensive but are also used in some cases.

The polyesters are traditionally orthophthalic, but isophthalic polyesters and some vinyl esters and epoxies are also used. Recent legislation in Europe on volatile organic compound (VOC) emissions[22] has resulted in the adoption of low-styrene polyesters. These resin formulations, based on DCPD (dicyclopentadiene) chemistry (low styrene content) or the inclusion of film forming components (low styrene emission) or a combination of the two, are more brittle than the traditional resins.[23,24] The surface film layers may also make secondary bonding more difficult.

Scantlings are determined by experience and this has progressively been integrated into classification society regulations (ABS, DNV, BV, Lloyds, etc.). A major recent development has been the consolidation of this experience in an ISO document, which will appear as ISO 12215–5.[25] This standard is based on equivalent pressures, and includes various safety factors that allow hull panel thicknesses to be determined using typical material properties defined by fibre content. The document also provides a database of material properties to be used if test data are not available. Typical hull laminate thicknesses for small pleasure boats are around 5 mm. This is very different from certain workboat hull thicknesses, which are dimensioned to resist underwater explosion. Smith[2] indicates that one option used by the Italian navy is a monolithic unstiffened design up to 150 mm thick.

Traditional marine composites have quite low fibre contents, 20–40% by volume. The recent development of infusion techniques is enabling higher fibre contents (>50%) and lower thicknesses to be obtained. This has important repercussions for ageing, as water diffusion rate is directly related to resin content and thickness. Ageing is not explicitly taken into account in the classification society rules but is included in long-term safety factors. Smith cites a partial safety factor for long-term degradation relating ultimate to permissible stress, to account for the influence of water absorption on matrix-dominated failure modes.[2]

12.2.2 Sandwich materials

Sandwich construction is an essential element of many marine vessels. Foam cores, mostly PVC and balsa wood, are widely used in small boats, and honeycomb is the standard core for racing yachts. When sandwich is employed, laminate facings are thinner and the effect of water on the core may be important. Few studies have examined this aspect of marine ageing but some results are discussed below.

12.2.3 Gel-coats

The presence of gel-coat layers which protect marine laminates is one of the major reasons why very few examples can be found where wet ageing has caused severe structural damage in service. Gel-coats are mostly isophthalic-NPG resins (resins in which a neo-pentyl glycol or similar has replaced the cheaper, traditional propylene glycol), a few tenths of a millimetre thick. The addition of thixotropic agents is necessary to enable the gel-coat to be applied to vertical moulds without sagging and various other additives including pigments, such as rutile (titanium dioxide), are also included. The resulting formulation is brushed or sprayed onto the mould. In the 1980s the widespread appearance of blisters on boat hulls resulted in many studies of osmosis and blister mechanics, and this will be discussed in more detail in Section 12.7 below.

12.2.4 Manufacture

The vast majority of marine structures are still produced by hand lay-up; either fabrics are impregnated manually with rollers or chopped fibres and resin are sprayed into moulds. This situation is evolving as health and safety regulations push towards closed-mould techniques such as infusion but manual impregnation remains the principal method today. As a result there are a number of different types of defect that can appear and which may play a significant role during subsequent service. Defects in the gel-coat layer are particularly critical and Rymill has listed a dozen common faults including pinholes, wrinkling, poor adhesion and crazing.[26] Resin suppliers provide similar examples[27] with advice on how to avoid them.

12.3 The marine environment

The marine environment is described in more detail in Chapter 18. The two main parameters that will influence the ageing behaviour of marine composites are the composition of the sea water and ambient temperature. The loading on a boat hull will depend on many parameters including sea conditions, and uncertainty in these loads is one of the main difficulties in design. For underwater applications, particularly in deep sea, hydrostatic pressure will play an important role, but this is directly related to immersion depth and hence is easier to predict. Biological factors (bio-film adhesion leading to marine fouling growth) may be important and the need to apply anti-fouling treatments regularly may affect the long-term effectiveness of the gel-coat. Weathering, exposure to sun, rain, ice, etc., may also affect surface layers but are not generally the cause of structural problems.

12.4 Recent published studies on marine ageing

The durability of composites is a popular topic and has been the subject of several major conferences such as the DURACOSYS series (Progress in Durability Analysis of Composite Systems) and a number of books (e.g. references 28–30) before the present one. A large part of the published literature on durability is concerned with fatigue,[31] aerospace structures are also well represented and relative humidity rather than immersion is the most popular environmental parameter studied. Ageing of marine structures has been seen as less critical and has received less attention, but Searle and Summerscales presented a review of the subject.[32]

Chapter 18 will describe the more general topic of wet ageing, and in the present chapter the literature review will be limited to studies that describe wet ageing of composites for marine vessels in sea water. Very few laboratory studies have considered marine composites under realistic marine conditions (gel-coated composites exposed on one surface to natural sea water) except in studies of blistering. Nevertheless, various studies have examined marine composites in sea water and the Section 12.4.1 will review some of the results, with particular emphasis on recent studies. In Section 12.4.2, the influence of load on ageing of these materials will be considered. In Section 12.4.3, ageing of marine sandwich materials will be discussed, and in Section 12.4.4 results from in-service experience will be presented.

12.4.1 Ageing of marine laminates

There are two types of mechanism involved in wet ageing of polyester composites: physical degradation – caused by plasticization and swelling, interfacial debonding and delamination – and chemical degradation involving matrix hydrolysis and, under particular conditions, fibre degradation. Early studies on moisture effects have been summarized by Smith[2] and will not be repeated here. More recent studies include a 2-year study of thin woven glass and carbon composites with polyester and vinyl ester matrix resins[33] aged in sea water at 30 °C. Large reductions in flexural strength (20–40%) were noted, but mode I fracture energy was not significantly affected. Vinyl ester composites degraded to the same extent as polyester in that study.

Several studies have been performed at ENSAM in Paris on the degradation mechanisms in polyester composites. For example, Gautier *et al.* presented results from a study on interface damage in woven glass reinforced polyester composites.[34] By determining characteristic times for mechanical degradation, τ_{ILSS} (interlaminar shear strength), diffusion rate, τ_D, and matrix hydrolysis, τ_H, measured by chemical analysis, they were able to show that for their materials (isophthalic and DCPD polyester resins) and ageing

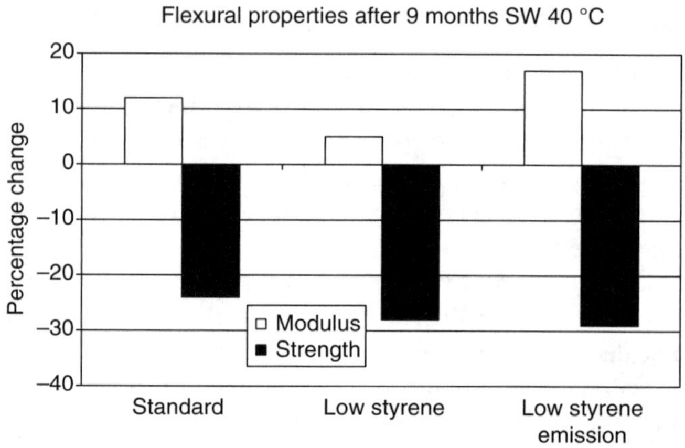

12.2 Influence of immersion for 9 months in sea water at 40 °C on flexural properties of standard orthophthalic and low-styrene polyester composites.

conditions from 30 to 100 °C that $\tau_D << \tau_{ILSS} << \tau_H$. They concluded that the macroscopic mechanical property losses noted in accelerated ageing tests must be caused by interfacial debonding, and not matrix hydrolysis. This is often stated in laboratory ageing studies but rarely verified.

Davies *et al.* presented results from a study of four marine composites, two polyesters, vinyl ester and epoxy and their composites aged in natural sea water and distilled water at 20 and 50 °C.[35] Samples were aged for 2 years and the particularity of this study was that damage mechanics parameters were used to follow mechanical behaviour, rather than the more usual simple short-beam shear or flexural tests. Tensile tests on 0/90° reinforced composites loaded at 45° with load–unload cycles enabled values to be obtained that can be introduced into more complex analyses than the simple laminate theory generally used today. More details will be given in Section 12.5 below.

Perrot and colleagues examined the ageing of two low-styrene polyester resins and a standard polyester and their composites in sea water.[36] They showed that the degradation kinetics were similar for all three materials, as shown in Fig 12.2. More brittle initially, the low-styrene resin composites remain more brittle after ageing.

12.4.2 Ageing under load

Marine vessels are rarely subjected to water exposure without also being loaded at the same time. Excepting storm conditions, loads are generally quite low, but it is still important to establish to what extent the load level affects water ingress, and more importantly the degradation in mechanical properties. Despite concerns expressed in the 1980s and 1990s over stress corrosion,[36–39] a loss in strength due to combined moisture and mechanical loading, this has rarely resulted in premature failure in service. Nevertheless, more highly loaded applications in the future may require attention to this failure mode. Improved glass fibres without boron in the glass composition are now widely available,[40] and these show significantly better strength than E-glass fibres in water under load. For example, the 50-year allowable constant stress in sea water is estimated to be 42% of ultimate strength for ECR (corrosion-resistant grade) fibres compared with only 30% for E-glass. There has been some recent work on the coupling between stress and wet ageing, using flexural[41] and tensile[42] loads.

Gellert and Turley studied flexural properties of polyester, vinyl ester and phenolic composites in natural sea water with and without loading to 20% of outer fibre flexural failure strain, for up to 9 months.[41] Only the phenolic composite showed an influence of loading on subsequent flexural properties; flexural strength dropped from –25% to –36% when load was applied during ageing. Pauchard and colleagues examined how tensile loading of unidirectional composites in water affects lifetime[42] and applied a stress corrosion model based on a statistical defect population to predict subcritical crack growth rate and stiffness. For their test conditions they also concluded that a stress corrosion mechanism controlled cyclic behaviour, as fatigue lifetimes depended on time under load and not number of cycles, for various loading frequencies.

Another combination of water and mechanical load is encountered in deep sea applications when the loading, hydrostatic pressure, is directly proportional to immersion depth. Pressure can affect weight gain, but there is little evidence of significant acceleration in degradation with pressure. Table 12.1 shows some results from a study on carbon/epoxy composites. Short-beam shear tests were used to measure apparent ILSSs after different exposure times and temperatures. Specimens were placed in water at 60 °C with and without a pressure of 100 bars (equivalent to 1000 m depth). Published studies have indicated both weight increases with pressure[43–47] and pressure independence. This topic is discussed in more detail in Chapter 18.

Table 12.1 Percentage loss in ILSS after immersion in distilled water for 3.5 years (tube samples) and 4 years (plate samples)[43]

Material	20 °C	60 °C	60 °C + P
Tube [±55°]	–7%	–16%	–14%
Plate [±45°]	–1%	–11%	–10%

P, pressure.

12.4.3 Ageing of marine sandwich materials

There are relatively few published studies of the ageing of marine composite sandwich materials. It might be argued that the critical step is the diffusion through the composite facings, and that once water reaches the core, particularly when the core is balsa wood, then the material should be replaced. However, this is not a practical approach and it is more important to understand how water affects the core. Wet balsa cores have been described in some detail by Strand.[48] In a study performed with core suppliers and a boatyard he examined three aspects: the influence of water on mechanical properties, decay of balsa wood and delamination caused by freeze/thaw cycling. Balsa is composed of a small amount of cellulose cells and a lot of free space (Fig. 12.3(a)).

The cellulose fibres saturate at about 28% by weight, but when all the free space within the cells is filled the weight gain may be nearer 300%. When the fibres are saturated shear strength may drop by 25%. Strand measured a 20% drop, but further drops were not noted at 100% weight increase and he concluded that wet balsa core may not be a problem for the majority of marine structures. The freeze/thaw issue is more complex. Volume changes due to water freezing in the core should not cause problems if there is still air in the cells but may result in core–facing debonding if water fills interfacial voids. Cantwell *et al.*[49,50] studied how sea water affected interfacial crack propagation between glass reinforced polyester skins and balsa cores, and revealed that water can actually increase crack resistance. Finally, fungal decay will cause loss of material and larger strength loss than moisture take-up alone, but analysis of samples in Strand's study indicated fungal activity in only 20% of tested samples of wet core.

(a) (b)

12.3 Core structures, same scale. (a) Balsa structure; (b) PVC closed cell foam.

When closed cell foam cores are used (Fig 12.3(b)) the water ingress may be limited to the surface cells. Li and Weitsman showed that, while water may improve foam core fracture toughness, its presence at the core–facing interface results in a drop in interfacial crack propagation resistance.[51] Ionita and Weitsman proposed a model for water ingress based on cell wall breakage, to explain the high weight gains measured.[52] Earl and Shenoi[53] used dynamic mechanical analysis to follow the progressive reduction in modulus of PVC foam due to plasticization by water.

Honeycomb used for racing yacht cores is generally Nomex™ of density in the range 48–96 kg/m^3. Moisture has been a concern in the use of honeycomb for aircraft structures[54] but little work has been published on marine immersion. As this type of core is limited to high-performance craft at present, ageing behaviour has not been considered to be a priority.

12.4.4 In-service experience

Fully documented studies of composite samples recovered after ageing in service are not easy to find. Early results from tests performed in 1962 and 1972 by Owens-Corning Fiberglas and the US Coastguard on samples from a 40-foot patrol boat were presented by Graner and Della Rocca.[55] Laminates were based on woven cloth and mat and were 19 mm thick in the hull bottom and 9 mm thick for the sides. Test results after 10 and 20 years showed that the structure was still in excellent condition.

Fried and Graner described results from tests on samples removed from early submarine casings.[56] The fairwater of USS *Halfbeak* was installed for trials in 1953, to compare its performance with that of aluminium. When it was replaced in 1965, after 11 years in service, mean properties were within the original specification. Williams described tests performed by Scott Bader on a 12-foot GRP dinghy built in the early 1950s, after 15 years in service.[10] He noted that the dinghy had 'lost only a small percentage of her original strength.' Based on these results and data from many laboratory tests, Smith concluded that maximum reductions to be expected after many years in service in Young's modulus and tensile, flexural and compressive strengths are, respectively, 20, 20, 30 and 35%.[2]

Other data have been obtained from samples taken from ship structures placed at sea but not in service. The French navy maintained two natural ageing sites, a minehunter section at Lorient and a site at Cherbourg, and removed samples periodically for testing over a period of more than 20 years. Gutierrez *et al.*[57] indicated that ageing in sea air is as severe as sea water immersion, and presented results suggesting that a drop (not quantified) in flexural and short-beam shear properties of samples exposed to 'natural ageing' after 15 years corresponds to accelerated ageing in artificial sea water at 70 °C for 1000 hours. Few details of test

Table 12.2 Estimations of percentage strength retention under flexure and short-beam shear loading of marine composites without post-cure after 30 years' exposure to sea air at 20 °C, based on reference 57

Material	Flexural strength retention (%)	ILSS retention (%)
Isophthalic polyester	80	75
Vinyl ester	85	87
Epoxy	80	65

Table 12.3 Mechanical properties after ageing in service, based on reference 58

Vessel	Ageing period (years)	ILSS (MPa)	Flexure strength (MPa)
Prototype minehunter	11		
Initial		20–24	220–310
After ageing		23–27	260–325
Minehunter	17–18		
Initial		23	285
After ageing		21–23	190–250
Submarine	5–6		
Initial		31–33	230–370
After ageing		24–29	170–214

conditions, materials, or damage mechanisms were given to validate these results, but Table 12.2 shows estimated retention of flexural and short-beam shear strengths, based on 21-year data extrapolated to 30 years' exposure to sea air. These results are again consistent with the values proposed by Smith.

More recent studies have been performed as part of the studies used to extend the life of MCMVs in the UK and France. Many other tests have been performed but few details of results have been published. One exception is a set of results from a French study, performed to examine the extension of the lifetime of marine vessels, which was presented recently.[58] Samples were taken from a prototype anti-mine vessel (the BAMO), a mine-hunter built in the 1980s and a submarine. Table 12.3 shows an example of the mechanical property results before and after ageing in service. Overall, very little change in properties was measured. The only exception was the submarine outer deck application, for which a drop in flexural strength was noted. Sections indicated the presence of damage in samples, in the form of cracks in the fibre–matrix interface region.

12.5 Example 1: glass-reinforced thermoset ageing

In order to illustrate the discussion above, some results from a study on marine resins and their composites are shown in this section.[35] First, data from weighing will be presented, then property changes will be briefly discussed. Two polyester resins, an orthophthalic and an isophthalic grade, a vinyl ester and an epoxy were studied. Resin dog-bone specimens were machined from cast resin plates and immersed in sea water at 20 and 50 °C and in distilled water at 50 °C. Composite samples were produced by reinforcing the same resins with a balanced 0/90° stitched glass cloth of 566 g/m^2. Samples were 4 mm thick and fibre content varied from 53 to 59% by weight.

12.5.1 Diffusion

Figure 12.4(a) shows an example of weight gain measurements for an isophthalic polyester resin, plotted in the usual way versus the square root of time divided by specimen thickness. The resin weight gain at 20 °C in sea water is approximately Fickian. At 50 °C the behaviour is initially Fickian but then deviates after about 3 months. After 650 days samples were removed from the 50 °C water bath and dried, and a net weight loss was measured. Figure 12.4(b) shows results for the composite of the same resin, aged in the same water bath. It is apparent that there is little effect of temperature on the composite response for this resin. The composite at 20 °C gains about 1% by weight whereas, if only the resin absorbs water, this should be limited to about 0.4%, indicating that the fibre–matrix interface is also active in the diffusion process. Figure 12.5 shows the same plots for the epoxy resin. Here the resin weight gain is much higher, nearly 3% at 50 °C, but the composite weight gain can be quite closely predicted from that of the resin by simply correcting for the quantity of resin (40% by weight). This suggests that water only diffuses into the matrix. There is also a strong influence of temperature on the rate of diffusion. These very different results from two composites with exactly the same reinforcement show the difficulty in generalizing about weight gains in marine composites.

12.5.2 Property changes

In this study, shear properties were measured by tensile tests on specimens cut at 45° to the fibre directions. Using a biaxial extensometer this enabled the shear stress–shear strain to be obtained, and Fig. 12.6 shows an example of shear–shear strain plots, together with shear modulus and strength losses after 2 years' ageing. Few shear data are available for comparison, but the

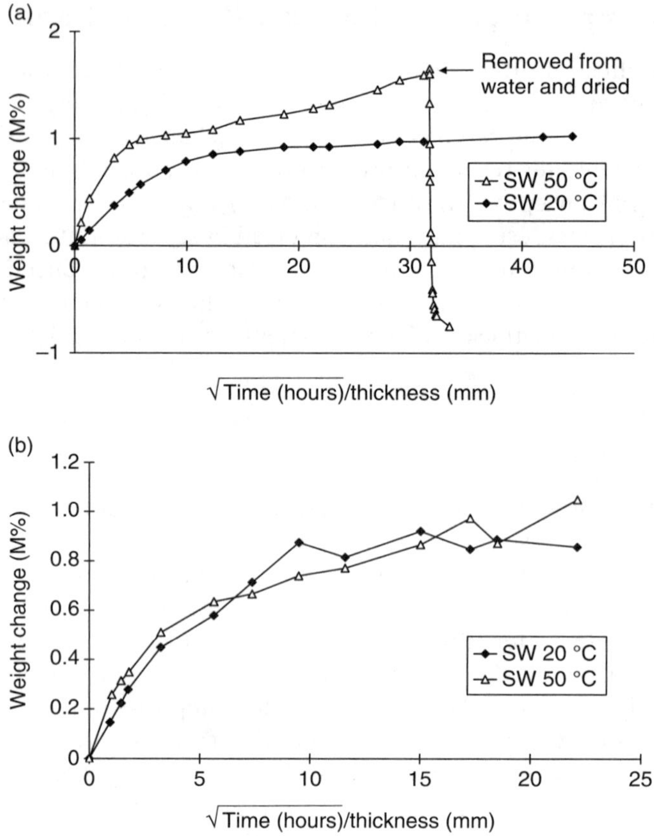

12.4 (a) Isophthalic polyester resin and (b) isophthalic polyester composite weight gains in sea water (SW).

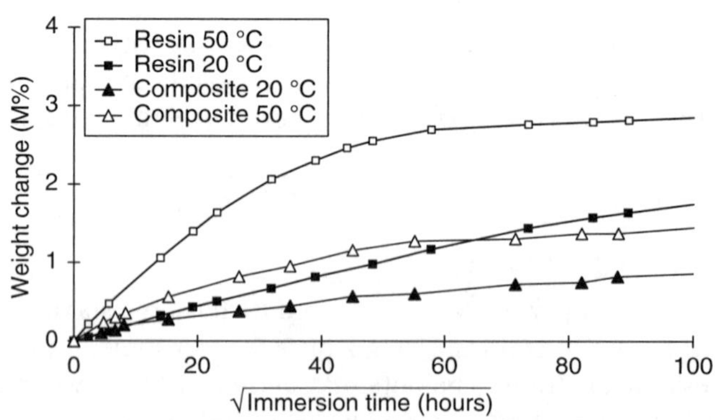

12.5 Epoxy resin and composite weight gains.

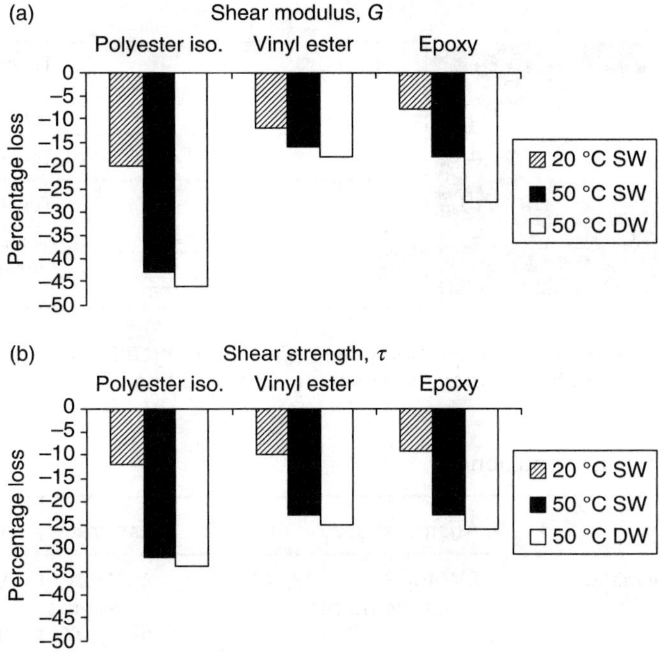

12.6 Influence of ageing on shear properties: (a) modulus, *G*; (b) strength, *τ*. iso., isophthalic.

strength losses are consistent with Smith's conclusions for strength under other loading conditions.

12.6 Example 2: ageing at sea

In this section one example of results from a study of sea ageing, performed at IFREMER, France on various marine composites and adhesively bonded assemblies, will be presented.

12.6.1 Materials and ageing conditions

A marine composite system consisting of isophthalic polyester resin reinforced with five layers of E-glass Rovimat (a 500 g/m^2 woven cloth lightly stitched to a 300 g/m^2 mat layer) was prepared in the form of panels 125 × 250 mm^2, of thickness 6 mm.

A first series of panels was immersed at sea at an IFREMER test site in the Brest Estuary for up to 2 years. Panels were fixed to immersion baskets, placed at 5 m depth, and periodically removed by divers (Fig. 12.7). A second series of panels was aged in the laboratory in distilled water at 50 °C.

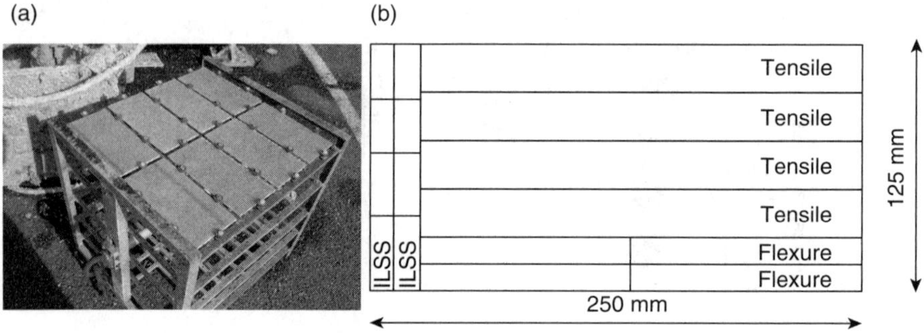

12.7 Ageing at sea. (a) Composite panels mounted in one of the baskets before immersion at sea. (b) Panel cutting plan.

Table 12.4 Ageing conditions

Material	Ageing at sea	Laboratory ageing
Rovimat/polyester	SW for 3, 6, 9, 12, 18 and 24 months	16 °C DW for 9 months 50 °C DW for 1, 2, 3 and 9 months

DW; distilled water; SW; sea water.

After ageing for different periods, as shown in Table 12.4, specimens were placed in a temperature (20 °C)- and relative humidity (50%)-controlled laboratory for 2 weeks to stabilize them, before testing in the same laboratory. The sea-aged specimens were covered with marine growth, and this was carefully removed before stabilization. The panels were always cut in the same way in order to obtain four tension, four flexure and eight short-beam shear specimens as shown in Fig. 12.7 (b).

12.6.2 Results and correlation with laboratory ageing

Figure 12.8(a) shows the tensile stiffness changes measured after different ageing periods for sea and laboratory ageing. The flexural modulus, Fig. 12.8(b), shows a very similar trend but with larger variability. This figure shows that there is a correlation between the stiffness loss in the accelerated test and the change after ageing at sea for up to a year, but beyond that the stiffness appears to stabilize and does not drop further. This indicates that even running an accelerated laboratory test for 9 months will seriously overestimate the drop in stiffness for longer periods. Figure 12.8(b) shows how flexural strength changes with immersion time. There is an

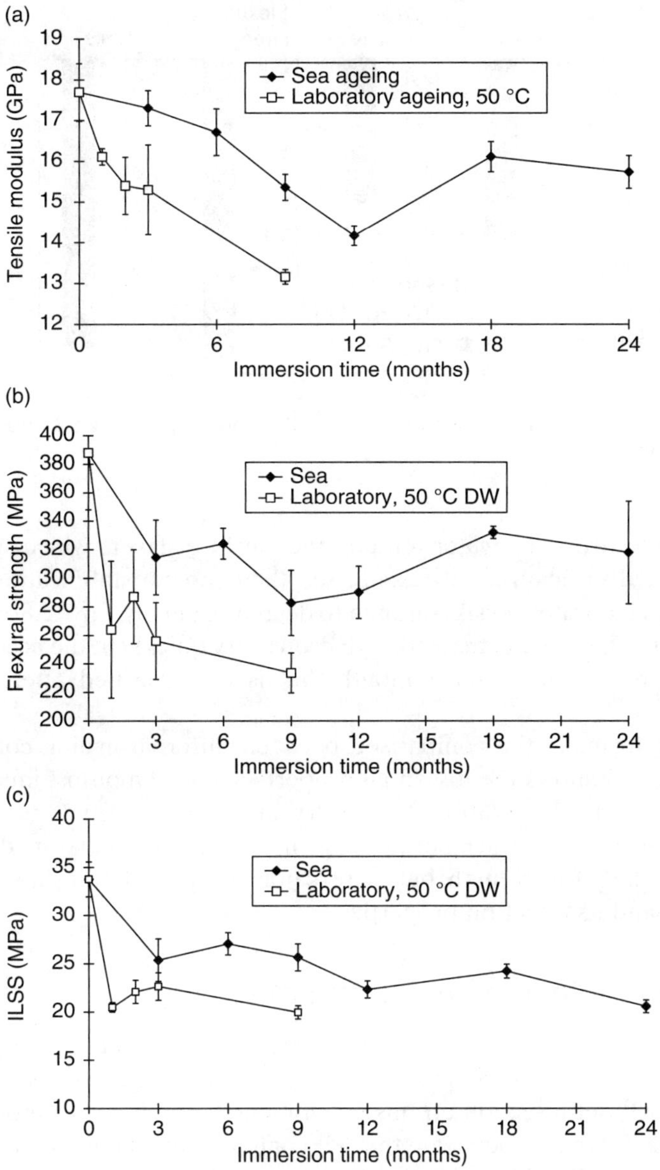

12.8 Comparison between ageing at sea and accelerated laboratory conditions for a Rovimat/isophthalic polyester. (a) Loss in tensile modulus versus immersion time; (b) loss of flexural strength versus immersion time; (c) ILSS versus immersion time.

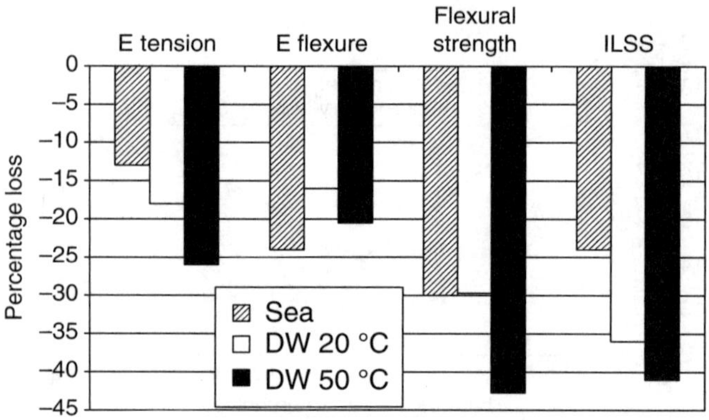

12.9 Loss in properties after 9 months', ageing. E tension, modulus in tension.

initial rapid drop as water enters the surface layer which is most highly loaded in flexure, but tests at sea then show a stabilization, while results from laboratory tests continue to decrease. Finally, Fig. 12.8(c) shows ILSS results. Once again, that strength drops very quickly in the accelerated test, reaching a value after 1 month that is only reached after 2 years at sea.

In order to make the comparison between different ageing conditions easier, Fig. 12.9 shows the loss in all properties after 9 months' immersion. Results for a third condition, laboratory immersion at 20 °C in distilled water, are also shown. The laboratory ageing conditions accelerate degradation, particularly for strength, but there is not a simple relationship between the media and its effect on properties.

12.7 Example 3: osmosis and blistering

The appearance of blisters on boat hulls caused much concern in the 1980s and early 1990s, and no chapter on the ageing of marine vessels would be complete without a section on this phenomenon, Fig. 12.10. Osmotic pressure at the interface between the gel-coat and the laminate has been observed to result in blistering. It is particularly likely when the gel-coat is more permeable than the underlying laminate as this results in a concentration gradient. Moisture reaches the interface and leaching out of products such as glycols appears to play a role in the process. Figure 12.11 shows a test set-up designed to examine this phenomenon; samples are subjected to hot water on one face and the time to appearance of blisters is noted. Castaing *et al.* studied this in detail[59–61] and were able to predict the kinetics of blister initiation and growth.

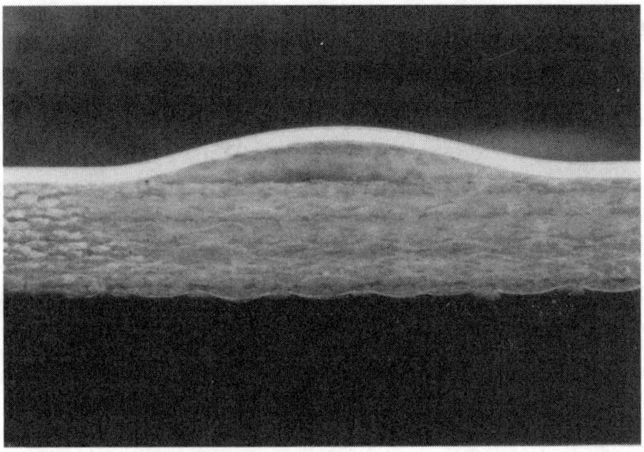

12.10 Section through blister.

12.11 Test set-up to measure blistering resistance of gel-coat/laminate combinations.

12.8 Relevance of accelerated tests

Laboratory ageing studies generally involve placing small samples in water baths at elevated temperature, weighing periodically and measuring residual properties after a few weeks or months. Water enters the samples through the faces but also from the edges which are usually unprotected. Flexural tests are most often used to measure residual behaviour and the sensitivity of this type of loading to the state of the surface further accentuates property losses. It is thus not unusual to find 50% strength reductions. As was described above in Section 12.4.4, such reductions have never been measured on samples from marine vessels, so the usefulness of such tests must be examined more closely.

12.8.1 Influence of gel-coat

The first observation is that marine composites are always protected by a gel-coat on the face in contact with water. This is not an impermeable layer but is resin-rich with a light glass veil which slows water ingress considerably. An example of the ability of this coating to protect the underlying laminate is shown in Fig. 12.12.

12.8.2 Water composition

Distilled water and tap water are widely used in laboratory ageing tests. It is well-known that distilled water diffuses into composites more quickly than sea water. Figure 12.13 shows the ratio between weight gains in distilled water and sea water for four composites with different matrix resins

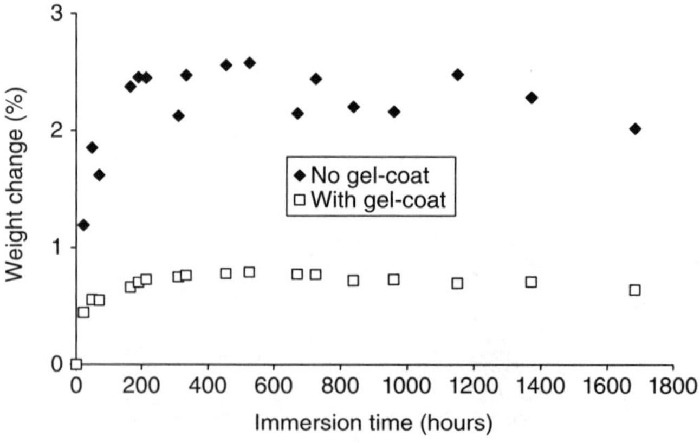

12.12 Weight gains for 1 mm thick carbon/epoxy laminates in sea water at 40 °C, completely covered by and without gel-coat.

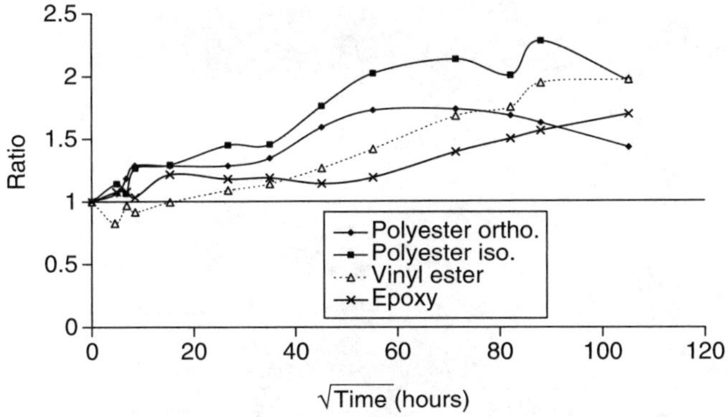

12.13 Ratio of weight gain in distilled water to that in sea water. Abbreviations: ortho., orthophthalic; iso., isophthalic.

and the same reinforcement. In all cases the weight gains are considerably higher in distilled water and the ratio tends to increase with immersion time.

This aspect of ageing is discussed in more detail in Chapter 18. However, another important point often neglected is that ageing of polyester samples in small water baths will in time lead to significant changes in water composition as products leach out of the composite. For this reason tests at IFREMER are performed in natural sea water pumped from the Brest Estuary into large tanks, with circulation and continuous water renewal (Fig. 12.14).

12.8.3 Influence of temperature

The dilemma faced by test laboratories is how to estimate long-term behaviour of marine composites from short-term test data. Increasing the water temperature and relying on an Arrhenius law to predict behaviour at lower temperatures is widely used, often with little justification. Miller examined a polyester composite used in sailboat construction. He showed that different moisture conditions during testing result in different strength losses compared with dry reference specimens[62] (Table 12.5). He concluded that the boil test, a common method for rapid screening, cannot be used to predict long-term exposure resistance.

Any increase in test temperature for polyesters that are only partly cured will promote further cross-linking. Use of high temperatures, above the glass transition temperature (typically around 70 °C for common polyesters) can induce degradation mechanisms that will never occur at lower temperatures. Nevertheless, Miyano *et al.* have based a long-term durability

12.14 Laboratory sea water ageing set-up, IFREMER.

Table 12.5 Loss in properties after different conditioning treatments, after reference 62

Ageing condition	Tension	Shear	Compression
15 months, tap water, room temperature	–20%	–11%	–16%
15 months, 100% RH, 24 h boiling water	–24%	–22%	–25%

prediction method on time–temperature superposition and show results for woven fibre reinforced glass[63] and carbon[64] marine composites. Using quasi-static and creep rupture failure tests at different loading rates and temperatures, they construct master curves that enable long-term strength to be predicted. Brinson *et al.*[65] also developed this approach, while Cardon and colleagues[66] applied a superposition principle but using moisture to accelerate tests. One can thus envisage a global life prediction tool based on superposition, but there are a number of assumptions that must be satisfied if such an approach is to be validated. First, the material must be stable at the temperatures employed. The reinforced vinyl ester materials tested by Miyano *et al.* were stabilized at 150 °C and accelerating temperatures up to 120 °C were applied. The under-cured polyesters currently used for most marine vessels are certainly not stabilized and such temperatures, necessary to obtain a useful acceleration factor, are much too high; as shown by Miller.[62] Other assumptions also need to be checked, namely that the same mechanisms occur for quasi-static, creep and fatigue failure and that the same time–temperature superposition can be applied to all three. Finally, a linear damage accumulation law was applied in the work of Miyano *et al.*, which may or may not be appropriate in polyester composites. Attempts to validate accelerated test conditions by referring to in-service experience are rare, as noted in Section 12.4.4, and the studies performed to compare ageing at sea and laboratory test results are often controversial. The results in Section 12.6 above show the risks in extrapolating from short-term data.

12.8.4 Orthotropy

Another difficulty with accelerated testing is that even without ageing the number of different tests required to fully characterize composite behaviour is very large. In addition to in-plane tension, shear and compression in two or three directions to obtain modulus and strength, through-thickness properties are also of interest, particularly when thick laminates such as those often used in marine applications are employed. In order to fully exploit the properties of these materials, damage parameters and fracture energy are also needed, together with some indicators for rapid (impact), slow (creep) and cyclic (fatigue) loading. If the influence of ageing on all these properties is to be completely understood this requires years of testing. As a result, for most testing campaigns small numbers of tests are applied and the majority of test results for marine composite ageing come from flexure and short-beam shear specimens. These are simple to test and require no extensometry but are not easy to analyse due to the complex stress state below the loading points.

12.8.5 Biological factors

A final consideration in ageing tests is that the marine environment is not sterile, but contains a multitude of living organisms. Figure 12.15 shows the appearance of the composite samples described in Section 12.6, after 11 months at sea in the Brest Estuary. The water temperatures in Brittany are

12.15 Eight of the polyester composite panels after 11 months' immersion.

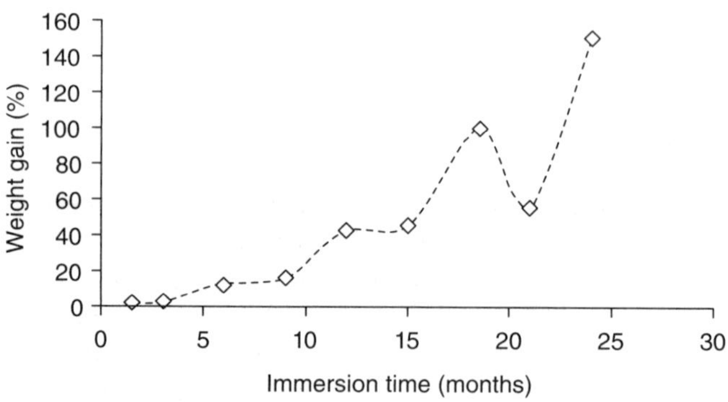

12.16 Weight gain of composite panels after different periods of immersion at sea.

far from tropical, typically in the range from 10 to 20 °C, but even after a short immersion period bio-films develop followed by marine fouling, and this results in a thick covering of living matter. Weight gains of composite plaques during ageing at sea are shown in Fig. 12.16, and a gradual increase over the first 9 months jumps to around 50% after a year and to over 100% after 2 years. After removal of the biomass the weight gains were all less than 1% of the original weight. The influence of this layer, which includes both soft and hard fouling, has never been studied in detail, to the authors' knowledge, but must affect residual properties and may contribute to the stabilization of property losses after a year at sea noted above (Fig. 12.8). Once again, the relevance of ageing samples without the protective coatings commonly applied to marine vessels (gel-coat, anti-fouling treatments) is open to discussion. There have been some studies of biological factors on composite behaviour[67] which showed little influence of sulphate-reducing, acid-producing or aerobic bacteria on the ageing behaviour of carbon/epoxy composites. The influence of fouling will vary with the season and with water properties. For deep sea applications it is not a problem.

12.9 Conclusions and future trends

Glass reinforced polyester composite materials have been widely adopted for marine vessels as they offer a good balance of mechanical properties at a reasonable cost. The experience from over 50 years' service has revealed very few cases of premature ageing and glass reinforced thermosets have generally performed extremely well. Manufacturing methods are slowly evolving away from hand lay-up and techniques such as infusion will become more popular, but the long-term future of these materials is less clear. New environmental regulations, uncertainty about the future availability and cost of oil-based resins, and the energy cost of producing glass fibres have prompted some interest in alternative materials. Thermoplastic matrix composites such as glass/polypropylene are being studied as they offer some possibilities for recycling, but they have the same sources and require elevated temperature tooling. Bio-composites, combinations of resins and fibres from natural sources, are also starting to attract interest, and may ultimately provide a more environmentally friendly route to build marine vessels. Guaranteeing the long-term durability of these bio-degradable materials is a real challenge, but one which is worth addressing. Most cellulose-based fibres are hydrophilic – as mentioned above for balsa – but wood, a natural composite, has been employed for centuries in naval construction and is still being used. An elegant approach would be to build bio-composite boats with fibres from marine plants such as sea-grass[68] reinforcing bacterial polyester resins produced by fermentation with marine bacteria.[69] This might appear to be a distant prospect, even though it is theoretically feasible today, it is certainly not economical. But then, 50 years ago high-speed composite sandwich ferries cruising at 50 knots would also have seemed highly improbable!

12.10 References

1 GIBBS & COX INC., *Marine Design Manual for Fibreglass Reinforced Plastics*, McGraw Hill, New York, 1960.
2 SMITH CS, ed. *Design of Marine Structures in Composite Materials*, Elsevier, Amsterdam, 1990.
3 GREENE E, ed. *Marine Composites*, 2nd edition, http://www.marinecomposites.com/.

4 SHENOI RA, WELLICOME JF, eds, *Composite Materials in Maritime Structures*, Cambridge Ocean Technology Series, Cambridge University Press, 1993.
5 MARSH G, 50 years of reinforced plastic boats, *Reinforced Plastics*, **50**, October 2006, 16–19.
6 BRITISH MARINE FEDERATION, UK leisure and small commercial marine industry, key performance indicators 2005/6.
7 BERREUR L, DE MAILLARD B, NOSPERGER S, L'industrie Française des matériaux composites, French Ministère de l'Economie, des Finances et de l'Industrie report 2002.
8 MINISTÈRE DE L'ECONOMIE, Les bateaux de plaisance, février 2006.
9 SPURR D, BÉNÉTEAU, *Professional Boatbuilder*, February–March 2005, 46–61.
10 WILLIAMS SG, Reinforced plastics in the marine industry, *Naval Architect*, November 1978, 197–201.
11 Scott Bader Technical leaflet 950, GRP for workboats, February 1977.
12 DIXON RH, RAMSEY BW, USHER PJ, Design and build of the GRP hull of HMS *Wilton*, in *Proceedings of the Symposium on GRP ship construction*, Royal Institution or Naval Architects, October 1972, pp. 1–32.
13 MOURITZ AP, GELLERT E, BURCHILL P, CHALLIS K, Review of advanced composite structures for naval ships and submarines, *Composite Structures*, **53**, 2001, 21–41.
14 REMEN W, The use of FRP sandwich, in *Proceedings of the International Conference On Nautical Construction with Composite Materials*, eds Davies P, Lemoine L, Paris, December 1992, IFREMER, paper 42, pp. 432–439.
15 HELLBRAT S-E, Design of dynamically supported craft, in *Composite Materials in Maritime Structures*, eds Shenoi RA, Wellicombe JF, Cambridge University Press, 1993, Chapter 2, pp 26–42.
16 MARSH G, Can composites become serious seagoers? *Reinforced Plastics*, **48**, October 2004, 20–24.
17 DNV Tentative rules for classification of high speed and light craft, 1991.
18 CASARI P, CHOQUEUSE D, DAVIES P, DEVAUX H, Applications marines des composites. Cas des voiliers de compétition, Techniques de l'Ingénieur, 2008, AM 5 655, http://www.techniques-ingenieur.fr (in French).
19 SUMMERSCALES J, Marine applications, in *Engineered Materials Handbook I*, ASM, Materials Park, OH, 1987, Chapter 12 G, pp 837–844.
20 GARNER WR, Marine applications, in *Handbook of Composites*, ed. Lubin G, van Nostrand Reinhold, New York, 1982. p. 699.
21 LEMIÈRE Y, The evolution of composite materials in submarine structures, in *Proceedings of the International Conference on Nautical Construction with Composite Materials*, eds Davies P, Lemoine L, Paris, December 1992, IFREMER, pp 441–449.
22 MARSH G, Reduced styrene content offers answer for LSE, *Reinforced Plastics*, **45**, December 2001, 24–30.
23 BALEY C, PERROT Y, DAVIES P, BOURMAUD A, GROHENS Y, Mechanical properties of composites based on low styrene emission polyester resins for marine applications, *Applied Composite Materials*, **13**(1), 2006, 1–22.
24 PERROT Y, BALEY C, GROHENS Y, DAVIES P, Damage resistance of composites based on glass fibre reinforced low styrene emission resins for marine applications, *Applied Composite Materials*, **14**(1), 2007, 67–87.

25 ISO 12215, Hull construction – Scantlings – Part 5 Design pressures for monohulls, 2006.
26 RYMILL RJ, Material case study – failures and their repairs, in Composite Materials in Maritime Structures, eds Shenoi RA, Wellicome JF, Cambridge University Press, 1993, Chapter 6, pp. 134–160.
27 CRAY VALLEY UNSATURATED POLYESTERS, *Application Guide*, Cray Valley, Paris, France, 1998.
28 SPRINGER GS, ed., *Environmental Effects on Composite Materials*, Volumes 1 and 2, Technomic Publishing, Lancaster, PA, 1981 and 1984.
29 MALLINSON JH, ed., *Corrosion Resistant Plastic Composites in Chemical Plant Design*, Marcel Dekker, New York, 1988.
30 PRITCHARD G, ed., *Reinforced Plastics Durability*, Woodhead/CRC, Abington, UK, 1999.
31 HARRIS B, ed., *Fatigue of Composites*, Woodhead Publishing, Abington, UK, 2003.
32 SEARLE TJ, SUMMERSCALES J, Review of the durability of marine laminates, in *Corrosion Resistant Plastic Composites in Chemical Plant Design*, Marcel Dekker, New York, 1999, Chapter 7, pp. 219–266.
33 KOOTSOOKOS A, MOURITZ A, Seawater durability of glass– and carbon–polymer composites, *Composites Science and Technology*, **64**, 2004, 1503–1511.
34 GAUTIER L, MORTAIGNE B, BELLENGER V, Interface damage study of hydrothermally aged glass-fiber reinforced polyester composites, *Composites Science and Technology*, **59**(16), 1999, 2329–2337.
35 DAVIES P, MAZEAS F, CASARI P, Sea water ageing of glass reinforced composites: shear behaviour and damage modelling, *Journal of Composite Materials*, **35**(15), 2001, 1343–1372.
36 PERROT Y, DAVIES P, BALEY C, Influence of polyester resins formulated to limit styrene emission on the ageing resistance of marine composites, in *Proceedings of the French Composites Conference JNC15*, Marseille, June 2007, pp. 1115–1122 (in French).
37 WHITE RJ, PHILLIPS MG, Environmental stress-rupture mechanisms in glass fibre/polyester laminates, in *Proceedings of ICCM5*, eds Harrison WC, Strike J, Dhingra AK, San Diego, 1985, pp. 1089–1099.
38 HOGG PJ, HULL D, Corrosion and environmental deterioration of GRP, in *Developments in GRP Technology 1*, ed. B. Harris, Applied Science, London, 1983, Chapter 2.
39 MELENNEC A, LAGRANGE A, JACQUEMET R, Influence of various stress conditions on the moisture diffusion of composites in distilled water and natural seawater, in *Durability of Polymer Composite Based Systems for Structural Applications*, eds Cardon A, Verchery G, Elsevier, Amsterdam, 1991, p. 385.
40 GREENWOOD M, Pultruded composites durability, in *Proceedings of Composites 2001*, Tampa, Florida, October 2001.
41 GELLERT EP, TURLEY DM, Seawater immersion of glass-fibre reinforced polymer laminates for marine applications, *Composites*, **30A**, 1999, 1259.
42 PAUCHARD V, GROSJEAN F, CAMPION-BOULHARTS H, CHATEAUMINOIS A, Application of a stress-corrosion cracking model to an analysis of the durability of glass/epoxy composites in wet environments, *Composites Science and Technology*, **62**, 2002, 493–498.

43 DAVIES P, CHOQUEUSE D, MAZEAS F, Composites underwater, in *Proceedings of DURACOSYS* 1997, eds Reifsnider KL, Dillard DA, Cardon AH, Blacksburg, VA, September 1997, Balkema Publishers, Rotterdam.
44 POLLARD A, BAGGOTT R, WOSTENHOLM GH, YATES B, GEORGE AP, Influence of hydrostatic pressure on the moisture absorption of glass fibre reinforced polyester, *Journal of Materials Science*, **24**, 1989, 1665–1669.
45 AVENA A, BUNSELL AR, Effect of hydrostatic pressure on the water absorption of glass fibre reinforced epoxy resin, *Composites*, **19**(5), 1988, 355–357.
46 TUCKER WC, BROWN R, Moisture absorption of graphite/polymer composites under 2000 feet of seawater, *Journal of Composite Materials*, **23**, 1989, 787–797.
47 CHOQUEUSE D, DAVIES P, MAZÉAS F, BAIZEAU R, Ageing of composites in water, ASTM STP 1302, 1997, pp. 73–96.
48 STRAND R, Wet balsa core, *Professional Boatbuilder*, August/September 2005, 16–35.
49 CANTWELL WJ, BROSTER G, DAVIES P, The influence of water immersion on skin-core debonding in GFRP-balsa sandwich structures, *Journal of Reinforced Plastics and Composites*, **15**(11), 1996, 1161–1172.
50 CANTWELL WJ, SCUDAMORE R, RATCLIFFE J, DAVIES P, Interfacial fracture in sandwich laminates, *Composites Science and Technology,* **59**, 1999, 2079–2085.
51 LI X, WEITSMAN J, Sea-water effects on foam-cored composite sandwich lay-ups, *Composites Part B*, **35**, 2004, 451–459.
52 IONITA A, WEITSMAN YJ, A model for fluid ingress in closed cell polymeric foams, *Mechanics of Materials*, **39**, 2007, 434–444.
53 EARL JS, SHENOI RA, Hygrothermal ageing effects on FRP laminate and structural foam materials, *Composites Part A*, **35**(11), 2004, 1237–1247.
54 LAPLANTE G, MARBLE AE, MACMILLAN B, LEE-SULLIVAN P, COLPITTS BG, BALCOM BJ, Detection of water ingress in composite sandwich structures: a magnetic resonance approach, *NDT & E International*, **38**(6), 2005, 501–507.
55 GRANER WR, DELLA ROCCA RJ, Evaluation of US Navy boats for material durability, in *Proceedings of the 26th Annual Technical Conference on Reinforced Plastics/Composites SPI*, 1971.
56 FRIED N, GRANER WR, Durability of reinforced plastic materials in marine service, *Marine Technology*, **3**(3), 1966, 321–327.
57 GUTIERREZ J, LELAY F, HOARAU P, A study of ageing of glass fibre-resin composites in a marine environment, in *Proceedings of the International Conference on 'Nautical Construction with Composite Materials*, Paris', 1992, IFREMER, p. 338.
58 ARTIGA-DUBOIS F, Durability and expertise in composites for naval applications, in *Proceedings of JNC15, 15th French Composites Conference*, Marseilles, 2007, pp. 1107–1114 (in French).
59 CASTAING P, Vieillissement des matériaux composites verre/polyester en milieu marin: délaminage d'origine osmotique, PhD Thesis, INPToulouse, 1992 (in French).
60 CASTAING P, LEMOINE L, GOURDENNE A, Mechanical modelling of blisters on gel-coated laminates I – Theoretical aspects, *Composite Structures*, **30**(2), 1995, 217–222.
61 CASTAING P, LEMOINE L, GOURDENNE A, Mechanical modelling of blisters on gel-coated laminates I – Experimental results, *Composite Structures*, **30**(2), 1995, 223–228.

62 MILLER PH, Effect of moisture absorption and test method on the properties of E glass/polyester hull laminates, *Journal of Composite Materials*, **36**(9), 2002, 1065–1078.

63 MIYANO Y, NAKADA M, SEKINE N, Accelerated testing for long-term durability of GFRP laminates for marine use, *Composites Part B*, **35,** 2004, 497–502.

64 MIYANO Y, NAKADA M, SEKINE N, Accelerated testing for long-term durability of FRP laminates for marine use, *Journal of Composite Materials*, **39**(5), 2005, 5–20.

65 BRINSON HF, Matrix dominated time dependent failure predictions in polymer matrix composites, *Composite Structures*, **47**(1–4), 1999, 445–456.

66 MIRANDA GUEDES R, MORAIS JLL, TORRES MARQUES A, CARDON AH, Prediction of long-term behaviour of composite materials, *Computers & Structures*, **76**(1–3), 2000, 183–194.

67 PUH JS, WAGNER P, LITTLE B, BRADLEY W, The effect of biofouling on graphite/epoxy composites, *Journal or Composites Technology and Research,* **20**(1), 1998, 59–67.

68 DAVIES P, MORVAN C, SIRE O, BALEY C, Structure and properties of fibres from sea-grass (Zostera marina), *Journal of Materials Science*, **42**(13), 2007, 4850–4857.

69 DOI Y, ed., *Microbial Polyesters*, VCH Publishers, New York, 1990.

Part III

Ageing of composites in non-transport applications

13
Ageing of polyethylene composite implants in medical devices

S. AFFATATO, Istituti Ortopedici Rizzoli, Italy

13.1 Definition of medical devices

The number of medical devices implanted every year in humans is very high. Some 20 million people in the world are estimated to have at least one medical-device implant and therefore billions of dollars are spent on medical devices every year.[1]

What does the term 'medical device' mean?

The European[2] and the American[3] definitions of a medical device are quite similar. Both definitions agree that a medical device is 'any instrument, apparatus, appliance, material or other article, whether used alone or in combination, including the software necessary for its proper application intended by the manufacturer to be used for human beings'. The purposes of using a medical device are:

- diagnosis, prevention, monitoring, treatment or alleviation of disease;
- diagnosis, monitoring, treatment, alleviation of or compensation for an injury or handicap;
- investigation, replacement or modification of the anatomy or of a physiological process;
- control of conception, for a device that does not achieve its principal intended action in or on the human body by pharmacological, immunological or metabolic means, but which may be assisted in its function by such means.

A risk classification scheme has been developed to categorise medical devices according to their potential risk.[2] The degree of regulation imposed on any device is proportional to its risk. The classification rules were designed to be straightforward and uscr-friendly, and the following indicators of risk were used to create the rules: degree of invasiveness, duration of contact, body system affected and local versus systemic effects. In contrast to other classification systems, in Europe the classification rules are set out in *Annex IX* of the directive CE Marking[2] for active implantable medical devices.

Annex IX classification is based on a risk assessment of the product, when used as intended. The 'risk' is composed of the duration of use and the level of invasiveness, as defined by the intended use stated by the manufacturer. It is for this reason that the manufacturer assumes the responsibility for a proper classification.

The classification of medical devices in the European Union is divided in four classes[2,4] (ranging from low to high risk). In particular:

- *Class I: low-risk devices.* All non-invasive devices that come into contact with injured skin and are intended to be used as a mechanical barrier, for compression or for absorption of exudates (i.e. wheelchairs, patient beds, non-invasive electrodes, sterile wound dressings, thermometers).
- *Class IIa: low- to medium-risk devices.* All non-invasive devices intended for channelling or storing blood, body liquids or tissues, liquids or gases for the purpose of eventual infusion, administration or introduction into the body (i.e. ultrasound diagnostic devices, respiration tubes, blood pressure monitoring devices, products for transfusion of blood).
- *Class IIb: low- to medium-risk devices.* All non-invasive devices intended for modifying the biological or chemical composition of blood, other body liquids or other liquids intended for infusion into the body (i.e. artificial joints, lithotripters).
- *Class III: high-risk devices.* All surgically invasive devices intended specifically to diagnose, monitor or correct a defect of the heart or of the central circulatory system through direct contact with these parts of the body (i.e. prosthetic heart valves, neurological catheters).

The difference between each Class depends on rules that involve the medical device's duration of body contact, its invasive character, its use of an energy source, its effect on the central circulation or nervous system, its diagnostic impact or its incorporation of a medicinal product. Certified medical devices should have the CE mark[2] on the packaging, insert leaflets, etc. These packagings should also show harmonised pictograms and International Organization for Standardization (ISO) standardised logos to indicate essential features such as instructions for use, expiry date, manufacturer, sterile, do not reuse, etc.

Many countries – such as Japan, People's Republic of China and Latin America – have their own classification scheme. In these countries, *in vitro* diagnostics (IVDs) are classified separately. For all practical purposes, the classification schemes of Canada and Australia/New Zealand (covered by the Trans-Tasman Agreement) resemble the European one so closely that a separate review makes little sense. For the United States, a difference exists. The American Federal Food and Drug Administration (FDA)[5] has recognised three classes of medical devices based on the level of control necessary to assure the safety and effectiveness of the device.[5] The classification procedures are briefly described here.

- *Class I.* Generally defined as *'General controls'* and present minimal potential for harm to the user. These devices are subject only to general controls that cover such issues as manufacturer registration with the FDA, good manufacturing techniques, proper branding and labelling, notification of the FDA before marketing the device and general reporting procedures.[3] (Most Class I devices are exempt from the good manufacturing practices and/or the FDA notification regulations.)[3]
- *Class II.* Generally defined as *'General controls with special controls'*; this is the class for which general controls alone are insufficient to assure safety and effectiveness, and additional existing methods are available to provide such assurances. Therefore, Class II devices are also subject to special controls; special controls may include special labelling requirements, mandatory performance standards, and postmarket surveillance.[3] Devices in this class are typically non-invasive.
- *Class III.* Generally defined as *'General controls and premarket Approval'* in which scientific reviewers ensure the device's safety and effectiveness, in addition to the general controls of Class I. Class III devices are described as those for which 'insufficient information exists to determine that general controls are sufficient to provide reasonable assurance of its safety and effectiveness or that application of special controls would provide such assurance and if, in addition, the device is life-supporting or life-sustaining, or for a use which is of substantial importance in preventing impairment of human health, or if the device presents a potential unreasonable risk of illness or injury'.[3]

Today, more than 500 000 arthroplasty procedures and total joint replacements (TJR) are performed annually in Europe. Nearly 20% of the hip replacement operations are revisions, repairs to the original implants, and as the population receiving such implants becomes younger relative to mortality, this number is likely to increase. Their increased activity plus longer usage is expected to result in a higher incidence of eventual failure of conventional hip and knee replacements. However, as it is necessary to remove bone surrounding the implant, generally only one revision surgery is possible, thus limiting current orthopaedic implant technology to older, less active individuals.

13.2 Brief history of polyethylene used in medical devices

Orthopaedic implantations are common procedures today: every year, approximately 1.4 million joint replacement procedures are performed worldwide.[6] Acetabular cups used in total hip replacement surgery have been made from a variety of materials throughout the years (metals, ceramics and polymers). The choice of polymer varies from manufacturer to manufacturer, but ultra-high molecular weight polyethylene (UHMWPE) is widely used and is now the major material used in artificial replacement of hip and knee joints. UHMWPE is known to have superior resistance to wear compared with other polymers.[7]

Polyethylene can be produced through radical polymerisation, anionic addition polymerisation, ion coordination polymerisation or cationic addition polymerisation.[8,9] UHMWPE, despite having relatively weak Van der Waals bonds between its molecules, derives ample strength from the length of each individual molecule.[10] It is made up of extremely long chains of polyethylene, which all align in the same direction. Each chain is bonded to the others with so many Van der Waals bonds that the whole can support great tensile loads. Polyethylene is classified into several different categories based mainly on its density and branching.[11,12] The mechanical properties of polyethylene depend significantly on variables such as the extent and type of branching, the crystal structure, and the molecular weight. UHMWPE molecules tend to have 100 000–250 000 monomers each, while other types of polyethylene have considerably fewer monomers; for example, high-density polyethylene (HDPE) molecules generally have between 700 and 1800 monomer units per molecule. Nowadays, medical-grade resins are described as GUR1020 or GUR1050 depending on their molecular weights: GUR1020 has an average molecular weight of 3.5 million Daltons and that of GUR1050 is above 5 million Daltons.[13,14]

The requirements for medical-grade UHMWPE powder are specified in ASTM standard F648 and ISO standard 5834–1. These standards characterise powders on molecular weight, trace impurities of titanium, aluminium and chlorine (residuals from catalysts), trace levels of calcium and ash content. UHMWPE is produced as a powder, which must be compacted in solid form in order to realise the final form, by the chemical process of polymerisation, which usually involves Ziegler catalysts such as titanium tetrachloride.[14–16]

The Ziegler–Natta catalyst[17,18] is synthesised by treating crystalline α-$TiCl_3$ with $(AlCl(C_2H_5)_2)_2$. Polymerisation occurs at special Ti centres located on the exterior of the crystallites. At the surface some Ti centres lack their full complement of six chloride ligands to give an octahedral structure. The alkene binds at these vacancies. The alkene converts to an alkyl ligand group.

Compaction of the powder of the polymer can be achieved by different processes such as ram extrusion, slab compression moulding or direct compression moulding.[13] It is left to the orthopaedic manufacturer to determine which conversion method is most suitable for orthopaedic applications, as there is still no consensus on which choice is best. The disadvantage of ram extrusion and slab compression moulding is that they may not produce uniform conditions in the compacted mass.[13] In the past, these techniques have used calcium stearate as an additive to stabilise the residue of the catalyser, in order to make the material flow more easily and to minimise corrosion of tools.[13,19] During the 1980s, the influence of calcium stearate on the properties and performance of UHMWPE total joint replacement was the subject of several studies; results showed an association of this additive with fusion defects and oxidation of UHMWPE. In fact, calcium stearate may accelerate oxidative degradation of polyethylene after γ-irradiation in air and may also cause poor consolidation of the powder.[20–22] In the meantime, polymerisation and processing technology evolved to the point that the additive was no longer necessary, consequently orthopaedic manufacturers began switching to UHMWPE resins without added calcium stearate.

In the cases of slab compression and ram extrusion, a solid block is produced from which a component is machined.[23] With direct compression moulding a component is formed that requires little or no further machining, but the technique is expensive and hence not suitable for prototypes.[14,24] The choice of conversion method is at least as important as the choice of resin for the properties of a UHMWPE component, because both of these factors introduce some change in the morphology of the consolidated polymer. However, there is still no consensus on which resin and conversion method would be universally proposed as the best choice for all orthopaedic applications and, as a result, it is the orthopaedic manufacturer who determines which conversion method is most suitable for orthopaedic applications.[25]

Since the molecular chains of UHMWPE are not static they can become mobile at elevated temperatures. When cooled below melt temperature the molecular chains have the tendency to rotate the C–C bonds creating chain folds.[10] These folds, in turn, enable chains to form a more ordered structure known as 'crystalline lamellae'. These lamellae are embedded within an amorphous (disordered) region and may communicate with surrounding lamellae.[26,27] The UHMWPE in the manufactured component, regardless of the process of compaction, is composed of two phases: crystalline and amorphous.[13,24] The degree and the orientation of crystalline regions within a polyethylene depend on many factors such as molecular weight, processing conditions, etc.

When air gains access inside a block of polyethylene, either by being trapped between the granules as they are compressed or by diffusion from ambient air during storage, the oxidation process can start. Oxidation of the polyethylene molecule is a chemical reaction that results in chain scission (fragmentation and shortening of the large polymer chains) and introduction of oxygen into the polymer, which is more rapid in amorphous regions than in crystalline lamellae.[28,29] If oxygen is present, it may react with free radicals, lowering the molecular weight of the polymer and reducing its yield strength, ultimate tensile strength, elongation to break and toughness, and increasing its density.[30,31]

Clinical and laboratory research have revealed that sterilisation methods can dramatically affect the *in vivo* performance (and life) of a UHMWPE component, particularly as far as the resistance to wear is concerned.[32] Historically, polyethylene components were sterilised by ethylene oxide gas supplanted by exposure to γ-irradiation. The γ-irradiation was typically carried out at doses ranging from 2.5 to 4.0 Mrads, with the components packaged in air.[33,34] However, since 1995 all the major orthopaedic manufacturers began to use γ-radiation in a reduced oxygen environment or to sterilise without ionising radiation (ethylene oxide or gas plasma). The evidence that γ-sterilisation, followed by long-term shelf storage, could accelerate oxidative chain scission and degradation of the mechanical and physical properties of UHMWPE, promoted the shift in sterilisation practice.[35,36] While these measures, taken in order to avoid oxidative degradation of UHMWPE prior to implantation, are universally accepted, it still remains unclear to what extent these processes mitigate the tendency of polyethylene components to oxidatively degrade during long-term implantation in the body.[37,38]

More recently, polymer scientists have proposed alternative or 'enhanced' varieties of UHMWPE to improve the wear resistance of the polymer. Among these attempts, the attention of various researchers has been focused on the development of methods that increase the level of cross-linking in polyethylene since the polymer microstructure for optimal wear resistance appeared to be a three-dimensional network with the polymeric chains connected by covalent bonds.[24] At the same time, new 'composite' biomaterials suitable for medical applications have been proposed. The word 'composite' refers to the combination of two or more materials, different in composition, morphology and physical properties.[39,40] These materials have been studied for the last 40 years and their use in different applications has progressively increased. The composite materials present some advantages in quality and improved properties that the single constituents do not have. Moreover, composite materials allow a flexible design so that their properties can be tailored to specific applications.[39]

Biocompatibility is a crucial point in developing new materials; this property cannot be compromised in favour of better mechanical properties.[41] Composite materials suitable for joint prostheses (hip or knee) must provide exceptional wear performance and reduced particle debris. The leading opinion is to develop a high-strength, reinforced polymeric matrix such as self-reinforced UHMWPE, UHMWPE reinforced with polymethylmethacrylate (PMMA) or carbon fibres, and polyether ether ketone (PEEK) reinforced with carbon fibres.[6,42–44] The self-reinforced UHMWPE is essentially a non-oriented matrix of UHMWPE in which reinforcement has been dispersed. The resulting polymer presents excellent biocompatibility, increased mechanical properties and the possibility to be sterilised and cross-linked in exactly the same manner as the conventional UHMWPE.[44]

Acetabular cups made from composite UHMWPE–PMMA could consist of two zones formed exclusively either by UHMWPE or PMMA, or could have a layer of UHMWPE forming an articulating surface and a body formed from a mixture of PMMA and UHMWPE.[45] Generally, acetabular cups could be screwed to the pelvi, or press-fitted, or cemented. In this way advantages or disadvantages are associated with each style of fixation method.[43] For this kind of composite there is still no consensus on the clinical behaviour and further investigation is necessary.

In order to increase the long-term performance of UHMWPE acetabular cups, reinforcement in the form of carbon fibres has been added in the polymeric matrix.[39,46] Clinical experience with these composite materials was not satisfactory; mainly due to the poor bonding strength between the carbon fibres and the UHMWPE matrix or to the poor creep resistance of this latter, which promotes debonding of the carbon fibre from the matrix under load.

A carbon fibre reinforced PEEK has been proposed as an alternative for UHMWPE and has been studied with great interest although is not currently used in clinical practice for total hip replacement.[47–49] For carbon fibre reinforced PEEK it is crucial to determine the optimal combination of carbon fibre type and weight, and the nature of the counterface (femoral head material). Many investigations[50–52] have even observed that excellent tribological behaviour of acetabular cups is obtained with a combination of 30-wt% pitch-based carbon fibre reinforced PEEK composite. This combination showed a reduction in the wear rate by almost two orders of magnitude compared with conventional UHMWPE–metal or UHMWPE–ceramic couples on a hip joint simulator. Although traditional materials continue to provide clinical benefits, many attempts to obtain improved composite biomaterials have been made and the goal of a new composite material with excellent mechanical, tribological, and biocompatibility properties, seems to be getting closer each day.

13.3 Improvements on polyethylene for medical devices

It is well known that the debris generated from a UHMWPE socket may cause adverse tissue biological reactions leading to bone loss or osteolysis.[53,54] A major consequence of the debris-induced osteolysis is the loosening of the implant inside the femur or acetabular cup, which often necessitates a revision surgery. It is therefore essential to find a way of developing improved materials in order to extend the lifetime of orthopaedic implants such as knees and hips up to a minimum of 30 years.

Several new and pending products aim to address this question. A ceramic-on-ceramic bearing for a hip replacement is undergoing pre-FDA approval studies in the United States, and has been in use in Europe. In contrast to the traditional metal-on-plastic articulation, ceramic surfaces are said to have better wettability and toughness. Ceramic components, however, are brittle and can fracture more easily that other materials.[55–57] A metal-on-metal bearing is considered an alternative of hip replacement but, however, concerns remain about metal ion release into the patient's body.[58,59]

The need to improve the UHMWPE polymer has been the subject of considerable interest in the scientific community. In fact, scientists have proposed an alternative variety of UHMWPE to improve the wear resistance of the polymer. Among these attempts, the attention of various researchers has been focused on: (a) the development of methods that increase the level of cross-linking in polyethylene and (b) the creation of polymer composites (i.e. UHMWPE with the inclusion of carbon fibre). A composite material has superior mechanical properties relative to non-filled UHMWPE, including higher strength and impact strength, increased creep resistance and improved modulus. These two aspects will be discussed below.

13.3.1 Cross-linked polyethylene

Cross-linked polyethylene, commonly abbreviated to PEX or XLPE, is a form of polyethylene with cross-links.[60,61] It is the most exciting development in articular technology and it is a process in which polyethylene molecules are bonded together to result in a stronger material, substantially improving the material's wear resistance. Covalent bonds are formed between the polymeric chains and therefore chain mobility, orientation and, as a result, wear are inhibited.[49] There are three main methods of cross-linking polyethylene: (a) using peroxide, (b) by moisture cross-linking and (c) using irradiation.

Peroxide

The peroxide method has been practised extensively by the wire and cable industry for many years. It requires special compounds containing an initiator, usually organic peroxide in its original unprocessed chemical structure, and special downstream cross-linking equipment.[62] The compound must be prepared and extruded at low temperatures, below the peroxide's decomposition temperature, then cross-linked in the downstream equipment at significantly higher temperatures and pressures to complete the process. Higher temperatures decompose the peroxide and liberate free radicals that can abstract a hydrogen atom from the polymer chain. This site then becomes a reactive radical that can form a cross-link bond with another polyethylene radical. This reaction occurs repetitively until all the peroxide is consumed or the temperatures fall below the decomposition point. One major disadvantage of this method is the limitation on the use of additives (such as antioxidants); this is because they can interfere with the reaction. As a result, stabilisation of the polymer is more involved and more difficult.

Moisture cross-linking[63]

There are two basic methods of cross-linking polyethylene by moisture, both involving the use of ethylene vinylsilane 'copolymers'. These copolymers can either be produced in a reactor by polymerising ethylene with vinylsilane, or by extruder grafting of polyethylene with the vinylsilane. The processes produce similar products, but with some very important differences. Both have become increasingly popular as replacements for 'old' technologies, due to the lower capital investment required and increased productivity.

Made directly in the polyethylene reactor, silane copolymers were introduced in the late 1980s.[64] This copolymer is supplied to the processor as the base component of a multi-part system, which also includes a catalyst and other additives as desired. Additives such as antioxidants and flame-retardants are not a problem as they can be added as one of the master batches. A major disadvantage of the reactor process is the limited types of product that can be made. The polymerisation is carried out in a high-pressure reactor, which results in a highly branched low-density product. The creation of higher density polymers is not possible with the current technology, so this method cannot be used to manufacture HDPE and consequently UHMWPE-, linear low-density polyethylene (LLDPE)-, ethylene propylene dimonomer (EPDM)-, or chlorinated polyethylene (CPE)-type products. Silane cross-linked polyethylene was introduced into clinical practice in a limited series of acetabular cups in 1986 but today it is of little importance for medical applications.

Irradiation[65]

The irradiation method involves bombarding the polyethylene with high-energy electrons that liberate free radicals and cause the subsequent cross-linking reaction; it is the most commonly used method for cross-linking UHMWPE for medical purposes. When polyethylene is irradiated with high-energy radiation, such as γ ray or electron beam (EB), it will lead to chain breakage and radical formation. The irradiation procedures are similar to sterilisation methods. In practice, the finished product is passed through a large irradiation unit one or more times to achieve the desired level of cure. Polyethylene can be cross-linked in this way without any chemical additions but usually accelerators or promoters are incorporated to speed up the reaction times and improve cross-linking efficiency. The high cost of installation and operation, the many regulations imposed on operators of irradiation units and the safety precautions needed when operating the equipment make irradiation unpopular. A final disadvantage of this method is the potential non-uniformity of the cross-link density compared with the previous two methods. Some of the advantages of this method are high extrusion speeds, because curing is done off-line, potentially lower raw material costs and the 'cleanliness' of the process.

Oxygen that was present in the polyethylene when it was irradiated, or that diffused into the polyethylene during shelf storage and/or *in vivo*, could react with the free radicals that were generated by the radiation, causing oxidative degradation that lowers the molecular weight, and reduces mechanical properties.[66] In order to achive cross-linking and to minimise the degradation due to chain breakage and oxidative degradation, the irradiation and post-treatment of the cups must be optimised.[67] The top layer of the material will be oxidised during the process, so instead of irradiating the end product (as is done for sterilisation), the material is cross-linked as blocks or prepregs and then machined into cups. During the machining, the oxidised top layer is machined off leaving only the cross-linked material underneath. Instead of the low irradiation dose used for sterilisation (about 25–40 kGy), higher doses are used for cross-linking (about 50–100 kGy) and after the irradiation the cups are heat-treated to quench any radicals that are still present in the material. Typical post-irradiation treatments are annealing heating below the melting point of UHMWPE and remelting heating above the melting point of UHMWPE.[68] After the heat treatment, the material is then sterilised, either without using irradiation (ethylene oxide or gas plasma) or by irradiating in an inert atmosphere.

There are at present six different types of cross-linked cups on the market.[60] They are all made using a range of irradiation doses and sources, and are heat-treated in different ways. Cross-linking has been reported to improve the wear characteristics with respect to non-cross-linked polyethylene in clinical studies[24,69] and laboratory tests using hip joint simulators. However, the reduction in the mechanical properties of polyethylene following certain methods used to produce cross-linking has been a concern. These reductions are known to result from the processes used to increase the cross-link density and could affect the device performance *in vivo*.[70]

13.3.2 Composite polyethylene

Composites of UHMWPE with added particles have shown great improvements in tribological characteristics and mechanics behaviours.[71] Composites are typically made from mixtures containing up to 5.0 wt% of lubricant and are prepared by compression moulding.[72] The lubricants used may be different, i.e. solid (such as molybdenum disulphide (MoS_2) and carbon black (CB)) or liquid (such as perfluoropolyether (PFPE)). The UHMWPE and the lubricants form three-dimensional networks, where the lubricant is evenly spread over the UHMWPE particles.

The development of new forms of composite UHMWPE is a common aim for the scientific community. Nowadays, in this class of materials we can mention composites of: UHMWPE reinforced with carbon fibre; UHMWPE enhanced with PMMA composites; UHMWPE reinforced with nano-sized hydroxyapatite (HA) particles; nano-Al_2O_3/UHMWPE; wollastonite fibre-filled UHMWPE; UHMWPE with nano-powder of SiO_2 fibre.[73,74] Many of these composites have resulted in an improvement in some mechanical properties; however, there is still no consensus about the *in vivo* performance of these new materials. Therefore, other clinical and *in vitro* studies are necessary to convince a sceptical scientific community of the improved quality of these composites.

13.4 Ageing of polyethylene

Even if UHMWPE was improved using other methods, UHMWPE components oxidatively degrade because of γ-radiation sterilisation and subsequent shelf-ageing in air producing elevated wear of the polymer. The oxidation of the material causes material degradation with changes in properties such as density, molecular weight, degree of crystallinity or toughness.[28,75,76] The oxidation of UHMWPE is in reality a complex sequence of various cascade reactions, which are not fully understood. The degradation induced is detected as a 'white band', which mostly appears concentrated in a subsurface zone 1–2 mm below the surface[77,78] and not on the surface where the exposure to oxidising species should be highest.

In order to evaluate the long-term stability of polymers, many accelerated ageing methods have been developed to test the relative oxidation of various types of polyethylene. Typically, all these procedures consist of the exposure of the material for some days in high-temperature and -pressure environments. In particular, standard artificial protocols are based on the procedures proposed by Sun *et al.*[79] and Sanford and Saum.[80] Sun *et al.*[79] have argued that for UHMWPE γ irradiated in air, 11 or 23 days of thermal ageing at 80 °C, with an initial heating rate of 0.63 °C/min or slower, was considered equivalent to 4–6 or 7–10 years of natural shelf-ageing, respectively. An alternative accelerated ageing method has been proposed by Sanford and Saum.[80] This procedure consists of using a pressurised vessel in order to put the UHMWPE specimens at 70 °C in an oxygen bomb (maintained at 5 atm of oxygen) for 2 weeks. Using this technique, 5–10 years of shelf-ageing could be simulated in as little as 1 week. The technique also produces the subsurface oxidation maximum that is often observed for shelf-aged components.

In order to fix a standard procedure, an ASTM recommendation[81] has recently been published. This guide outlines two procedures for accelerated ageing to evaluate the oxidative stability of UHMWPE. The procedures have been validated based on certain components that were γ-irradiated in air and subsequently shelf-aged in air. However, the procedures do not replicate the natural ageing of UHMWPE irradiated in an inert environment. In particular, the ASTM recommendation incorporates both the above-mentioned methods.[79,80] However, interlaboratory studies have identified that, although materials can be ranked successfully by both of these methods, there is poor interlaboratory reproducibility (especially for the method proposed by Sanford and Saum[79]). Various studies have been carried out to evaluate if the UHMWPE specimens treated according to these two protocols really resemble shelf-aged and retrieved components.[24] Mazzucco and co-workers[82] investigated the effects of temperature, solute and oxygen in aqueous media on the oxidation of γ-sterilised UHMWPE. In these studies, they evaluated polyethylene γ-irradiated in air and non-irradiated. Following this, these polyethylenes were aged using oxygenated pressure at 5 atm of pure oxygen and at 70 °C for 2 weeks. The authors found that this accelerated shelf-ageing caused the greatest degradation for both irradiated and non-irradiated polyethylene specimens. Edidin *et al.*[83] confirmed the same results in their experiments.

Kurtz *et al.*[14] have performed an interlaboratory study to quantify the repeatability and reproducibility of standard accelerated ageing methods for UHMWPE; they have concluded that, although differently sterilised materials can be ranked successfully by both protocols, there is poor interlaboratory reproducibility. Actually, even though ageing protocols are widely used for preconditioning UHMWPE prior to mechanical testing, the question remains as to whether or not the thermal techniques precisely recreate the morphology and mechanical properties of shelf-ageed UHMWPE. Recent research has highlighted certain limitations of thermal ageing techniques. The mechanical behaviour of UHMWPE evolves during natural (shelf) ageing after γ-irradiation in air, but the kinetics and characteristics of mechanical degradation remain poorly understood, due largely to previous emphasis on indirect measurement techniques.[85,86]

Today, all the ageing protocols used should not be interpreted as a complete reproduction of all the characteristics associated with natural ageing.

13.5 Future trends

Even if problems of oxidative degradation and wear are still unsolved, UHMWPE remains the gold standard as a bearing surface for total arthroplasty. The categorical factors limiting the function and longevity of total hip replacement are surgical technique, fixation of the implant, osteolysis, fatigue failure and long-term skeletal remodelling. The challenge is to obtain improvements in fixation and durability, which in UHMWPE is widely influenced by oxidative degradation and wear.

Attempts to solve these problems have involved new sterilisation techniques and new forms of more highly cross-linked polyethylene. In fact, multi-walled carbon nanotubes (MWCNTs)[87] or UHMWPE with PMMA composites[16] have been studied in order to improve the properties of the polyethylene. In particular, a drastically enhanced toughness in UHMWPE films, due to the addition of 1 wt% MWCNTs, has been shown. Raman spectroscopic measures revealed that the presence of MWCNTs in the composites could lead to a 150% increase in strain energy density accompanied by an increase of 140% in ductility and up to 25% in tensile strength, in comparison with pure UHMWPE. UHMWPE enhanced with PMMA composites consists of three-dimensional braided UHMWPE fibre reinforced PMMA resin and is characterised by excellent impact strength.

With increasing patient longevity and activity levels, the search for the ultimate polymer is important, even if as yet these 'new age polyethylenes' offer no direct clinical evidence to demonstrate their efficacy. The merits of these potential solutions have to be investigated through clinical performance and *in vitro* studies in order to confirm their better performances with respect to previous polyethylenes.

13.6 Acknowledgements

Thanks are due to Ms Panagiota Dimopoulou and MariaChiara Scandellari for their editorial assistance and help.

13.7 References

1 *An Introduction to Biomaterials.* Available at: http://www.uweb.engr.washington.edu/research/tutorials/introbiomat.html, 21 June 2007.
2 CE MARKING DIRECTIVE, 1993.
3 Medical device. Available at: http://en.wikipedia.org/wiki/Medical_device, 21 June 2007.
4 COUNCIL DIRECTIVE 93/42/EEC of 14 June 1993 concerning medical devices, 1993.
5 Device Classes: FDA regulation. Available at: http://www.fda.gov/cdrh/devadvice/3132.html, 21 June 2007.
6 Developing PEEK polymer as a bearing material for implants. Available at: http://www.devicelink.com/mddi/archive/06/05/007.html, 22 June 2007.
7 BELLARE A., BISTOLFI A., SIMIS K., PRUITT L. Physico-chemical characteristics of UHMWPE before and after processing. In *UHMWPE for Arthroplasty: Past, Present and Future.* UHMWPE conference, Turin, 2003.
8 Radical polymerization of polyethylene. Report. Available at: http://plastics-engineering.blogspot.com/, Feb. 2008.
9 Polyethylene. Available at: http://www.sciencedaily.com/articles/p/polyethylene.htm, Feb. 2008.
10 DE JESÚS W., ECHEVARRÍA L., RODRÍGUEZ J., VARGAS A. Biomechanics of elbow prostheses. Available at: http://www.google.it/search?hl=it&q=Biomechanics+of+elbow+prostheses&btnG=Cerca&meta=, 2004.
11 KOSIOR E., BRAGANCA R.M., FOWLER P. Lightweight compostable packaging: literature review. Available at: http://www.google.it/search?hl=it&q=Lightweight+compostable+packaging%3A+literature+review&btnG=Cerca&meta=, 2006.
12 Polyethylene. Available at: http://en.wikipedia.org/wiki/Polyethylene, Feb. 2008.
13 BLUNN G., BRACH DEL PREVER E.M., COSTA L., FISHER J., FREEMAN M.A.R. Ultra high molecular-weight polyethylene (uhmwpe) in total knee replacement: fabrication, sterilisation and wear. *J Bone J Surg [Br].* 2002; **84-B**:946–949.
14 LI S., BURSTEIN A.H. Ultra-high molecular weight polyethylene: the material and its use in total joint implants. *J Bone J Surg [Am].* 1994; **76-A**:1080–1089.
15 COOK J.T.E., KLEIN P.G., WARD I.M., BRAIN A.A., FARRAR D.F., ROSE J. The morphology of nascent and moulded ultra-high molecular weight polyethylene: insights from solid state NMR nitric acid etching GPC and DSC. *Polymer.* 2000; **41**:8615–8623.
16 FISHER J., REEVES E.A., ISAAC G.H., SAUM K.A., SANDORD W.M. Comparison of the wear of aged and non-aged ultrahigh molecular weight polyethylene sterilized by gamma irradiation and by gas plasma. *J Mater Sci: Mater Med.* 1997; **8**:375–378.

17 COUTINHO F.M.B., COSTA M.A.S., SANTOS A.L.S.S., COSTA T.H.S., SANTA MARIA L.C., PEREIRA R.A. Characterization of Ziegler–Natta catalysts based on $TiCl_3$ synthesized by different methods. *J Anal Chem*. 1992; **344**:514–516.

18 Process For Preparation Of Polyacetylene Film. US Patent 5008040. Available at: http://www.patentstorm.us/patents/5008040-description.html, Feb. 2008.

19 Method for processing thermoplastics, thermosets and elastomers. US Patent 5830396. Available at: http://www.patentstorm.us/patents/5830396.html, Feb. 2008.

20 JACOB R.J., PIENKOWSKI D., LEE K.Y., HAMILTON D.M., SCHROEDER D., HIGGINS J. Time and depth dependent changes in cross-linking and oxidation of shelf-aged polyethylene acetabular liners. *J Biomed Mater Res*. 2001; **56**:168–176.

21 RIMNAC C.M., KLEIN R.W., BETTS F., WRIGHT T.M. Post-irradiation aging of Ultra-High Molecular Weight Polyethylene. *J Bone Jt Surg [Am]*. 1994; **76**(7): 1052–1056.

22 SCHMIDT M.B., HAMILTON J.V. Calcium stearate-free UHMWPE therefore appears to be a superior choice for use in total joint replacement applications. In 42nd ORS, Atlanta, GA, 19–22 Feb. 1996.

23 Compression molded flame retardant and high impact strength ultra high molecular weight polyethylene composition. European Patent 0414110. Available at: http://www.freepatentsonline.com/EP0414110.html, Feb. 2008.

24 KURTZ S.M., MURATOGLU O.K., EVANS M., EDIDIN A.A. Advances in the processing sterilization, and crosslinking of Ultra-High Molecular Weight Polyethylene for total joint arthroplasty. *Biomaterials*. 1999; **20**:1659–1688.

25 Method of generating chemical compounds having desired properties. US Patent 6434490. Available at: http://www.freepatentsonline.com/6434490.html, Feb. 2008.

26 LOTZ B. Phase transitions and structure of crystalline polymers. Available at: http://www.google.it/search?hl=it&q=Phase+transitions+and+structure+of+crystalline+polymers&as_q=Lotz+B.+&btnG=Cerca%C2%A0tra%C2%A0i%C2%A0risultati, 1996.

27 LOTZ B., WITTMANN J.C., LOVINGER A.J. Structure and morphology of poly(propylenes): a molecular analysis. *Polymer*. 1996; **37**:4979–4992.

28 GOLDMAN M., LEE M., GRONSKY R. Oxidation of UHMWPE characterized by Fourier Transforma Infrared Spectometry. *J Biomed Mater Res*. 1997; **37**:43–50.

29 PREMNATH V., HARRIS W.H., JASTY M., MERRILL E.W. Gamma sterilization of UHMWPE articular implants: an analysis of the oxidation problem. *Biomaterials*. 1996; **17**:1741–1753.

30 BRAUN D. Recycling of PVC. *Prog in Polym Sci*. 2002; **27**:2171–2195.

31 Polymer material and method of making same utilizing inert atmosphere. Available at: http://www.patentstorm.us/patents/6017586-description.html, US Patent 6017586. Feb. 2008.

32 HEISEL C., SILVA M., SCHMALZRIED P. Bearing surface options for total hip replacement in young patients. *J Bone Jt Surg [Am]*. 2003; **85**:1367–1369.

33 MCKELLOP H. Does gamma irradiation speed or slow wear? *AAOS Bull*. 1996; **44**:10–11.

34 MCKELLOP H., SHEN F.W., LU B., CAMPBELL P., SALOVEY R. Effect of sterilization method and other modifications on the wear resistance of acetabular cups made of ultra-high molecular weight polyethylene. A hip-simulator study. *J Bone Jt Surg [Am]*. 2000; **82-A**:1708–1725.

35 MCKELLOP H.A., SHEN F.W., CAMPBELL P., OTA T. Effect of molecular weight, calcium stearate, and sterilization methods on the wear of uhmwpe acetabular cups in a hip joint simulator. *J Orthop Res*. 1999; **17**:329–339.
36 FISHER J, CHAN KL, HAILEY JL, SHAW D, STONE M. Preliminary study of the effect of aging following irradiation on the wear of ultra-high molecular weight polyethylene. *J Arthroplasty*. 1995; **10**(5):689–692.
37 KURTZ S.M., HOZACK W., MARCOLONGO M., TURNER J., RIMNAC C., EDIDIN A.A. Degradation of mechanical properties of UHMWPE acetabular liners following long-term implantation. *J Arthroplasty*. 2003; **18**:68–78.
38 KURTZ S.M., RIMNAC C.M., HOZACK W.J., TURNER J., MARCOLONGO M., GOLDBERG V.M., KRAAY M.J., EDIDIN A.A. In vivo degradation of polyethylene liners after gamma sterilization in air. *J Bone J Surg [Am]*. 2005; **87**:815–823.
39 SALERNITANO E., MIGLIARESI C. Composite materials for biomedical application: a review. *J Appl Biomater Biomech*. 2003; **1**:3–18.
40 Fused nanostructure material. Available at: http://www.wipo.int/pctdb/en/wo.jsp?IA=WO2005047181&DISPLAY=DESC, Feb. 2008.
41 NEUMANN A., RAGO C., MAIER H.R., JAHNKE K. Biomaterials: fundamentals and clinical applications. In *6th Interdisciplinary Essen-Symposium on Biomaterials*, 2003.
42 *Composite Materials Handbook*. Available at: http://snebulos.mit.edu/projects/reference/MIL-STD/MIL-HDBK-17-3F.pdf., USA Department of Defense, Feb. 2008.
43 Polymer composite implant and method of making the same. US Patent 5645594. Available at: http://www.patentstorm.us/patents/5645594-fulltext.html.
44 Self-reinforced ultra-high molecular weight polyethylene composite medical implants. US Patent 5834113. Available at: http://www.patentstorm.us/patents/5834113-fulltext.html.
45 KURTZ S.M. *The UHMWPE Handbook: Ultra-High Molecular Weight Polyethylene in Total Joint Replacement*, Adobe eBook, 2004.
46 CHOWDHURY S.K.R., MISHRA A., PRADHAN B., SAHA D. Wear characteristic and biocompatibility of some polymer composite acetabular cups. *Wear*. 2004; **256**: 1026–1036.
47 Materials Sciences at ANL. Report. Available at: http://www.scied.science.doe.gov/scied/Abstracts2004/ANLms.htm, Feb. 2008.
48 KURTZ S.M., DEVINE J.N. PEEK biomaterials in trauma, orthopedic, and spinal implants. *Biomaterials*. 2007; **28**:4845–4869.
49 MANO J.F., SOUSA R.A., BOESEL L.F., NEVES N.M., REIS R.L. Bioinert, biodegradable and injectable polymeric matrix composites for hard tissue replacement: state of the art and recent developments. *Composites Sci Technol* 2004; **64**:789–817.
50 Goodfellow. Carbon fibre/polyetheretherketone (PEEK) matrix composite sheet – properties. Supplier data from Goodfellow. Available at: http://www.azom.com/details.asp?ArticleID=2002, Feb. 2008.
51 JIA J., CHEN J., ZHOU H., HU L. Comparative study on tribological behaviors of polyetheretherketone composite reinforced with carbon fiber and polytetrafluoroethylene under water-lubricated and dry-sliding against stainless steel. *Tribology Lett*. 2004; **17**:231–238.
52 WANG A., LIN R., POLINENI V.K., ESSNER A., STARK C., DUMBLETON J.H. Carbon fiber reinforced polyether ether ketone composite as a bearing surface for total hip replacement *Tribol Int*. 1998; **11**:661–666.

53 INGHAM E., FISHER J. Biological reactions to wear debris in total joint replacement. *Proc Inst–Mech Eng(s) Part H: J Engng Medicine*. 2000; **214**:21–37.

54 INGHAM E., FISHER J. The role of macrophages in osteolysis of total joint replacement. *Biomaterials*. 2005; **26**:1271–1286.

55 BIZOT P., NIZARD R., LEROUGE S., PRUDHOMMEAUX F., SEDEL L. Ceramic/ceramic total hip arthroplasty. *J Orthop Sci*. 2004; **5**:622–627.

56 CALE B. Zirconia as a sliding material: histologic, laboratory, and clinical data. *Clin Orthop*. 2000; **379**:94.

57 WILLMANN G. Ceramics for total hip replacement: what a surgeon should know. *Orthopedics*. 1998; **21**:173.

58 CLARKE M.T., LEE P.T.H., ARORA A., VILLAR R.N. Levels of metal ions after small- and large-diameter metal-on-metal hip arthroplasty. *J Bone J Surg [Am]*. 2003; **85**: 913–917.

59 MACDONALD S.J., BRODNER W., JACOBS J.J. A consensus paper on metal ions in metal-on-metal hip arthroplasties. *J Arthroplasty*. 2004; **19**:12–16.

60 Cross-linked ultra-high molecular weight polyethylene. Available at: http://www.uhmwpe.unito.it/atti/07%20Jacobson.pdf, Feb. 2008.

61 Cross-linked polyethylene. Available at: http://en.wikipedia.org/wiki/PEX, 25 June 2007.

62 Peroxide. Available at: http://en.wikipedia.org/wiki/Peroxide.

63 BERGMAN F.A.C. *Moisture crosslinkable polymers: studies on the synthesis, crosslinking and rheology of methoxysilane functional poly(vinyl esters)*. PhD thesis, Technical University Eindhoven, 2001.

64 Silane-grafted moisture-crosslinkable polyethylene. Available at: http://www.geon.com/downloads/pdfs/TSR/TSR66Revised.pdf.

65 Electron-beam processing of plastics: an alternative to chemical additives. Available at: http://www.ebeamservices.com/ebeam_spe_antec.htm, Feb. 2008.

66 SHEN F.W., MCKELLOP H.A. Interaction of oxidation and crosslinking in gamma-irradiated ultrahigh molecular-weight polyethylene. *J Biomed Mater Res*. 2002; **61**:430–439.

67 CURRIER B.H., CURRIER J.H., MAYOR M.B., LYFORD K.A., VAN CITTERS D.W., COLLIER J.P. In vivo oxidation of γ-barrier-sterilized ultra-high-molecular-weight polyethylene bearings. *J Arthroplasty*. 2007; **22**:721–731.

68 KURTZ S.M., VILLARRAGA M.L., HERR M.P., BERGSTRÖM J.S., RIMNAC C.M., EDIDIN A.A. Thermomechanical behavior of virgin and highly crosslinked ultra-high molecular weight polyethylene used in total joint replacements. *Biomaterials*. 2002; **23**:3681–3697.

69 OONISHI H. Long term clinical results of THR: clinical results of THR of an alumina head with a cross-linked UHMWPE cup. *Orthop Surg Traumatol*. 1995; **38**:1255–1264.

70 MURATOGLU O.K., BRAGDON C.R., O'CONNOR D.O., JASTY M., HARRIS W.H. A novel method of cross-linking ultra-high-molecular-weight polyethylene to improve wear, reduce oxidation, and retain mechanical properties. *J Arthroplasty*. 2001; **16**:149–160.

71 CHAND N., SHARMA M.K. Development and tribological behaviour of UHMWPE filled epoxy gradient composites. *Wear*. 2007; **262**:184–190.

72 PUUKILAINEN E., SAARENPÄÄ H., PAKKANEN T.A. Compression-molded, lubricant-treated UHMWPE composites. *J Appl Polym Sci*. 2007; **104**:1762–1768.

73 FANG L., LENG Y., GAO P. Processing of hydroxyapatite reinforced ultrahigh molecular weight polyethylene for biomedical applications. *Biomaterials*. 2005; **26**: 3471–3478.

74 TONG J., MA Y., JIANG M. Effects of the wollastonite fiber modification on the sliding wear behavior of the UHMWPE composites. *Wear*. 2003; **255**:734–741.

75 COSTA L., LUDA M.P., TROSSARELLI L., BRACH DEL PREVER E.M., CROVA M., GALLINARO P. Oxidation in orthopaedic UHMWPE sterilized by gamma-radiation and ethylene oxide. *Biomaterials*. 1998; **19**:659–668.

76 CURRIER B.H., CURRIER J.H., COLLIER J.P., MAYOR M.B., SCOTT R.D. Shelf life and in vivo duration. Impacts on performance of tibial bearings. *Clin Orthop Rel Res*. 1997; **342**:111–122.

77 BOSTROM M.P., BENNETT A.P., RIMNAC C.M., WRIGHT T.M. The natural history of UHMWPE. *Clin Orthop Rel Res*. 1994; **309**:11–19.

78 SUTULA L.C., COLLIER J.P., SAUM K.A., CURRIER B.H., CURRIER J.H., SANDORF W.M., MAYOR M.B., WOODING R.E., SPERLING D.K., WILLIAMS I.R., KASPRZAK D.J., SURPRENANT V.A. Impact of gamma sterilization on clinical performance of polyethylene in the hip. *Clin Orthop Rel Res*. 1995; **319**:28–40.

79 SUN D.C., SCMIDIG G., STARK C., DUMBLETON J.H. A simple accelerated aging method for simulations of long-term oxidative aging effects in UHMWPE implants. In 42nd ORS, Atlanta, GA, 19–22 Feb. 1996.

80 SANFORD W.M., SAUM K.A. Accelerated oxidative aging testing of UHMWPE. In 41st ORS, Orlando, FL, 13–16 Feb. 1995.

81 ASTM F2003-00 A. Standard Guide for Accelerated Aging of UHMWPE, 2003.

82 MAZZUCCO D.C., DUMBLETON J., KURTZ S.M. Can accelerated aqueous aging simulate in vivo oxidation of gamma-sterilized UHMWPE? *J Biomed Mater Res B*. 2006; **79**:79–85.

83 EDIDIN A.A., VILLARRAGA M.L., HERR M.P., MUTH J., YAU S.S., KURTZ S.M. Accelerated aging studies of UHMWPE. II. Virgin UHMWPE is not immune to oxidative degradation. *J Biomed Mater Res*. 2002; **61**:323–329.

84 KURTZ S.M., MURATOGLU O.K., BUCHANAN F., CURRIER B., GSELL R., GREER K., GUALTIERI G., JOHNSON R., SCHAFFNER S., SEVO K., SPIEGELBERG S., SHEN F.W., YAU S.S. Interlaboratory reproducibility of standard accelerated aging methods for oxidation of UHMWPE. *Biomaterials*. 2001; **22**:1731–1737.

85 EDIDIN A.A., JEWETT C.W., KALINOWSKI A., KWARTENG K., KURTZ S.M. Degradation of mechanical behavior in UHMWPE after natural and accelerated aging. *Biomaterials*. 2000; **21**:1451–1460.

86 KURTZ S.M., RIMNAC C.M., PRUITT L., JEWETT C.W., GOLDBERG V., EDIDIN A.A. The relationship between the clinical performance and large deformation mechanical behavior of retrieved UHMWPE tibial inserts. *Biomaterials.* 2000; **21**(2): 283–291.

87 RUAN S.L., GAO P., YANG X.G., YU T.X. Toughening high performance ultrahigh molecular weight polyethylene using multiwalled carbon nanotubes. *Polymer.* 2003; **44**:5643–5654.

88 ZHANG Y.D., WANG Y.L., HUANG Y., WAN Y.Z. Preparation and properties of three-dimensional braided UHMWPE fiber reinforced PMMA composites. *J. Reinforced Plastics Composites.* 2003; **25**:1601–1609.

14
Ageing of composites in oil and gas applications

S. FROST, ESR Technology Ltd, UK

14.1 Introduction

Composite materials, which for the purposes of this chapter are defined as and limited to, fibre reinforced thermosetting matrix (or resin) systems, have several features that make them attractive for use in the Oil and Gas industry, namely light weight and good corrosion resistance. The primary fibre used is glass, although carbon and aramid are used in limited applications. The primary resin system (thermosetting) used is epoxy or polyester, although vinyl ester, polyurethane and furane are also used to a limited extent. The method of manufacture is predominantly filament winding implying continuous fibre composites, although some components are pultruded, resin transfer moulded or made by hand lay-up.

The primary applications of composites within the Oil and Gas industry include:

- pipelines, risers and piping systems;
- tubings, casings;
- process equipment;
- tanks and vessels;
- structures;
- access equipment (stairs, gratings);
- lifeboats;
- mudmats;
- protective covers.

Corrosion resistance, light weight and in some cases flexibility and continuous manufacture are the primary business drivers, which when used to advantage in design, can lead to either reduced life-cycle costs or improved safety. Generally speaking, the major use of composite components is in containment applications and current applications can be divided into three areas: on-shore, off-shore and downhole.

- On-shore, the main application is pipelines and piping systems (with some tanks and vessels), e.g. glass reinforced epoxy (GRP) pipes. A typical application of GRP piping is shown in Fig. 14.1 The business driver is reduced life-cycle costs through minimum maintenance (corrosion resistance).
- Off-shore applications are more diverse and include pipework, e.g. fire water mains systems, water injection systems, access structures and components of flexible risers. The business drivers are corrosion resistance and light weight (both ease of handling and reduced overall structural weight are also important).
- Downhole applications include tubings and lined tubings. As for on-shore systems, the business driver is reduced life-cycle costs through corrosion resistance.

In terms of volume (or weight) usage, on-shore applications far out-weigh both off-shore and downhole usage.

In its broadest definition ageing can be defined as the reduction in performance of a component as a function of the applied conditions. This is the definition that will be used in this chapter. The three primary causes of ageing for composite components in the Oil and Gas industry are through chemical species ingress, elevated operating temperature and length of time of load application.

As a significant number of applications of composite components are pressure containment and given the fact that internal polymeric liners are not commonly used, then the principal failure mechanism of concern is ply matrix cracking linked with inter-ply delaminations. As the load is increased or as time progresses, the number or density of these matrix cracks increases until they join together in a convoluted arrangement, creating a fluid path through the composite. The failure mode is often termed 'weepage'. Figure 14.2 is a photograph of the microstructure of a failed composite GRP pipe

14.1 Application of GRP pipework in a process plant.

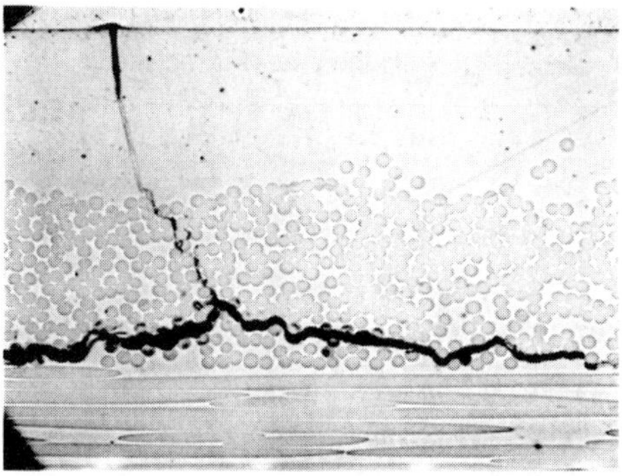

14.2 Photograph of the microstructure of a failed GRP pipe showing ply matrix cracking and inter-ply delamination. Fibre diameter is between 10 and 20 μm.

showing both ply matrix cracking and inter-ply delaminations. The other failure mode of concern is fibre failure. Generally, this occurs at the ultimate load-bearing capacity of the composite component and results in gross failure. The ageing process accelerates the failure process, be it increasing the density of micro-cracks or reducing the strength of fibres.

The following sections contain initially the development of a model to predict ageing and damage within composite components through matrix cracking followed by a detailed discussion on the consequence of the three causes of ageing on composite component performance. One of the most commonly used composite structures is the filament wound glass fibre reinforced thermosetting matrix, often epoxy (GRP) pipe. Typically, these pipes range in diameter from 50 to 4000 mm. Pressure ratings range from 5 to 120 bar, the higher pressure ratings only applicable to smaller pipe diameters. In order to simplify the discussion, GRP pipes will be used as the composite component example to illustrate the consequences of ageing.

14.2 Modelling of damage

GRP pipes are constructed from uni-directional plies angled sequentially at [±55°] to the axial pipe direction. At weepage, the fluid path through the pipe wall is a combination of mostly through-thickness matrix cracks running parallel to fibres and some delaminations (usually more prevalent during long-term tests). Both delaminations and through-thickness cracks result from the coalescence of matrix micro-cracks and interfacial debonding. Throughout this failure process the reinforcing fibres remain intact. The stress–strain behaviour of GRP pipes under internal pressure loading is initially linear elastic followed by a non-linear region until weepage

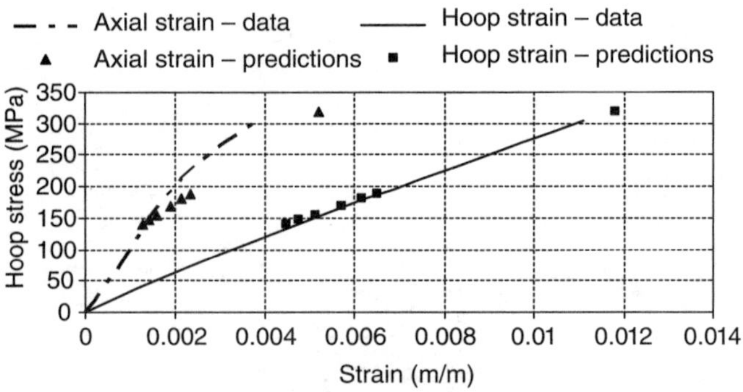

14.3 Comparisons between measurements and predictions of the stress–strain response of a GRP pipe loaded with a stress ratio of 2:1.

failure. Figure 14.3 presents a plot of the axial and hoop strain as a function of hoop stress, clearly showing the non-linear stress–strain behaviour (Frost and Cervenka[1]). The non-linear region is a consequence of matrix cracking within each ply, the density of which increases with increasing load, with the non-linearity becoming more severe at higher loads. In order to develop a model capable of predicting weepage, an allowance for the growth of damage as a function of applied load must be included.

The application of damage mechanics allows for the change in stiffness matrix to be quantified in terms of the amount of damage. For GRP pipes this damage is in the form of matrix cracking and is quantified in terms of density of, or spacing between, cracks. The stiffness matrix of the composite is derived from two components: the undamaged stiffness matrix, $\mathbf{Q}_{ij}^{\text{elastic}}$, and the damage stiffness matrix, $\mathbf{Q}_{ij}^{\text{damage}}$, i.e.

$$\mathbf{Q}_{ij}^{\text{total}} = \mathbf{Q}_{ij}^{\text{elastic}} - \mathbf{Q}_{ij}^{\text{damage}} \qquad [14.1]$$

Talreja[2] has derived the damage stiffness matrix for a uni-directional, transversely isotropic, ply with matrix cracking orientated parallel to the fibre direction. In its most general form, $\mathbf{Q}_{ij}^{\text{damage}}$ is given by

$$\mathbf{Q}_{ij}^{\text{damage}} = \begin{bmatrix} a_1 & a_2 & 0 \\ & a_3 & 0 \\ & & a_4 \end{bmatrix} \qquad [14.2]$$

In total, there are four constants (a_i, $i = 1$ to 4) defining the state of damage. Assuming that each ply within the pipe wall has the same damage state, then the stiffness matrix of the pipe can be calculated from summing the individual ply stiffness matrices, $\mathbf{Q}_{ij}^{\text{total}}$, using classical lamination theory.[3]

To simplify and reduce the number of damage constants, two assumptions are made, namely that the following ply material properties are independent of damage state:

(a) Young's modulus of the ply in the fibre (subscript 'f') direction;
(b) Poisson's ratio of the ply in the fibre direction.

Using these assumptions, Equation [14.2] simplifies to

$$\mathbf{Q}_{ij}^{\text{damage}} = \begin{bmatrix} \upsilon_{\text{ft}}^2 a_2 & \upsilon_{\text{ft}} a_2 & 0 \\ & a_2 & 0 \\ & & a_4 \end{bmatrix} \qquad [14.3]$$

where υ_{ft} is the in-plane Poisson's ratio (load in fibre direction, response in transverse direction). It is possible to derive values for a_2 and a_4 from analytical considerations of crack formation and geometry.[4] These are given by

$$a_2 = C_2 \frac{1}{s}\left(\frac{1-\nu_{\text{ft}}\nu_{\text{tf}}}{E_{\text{t}}}\right)\sum_{n=1}^{10}\frac{a_n}{(1+1/s)^n}$$
$$a_4 = C_4 \frac{s}{G_{\text{ft}}} \ln\cosh\left(\frac{\pi}{2}\frac{1}{s}\right) \qquad [14.4]$$

where s is the normalised crack spacing with respect to ply thickness and a_n are constants. Also in Equation [14.4], E_{t} is the transverse ply modulus, G_{ft} the in-plane shear modulus and υ_{tf} is the in-plane Poisson's ratio (load in transverse fibre direction, response in fibre direction). The constants C_2 and C_4 are determined by calibrating predictions against a control specimen under a specific loading condition. Highsmith and Reifsnider[5] measured Young's modulus (in the zero degree direction) as a function of crack spacing under tensile fatigue loading for a cross-ply $[0,90_3]_s$ glass–epoxy laminate. Figure 14.4 compares predictions with measurements of normalised modulus. A best fit is achieved when C_2 and C_4 are both set to 1.6.

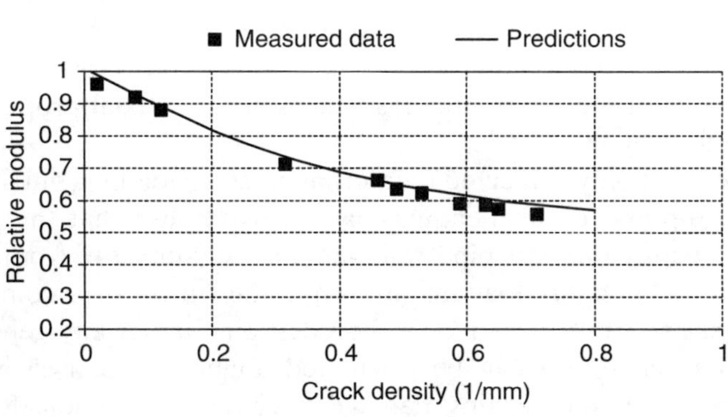

14.4 Calibration of damage constant, C_2 and C_4, using data from Highsmith and Reifsnider.[5]

Table 14.1 Undamaged and damaged moduli for a filament wound GRP pipe

Constituents (fibre and matrix)		Ply		Laminate (damaged)		Laminate (damaged)	
E_{fibre}	74 GPa	E_f	45.8 GPa	E_{ax}	15.77 GPa	E_{ax}	15.77–16.3 GPa
E_{matrix}	3.39 GPa	E_t	18.6 GPa	E_{ho}	23.07 GPa	E_{ho}	23.07–12.9 GPa
v_{fibre}	0.2	ν_{ft}	0.26	$\nu_{ax,ho}$	0.36	$\nu_{ax,ho}$	0.36+0.0008/s
v_{matrix}	0.35	ν_{tf}	0.11	$\nu_{ho,ax}$	0.52	$\nu_{ho,ax}$	0.52+0.44/s
G_{fibre}	30.8 GPa	G_{tf}	7.1 Gpa	G_{ah}	11.87 GPa	G_{ah}	11.87–3.5 GPa/s
G_{matrix}	1.26 GPa						

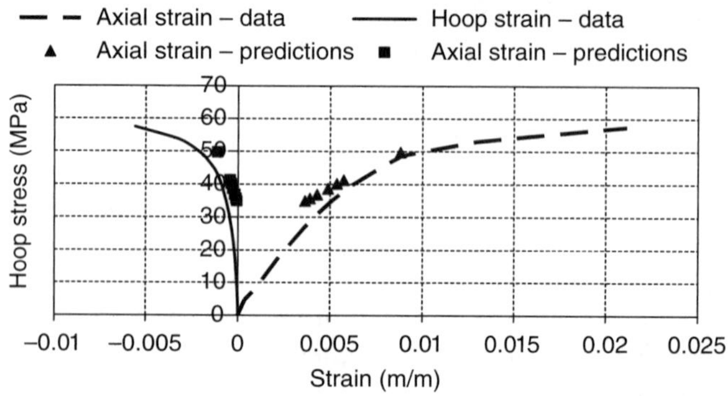

14.5 Comparisons between measurements and predictions of the stress–strain response of a GRP pipe loaded with the stress ratio 1:2.

Table 14.1 presents fibre and matrix, undamaged ply and laminate moduli, and damaged moduli for an anti-symmetric [±55°] laminate, i.e. a GRP pipe, and a simple prediction of laminate moduli as a function of crack spacing, *s*. From Table 14.1, the modulus in the axial direction is more sensitive to crack spacing than the hoop modulus. This is because axial properties are more influenced by the matrix.

Pipework is not only subjected to internal pressure loading, but also axial tensile or compressive and bending loads. This implies that the hoop to axial stress ratio within the pipe wall can vary significantly from the 2:1 ratio, i.e. internal pressure loading only. From the stiffness (or compliance) matrix, the stress–strain response of the GRP pipe under any general (in-plane) loading condition can be calculated. Figures 14.3, 14.5 and 14.6 present stress–strain data against damage mechanics predictions for short-term increasing load until failure. Figure 14.3 presents internal pressure test data (closed, free end test), applied stress ratios of 2:1 (hoop to axial),

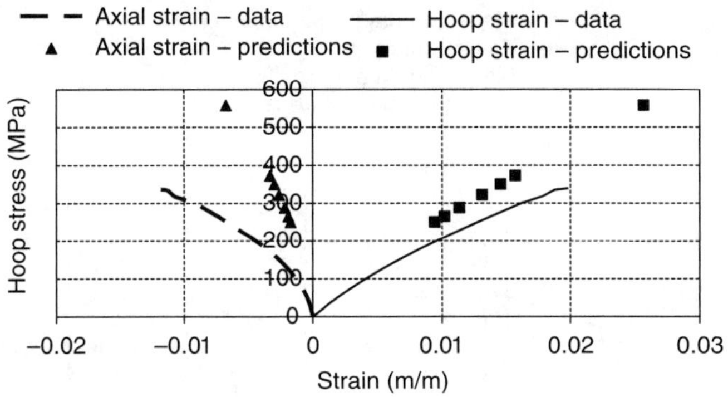

14.6 Comparisons between measurements and predictions of the stress–strain response of a GRP pipe loaded with the stress ratio 4:1.

whereas Figs 14.5 and 14.6 present data for an applied stress ratio of 1:2 (internal pressure plus tension) and 4:1 (internal pressure plus compression), respectively. The experimental procedure was reported in Frost and Cervenka.[1]

The general shape of the damage mechanics predictions are in overall agreement with measurements. For the 1:2 and 2:1 cases the ply transverse tensile stresses dominate where for the 4:1 cases ply shear stresses are more important. Strain predictions were made up to the measured weepage stress. For all the tests performed (refer to reference 1 for more details) pipe joints were present. For the 4:1 case the pipe joint failed rather then the pipe body.

The previous discussion on the damage mechanics model has demonstrated that the stress–strain response as a function of crack spacing can be predicted. However, a failure criterion is required that can predict the failure of the composite component as a function of the applied load. It is assumed that the failure criterion is related to the ply stresses that contribute to the crack formation, the transverse and shear ply stresses to the fibre direction. The relation between crack spacing and applied stress is given by Roberts *et al.*:[4]

$$1+\left(\frac{1}{s\kappa_{\text{ply}}}\right)^2 = f(\sigma_{\text{t}}, \sigma_{\text{sh}}) \qquad [14.5]$$

where

$$\kappa_{\text{ply}} = \sqrt{\frac{(E_{\text{f}}+E_{\text{t}})G_{\text{ft}}}{E_{\text{f}}E_{\text{t}}}}$$

where σ_t and σ_{sh} are the transverse and shear ply stresses, respectively.

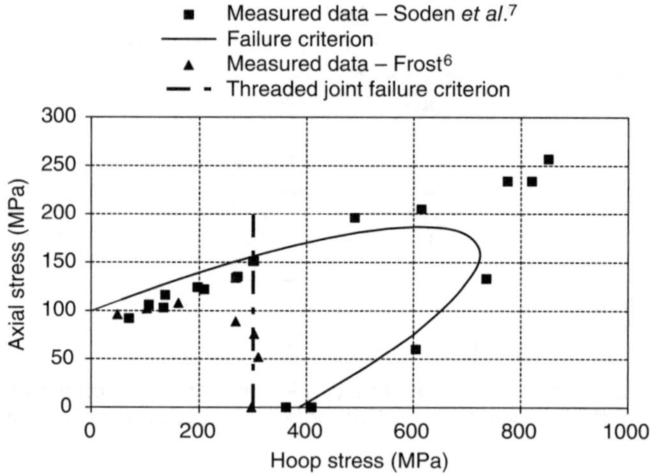

14.7 Measured and predicted short-term failure envelope for a 55° filament wound pipe; crack spacing = 2.2 mm.

A second-order polynomial criterion based on these two stress components is developed by fitting the function, *f*, to available data. The data presented by Frost[6] and Soden *et al.*[7] are best fit with a second-order polynomial expressed as

$$\left(\frac{\sigma_t}{\sigma_{t,fail}}\right)^2 + \left(\frac{\sigma_{sh}}{\sigma_{sh,fail}}\right)^2 - \frac{\sigma_t}{\sigma_{t,fail}}\frac{\sigma_{sh}}{\sigma_{sh,fail}} = f(\sigma_t, \sigma_{sh}) = C \qquad [14.6]$$

where $\sigma_{t,fail}$ and $\sigma_{sh,fail}$ are the respective transverse and shear ply failure stresses.

C is the failure parameter and can be considered to be a function of time, temperature and chemical species, i.e. ageing, and can be represented by

$$C = A_t A_T A_C A_{cyc} \qquad [14.7]$$

A_t is the partial factor for constant loading (time) and can be related to the regression gradient of the GRP pipe. It can conservatively be set to 0.5 for a 20-year design lifetime. Partial factors A_T, A_C and A_{cyc} account for temperature, chemical species and cyclic effects and are equivalent to the partial factors A_1, A_2 and A_3 as defined in ISO 14692.[8]

Figure 14.7 presents the measured failure envelope (tensile quadrant)[6,7] for a [±55°] filament wound GRP pipe against damage mechanics predictions based on values of ply strengths $\sigma_{t,fail}$ 50 Mpa, $\sigma_{sh,fail}$, 120 MPa. As the data from Frost[6]and Soden *et al.*[7] are short term and measured on unaged test samples then *C* is set to unity. In general, the agreement between predictions and experiments is good when a final crack spacing of 2.2 mm is

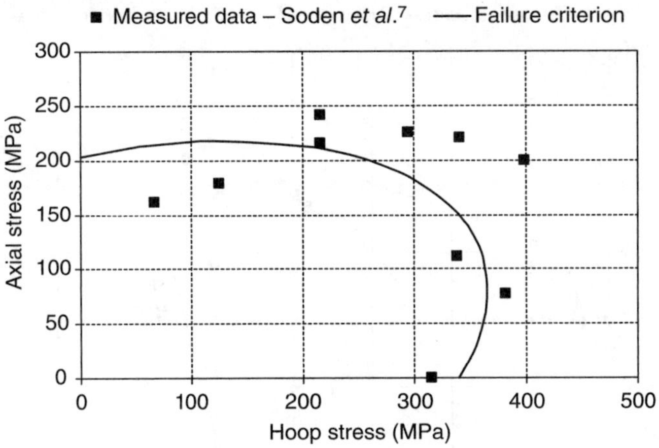

14.8 Measured and predicted short-term failure envelope for a 45° filament wound pipe; crack spacing = 2.2 mm.

assumed (corresponding to approximately eight ply thicknesses). The general shape of the failure envelope is skewed towards applied stress ratios of 3:1 to 4:1. This implies that, at this winding angle for a 2:1 applied stress ratio, i.e. internal pressure, closed free end test, the pipe is not optimally designed. Also plotted in Fig. 14.7 is a prediction of threaded joint failure. Essentially, the failure of threaded joints (the most common joint in higher pressure GRP piping) is governed by hoop stress. Above a certain limit the interlocking threads are opened sufficiently for fluid to escape. Figures 14.8 and 14.9 present a comparison between measured and predicted failure envelopes,[7] for 45° and 75° filament wound pipes respectively, for the same crack spacings as Fig. 14.7. The agreement for the 45° pipe is good, but what is surprising is that the envelope is not symmetrical. The reason for this non-symmetry is unclear. For the 75° envelope again the agreement is good when failure is controlled by weepage. For a large part of the envelope, failure is controlled by the fibres. This failure mode is predicted by comparing the in-plane ply fibre stresses with the volume fraction weighted average strength of glass fibres. The agreement between predictions and measurements is satisfactory.

These comparisons between predictions and measurements have demonstrated the validity of applying a damage mechanics solution to the short-term failure of composite components. The next section addresses the question of predicting the long-term failure of aged composite components. However, it should be noted that there have been very few studies on the combined effects of pressure, temperature and chemical species on the long-term performance of GRP pipes and, consequently, there is little experimental data to assess ageing predictive models.

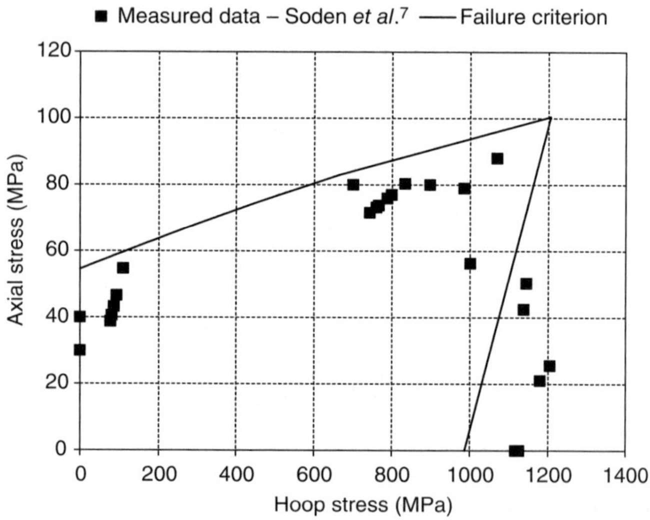

14.9 Measured and predicted short-term failure envelope for a 75° filament wound pipe; crack spacing = 2.2 mm.

In order to predict the influence of ageing within a composite component, the assumption taken is that fibres remain intact and that only the matrix or resin degrades. This degradation is time-dependent and is influenced by the concentration of species ingress, the temperature and the applied load. It is further assumed that ageing due to temperature or chemical species is independent of pressure. Glass fibres do suffer from stress corrosion in both low- and high-pH conditions. However, the operating strain of most composite components in the Oil and Gas industry is typically no greater than 0.35% strain, well below that required to initiate stress corrosion of the fibres.

14.3 Ageing due to temperature

Hale *et al.*[9] presented data on the short-term performance of GRP pipes under the combined effects of short-term pressure and temperature. Figure 14.10 presents the short-term failure envelope (in terms of hoop stress against axial stress) for a GRP pipe as a function of temperature. The highest test temperature was 160 °C, which is above the measured glass transition temperature (T_g) of the resin system (measured at 130 °C). The test temperature of 120 °C is within 10 °C of the measured T_g of the resin system. Note that ISO 14692[8] stipulates that the maximum operating temperature of a GRP pipe must be at least 30 °C less than the T_g.

The failure mode of all tests was weepage. There is a significant reduction in strength of the pipe at the test temperature of 160 °C and also 120 °C.

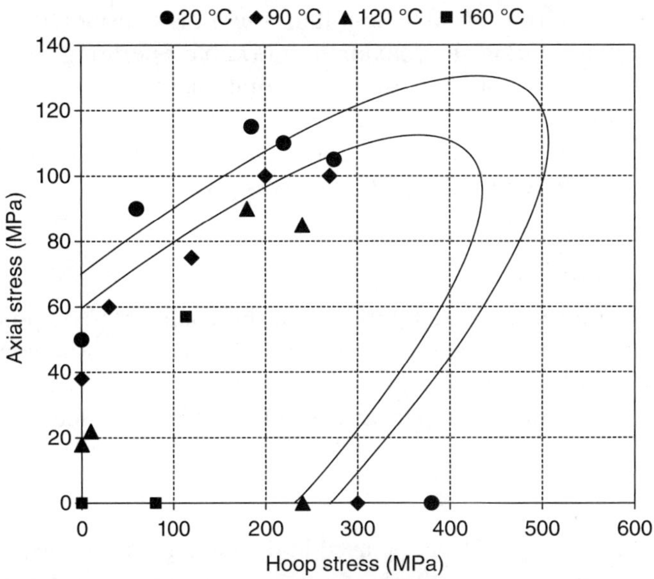

14.10 Failure envelopes (hoop stress against axial stress) as a function or temperature for a GRP pipe.

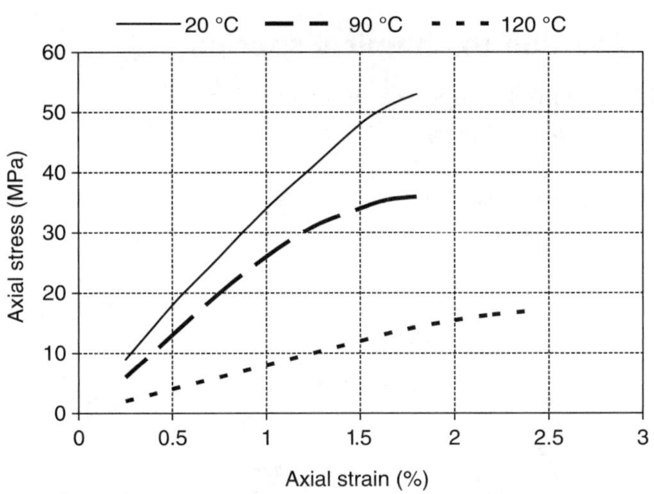

14.11 Axial strain to failure as a function of temperature.

However, at a test temperature of 90 °C there is on average a strength reduction of approximately 15–20%. Figure 14.11 presents the axial strain to failure as a function of pressure under the application of internal pressure. For temperatures below the ISO 14692 maximum operating temperature criterion, the axial strain to failure is independent of temperature.

Therefore, the criterion of 30 °C less than the T_g is considered a sensible criterion for determining the maximum allowable operating temperature of a thermosetting resin-based composite component.

In order to predict the ageing influence of temperature, Equation [14.7], an estimate of the parameter A_T is required. In general, the relationship between the design temperature, T, and A_T is given by

$$A_T = \left(\frac{T - T_g}{T_0 - T_g} \right)^n \qquad [14.8]$$

where T_0 is the installation temperature of the composite component and the exponent n is usually taken as 0.5. Similarly, for composite component material properties that are dominated by the matrix (or resin), the reduced property as a function of temperature, $P(T)$, can be expressed as

$$P(T) = P(T_0)\sqrt{\left(\frac{T - T_g}{T_0 - T_g} \right)} \qquad [14.9]$$

Equations [14.8] and [14.9] can be used in the damage mechanics model to estimate the reduction is strength of a composite component due to elevated temperature. The predictions for the two lower temperatures are presented in Fig. 14.10 where the agreement is acceptable.

14.4 Ageing due to chemical species

Figure 14.12 presents the up-take of species (or fluid) into a ring section cut from a GRP pipe (up-take defined as weight percent.[1]) The species

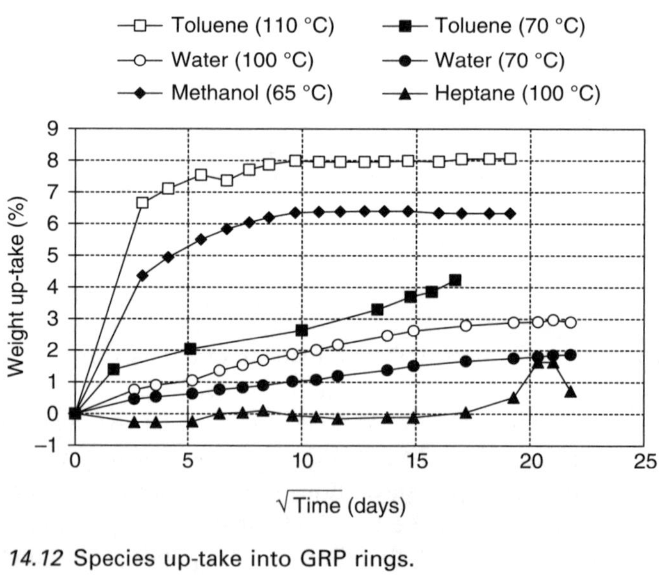

14.12 Species up-take into GRP rings.

considered are typical or representative of those occurring in the Oil and Gas industry, specifically hydrocarbon transport. In these ring tests, species concentration was 100% whereas in practice, mixtures of these species occur. As a first approximation, a weight fractioned average of the individual species' influence is probably the best solution to dealing with assessing the influence of fluid mixtures on performance. The temperatures (or higher temperature tests) were set to the atmospheric boiling point of the individual fluids.

Of the species under investigation, the largest weight up-take was for toluene (8%) at 100 °C, followed by methanol (6%) at 65 °C. In these experiments, the rings were completely immersed and the ends were not sealed. Therefore the time to reach saturation will be much shorter than for an in-service GRP pipe or component where the ingress of species is only from the inside pipe wall, i.e. one sided. Other species were absorbed to a lesser amount. Water up-take was 3.5% at 100 °C compared with 1.5% at 70 °C. Heptane was essentially not absorbed.

The reduction in hoop modulus caused by the ingress of species is presented in Fig. 14.13.[1] The experimental procedure for determining the hoop modulus was to remove the GRP ring from the ageing bath, allow the ring to cool, typically 10 minutes waiting time, then measure the modulus in a standard hydraulic test machine. For all tests, except toluene at 110 °C, the hoop modulus degrades to between 80 and 85% of its original value. It appears from these tests that the degradation in GRP stiffness is not that

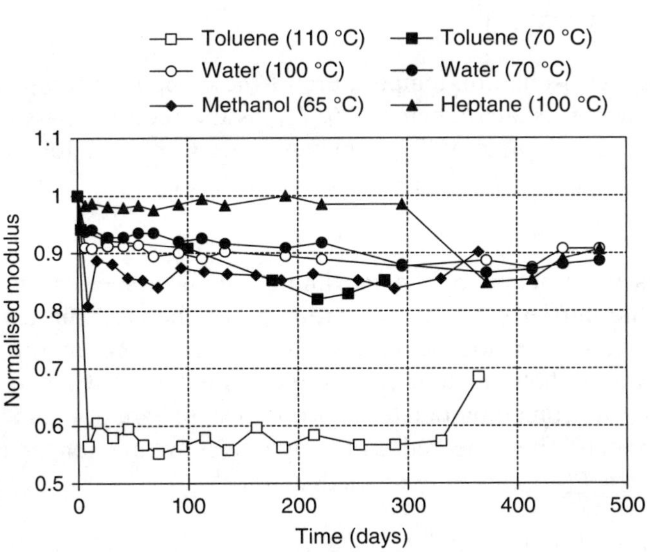

14.13 Modulus reduction of GRP rings as a function of species up-take.

sensitive to the species type or the up-take of the species into the ring, assuming that the ageing temperature is less than the aged T_g of the resin, measured at 100 °C. For the high-temperature toluene test, the modulus reduced to 60% of its original value.

Many applications of composite components in the Oil and Gas industry involve water. It is important to realise that water is a relatively aggressive species when considering the degradation of GRP pipes. It is the [OH] group within the water molecule that causes the degradation. In general, thermosetting resin-based composite components are compatible with a wide range of environments common in the Oil and Gas industry, but consideration is required when the environment is strongly acidic (pH < 3.5), strongly alkaline (pH > 11) or contains a strong solvent, e.g. methanol or toluene in concentrations greater than 25%.

Ultraviolet degradation is generally not a concern as most resin systems nowadays contain an inhibitor and this degradation mechanism is considered minor as operating experience in hot, sunny environments, e.g. the Middle East, extends to at least 15 years without recorded problems.

Absorption of chemical species causes a reduction in the T_g of the composite. Therefore, to account for the ageing influence of chemical species in the design process an estimate of the parameter A_C in Equation (14.7) is required. A similar approach to that for the temperature ageing factor, A_T, is proposed but using the reduced T_g. In general, the relationship between chemical species ingress, the design temperature, T, and A_C is given by

$$A_C = \left(\frac{T - T_{g,red}}{T_0 - T_{g,red}} \right) \qquad [14.10]$$

where T_0 is the installation temperature of the composite component and the exponent n is usually taken as 0.5. $T_{g,red}$ is the reduced glass transition temperature. An estimate of $T_{g,red}$ can be deduced from the following relationship linking chemical species ingress to a shift in T_g:

$$T_{g,red} = T_g - mT_{g,red,100\%} \qquad [14.11]$$

where m is the fraction of the maximum chemical species ingress and $T_{g,red,100\%}$ is the shift in T_g for full saturation of the chemical species of interest. To provide some guidance on the shift in T_g for GRP pipes fully saturated with water, the reduction in T_g is approximately 15 °C.

For composite component material properties that are dominated by the matrix (or resin) then the reduced property as a function of chemical species ingress, $P(C)$, can be expressed as

$$P(C) = P(C_0)\sqrt{\left(\frac{C - C_0}{C_{sat} - C_0} \right)} \qquad [14.12]$$

where C is the concentration of ingress species and the subscripts '0' and 'sat' refer to the initial and fully saturated conditions. It is generally assumed that Fickian diffusion is applicable for estimating the concentration profile of chemical species within the wall of the composite component. Equations [14.10] and [14.12] can be used in the damage mechanics model to estimate the reduction in strength of a composite component due to chemical species ingress.

14.5 Ageing due to applied load

The basis for applying damage mechanics to predict the long-term behaviour of composite components under long-term applied loads is an adapted form of the Paris Law, which relates the growth in damage (or decrease in crack spacing) as a result of time or fatigue cycles to a power law in terms of the failure criterion of the ply stresses controlling damage crack growth. To extrapolate from short- to long-term behaviour, the failure mechanism common to both times must be similar, which is the case for most composite components, e.g. GRP pipes.

The adapted form of the Paris Law is expressed as

$$\frac{\mathrm{d}}{\mathrm{d}t}\left(\frac{1}{s}\right) = AC^n \qquad [14.13]$$

where t is time and A and n are constants,[6] and C is defined in Equation [14.6]. Equation [14.13] can equally be written in terms of number of cycles rather than time, but with a different exponent, n. Integrating Equation [14.13] and assuming that the influence of damage on the shear stress term is comparable to that of the transverse tensile stress component, then:

$$\begin{aligned} t &= \frac{1}{A}\int C^n \mathrm{d}\left(\frac{1}{s}\right) \\ &= \frac{C^n}{As_{\text{final}}} \end{aligned} \qquad [14.14]$$

The time (or number of cycles) is proportional to C^n which, from Equation [14.6], implies that it is also proportional to the applied pressure to the power 1/2n as the ply stresses are related directly to the applied pressure. Therefore:

$$P \propto \frac{1}{t^{1/2n}} \qquad [14.15]$$

where the constant of proportionality is predicted from short-term measurements. Note that, in the qualification of GRP pipes using ISO 14692, long-term testing is used to infer the regression gradient which represents the gradient of the reduction in failure pressure as a function of time. Therefore, the constant A_t, (Equation [14.7]) is given by

$$A_t = \frac{1}{t^{1/2n}} \quad [14.16]$$

and the regression gradient, G, is equivalent to 1/2n. In the above derivation it is assumed that the damage at failure (in terms of crack spacing or density) is similar and independent of the cycles or time to failure, and this is supported by experimental evidence.[6]

Figure 14.14 presents predictions against measurements of a static fatigue failure.[10] The measurements were performed on GRP pipes according to ASTM D 2992 under internal pressure. Agreement between prediction and experiment is good when the exponent n is set to 7. This corresponds to a value for the regression gradient, G, i.e. the slope of the best fit line through the measured data points, of 0.071. To help put the data presented in Fig. 14.14 into perspective, and also to provide insight into the effect of constant load on the reduction in strength of a composite component, then a conservative estimate of the reduction in strength from short term to long term, where long term is defined as 20 years, is 0.5. Therefore, for a 20-year design life the degradation constant, A_t, can be set conservatively to 0.5.

Constant pressure tests have some drawbacks as a means of generating data for the assessment of long-term performance. The principal difficulty arises when the magnitude of the regression gradient is relatively small, i.e. the slope of the curve is almost flat, close to horizontal, which is the case for many composite components. In this case small statistical variations in the test and test sample can produce large variations in the time to failure, with the effect that it is difficult to estimate accurately the time to failure of a composite component under a given applied load. In fact, the potential variation in failure time can be greater than one order of magnitude in

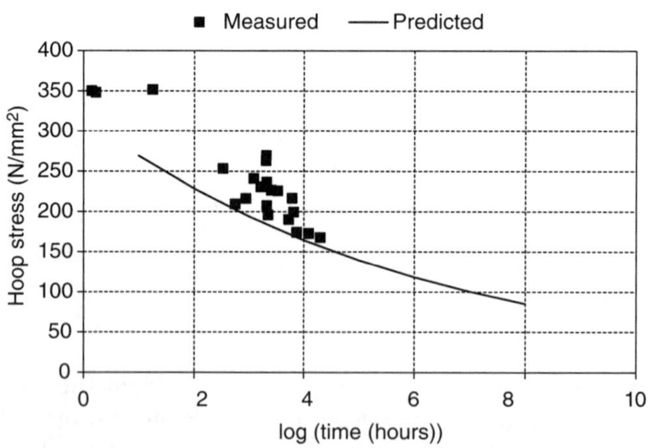

14.14 Hoop stress against time to failure.

terms of the logarithm of time. This variation can be as high as 5000 hours, impractical for testing purposes.

Rather than applying a constant pressure, an alternative loading option is to apply the pressure as a linear function of time. This low-speed loading rate (LSL) test provides an improved test in terms of reducing the scatter in time to failure. Further details of this test method can be found in Gibson *et al.*[11]. The analogy that can be used to illustrate the point is that of intersecting lines. Two lines almost parallel have a large potential intersection zone (constant pressure test and regression curve), whereas lines that are close to being perpendicular (LSL test and regression curve) have a well-defined intersection point. Figure 14.15 shows this point schematically. The disadvantage of the LSL test is that a degree of uncertainty is introduced into the measured failure pressure and a more complicated pressure delivery system is required.

It is possible to convert measured failure pressures directly from a linear LSL test to a constant load test. The relationship between the pressures is given by

$$P_{\text{contant pressure}} = \left(\frac{G}{G+1}\right)^{G} p_{\text{LSL}} \qquad [14.17]$$

where G is the regression gradient. Figure 14.16 presents a plot of the axial strain against hoop stress during an LSL test. The pipe tested in Fig. 14.16 is identical to the pipe tested in the short-term test presented in Fig. 14.3. Each point in Fig. 14.16 represents the strain at a particular time during the test. If the pressure is scaled according the regression relationship (Equation [14.12]), then a comparison between the scaled axial strain results from

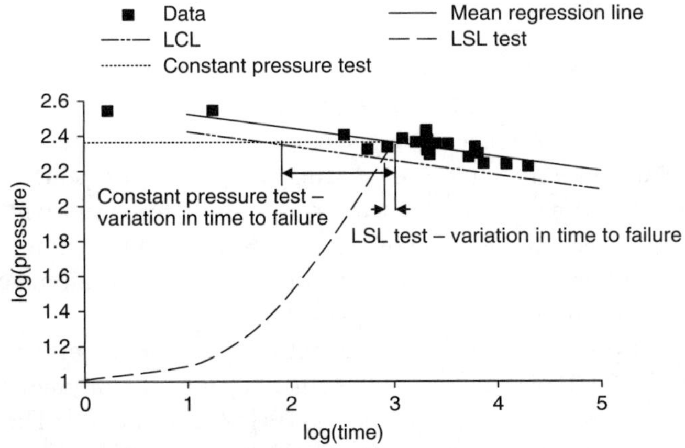

14.15 LSL pressure–time plot.

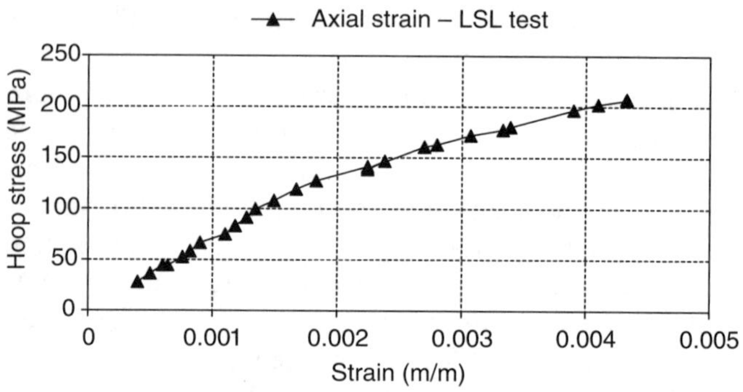

14.16 Axial strain response to internal pressure during an LSL test.

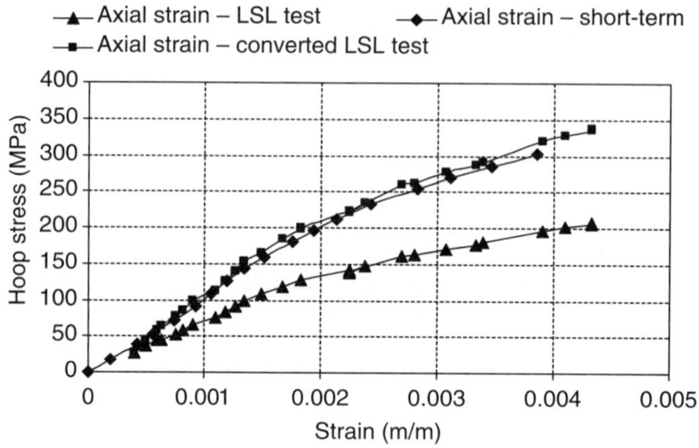

14.17 Comparison between scaled axial strain results from an LSL test and a short-term internal pressure test.

the LSL test and the short-term internal pressure test can be made. This comparison is shown in Fig. 14.17. The agreement between these two curves shows that, if the strain response is measured in both short-term and LSL tests, then the regression gradient of the composite component can be inferred directly from these tests by interpretation. This result will be used in the proposed assessment procedure, discussed later in section 14.7, as a means of experimentally quantifying the amount of degradation a composite component has undergone.

Figure 14.18 presents a comparison between measured and predicted hoop stress of a GRP pipe under cyclic fatigue against the number of cycles to failure for ratios of hoop to axial stress of 1:2, 2:1 and 4:1. The predictions are based on an exponent, $n = 6$. It has been shown that this exponent is also the same as for fatigue crack growth in pure epoxy resin.[6] The con-

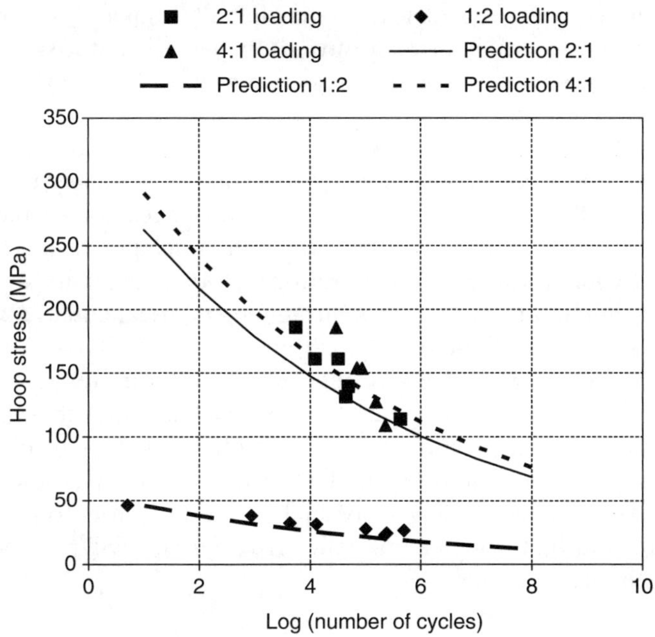

14.18 Hoop stress against cycles to failure.

stant of proportionality was deduced by matching short-term failure predictions to those of the first cycle, $N = 1$, i.e. the constant of proportionality was set to the relevant short-term failure pressure. The linkage between the fatigue crack growth in the pure resin and the GRP pipe demonstrates the effectiveness of the proposed theory for predicting ageing.

For cyclic load fatigue effects, which are more severe than constant load effects, the following correlation from experimental data[6] and also quoted in ISO 14692,[8] is used to estimate the degradation constant, A_{cyc}:

$$A_{cyc} = \sqrt{\left(R_c^2 + \frac{1}{2.888\log(N) - 7.108}(1 - R_c^2)\right)} \qquad [4.18]$$

where N is the number of cycles and R_c is the ratio of the minimum to maximum of the load cycle. For a large number of cycles, A_{cyc} reduces to 0.25.

14.6 Design against ageing

Design standards for composite components generically fall within two approaches, either performance based or design allowables. Performance-based standards rely on testing to demonstrate performance whereas design allowable approaches use conservative default values for composite strengths (or failure strains). Irrespective of the generic nature of the standard, the influence of ageing on the long-term performance of the composite component must be accounted for in the design.

The general approach to account for ageing is to derive individual values for the four de-rating constants listed in Equation [14.7]. Ideally, this derivation should be through measurement of performance under the actual ageing conditions. In most cases this approach is not practical. The usual approach is to derive values of the de-rating factors individually, sometimes through experiment but mostly through simple empirical equations as presented in, for example, Equations [14.8] and [14.10]. ISO 14692 provides some guidance on how to measure the de-rating factors for temperature and chemical degradation. However, this approach assumes that the effect of the individual ageing factors can be multiplied together to obtain the overall ageing de-rating factor without allowance for any possible interaction of effects. In general, using individual de-rating factors to predict the effect of degradation through a multiplicative approach will provide a conservative estimate of ageing.

14.7 Assessment of ageing

Composite components within the Oil and Gas industry have been in service for up to 20 years. With such long service times, assessment of the condition of composite components is becoming part of the integrity management process within many Oil and Gas plants. However, there is little guidance in the open literature or standards on the recommended practice of assessing that integrity. When attempting to perform an assessment of the integrity or fitness for purpose of a composite component, the question that is asked is usually along the lines of: 'Will the component remain fit for service for the remaining design life or how many more years will the component remain fit for service?' In order to answer these questions, an assessment procedure is presented for the degradation mode of failure, matrix cracking. This assessment procedure will not be relevant for other failure modes, e.g. delaminations.

The mechanical and physical properties of the 'damaged' component that are required for the assessment procedure are:

(i) glass transition temperature;
(ii) density or spacing of matrix micro-cracking;
(iii) regression gradient.

Mechanical data of the undamaged composite component are also required and include:

(a) hoop and axial modulus, and Poisson's ratio;
(b) short-term stress–strain data of the component to failure under the same loading conditions as the component is subjected to in-service;
(c) glass transition temperature.

These latter data for the undamaged component should be available from the composite component supplier.

The proposed assessment procedure consists of the following steps.

1 Mechanical data from steps (a) and (b) of the undamaged component are used to calibrate the damage mechanics model (Equations [14.3] and [14.4]) from short-term tests to predict crack spacing at failure.
2 The regression gradient from step (iii) is used to calibrate the partial factor, A_t, in Equation [14.7]. If cyclic fatigue is to be considered, then the regression gradient will be required for this type of loading which can then be used to calibrate the product of the partial factors, $A_t A_{cyc}$ in Equation [14.7].
3 The T_g, of the damaged component is required. If it cannot be measured directly, then a conservative assumed value of the undamaged T_g minus 40 °C should be used. The T_g is used to estimate the product of the constants $A_T A_C$ in Equation [14.7] through use of Equations [14.8] and/or [14.10]. This value along with the value of the partial factors from step 2 can be used to derive the constant, C, in Equation [14.7].
4 The amount of damage or crack density within the damaged component is required. If it can be measured destructively, then the crack density can be determined from performing two LSL tests (for an estimate of repeatability) for a minimum duration of 1000 hours. The damage mechanics analysis is used to infer the regression gradient by comparison and scaling the stress–strain data from the LSL test with the short-term undamaged stress–strain data. From the regression gradient an estimate of the remaining lifetime of the component can simply be estimated based on the current state of damage and extrapolation to the predicted long-term failure pressure. If only NDT methods are available for determining the crack density, then the amount of damage within the composite component needs to be measured. For GRP pipes, the measurement of axial velocity of an ultrasonic signal and subsequent inference of the axial modulus is one method of assessing and measuring the crack density. Figure 14.19 shows the measured ultrasonic velocity as a function of axial strain measurement for the same GRP pipe as shown in Fig. 14.3.

The reduction in velocity as a function of increasing axial strain and therefore crack density is clearly demonstrated. However, the relationship

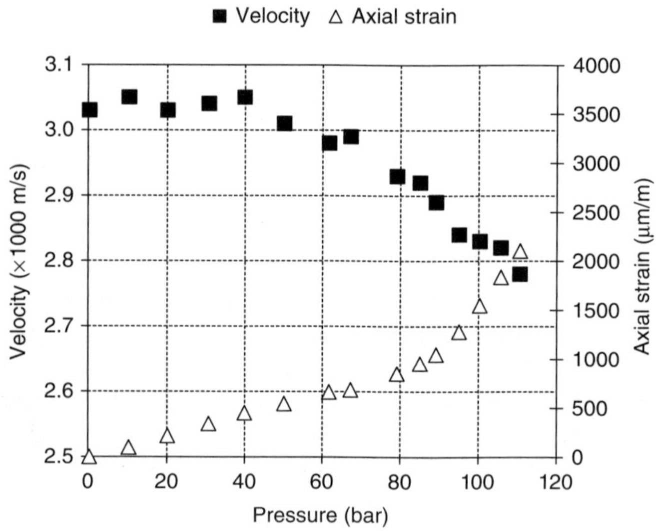

14.19 Velocity against axial strain.

between axial modulus and measured velocity requires calibration from a test sample, i.e. an equivalent undamaged pipe sample. From the axial velocity ultrasound measurement of the damaged sample, the crack spacing within the component can be inferred. From this crack spacing, the damaged stress–strain curve can be predicted from the damage mechanics model. Comparing this predicted stress–strain curve with the undamaged stress, strain curve as discussed previously provides confirmation of the current state of damage and an estimate of remaining life.

The aim of these destructive or non-destructive tests is to measure the degradation that the GRP pipe has suffered during its operational life. The procedure either infers or measures the amount of damage and infers from this the actual regression gradient of the damaged component. The outcome of these tests and the damage mechanics analysis is a quantitative measure of the degradation the component has suffered. Using predictions from the damage mechanics model and the regression gradient, an estimate of the future remaining life of the composite component can be made given the anticipated future operating conditions.

The above procedure for assessing the condition of degradation within the wall of a composite component is considered a practical approach. The destructive procedure requires sections of the composite component to be removed from the field and returned to the laboratory for testing, which may not be practical. The non-destructive approach requires calibration of the measured axial velocity against axial strain and crack spacing.

This assessment procedure is not fully tested or verified but it does provide an insight into how to assess ageing and predict the future remaining life of a damaged composite component assuming that the degradation mechanism is matrix cracking. This assessment procedure is new and further testing and trials of the approach are on-going. Based on these trials further refinements will no doubt result. Currently, no other assessment procedure for composite components exists.

14.8 Examples of ageing

The following three paragraphs provide a short pictorial description of common ageing effects seen within composite components.

1 *Matrix cracking due to applied load.* Figure 14.20 presents a micrograph of a GRP pipe which has suffered matrix micro-cracking as a consequence of continuous applied load. The plies within the pipe wall can be clearly identified and the matrix crack can also be clearly seen in the top, centre of the photograph. The dark regions, for information, are voids. This microstructure is typical of the weepage failure mode.
2 *Elevated temperature.* Figure 14.21 presents a photograph of a failed GRP due to elevated temperature and axial load. Due to the elevated temperature there was significant reduction in the axial (resin-dominated) mechanical properties of the pipe which resulted in significant fibre rotation.[12]
3 *Chemical species.* Figure 14.22 is a photograph of a GRP pipe that has failed due to chemical species ingress. The photograph shows the inner

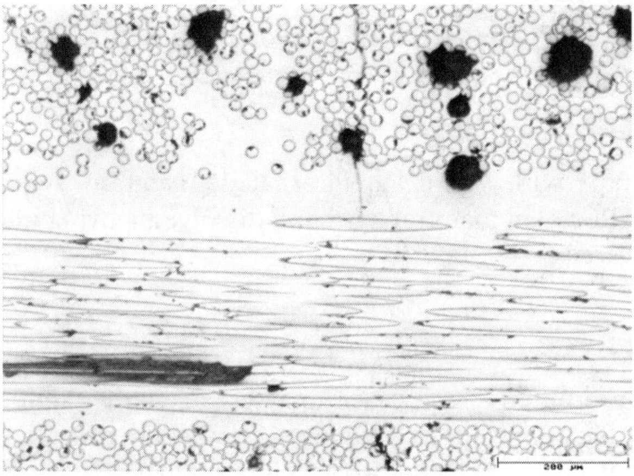

14.20 Microstructure of a GRP pipe showing matrix cracking due to the application of continuous load.

14.21 Photograph of a failed GRP pipe at elevated temperature showing matrix cracking and fibre rotation.[12]

14.22 Photograph of a GRP pipe that has due to chemical species ingress with significant inner liner degradation.

liner of the pipe with significant internal degradation. A significant proportion of the resin material has been dissolved leaving many bare, exposed fibre areas.

14.9 Conclusions

This chapter has presented an approach to assess ageing within fibre reinforced thermosetting composite components. The ageing process has been limited to the matrix cracking failure mechanism. This mechanism is a common failure mode within GRP pipes, one of the most common composite components within the Oil and Gas industry.

Three ageing influences have been discussed: temperature, chemical species and time of application of load. For all three ageing influences, their consequence on the failure mechanism has been presented in terms of a physical model that can quantify their effect on the failure mechanism, i.e. reduction in performance. Examples of the predictive capability of the model have been presented.

The design of composite components must include or account for the anticipated ageing effects that are likely to be present during the lifetime of the component. This implies that the designer must be aware of the failure mechanisms that can occur in the component and also the relationship between the ageing process and the failure mechanism. In addition for suppliers of composite components, the ageing process must also be considered in the design of long-term qualification tests used to demonstrate performance.

Much of the chapter has taken information from previous activities associated with studies on the ageing processes of composite components. However, a new procedure for assessment of ageing has been presented, which enables owners of the composite component to demonstrate the integrity of that component and also provide an estimate of the remaining life. However, this procedure is not generic and is specific to the failure process of matrix cracking. Further development of the assessment procedure for the other common failure mechanisms of composite components – i.e. delaminations – is required and this is on-going.

14.10 References

1 S.R. FROST and A. CERVENKA, Glass fibre reinforced epoxy matrix filament wound pipes for use in the oil industry. *Composites Manufacturing*, **5**, 73–82, 1994.

2 R. TALREJA, Stiffness properties of composite laminates with matrix cracking and interior delamination. *Engineering Fracture Mechanics*, **25**, 751–762, 1986.

3 S.W. TSAI and H.T. HAHN, *Introduction to Composite Materials*, Technomic Publishing Co., Inc., Lancaster, PA, 1980.

4 S.J. ROBERTS, J.T. EVANS, S.R. FROST and A.G. GIBSON, Strain from matrix microcracking in fibre composite laminated tubes, *Journal of Composite Materials*, **37** (17), 1509–1524, 2003.

5 A.L. HIGHSMITH and K.L. REIFSNIDER, in *Damage in Composite Materials* (Ed. K.L. Reifsnider), ASTM STP 775, American Society for Testing and Materials, Philadelphia, PA, pp. 103–117, 1982.

6 S.R. FROST, Predicting the long term fatigue behaviour of filament wound Glass fibre/Epoxy matrix pipes, in *10th International Conference on Composite Materials (ICCM/10)*, Whistler, BC, July 1995.

7 P.D. SODEN, R. KITCHING, P.C. TSE, Y. TSAVALAS and M.J. HINTON, Influence of winding angle on the strength of filament wound composite tubes subjected to uni-axial and bi-axial loads, *Composites Science and Technology*, **46**, 363–378, 1993.

8 ISO 14692 – Petroleum and natural gas industries – Glass-reinforced plastic (GRP) piping, International Organization for Standardization, Geneva, 2002.

9 J.M. HALE, B.A. SHAW, S.D. SPEAKE and A.G. GIBSON, High temperature failure envelopes for thermosetting composite pipes in water, *Plastics, Rubber and Composites*, **29** (10), 539–548, 2000.

10 Ameron International Fiberglass Limited, Product literature, Ameron International, Houston, TX.

11 A.G. GIBSON, N. DODDS and S.R. FROST. Use of Miner's law in the short and long term qualification testing of non-metallic pipe systems, in *4th International Conference on Composite Materials in Offshore Operations (CMOO-4)*, CEAC, Houston, TX, 2005.

12 R.O. SAIED, *Failure envelopes for filament wound composite tubes in water at elevated temperatures*, PhD Thesis, Newcastle University, 2004.

15
Ageing of composites in the construction industry

S. HALLIWELL, NetComposites Ltd, UK

15.1 Introduction

Fibre-reinforced polymers (FRPs) have been used successfully over the past 50 years in a wide range of applications in the civil engineering sector including pipes, tanks, slabs, walkways, bridge decks, gratings, column reinforcing wraps and reinforcing bars for concrete. In many of these applications FRPs are exposed to one or more environmental influences. All FRPs are durable inasmuch as they are water resistant, thermally stable and cannot rust. Particular grades of high-durability (5–20 year lifetime) FRPs are available for specific applications, e.g. FRPs for concrete re-bars incorporating alkali-resistant glass fibres in order to resist fibre attack by pore water. In almost all applications, the durability of an FRP may be enhanced by imposing a conservative safety factor (2–4) on the design, and in many such cases additional durability may be achieved by the use of a protective coating and/or the incorporation of light stabilisers and antioxidants.

FRP components have good durability:

- FRP structures where appropriately designed perform exceptionally well;
- several structures in the United Kingdom have given over 35 years of service, and are still meeting performance requirements;
- where reported, failures are not due to the material, but due to a lack of understanding of material properties at the initial design stage or poor detailing in some prefabricated sections.

Any deterioration caused by weathering is restricted to the surface of the FRP component. This does not generally affect the structural performance of the component or building, but may be significant for certain applications where aesthetics is important. Regular inspection and correct cleaning will ensure such effects are spotted at an early stage and are treated to restore the FRP. Improvements in resin and manufacturing technology over the last 10 years will lead to improved durability of FRP components enabling design lives of 60–100 years to be realised. FRPs already have a market in the construction industry. Their share of this market is set to expand rapidly in the coming decade as construction demands improved performance and cleaner processes.

15.2 Use of fibre-reinforced polymers in construction

FRPs offer the designer a combination of properties not available in traditional materials. The key advantages offered by the material are the ability to tailor-make properties, lightness, resistance to corrosion, resilience, translucency and greater efficiency in construction compared with the more conventional materials.

FRPs were first developed during the 1940s for military and aerospace applications. Considerable advances have been made since then in the use of this material and applications developed in the construction sector. FRPs have been successfully used in many construction applications including load-bearing and infill panels, pressure pipes, tank liners, roofs, and complete structures where FRP units are connected together to form the complete system in which the shape provides the rigidity. FRP materials have good durability and there are many examples in the United Kingdom to demonstrate their long-term performance in service (Fig. 15.1).

In the last decade, FRPs have found application in the construction sector in other areas such as bridge repair, bridge design, mooring cables, structural strengthening and stand-alone components. These materials are

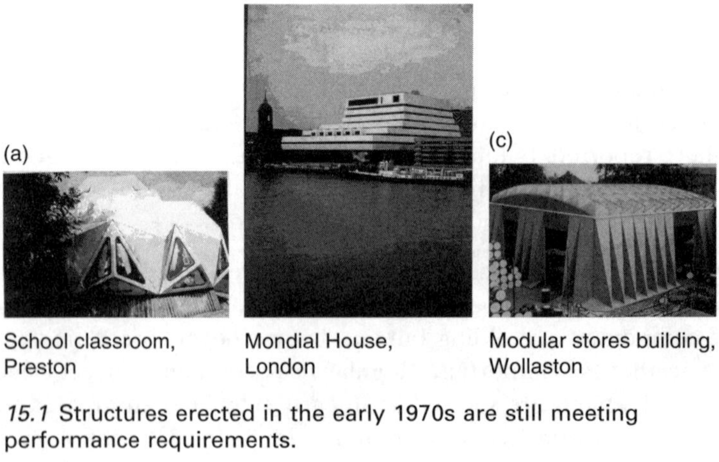

(a) School classroom, Preston

(b) Mondial House, London

(c) Modular stores building, Wollaston

15.1 Structures erected in the early 1970s are still meeting performance requirements.

often referred to as 'advanced composites' and have properties considerably superior to those of earlier materials. In FRPs or 'advanced composites', fibre with high strength and stiffness is used in relatively high volume fractions, while the orientation of the fibre is controlled to enable high mechanical stresses to be carried safely. The major advantage of these materials lies in their anisotropic nature. The reinforcement can be tailored and orientated to follow the stress patterns in the component, leading to much greater design economy than can be achieved with traditional isotropic materials.

15.2.1 Materials used

Thermosetting resins are most widely used in FRPs for the construction industry, the most common being the unsaturated polyesters, epoxides and phenolics.

- Polyester resins are relatively inexpensive, easy to process, allow room-temperature cure and have a good balance of mechanical properties and environmental/chemical resistance.
- Epoxy resins are used for the majority of high-performance FRP structures. They have excellent environmental and chemical resistance and superior resistance to hot–wet conditions. Compared with polyesters, they are more expensive and require more careful processing; however, they give better mechanical properties and better performance at high temperatures.
- Phenolic resins find specific application in construction due to their flame-retardant properties, low smoke generation, dimensional stability at high temperature and excellent resistance to environmental degradation.

A wide range of amorphous and crystalline materials can be used as the fibre. In the construction industry glass fibre is most widely used, mainly from economic considerations. There are four classes of glass fibre: E-glass, AR-glass, A-glass and high-strength glass, but E-glass tends to dominate the reinforcement sector. Carbon fibre, of which there are three types (Types I, II and III), can be used separately or in conjunction with glass fibre as a hybrid to increase the stiffness of a structural member or the area within a structure. The stiffness obtained exceeds the value possible using only glass fibre. Ultra-high modulus (UHM) carbon is used for steel reinforcement. Aramid fibres can be used instead of glass fibres to give increased stiffness to the component.

For structural applications it is mandatory to achieve some degree of flame retardance. Fire retardants are usually incorporated in the resin itself or as an applied gel-coat. Fillers and pigments are also used in resins for a

Table 15.1 Fabrication of FRP parts for the construction industry

Processing route	Typical components
Open moulding – hand and spray lamination	Cladding panels, radomes, garage doors, caravan parts
Vacuum infusion	Strengthening of components, masts
Pre-pregs	Architectural mouldings, infrastructure repair
Compression moulding of SMC	Electrical cabinets, sectional water storage panels, modular components
Pultrusion	Standard sections, access ladders, reinforcement bars, roof trusses, space frames, window profiles
Filament winding	Pipes, wrapping of columns
Continuous sheeting	Cladding, roofing
Centrifugal casting	Pipes, tanks, masts, poles

variety of purposes, the former principally to improve mechanical properties and the latter for appearance and protective action.

15.2.2 Fabrication

A wide range of different processes have developed for moulding of FRP parts, ranging from very simple manual processes such as hand lay-up to very sophisticated highly industrialised processes such as sheet moulding compound (SMC) and pultrusion. Each process has its own particular benefits and limitations making it applicable for certain applications. The choice of process is important in order to achieve the required technical performance at an economic cost.

The main technical factors that govern the choice of process are the size and shape of the part, the mechanical and environmental performance, and aesthetics. The main economic factor is the number of identical parts required or run length. This is because FRP parts do not generally come as standard components but are custom-designed for a particular application. Pultrusion and continuous sheeting are exceptions but most processes will have an initial investment or set-up cost that must be amortised over the length of the project. This is a major factor in the choice of process and is one of the reasons for the proliferation in processing methods. Table 15.1 summarises the types of component produced from common manufacturing processes. For further information on manufacturing, see the Network Group for Composites in Construction website: www.ngcc.org.uk.

15.3 Benefits of fibre-reinforced polymers for construction

FRPs have a number of advantages when compared with traditional construction materials such as steel and concrete. FRP materials have been utilised in small quantities in the building and construction industry for decades. However, because of the need to repair and retrofit rapidly deteriorating infrastructure in recent years, the potential market for using FRP materials for repair (and for a wider range of applications) is now being realised to a much greater extent. Numerous successful applications using FRP materials in the construction industry – such as bridges, piers, building panels, walkways, pipelines, and offshore structures – to mention only a few – have been reported.[1]

FRPs offer excellent corrosion resistance to environmental agents as well as the advantages of high stiffness-to-weight and strength-to-weight ratios when compared with conventional construction materials. For instance, such ratios for carbon-fibre-reinforced composite (CFRP) are 10–15 times higher than those of steel. Other advantages of FRPs include low thermal expansion, good fatigue performance and damage tolerance, non-magnetic properties, ease of transportation and handling, low energy consumption during fabrication of raw material and structure, and the potential for real-time monitoring.[2] Perhaps the biggest advantage of FRPs is tailorability. Reinforcement can be arranged according to the loading conditions so that an FRP structure or component can be optimised for performance.

The apparent high cost of FRPs compared with conventional materials has been a major restraint on their use. However, a direct comparison on a unit price basis may not be appropriate. When installation is included in the cost comparison, FRPs can be competitive with conventional materials. The light weight of FRPs allows some prefabrication to take place at the factory which reduces time at the job site. For example, Meier *et al.*[3] reported that, in repairing the Ibach bridge, 175 kg of steel can be replaced by 6.2 kg of carbon-FRP because of the light weight. Easy transportation and low installation cost (for instance there is no need for scaffolding) for FRP structures implies significant reduction in labour cost. In a recent geodesic dome structure constructed using glass-FRP in Reno, Nevada, it was reported that a substantial cost saving of 20% was estimated in comparison with a metal design.[4] If the comparisons include life-cycle costs, FRPs can have significant advantage. The unique properties of FRPs, such as high corrosion resistance, make the life-cycle costs lower than they are with conventional materials. In many cases, an FRP structure can last much longer than conventional materials, thus assuring a lower life-cycle cost.[5] In addition, the cost of FRPs will be driven down by increasing demand. The current demand for FRP in the construction industry continues to increase at a rate of 2.5% per year.[6] As FRPs are gradually accepted by civil engineers and designers, and confidence in using them is gained through research and experience, such demand will certainly escalate in the future.

Beyond cost issues, the most significant technical obstacle preventing the extensive use of such materials is a lack of long-term durability and performance data comparable with the data available for more traditional construction materials like steel and concrete. Although there have been numerous studies in the areas of creep, stress corrosion, fatigue and environmental fatigue, chemical and physical ageing, and natural weathering of FRPs in the past four decades, most of these are not aimed at applications for the construction industry. The expected service life of a structure is much longer in infrastructure applications. For instance, bridges are designed to last for over 120 years, and buildings in the region of 60 years. Additionally, at this time, the construction industry has focused predominantly on lower-cost glass reinforcement rather than the carbon fibre reinforcement used in aerospace applications. Most data for glass reinforcement are from work with short fibre materials and not at the loadings associated with structural applications. Hence the infrastructure community must be concerned with longer-term behaviour as well as with different materials and service environments compared with the aerospace industry. As a result, although data and experience gained from the past may serve as a general guideline, new studies and data pertaining to infrastructure applications are in great demand, especially for composites produced by low-cost, large-volume processing methods such as pultrusion.

15.4 Performance requirements

15.4.1 Service life

Most construction materials have a finite life. Metals can corrode and can suffer from fatigue. Wood can rot, even preservative-treated timber can rot eventually in a severe hazard. Concrete can crack or suffer from various chemical degradation processes. Natural rubber can perish as a result of ozone attack. All these materials have been around long enough for us to know and make allowances for their weaknesses.

While FRPs are no exception to deterioration, they can easily be designed to meet even the most challenging service environment. FRPs are being specified for applications in service environments ranging from the Middle East to Antarctica. In addition, there are continuous improvements in resin technology (new improved varieties of resin tend to be developed around every 7 years). FRPs are now being specified for applications designed to last for 40 or even 60 years without loss of functional effectiveness. The accelerating trend towards using FRPs in bridges and buildings means a further extension of the required lifetime, possibly to almost a century.

15.4.2 When does a fibre-reinforced polymer product have to be replaced?

It is sometimes difficult to determine the end of life of a product. There are three key factors:

- the product must remain safe to use despite the stresses and the external weathering they experience over decades;
- it must not become too expensive to maintain;
- it must continue to meet performance requirements – structural or aesthetic.

The possibility of repair is an attractive feature of FRPs. Their useful life can often be extended because they are more easily repaired than some other materials. Additionally, they can be used to extend the life of structures originally made from another material, such as concrete or metals.

15.5 Performance in service

15.5.1 Causes of deterioration

All construction materials are subject to deterioration in service due to exposure to certain environmental elements. Material deterioration may begin through one or more of the following influences:[7]

- mechanical stresses – including static loading, fatigue, repeated minor impact, erosion (including water erosion) and abrasion;
- chemicals (water, solvents, fuels, oils, acids, cleaning liquids, atmospheric oxygen, oxidising agents, caustic alkalis, etc.);
- radiation (including sunlight);
- heat – including high temperatures and large and rapid fluctuations in temperature;
- biological attack from bacteria, fungi, insects and marine borers.

Outdoor weathering can involve all five factors simultaneously. Materials can often survive individual threats such as ultraviolet light or a specific solvent, but they can still succumb to a combination of influences.

Biodegradation through micro-organisms, while a major factor for other materials (timber in particular), has very little importance in the degradation of most polymers, whether reinforced or not. Most FRPs can be buried safely underground for decades without rotting.

15.5.2 Fabrication of products

There are obvious links between fabrication procedures, inspection methods and subsequent product durability. Some methods of manufacturing FRPs produce better quality products than others. They introduce fewer defects, allow better control over fibre placement and orientation, enable a higher volume fraction of fibre reinforcement to be used, or lend themselves to better quality control.

Quality control is an important element in the optimisation of material durability. Achieving consistent output involves sound operator training, regular routine screening of raw materials, competent maintenance of processing machinery and good mould tool design. It also requires vigilant oversight of the actual processing operation, including atmospheric conditions such as humidity, temperature and dust content. Furthermore, it requires the use of up-to-date inspection methods on the final products.

15.5.3 Maintenance

FRP structures require little maintenance. Panels facing prevailing winds are generally 'self-cleaning' while those in the shadow will become contaminated and require periodic cleaning with a mild detergent solution[8] and the application of a wax specially formulated for the upkeep of FRP products. Incorrect maintenance of an FRP structure can deleteriously affect the durability. The use of inappropriate cleaning agents based on strong alkalis, solvents or abrasives can damage the surface.

15.5.4 Property retention as a guide to durability

After several decades in service, the initial properties of a component will have changed, even in the absence of obvious mechanical damage. It is customary to cite the change in properties with time as a measure of the extent of deterioration, and '% retention' has become by implication a measure of durability.[9] Not all properties change equally rapidly and the selection of significant properties requires careful consideration. Where there are no visible signs of deterioration, any internal changes may be detected using advanced diagnostic equipment, such as thermography, acoustic emission equipment, ultrasonic instrumentation, an electron microscope or spectrometers for chemical analysis.

15.5.5 Changes caused by weathering

All materials of construction change in appearance on extended exposure to the weather. Many materials such as brick, stone, exposed hardwood, slate and copper are judged to improve aesthetically on weathering; whereas many such as concrete, painted timber and polymers (including FRPs) become less pleasing to the eye. Whether or not this factor is important depends largely on the application of the material.

The changes that take place when polymers are exposed to weathering arise mainly from the effects of a combination of ultraviolet radiation and heat from the sun, together with the effects of moisture (precipitation and condensation) and oxygen from the atmosphere. Resins on their own vary a great deal in their ability to withstand outdoor use for long periods. Poor performance can sometimes be completely transformed by trace additives, so the solution becomes one of using the right grade of resin and appropriate additives.

The effects of outdoor use on structural FRPs such as glass/polyester or carbon/epoxy laminates are confined to the surface and do not often involve a serious threat to their structural integrity. The effects are mainly cosmetic, including:[10]

- fading and darkening;
- yellowing;
- blooming;
- loss of gloss and chalking.

Figure 15.2 shows the deterioration that may be observed in extreme cases. This can be prevented by the correct choice of resin and additives. Colour fading or darkening without loss of gloss can be due to the use of unstable pigments or pigment combinations that change colour after exposure. This can be overcome by the appropriate choice of pigment. Yellowing is usually due to the darkening of the base gel-coat resin, especially in whites. This can be overcome by using a more ultraviolet (UV)-resistant resin and better UV-absorbant additives, and by ensuring good cure of the resin. Blooming is caused by migration of an incompatible pigment or additive to the surface of a gel-coat to give a mat, faded appearance. Certain organic pigments can be the cause of this. Bloom can be removed by polishing, but this is only a short-term solution. Judicious choice of the pigment should overcome this problem.

(a) (b) (c)

Crazing of surface | Deterioration of gel-coat | Uneven discolouration

15.2 Changes induced by weathering.

Loss of gloss is normally brought about by erosion of the surface layer of the gel-coat due to chemical and/or physical damage. The colour of the gel-coat then appears to whiten, due to the diffused reflection of light from the matt surface. This is most serious in mouldings with strong bright colours where the phenomenon is most easily observed. On paler colours the effect is less noticeable; indeed the whiteness of a white structure can even improve by this means because surface dirt is shed, leaving a fresh exposure of white pigment; this phenomenon is termed 'chalking'. The erosion of gel-coat after many years' service with no treatment or repair can bring about the eventual mechanical failure of a laminate by exposing the reinforcement underneath. It should be noted that the onset of loss of gloss or chalking does not presage the immediate disappearance of the gel-coat which normally lasts for many years longer.

15.5.6 Climate effects

The severity of the weathering effects depends on the climate to which a material is exposed. Climates can be classified into a number of types such as temperate, sub-tropical, desert, arctic and Mediterranean, as well as industrial, rural or marine. In addition, variations occur from season to season and from year to year. It is thus complicated to compare the weathering performance of one material with that of another and to predict accurately the performance in service of a new component. Despite this, FRPs can be designed to meet specific climatic conditions, however severe. For example, they are finding widespread application in the Middle East (Fig. 15.3).

15.5.7 Prediction of durability

Predicting the weathering performance of building materials (including FRPs) can be carried out based on artificially accelerated laboratory weathering experiments and field trials. The latter take several years and relatively few organisations have been able to generate large data banks. However, there are now sufficient case histories of FRP products to give us performance data extending over three decades or more.[1–11] Accelerated methods can be undertaken indoors or outdoors.

In general, UV exposure tests for FRPs can be grouped into three main categories:

- Outdoor testing – specimens are directly exposed to outdoor conditions at a fixed angle relative to the horizontal and in a fixed direction.
- Accelerated outdoor weathering – UV radiation is concentrated onto the test specimen using special mirrors.

(b)

(a)

Sharjah Airport Dome, 1974

Burj Al Arab hotel cladding, 1990

15.3 Application of FRPs in extreme climates.

15.4 Outdoor test site. Courtesy of Atlas.

- Accelerated laboratory testing – specimens are exposed to UV radiation from a variety of UV light sources; filters are often employed to remove wavelengths that fall outside of the solar spectral range of interest. Equipment is also fitted with rain or moisture cycles to mimic the effect of rain.

Natural weathering in Florida is often used as a means of forecasting more rapidly those effects that will occur in temperate climates. For example, one year's exposure in Florida is taken to be roughly equivalent to 4–5 years in central Europe. Such comparisons are necessarily approximate. However, Florida exposure often forms part of the specifications for weather resistance throughout the rest of the world. Figure 15.4 illustrates an outdoor test site. The effects of weathering at locations such as Florida and Arizona can be accelerated by using panel mountings that track the sun, combined with Fresnel reflectors to concentrate the rays of the sun, water sprays and cooling devices. This method is known as EMMAQUA, which stands for equatorial mounted mirrors with water spray (AQUA) as shown in Fig. 15.5. By such means, 1 year of normal Florida weathering can

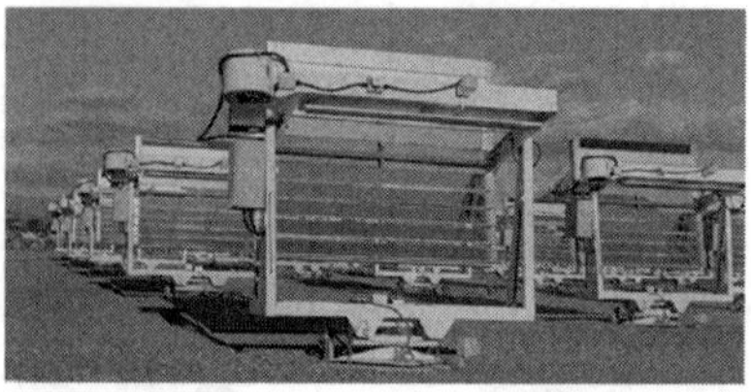

15.5 EMMAQUA test rigs. Courtesy of Atlas.

be achieved in approximately 40–45 days. Therefore, this is a very cost-effective method for testing materials needing a long service life.

Accelerated laboratory (artificial) weathering tests are widely used to investigate the durability of FRPs and other polymeric materials. These tests are based on the use of UV light, which is the principal agent responsible for the degradation of polymers. There are currently three key methods used for artificial weathering:

- carbon arc;
- xenon arc;
- fluorescent lamps.

Carbon arc

For the enclosed carbon arc, two strong emission bands peaking at 358 and 386 nm are much more intense than natural sunlight. This type of light source can be expected to have a weaker effect than solar radiation on materials that absorb only short-wavelength UV radiation, because there is very little irradiance below 310 nm. This technology has largely been replaced with fluorescent UV or xenon arc systems.

Xenon arc

Xenon arc lamps give a broad spectrum of light that matches the solar spectrum quite closely. At the short-wavelength end of the spectrum, the lamps produce a small amount of short-wavelength UV light that is not seen in sunlight because it is filtered out by the earth's atmosphere. At long wavelengths, xenon arc lamps produce a larger proportion of near-infrared light than is present in sunlight. However, filters can be used to correct for the greater part of these imbalances, producing a close match to a selected solar spectrum.

Fluorescent lamps

Fluorescent lamps have special phosphors selected to emit UV light at a particular waveband. Several types of lamp are available, concentrating their radiation either in the UV-A (400–315 nm) or UV-B (315–280 nm) wavebands. With either type of device, exposure to light is carried out under relatively controlled conditions of temperature and moisture.

The inherent variability in both laboratory and outdoor exposure testing makes it extremely difficult to obtain acceleration factors for correlating laboratory and outdoor test results (i.e. *x* hours in accelerated test equals *y* years in an outdoor environment). Thus any correlations must be qualitative.

The use of reliable methods makes it possible to predict service lifetimes more accurately. Such methods involve the precise measurement of temperature, time-of-wetness, and spectral UV radiation (in outdoor and laboratory accelerated testing). In particular, the calculation of total effective UV dosage (the actual amount of UV radiation that induces photochemical reactions in a material) takes into account the fact that not all UV radiation is absorbed, and that not all absorbed UV radiation causes a photochemical reaction to occur.

15.5.8 Controlling weathering performance with additives

UV absorbers are commonly used in the gel-coats on FRP components to absorb UV light and dissipate the absorbed energy. In the first few years of life in temperate climates, the use of a UV agent makes little difference to the weathering properties of a good-quality pigmented gel-coat. However, experience shows that patchy yellowing of white gel-coats, which sometimes occurs in these climates, can be overcome by such means. It is thought that this type of discolouration is caused by areas of the gel-coat having a lower pigment content. UV absorbers help to overcome such yellowing of the resin.

In the longer term or under conditions of high levels of radiation (such as sub-tropical climates), UV agents can be shown to improve gloss retention and colour stability. Traditional UV absorbers that have been used for many years are benzotriazole and hydroxybenzophenone derivatives. Other UV agents known as hindered amine light stabilisers (HALS) do not absorb UV radiation, but act by absorbing any free radicals that have been formed. The use of fire-retardant additives in gel-coats has a detrimental effect on weathering properties. Thus it is good practice to achieve fire retardancy by using a highly fire-retardant resin behind a non-fire-retardant gel-coat.

15.5.9 Durability in liquid environments

All resins and organic reinforcing fibres (but not glass or carbon) absorb water to varying extents, usually at a very low level, and are water permeable. Water absorption into glass or carbon FRPs is slow and the interfacial adhesion is protected by silane treatments included in surface 'sizings' applied during their production. Where absorption occurs, moisture migrates through the resin and eventually reaches the fibre–resin interfaces. It is often said that moisture migrates by 'wicking' along the interface by capillary action, starting from exposed fibre ends. However, hard evidence is usually absent, and a well-bonded fibre is not so easily separated from the resin that it would allow migration at the interface of a long, continuous section of its length.

The effects of moisture, once absorbed, are complex. Changes in the appearance and properties of the FRP product may be slight or severe, chemical or physical, permanent or reversible. The more moisture absorbed, the more deterioration in properties is likely to be found and the less reversible are the changes on drying. Reductions in strength and modulus are observed. An initial increase in strength is possible, because of the relief in internal stresses, which is followed by a decline after further absorption.

The more susceptible resins (polyester, polyester urethanes, some epoxies) are attacked by boiling water fairly quickly, but will resist cold water for very long periods. Other resins, with different chemical structures, are unaffected at temperatures within their normal range of use. Thus it is important to specify the correct resin formulation for the particular application.

Thick laminates are much less affected than thin ones in a given period and this explains the durability of many early FRP structures. It has been calculated that an epoxy-based FRP with a typical diffusivity towards moisture of 10^{-13} m^2s^{-1} would require 13 months to reach saturation if left in a tropical climate at 35 °C and 95% relative humidity (RH) if the thickness was 2 mm, but a 90 mm thick section would need 1342 years. During the approach to saturation, there is a through-the-thickness variation in moisture content and therefore in properties.[12,13]

Despite these reported effects of moisture, careful selection of material and component design can overcome any potential problems. FRP components, being tailor-made parts, are designed to prevent moisture absorption. Cutting or drilling on site exposes fibre and resin which could affect the absorption properties of the component and is strongly discouraged. FRPs have been used in the marine industry for many decades with very few reports of moisture ingress problems.

15.5.10 Chemicals

A surprising number of FRP applications involve occasional or prolonged contact with chemicals other than water. Many FRP articles are routinely placed in contact with detergents, cleaning solvents, acids, alkalis, strong oxidising agents, bleach, cleaning and degreasing agents, fuels, hydraulic and brake fluids, de-icers, paint strippers (methylene chloride is known to be damaging), lubricants, etching chemicals, flue gases, or food and drink.

It must be stressed that the resistance of FRP to highly reactive chemicals is generally very good. This explains their widespread use in the chemical process equipment industry, where it is often difficult to find any other affordable, processable materials capable of withstanding the very harsh conditions. It is rare for FRP articles to be attacked as rapidly as some common metals when placed in contact with acids. A few chemicals that are handled in chemical factories – such as powerful oxidising agents, strong caustic alkalis, bromine and wet chlorine – still pose severe problems for general-purpose resins. Otherwise, the well-informed selection of materials, in consultation with the suppliers and after reference to the relevant data banks, means that problems with chemical attack can be avoided.

15.5.11 Effect of temperature on performance

Maximum temperatures for use of FRPs are governed by two main factors: the resin's glass transition temperature (T_g) and the temperature at which chemical decomposition starts to become significant. Decomposition temperatures are seldom actually reached in service life. FRPs are pre-eminently load-bearing materials, and it is their temperature-dependent mechanical properties, such as T_g, or the closely related heat distortion temperature, that usually determine the maximum use temperature. Strength, yield stress and modulus all decline with increasing temperature, reflecting the increasing mobility of the molecular structure. Unacceptable levels of loss of physical properties will often occur well before the onset of thermal or thermo-oxidative degradation. Most resins have only a limited ability to withstand high temperatures; however, high-temperature resins are available and these have superior heat resistance.[14,15] The fibres themselves are generally thermally stable materials that give no anxiety to users of FRPs and can withstand a higher temperature than any of the current generation of commercial resins.

When considering FRPs for use in extreme temperatures it is necessary to consider the following factors:

- whether the heating is continuous or intermittent;
- the maximum and minimum temperatures in a working cycle;
- the heating and cooling rates;
- other factors such as mechanical stress, or fluids;
- the required lifetime, which may be a matter of decades.

Engineers and designers must be careful to ensure that the thermal properties of a particular material are completely understood before recommending its use for a specific application. Again, correct material selection will ensure the FRP performs as required.

15.5.12 Mechanical stress

Correct material selection and design will ensure that any FRP component or structure has the mechanical properties to meet performance requirements.

Fibre, matrix and interface roles

FRPs containing continuous fibres rely on the load being carried almost entirely by the fibres. The direct contributions of the matrix to the tensile or flexural strength and modulus of the material are trivial in comparison. Most reinforcing fibres have excellent durability towards stress in a wide range of conditions. Mechanical durability is maintained through choice of appropriate resin – this must continue to facilitate load transfer between fibres and must protect individual fibres from mechanical abrasion, as well as penetrating fluids. Surface weathering of the matrix could expose fibres to mechanical damage and a soft matrix is easily eroded or scratched. Creep rate can be controlled through the use of 0° plies and utilisation of long-fibre composites. For other fibre types, the choice of resin is critical to control this property.

Minor impact damage

A common hazard for FRPs is minor impact damage resulting from scratching or collision with small objects. The resulting damage is often difficult to see with the naked eye, but it can include delamination, matrix cracking, fibre debonding and, in severe cases, fibre fracture. Most impacts occur in practice at an oblique angle, which tends to reduce the severity of normal incidence, regardless of whether damage is measured by the damage area, indentation depth or residual strength.[16] The fact that there is scope for on-site repair of impact damage in FRPs, even in remote areas, is an important favourable consideration in their durability.

Fatigue

Fatigue 'life' is usually measured as the number of cycles to failure for a given applied load. The degradation and failure of bridges, highways and service piping is nearly always associated with cyclic and dynamic loading. The loading may be mechanical (due to vehicle traffic, for example), thermal (due to changes in temperature) or chemical (from seasonal road treatments, oxidation, water, etc.).

In FRP components with aligned or randomly distributed short fibres, cracks can initiate at flaws – such as pores or in resin-rich areas with local strain inhomogeneities caused by improper fibre alignment – or at fibre ends. The local load transfer from the fibre into the matrix can lead to an overstressing of the matrix or a fibre–matrix debonding and then crack propagation may occur.[17]

In continuous-fibre FRPs, the fatigue process is characterised by the initiation and multiplication, rather than propagation, of cracks. Crack initiation occurs early in fatigue life, and coincides with the first ply failure in the laminates, that is the first cracking of the weakest ply. While in metals crack growth accelerates during fatigue, crack multiplication in FRPs decelerates, resulting in uncontrolled final rupture of the FRP.

Reasonable data have been produced for the fatigue of glass/vinylester, polyester and epoxy FRPs produced from low-cost fabrication methods, allowing the longevity of these systems to be predicted with some degree of confidence. Research continues on a global basis to understand the fatigue behaviour of FRPs in order to enable the prediction of lifetimes of structures that are designed for extended service conditions.

15.6 Joints

Joints are necessary in large FRP structures because of production and design considerations. These components are too large to be fabricated in one piece, so several parts have to be joined and stiffeners are necessary. Joints are potential failure sites.[18] This applies whether they are adhesively bonded or mechanically fastened, and whether they join two FRP sections, or one FRP component and one constructed from another material. In the construction industry, joint failure in FRPs is likely to mean leakage of water into the building, rather than structural collapse. It is thus recommended that joints should be located well away from supplies of water. On roofs, they should be on ridges and not in gutters. Adequate seal pressure is necessary. Joints should be readily accessible for inspection and replacement.

15.6.1 Adhesively bonded joints

Adhesive joints in FRP structures are capable of achieving higher strength than mechanical ones and may be preferred for that reason.[19] Their durability depends more on the flexibility and toughness of the resin used in the adhesive than on its strength. Prediction of joint strength can be carried out by performing a stress–strain analysis and applying an appropriate failure criterion. Stresses in the adhesive bonds can be predicted using finite element analysis and closed form or continuum mechanics.

15.6.2 Mechanically fastened joints

Mechanically fastened joints have the advantage that they can be disconnected if desired. Bolts offer the greatest mechanical strength obtainable without adhesives, especially when the bolt is a good fit to the hole. Metal bolts must be protected against corrosion and the use of special materials such as stainless steel can be cost-effective. The edges of drilled holes need to be coated if the joint is exposed to liquids that attack the fibres.

15.7 Repair of degraded fibre-reinforced polymer composite structures

Mechanical damage, during erection or in service, can be patched[8] and an attempt made to match the original, but site conditions are not usually favourable for proper curing of the resin and a patch is likely to show after a period of weathering. The quality of the repair will also be influenced by the level of surface preparation, the correct use of repair products and the final finishing of the repair. Many of these aspects depend on the skill of the individual conducting the repair and can affect the way in which the repaired part weathers. It is better to replace a complete panel or member if this is practicable.

Restoration of degraded surfaces can be difficult. In the extreme case of the glass fibres becoming badly exposed, they must be scrubbed off completely before any new surface treatment is applied. When coating FRP, as with the repainting of all exterior construction substrates, attention should be paid to surface preparation, the manufacturer's instructions for the specific paints and the supervision of trained labour. Surface preparation, prevailing weather conditions during application and the skill of the painter will all have an influence on the service life of the coating, regardless of paint type.

15.8 Summary

FRPs offer the ability to tailor-make components with the properties needed to meet performance requirements of a particular situation. Correct material selection and design means that FRPs can perform in the most demanding of service environments. FRPs offer good durability, their performance enhanced by incorporation of additives and correct maintenance procedures. FRP components have demonstrated service lives of over 35 years to date in construction applications in a variety of different environmental conditions. Advances in resin and additive technology mean that design lives of 60–100 years are possible.

15.9 Sources of further information and advice

- Network Group for Composites in Construction (NGCC) (http://www.ngcc.org.uk). Technical sheets on durability and case study information.
- National Composites Network (NCN) (http://www.ncn-uk.co.uk). Best practice guides – repair of FRP structures.
- European Pultrusion Technology Association (EPTA) (http://www.pultruders.com).
- Composites Fabricators Association (CFA) (http://www.cfa-hq.org).
- Infrastructure Applications of Composites Materials (http://iti.acns.nwu.edu/projects/comp.html).
- Analysis, testing and fabrication of structural composites (http://acatc.ame.arizona.edu).
- *Composites Materials Handbook* (http://mil-17.udel.edu/links.htm).
- International Research on Advanced Composites in Construction (IRACC) (http://www.iper.net/co-force/iracc.htm).

15.10 References

1 HALLIWELL S M. (2000) BRE Report 405: Composites in Construction.

2 LIAO K, SCHULTHEISZ, C R, HUNSTON D L and BRINSON L C. (1998) 'Long-term durability of fibre-reinforced polymer-matrix composite materials for infrastructure applications', *Journal of Advanced Materials*, **36**, 3–40.

3 MEIER U, DEURING M, MEIER H and SCHWEGLER C. (1992) 'Strengthening structures with CFRP laminates: research and applications in Switzerland', in *Advanced Composite Materials in Bridges and Structures*, K W Neals and P Labossiere (eds), (Montreal, Quebec, Canadian Society for Civil Engineering), pp. 243–251.

4 DAWSON D K. (1996) 'Architectural composites are now primary structures', *Composites Technology*, Sept/Oct, 48–50.

5 CHIU A and FRANCO R J. (1990) 'FRP pipeline for oil and gas production', *Modern Plastics*, June, 21–27.

6 COMPOSITES TECHNOLOGY (1996) News article, Sept/Oct, 8.

7 NGCC TECHNICAL SHEET 06/01 (2006), Durability of Fibre Reinforced Polymers in Construction, www.ngcc.org.uk.

8 NCN BEST PRACTICE GUIDE (2007) Repair and Maintenance of FRP Structures.

9 PRITCHARD G. (ed.). (1999) *Reinforced Plastics Durability* (Cambridge, UK, Woodhead Publishing).

10 HALLIWELL S M. (1999) 'Reinforced plastics cladding panels', in Proceedings of Composites and Plastics in Construction, BRE, Watford, UK, Paper 20.

11 NGCC TECHNICAL SHEET 06/02 (2006) Predicting Durability of FRP, www.ngcc.org.uk.

12 KASTURIARACHCHI K A. (1975) Hygrothermal degradation of fibre reinforced epoxide resins under stress, PhD thesis, Kingston Polytechnic, UK.

13 KOSURI R and WEITSMAN Y. (1995) 'Sorption process and immersed-fatigue response on graphite/epoxy composites in seawater', in *Proceedings of ICCM-10*, Whistler, BC, Canada (Cambridge, UK, Woodhead Publishing), Vol. 6, pp. 117–184.
14 PARKER J A, KOURTIDES D A and FOHLEN G M. (1987) 'Bismaleimides and related maleimideo polymers as matrix resins for high temperature environments', in *High Performance Polymer Matrix Composites*, T T Serafini (ed.), (Park Ridge, NJ, Noyes Data Corporation), pp. 54–75.
15 LEACH D C. (1989) 'Continuous fibre reinforced thermoplastic matrix composites', in *Advanced Composites*, I K Partridge (ed.), (Barking, UK, Elsevier Applied Science), Chapter 2.
16 MADJIDI S, ARNOLD W S and MARSHALL I M. (1996) 'Damage tolerance of CSM laminates subject to low velocity oblique impacts', *Composite Structures*, **34**, 101–116.
17 FRIEDRICH K, SCHULTE K, HORSTENKAMP, G and CHOU T-W. (1985) 'Fatigue behaviour of aligned short carbon-fibre reinforced PI and PES composites', *Journal of Materials Science*, **20**, 3353–3364.
18 LEGGATT A. (1984) *GRP and Buildings, a Design Guide for Architects and Engineers* (London, Butterworth).
19 LEE R J and MCCARTHY J C. (1989) 'Design of bonded structures', in *Advanced Composites*, I K Partridge (ed.), (Barking, UK, Elsevier Applied Science), Chapter 8.

16

Ageing of composite insulators

S. M. GUBANSKI, Chalmers University of Technology, Sweden

16.1 High-voltage insulators

Changes introduced by ageing that cause decay of material properties have been challenging designers of electrical devices since the very beginning of this engineering discipline. All materials, including components of electric insulation, have therefore always been selected with great care and some of the solutions adopted in the past appeared to be excellent choices. Records of lifetime lengths of 50+ years have not been unique. This also refers to insulators for applications in electric power transmission and distribution, for which the selection and acceptance rely on the assurance of well-specified and long-term stability properties. Insulators for applications in outdoor environments appeared first with the introduction of telegraph lines in the middle of the nineteenth century. The materials used for their manufacture were at that time porcelain and glass. The same materials were thereafter adopted to insulate electric power lines and substations. Consequently, the ceramic insulators have completely dominated the market for over 100 years.

Outdoor high-voltage insulators fulfil two main basic functions. They mechanically hold parts of electric installations exposed to high electric potentials as well as insulating them electrically from the ground potential, which is also the case of line insulators supporting energised conductors in electric power lines. Insulating shells, called hollow-core insulators, are also used for encapsulating different types of high-voltage apparatus to protect them from influences imposed by outdoor environmental factors. The most severe failure of an insulator carrying a power line happens when it breaks mechanically causing the line to drop. When falling, the line not only becomes disconnected, but the freely hanging energised conductors put people's lives at risk. The second unwanted situation is when insulators fail electrically, either due to an electric flashover along their surface or due to damage, both causing short- or long-lasting disconnections in the network. Consequently, research on insulator quality and performance has become an important field within high-voltage engineering.

Earlier investigations concentrated mainly on improving the electrical and mechanical properties of porcelain and glass used for manufacturing insulators. Attempts were also made to optimise their geometrical design. The main goal in this long development concentrated on optimising insulator performance under diverse outdoor conditions. The environmental factors that affect this performance include weather conditions (temperature, humidity, wind, solar radiation, rain and snow precipitation, etc.) and the presence of airborne pollutants at installation sites; the latter originating from marine, industrial, agricultural or desert sources and leading to build-up of a highly conductive layer on insulator surfaces. In order to combat the pollution problem, insulator geometry was often developed to either enlarge the length along the surface, so-called 'creepage', or increase self-cleaning ability under wind and rain. In other approaches, combating the discharge (arcing) activity on insulator surfaces was researched. In parallel, many efforts concentrated on improving material quality and the effectiveness of manufacturing technologies. This has led to a state in which the traditional ceramic insulators have been considered as very reliable.[1] In the meantime a need arose for new developments, which led to the application of polymeric materials in high-voltage insulator design, and a new family of insulators – called 'composite insulators' – were introduced. Composite insulators allow a reduction in insulator weight and an improvement in their performance under polluted conditions, especially when materials with long-lasting, water-repellent (hydrophobic) surface properties were used in their manufacture. Additional features making composite insulators a desirable alternative were: being an unattractive target for vandalism; mechanical flexibility providing good seismic performance; explosion safety in applications such as apparatus insulators.

With the new materials came new production techniques and new designs. In contrast to the solid ceramic counterparts, a typical modern composite insulator consists of a glass fibre reinforced (GFR), resin-bonded core (rod or pipe) onto which two metal end-fittings are attached. This is the mechanical supporting structure and its modulus-to-weight ratio is extremely high.[2] The resistance of such a core to environmental stresses is, however, not good and the material degrades rapidly when exposed to moisture and pollution in combination with high electric field strength. In order to protect the core from these environmental stresses, it is covered with a polymeric protective layer called a housing, which has no mechanical function. In addition to the protection from weather-related factors and pollution, the housing also provides the extra creepage length needed to obtain the desired pollution resistance level, which can either be achieved through varying the shed diameter and/or the number of sheds.

A cross-section of a line composite insulator is depicted in Fig. 16.1. In hollow-core insulators, the insulator core is made of a GFR pipe which

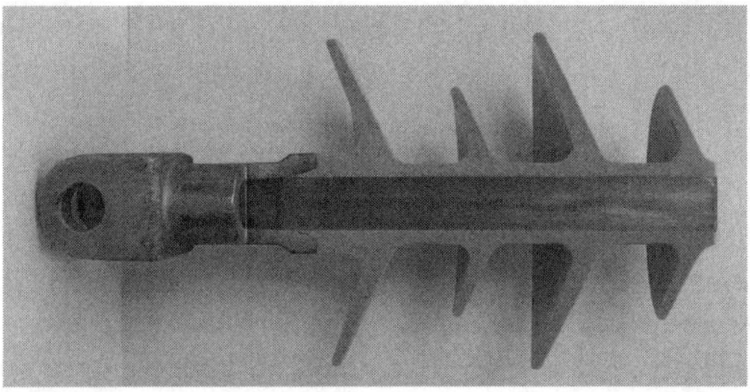

16.1 Cross-section of a modern composite insulator showing the three main components of its construction: GFR core, external housing and end-fitting.

allows different apparatus to be placed inside the insulator – such as bushings, circuit breakers, surge arresters, and current and voltage transformers. Special demands are made on the pipe design, since it may be stressed mechanically by high pressure from the inside. In cases where faults lead to explosions, the use of hollow-core insulators considerably lowers the risk of damage to people and property.[3]

16.2 Materials and manufacturing techniques

Epoxy composites were applied first as a housing material in the mid 1950s, but they quickly exhibited a relatively weak resistance against ageing under external environmental stresses and consequently their use today is limited almost entirely to indoor applications. Later on, different types of elastomeric materials replaced epoxies. In particular, those based on polydimethylsiloxanes (PDMS), commonly known as silicone rubber (SIR), have become attractive. Other materials used have been ethylene–propylene monomer (EPR)- or ethylene–propylene–diene monomer (EPDM)-based rubber blends, ethylene–vinyl-acetate (EVA) and so-called 'alloy rubbers', the latter being blends of EPDM and silicone resins.

Housings made of SIR provide high resistance to environmental stresses and long-term hydrophobic properties on the insulator surfaces. Three types of SIR are used in high-voltage insulation applications: high-temperature vulcanising (HTV) silicone rubber, room-temperature vulcanising (RTV) silicone rubber and liquid silicone rubber (LSR).[4] HTV is cured at high temperature and pressure and is catalysed by peroxide-induced free radicals or by hydrosilylation catalysed by a noble metal, e.g. platinum. RTV is cured at lower temperatures, i.e. around room temperature, by a condensation reaction as a one-component system or by hydrosilylation as a two-component system. The one-component system is cured by diffusion of moisture from the surrounding air into the material and nowadays is rarely used for the production of insulators. LSR is always a two-component system, and is vulcanised by hydrosilylation catalysed by platinum; LSR is cured at elevated temperatures. Fillers are added to the rubbers to control different properties of the material – such as mechanical stability and resistance to tracking – as well as to reduce the cost. The addition of fumed silica is necessary to achieve good mechanical properties during processing, and alumina trihydrate (ATH) is added as a flame-retardant.[5] Adding ATH also has the positive effect of improving the dielectric strength and tracking resistance.

Different techniques have been used over recent years to manufacture composite insulators. One method is to manufacture the sheds separately and to then push them onto the core.[1,2] This technique has been abandoned since insulators manufactured in this way created many problems. The weak spots were all the interfaces between the sheds where moisture could penetrate into the insulator causing internal tracking. A better technique has therefore been to first cover the core with a sheath, adding the sheds onto it and vulcanising the parts together afterwards. This has allowed a reduction in the number of interfaces where moisture can penetrate to the GFR core. However, today the most commonly used technique is one-shot moulding,[2] where the whole insulator housing is injection moulded in one piece directly around a pre-assembled core. In this way the housing can be chemically bonded to the core and the number of interfaces allowing moisture to penetrate inside is minimised. In addition, the technique is attractive to manufacturers because the number and time requirement of processing steps involved is minimised. There is, however, a problem connected with the use of two-part moulds – they give rise to so-called 'mould lines', which appear parallel to the electric field and may be weak spots with respect to tracking and erosion of the housing.[6] Insulator housings can also be made by means of spiral winding techniques. This is a very flexible technique, enabling large variations in insulator diameters and lengths without involving costly manufacturing tools.

The most common metallic materials for end-fittings of composite insulators are cast, forged or machined aluminium and forged iron or steel.[2,7] The end-fittings can be attached to the core by different methods. Today, the most commonly used technique is swaging (crimping) and gluing,[8] but the use of wedges also occurs in some applications. Swaging is the strongest type of attachment per unit area of the core.

16.3 Practical experiences with composite insulators

Experiences with first-generation composite insulators exposed to natural service conditions and in test stations allowed the identification of a number of problems that had to be solved before general market acceptance could be achieved. The initially limited range of applications led to a situation when the choice of material formulations and the geometrical designs were most frequently dictated by technological and manufacturing possibilities rather than by the optimum performance; the latter meaning good resistance to erosion and tracking, the ability to recover hydrophobic properties of the housing, high mechanical strength and protection of the GFR core against so-called 'brittle fracture', water sealing of the fittings, good adhesion at the interfaces, self-washable geometry of the housing and geometrical shape allowing the minimisation of the electric field distribution at the insulator surface, etc.

In parallel, different diagnostic techniques have been developed in order to identify faulty insulators or undesired stresses to which insulators can be exposed in service. The most common include visual inspection, infrared thermography, directional acoustic emission, ultraviolet image intensification and electric field measurements, and they are primarily used for inspecting insulators from towers, the ground or from helicopters. However, none of these methods alone is able to detect all types of defects. It is therefore recommended that a combination of at least two different techniques is used.

For the earlier generation of composite insulators, the typically observed defects were tracking or erosion damages caused by surface discharge activity on the insulator housing surfaces and at interfaces between housing and core. A tracked path usually contains carbonised by-products of polymer decomposition and is characterised by increased electrical conductivity, which, if left long enough, may lead to a flashover. On the other hand, erosion damages, though not as highly conductive as the tracking paths, may become severe enough to puncture sheds or to expose the core. Both types of damage are material specific. This is especially the case for mould lines along the insulator, which may be a preferential erosion site. Long-term exposure to ultraviolet (UV) radiation also yielded erosion changes in some types of the earlier generation of housing surfaces, which appeared as chalking, crazing and cracking, as illustrated in Fig. 16.2. Much research work has been dedicated to optimise housing material formulations[9] and such erosion changes are rare in the insulators produced today. These changes do not appear on SIR housing materials and have been greatly reduced on EPDM and rubber alloy formulations by adding UV stabilisers.[10]

Exposure to prolonged electric discharges in the form of dry band arcing or corona causes SIR housing materials to lose hydrophobic properties and become temporarily wettable,[11,12] as illustrated in Fig. 16.3. Fortunately, this property can recover when the materials are left to rest.[13,14] Migration of low molecular weight (LMW) polymer chains from the bulk to the surface is considered to be the main mechanism behind this recovery process.[5,14,15]

(a)

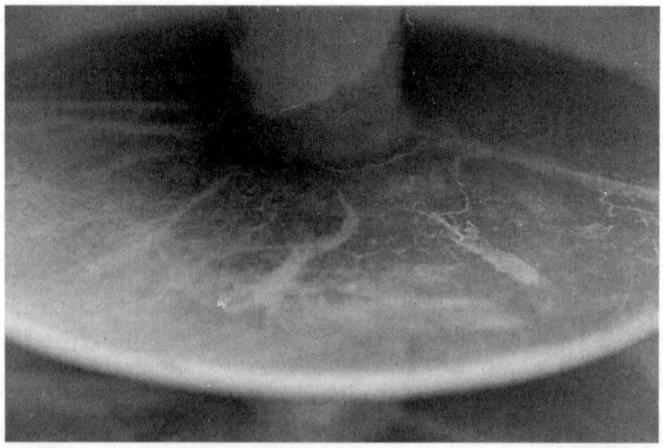

(b)

16.2 Examples of erosive ageing of composite insulator sheds due to (a) leakage current activity or (b) UV irradiation.

Moreover, long-term contact with water has been shown to reduce hydrophobic properties of SIR, and the effect is attributed to reorientation of methyl groups.[5,15,16] On the other hand, solar irradiation and increased temperature have been reported to enhance hydrophobicity and its recovery rate of polluted SIR in tropical climates.[17] In temperate climatic conditions, SIR insulators were reported to become less hydrophobic during autumn and winter, than during summer.[6] Areas of sheds protected from sunlight showed reduced hydrophobic properties as compared with the exposed parts.

Slipping of the end-fitting was a problem in the first generation of composite insulators,[7] but today it is one of the least common types of failure.[18–20] Swaging, which is the most common method of attaching end-fittings to a solid GFR core, works fine as long as the tools used for the

(a) (b)

16.3 Hydrophobic properties of insulator sheds (a) may be lost (b) and subsequently recovered.

joining are in good condition and the tolerance between the diameters of the core and the hardware remains within set limits – worn tools may give rise to cracks in the core and too large a difference in the diameters may yield hardware cracking.[2] Water ingress into damaged cores was reported to cause punctures and breakdowns.[21] Dissections revealed the existence of carbonised tracks along the rod/housing interface or inside the fibre rod.

Despite some ongoing concerns, a general acceptance of composite insulators has been achieved over recent years. A broad range of insulator types is available, including line insulators and hollow-core insulators for voltage levels ranging from distribution voltages of a few kV to the highest transmission voltages at 800–1000 kV[22] and their share of the market continues to grow. Polymeric housings have also found applications in cable terminations and joints. Records on good performance are available for composite insulators installed 30 years ago. Most of the composite insulator failures seen so far have mainly been due to bad insulator design, to selection of inferior materials or to poor quality control. It seems that most manufacturers are over this stage and produce far better products, and the problems observed in service conditions today are related to: (a) human errors during storage, handling, installation, (b) rodent and biological attacks as well as (c) inferior quality control.[2] If this is true, future failures will mainly depend on ageing of the materials.

16.4 Ageing of insulator housing

Ageing of polymeric insulators is to a large extent controlled by the climatic conditions prevailing at the site of installation, but insulator design and the presence of pollution are also very important factors. Different steps in the development of ageing and flashover processes on SIR composite insulators have been identified[3,23,24] as summarised below. A new composite insulator made of SIR performs well due to its excellent hydrophobic properties, but with time it collects hydrophilic pollution, which later becomes encapsulated in the LMW polymer, making the pollution layer hydrophobic. Water drops localised on the surface give rise to corona in high electric field regions, causing local loss of hydrophobicity. Here some of the pollution is dissolved into the water and forms a thin conductive layer, which changes the electric field distribution and increases the field strength in new regions, initiating new corona discharges. This process continues, and a conductive wet path between the insulator ends develops. The current in this layer causes it to dry in regions where the current density and/or the layer resistivity is high. At these dry areas, dry-band arcing can occur, which causes further loss of hydrophobicity, erosion of the surface and increase in leakage current. When the wet period ends, the insulator will dry and start to recover its hydrophobicity and if it is allowed to recover completely before the next wet period comes, the process will start from the beginning. If not, the process will resume. Depending on the insulator design and choice of material, this process may go on for many cycles and eventually result in a flashover along the insulator surface. In cases of bad design or inferior quality of the housing material, severe erosion of both sheath and sheds, leading in time to electrical or mechanical failure, may also be expected.

In order to better understand the mechanisms of hydrophobic recovery and material ageing in composite insulators, it was necessary to concentrate the investigations on well-defined materials and, in addition to studies in natural conditions, to develop acceptable accelerated treatment methods that imitate the natural exposure.[25] Exposure of energised insulators or material samples in fog chambers (salt fog and clean fog) (IEC standards 60507 and 61109) has been used most commonly. Other fog chamber procedures include a combination of different climatic stresses, including rain, UV radiation, and deposition of dry salt and dust layers. Among other test procedures, the rotating wheel dip test (IEC standard 61302) and the inclined plane test (IEC standard 60587) are available.

Nowadays, it is generally accepted that the main mechanism for hydrophobic recovery is the migration of mobile LMW siloxanes from the bulk of the rubber to the surface, which encapsulates pollution layers or covers hydrophilic (wettable) regions of the surface. However, it has been important to understand which of the stresses appearing under natural conditions have the strongest influence on the dynamics and long-term behaviour of the hydrophobic recovery and ageing. It has also been important to find how other physical properties of SIR – for example, cross-linking density, amount of fillers and content of stabilising additives – influence the ability to withstand the stresses. A diversified approach to the research was necessary to answer these questions; starting with studies on the molecular structure of the silicones and proceeding all the way through to the development of proper procedures for evaluating final products.

16.4.1 Exposure to corona discharges

Experiments performed on real insulators, in which corona from insulator hardware and eventually from water droplets on insulator surfaces appeared, showed that corona discharge is one of the main reasons for material ageing.[24,26,27] The first and most important observation was that corona reduced the hydrophobic properties of SIR. Discoloration, erosion marks and shed cutting have also been observed.[28] The damages were attributed to the influence of ions and gases from the discharge, ozone and nitrogen oxides, and possibly UV radiation.[29,30] One way to protect insulator surfaces is to minimise the risk of corona discharges by controlling the field strength along composite insulators, which it is nowadays believed should not exceed 4–6 kV/cm. This imposes special demands when designing insulator hardware, especially at voltage levels above 100 kV.

The effects of exposing pure and antioxidant-stabilised PDMS networks to corona discharges under well-controlled laboratory conditions were extensively studied in references 31 and 32. In Fig. 16.4 the appearance of corona in service as well as in electrode arrangements used for laboratory treatments is illustrated. Directly after a prolonged exposure to corona, PDMS surfaces become completely wettable. Receding contact angles[33] are lower than 10°, but the hydrophobic property gradually recovers with time after the exposure. Analyses of the treated surfaces by X-ray photoelectron spectroscopy (XPS) show that the corona causes a gradual oxidation of the surface region. The amount of oxygen, seen on the surface, increases and the amount of carbon decreases with increasing exposure time, whereas the silicon content remains essentially constant. A similar type of degradation was observed on insulators exposed to corona in field conditions.[34] The oxidation process results in the loss of hydrophobic properties of the surface. The oxygen becomes incorporated in the network by forming additional bridges between silicon atoms; thereby gradually forming a silica-like structure (SiO_x), which is a mixture of oxidised and virgin silicone species. A full conversion into pure non-organic silica (SiO_2) has never been observed, even after 200 h of continuous corona treatment.[35] The depth of the oxidised region can be estimated by means of neutron reflectivity

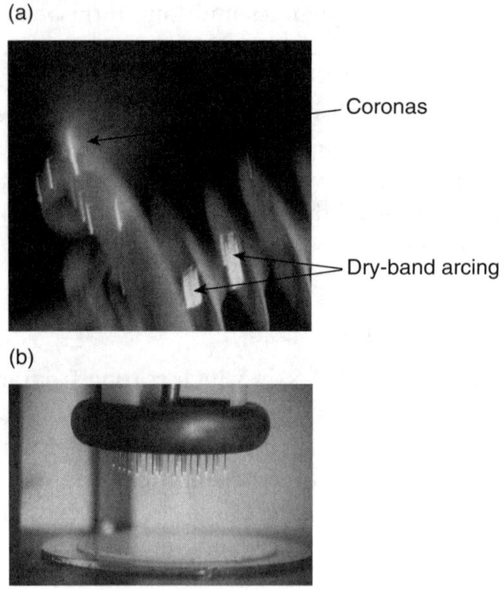

16.4 (a) Corona from insulator hardware and dry-band arcing on the insulator sheath of a contaminated composite insulator photographed at night under humid conditions. (b) Exposure of SIR samples to corona discharges during laboratory testing.

measurements, and is of the order of 150 nm.[36] Within this oxidised region, the silica-like layer gradually develops with increasing exposure time. Moreover, the degree of oxidation increases with the degree of cross-linking in the PDMS network, meaning that the cross-link points (consisting of ethylene bridges) appear to be more sensitive towards oxidation compared with the polymer backbone.

A typical indication of the transformation of the initially formed oxidised layer into a silica-like structure is the appearance of surface cracks upon elongation of the exposed materials. Such cracks can be observed using optical microscopy or atomic force microscopy. The surface cracking also occurs spontaneously at high enough doses of corona discharges. The depth of the cracks is initially of the same size range as the oxidised surface layer, but increases with increasing corona exposure time to several micrometres, meaning that the cracks probably penetrate into the unoxidised bulk. The cracks and the process of their propagation are schematically illustrated in Fig. 16.5. By adding stabilising agents (antioxidants) to SIR formulations the build-up of the silica-like layer can be impeded.[32] However, the process cannot be stopped completely.

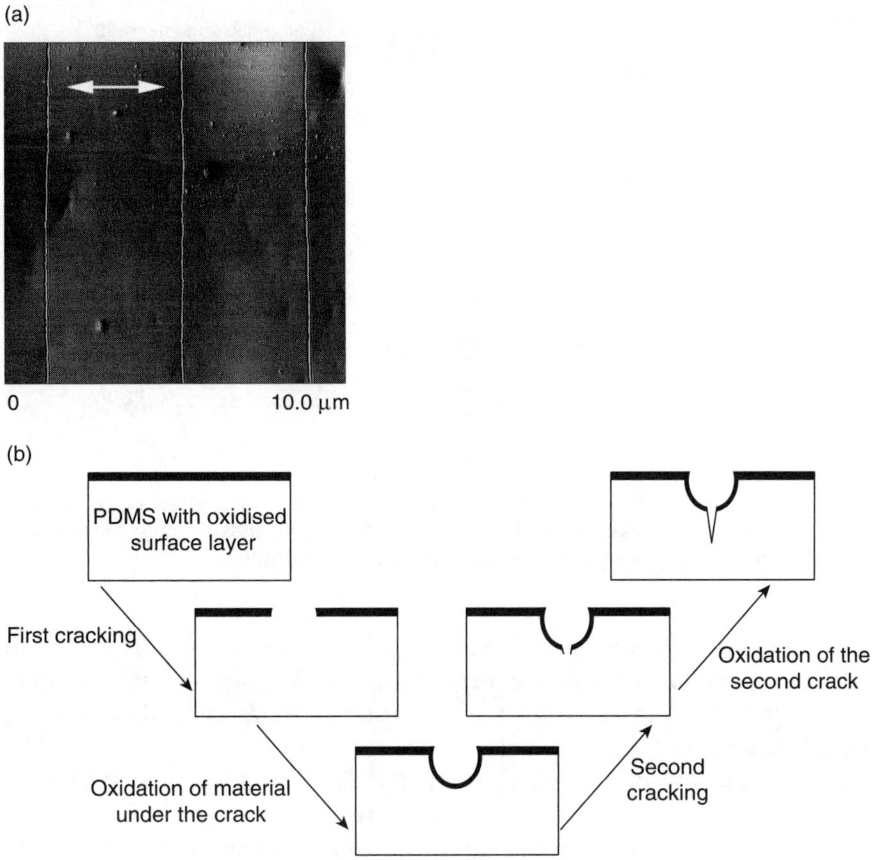

16.5 (a) Atomic force micrograph of cracks in the silica-like surface layer formed after 1 h of corona treatment – arrow indicates the direction of applied strain. (b) Schematic representation of crack propagation under prolonged action of corona discharges.

The investigations also revealed that the mechanical properties of the silica-like surface layers formed influence the hydrophobic recovery rates.[31] Undeformed material samples without cracks recover hydrophobicity at a low rate over several thousand hours, whereas stretched samples containing cracks recover hydrophobicity within tens of hours, as illustrated schematically in Fig. 16.6. The migrating low-molar siloxanes, responsible for the hydrophobic recovery and identified by gas chromatography–mass spectrometry (GC-MS), contain a homologous series of cyclic siloxane molecules ($D_n = [(CH_3)_2Si—O]_n$) with the same chemical composition as PDMS. They mainly consist of five to eight repeating units (D_5–D_8). This composition seems to be independent of the network structure and can be found in

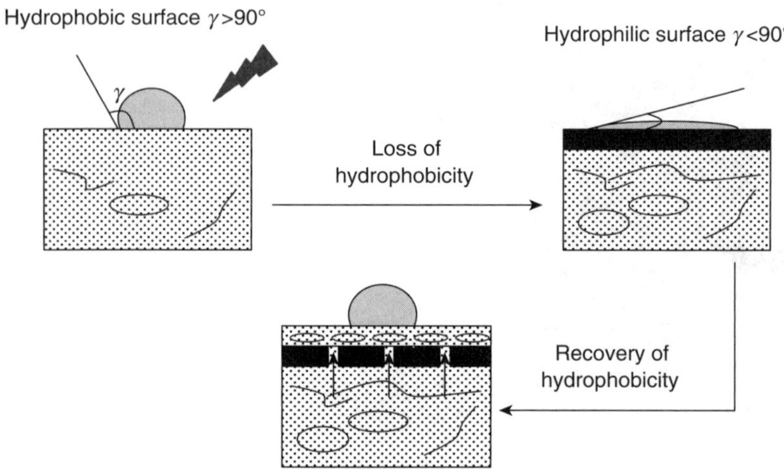

16.6 Mechanism of hydrophobic recovery of PDMS after exposure to corona – cracked silica-like (SiO_x) surface layer allows easy migration of cyclic siloxanes to the surface. γ, contact angle.

both exposed and unexposed networks. Moreover, during exposure to corona, additional amounts are formed. This strongly suggests that a depletion of the LWM reservoir of the migrating siloxanes with time is not a likely scenario.

From the results of the investigations, it is clear that formation of a brittle silica-like surface layer and its subsequent cracking occurs on SIR insulators exposed to corona in service. It results in increasing surface roughness of insulator sheds and sheaths. The formation of this layer should therefore be avoided since the increasing surface roughness makes trapping of moisture and pollution less difficult, leading to an overall decreased insulator performance, even though the surface can recover hydrophobicity well. The other unwanted phenomenon is that the surface cracking may initiate material cuts, which allow water to penetrate into the insulator core.

If corona exposure is indeed the main factor behind the onset of ageing and considering the fact that corona activity can be suppressed by reducing electric field by proper design of insulator hardware, it should then become quite possible to use insulators with lower creepages than are commonly found today. Once the corona is suppressed, the hydrophobic property should be more stable, being lost only during periods of very rapid pollution deposition. This also implies that, for the corona scenario, material formulations could ideally be optimised for oxidative stability as well as for high rates of hydrophobic recovery and, at the same time, less thermal protection would be needed. One should therefore pay attention to selecting material formulations in which the degree of cross-linking is controlled well.

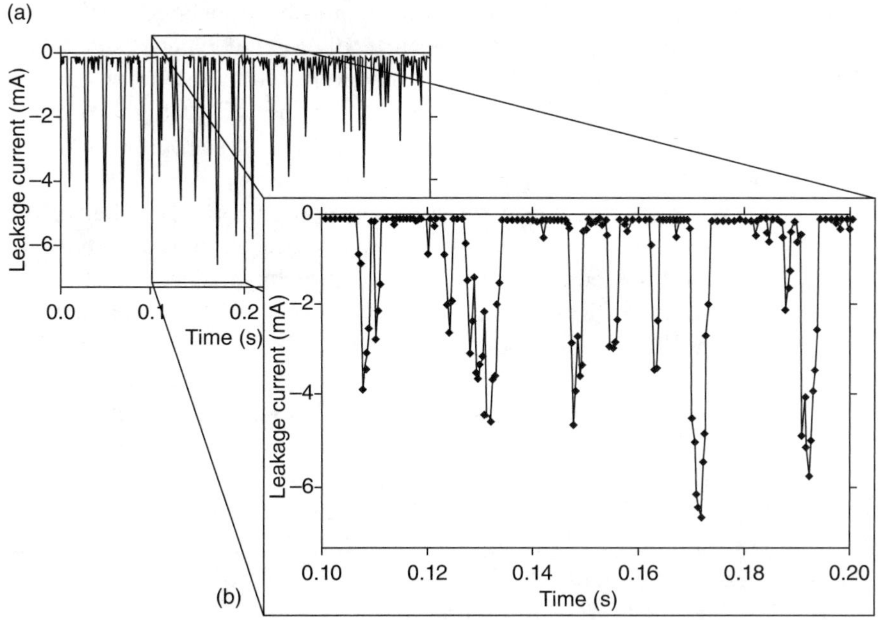

16.7 Record of a typical discharge current pattern recorded on a DC-energised sample (a) and magnification of selected time window (b). The appearance of such leakage current peaks is associated with the presence of arcing discharges on the surfaces of insulating materials energised under contaminated conditions.

16.4.2 Exposure to dry-band discharges

The reduction of hydrophobicity of SIR due to arcing (dry-band discharges or flashover arcs) and surface leakage currents seems to be less severe than that due to corona;[24,27] this is attributed to the fact that the energy spectrum in arcs is very different from that of corona. Experimental evidence was required to demonstrate that the mechanisms of surface ageing are different in the absence of corona. Numerous investigations have been dedicated to elucidating this problem (as exemplified in references 9 and 37–43) through assessing the ageing behaviour of material samples and insulators in both field and laboratory conditions.

In references 39 and 40, model samples and insulators, made of well-defined material compositions, were investigated under exposure to a marine (salt-contaminated) environment. The electrical performance was quantified by analysing the behaviour of leakage current peaks measured during the exposure, as shown in Fig. 16.7. The current peaks appearing could, at their lower level of about 1 mA, be well correlated with the condition of the housing material surface – such as loss and recovery of hydrophobicity as well as its weak degradation. On the other hand, higher current peaks, those exceeding 4 mA in amplitude, could be associated with arcing activity and local erosion spots. The type of applied electric stress, i.e. alternating (AC) or direct (DC) current voltages, was also important. A higher severity of degradation was found under the DC voltage exposure, as it resulted mainly from the extra pollution build-up brought about by electrostatic attraction.

An intrinsic effect resulting from a higher content of ATH filler in the housing material is an increase in surface electric discharge activity. In contrast, formulations containing less ATH filler have a prolonged silent (early ageing) period. However, once this early ageing state is passed, the pace of degradation becomes much faster than that for the highly filled materials. In other words, erosion develops faster, once initiated, on the low-filled material and electrical activity is strongly enhanced by this erosion. Infrared spectroscopy studies revealed that consumption of ATH filler takes place under the heat produced by surface discharges. For surfaces with higher leakage current activity, meaning higher discharge activity, the consumption of the ATH is higher. For low-filled materials, characterised by the high pace of degradation, an almost complete loss of the ATH on their surfaces can be found.[40]

Different chemical and structural analyses performed during the reported investigations confirmed an earlier formulated hypothesis[44,45] that the LMW fraction, responsible for the hydrophobic recovery, can actually be continuously regenerated through depolymerisation of longer silicone molecules during the life of the insulator. Since SIR is a very stable material, certain conditions must be fulfilled for such depolymerisation to occur. The heat released during the discharge activity, while having a detrimental impact on the top layer of an SIR surface, can at the same time increase the potential for hydrophobic recovery, thus stabilising this critical property of the material. Under hot conditions, at the SIR surface the smallest LMW molecules are generated, as revealed by GC-MS, among which three-unit cyclic oligomers (D_3) dominate. This process is well illustrated in Fig. 16.8, showing a comparison between percentage distributions of siloxanes in new as well as corona- and arcing-aged samples. The siloxane distribution of the arcing-aged samples differs significantly from new sample extractions, the latter consisting mainly of D_6 and larger cyclic oligomers. In addition, a comparison is also made between contaminated and clean, aged samples. It can be observed that the contaminated aged areas generally contain higher amounts of D_3 and D_4 siloxane cyclics. In contrast, under the action of corona, larger oligomer rings dominate.

Regarding the surface ageing under arcing, practically no signs of oxidation are found as compared with the action of corona, which can be observed in Fig. 16.9. The absence of the oxidative process indicates that surface discharges should be considered here as the principal initiator of ageing.

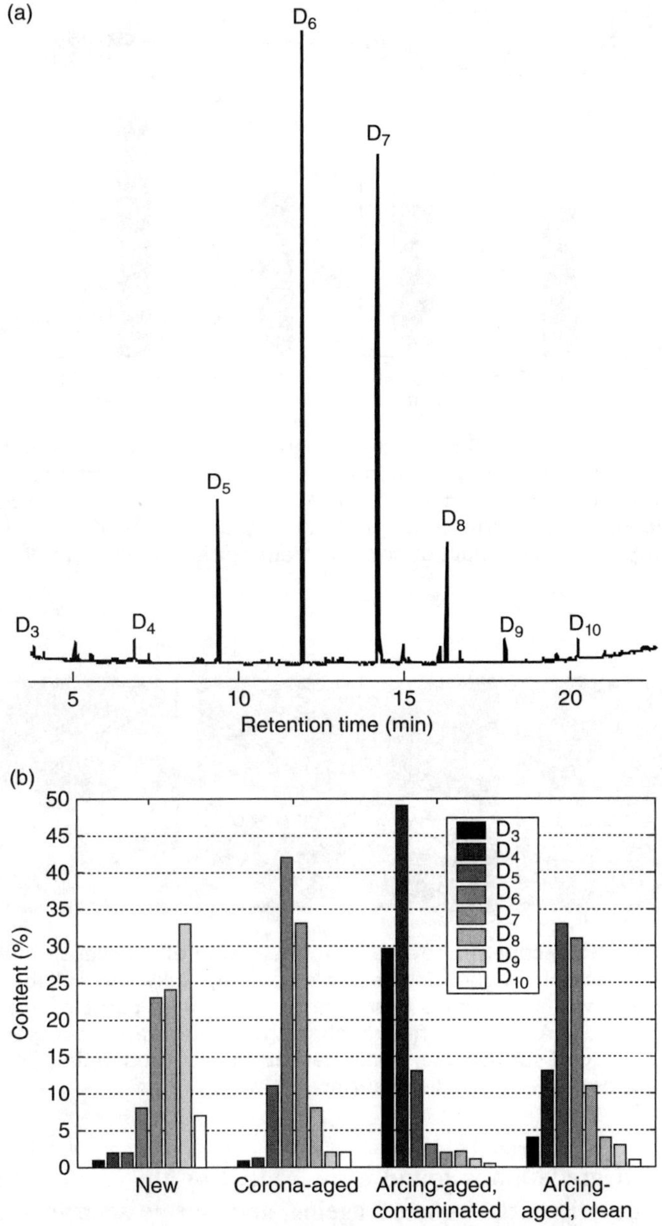

16.8 (a) Gas chromatograph showing identified migrating cyclic siloxanes (D_3–D_{10}) responsible for the hydrophobic recovery of SIR after exposure to corona. (b) Comparison of siloxane distributions found on new, corona-aged and arcing-aged SIR samples.

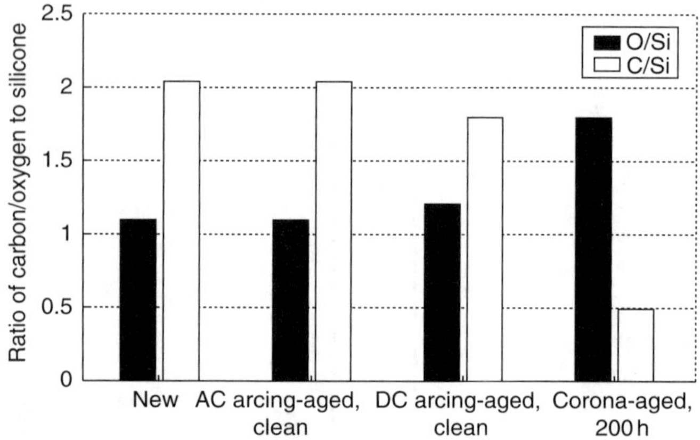

16.9 Results of XPS analyses indicate that oxidation is insignificant as the operating ageing mechanism for arcing-treated surfaces. The bars show the relative content of oxygen and carbon with reference to silicone on the surfaces of virgin and aged (AC and DC) samples. For comparison, the effect of corona treatment is also illustrated.

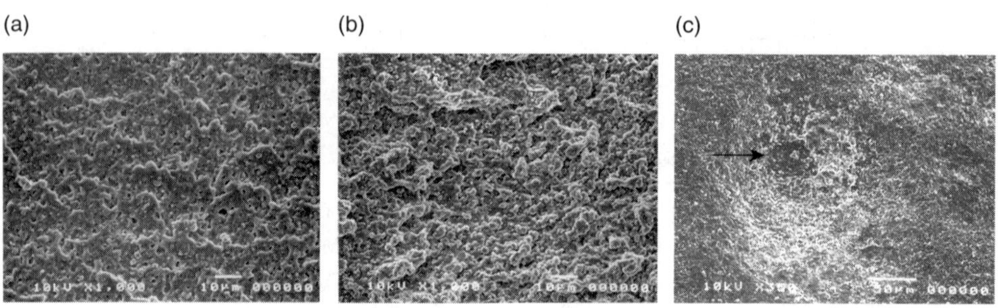

16.10 Microscope photographs of the ageing effects caused by arcing: (a) surface of DC-energised sample showing flaky, but polished, pattern with some holes from removed ATH filler particles; (b) surface aged under AC voltage has much rougher topography compared with the DC-aged surfaces; (c) crater caused by localised arcing (indicated by arrow) on AC-energised surface.

The type of degradation is found to be caused by the discharge activity, defined as spot-discharge-initiated ageing, and can be seen in Fig. 16.10. The spot discharges stress the surface locally by exposing it to high temperatures, estimated at above 700 K, where photons and radicals spread from the arc discharge channel and attack the material surface. Evidence of such thermal activity (see Fig. 16.10(c)) is clearly seen as loss of ATH filler. The loss of filler is also demonstrated by infrared spectroscopy investigation.

Results of the investigation presented above indicate that under natural field conditions in contaminated environments and given proper control of electric field distribution along insulator surfaces that eliminate corona, thermal depolymerisation becomes the dominant ageing mechanism. Electrical discharges appearing on the material surface are the source of thermal stressing at localised spots. Micro-craters are formed, ATH is consumed and volatile low molar mass siloxanes are produced. Therefore, while the surface becomes eroded, the hydrophobic recovery is actually enhanced.

16.4.3 Biological growth

The fact that microbiological colonisation of ceramic as well as composite insulators can take place in all parts of the world has been a matter of concerns.[8,46–50] There is clear evidence that biological contamination on composite and porcelain insulators leads to a reduction of their wet flashover voltage.[47] Insulators made of SIR, epoxies and blends of silicones and EPDMs have been found to support growth of bacteria, algae, fungi and lichen. In order to evaluate the susceptibility of EPDM and epoxy insulators to fungal growth, a set of tests was performed[51] where new and aged insulators were exposed to different strains of fungi and incubated under varying environmental conditions. Very little growth was observed, except for the samples incubated under high humidity.

Composite insulators, unlike ceramic ones, contain organic substances that could eventually be digested as nutrients by microorganisms. SIR is known to exhibit high resistance to biological degradation. One of the reasons is that the material consists of inorganic as well as organic components and that microorganisms like fungi cannot digest the inorganic parts.[52]

The microorganisms colonising bio-resistant substrates tend to form a film on the surface of the material. Such a biofilm consists of a mixture of different species embedded in a highly hydrated matrix of extracellular polymeric substances, mainly polysaccharides and proteins. The composition of the biofilm affects its interaction with the support material. Therefore, it is important to know the composition of the biofilms colonising SIR insulators in order to decide their effects on the properties of the housing material.

Three different biofilms, collected from SIR insulators exposed to outdoor environments in Sweden, Sri Lanka and Tanzania, were analysed in references 53 and 54. Even though the insulators were collected from three different continents, the compositions of the biofilms studied were remarkably similar. The similarity between the different biofilms indicates that the mechanism of bio-fouling of SIR insulators may be the same all over the

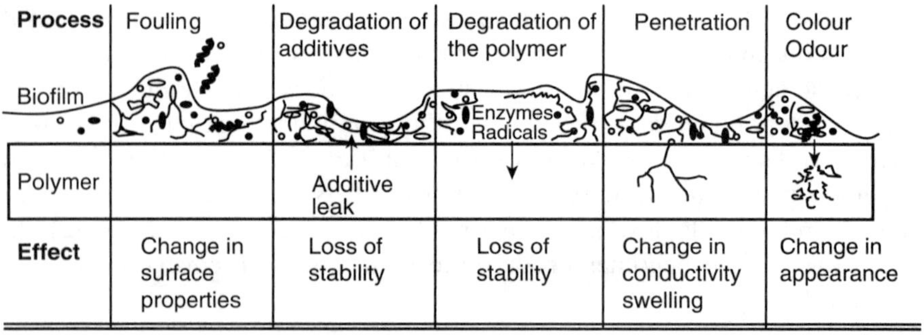

16.11 Potential undesirable effects of a biofilm evolving on a polymeric material surface.

world and this similarity should make it easier to develop a bio-resistant material that is useful world-wide.

Structure and function of a polymeric material can both be damaged by biofilms in various ways. The undesired effects range from disturbance of surface properties caused by the mere presence of accumulated biomass, via the degradation of leaking components to direct attack on the polymer matrix, as illustrated schematically in Fig. 16.11.

It has been noted that not all insulators in the field are affected by biological growth, which may indicate that some formulations are protected. In an attempt to identify the source of the observed differences in sensitivity, it was decided to analyse some of the materials used in practice. The investigation was focused on the effect of the flame-retardant fillers and different model SIR formulations containing ATH and zinc borate were prepared and tested. In addition, a commercial formulation was tested in parallel. The tests were performed in microenvironment chambers containing the biofilms studied and showed different sensitivity to the influence of microorganisms. Algae cells immediately started to grow on materials not containing zinc borate. After several weeks, fungal growth started to develop on the materials infected by the algae, and biofilms similar to those formed on insulators in the field appeared. The zinc borate-filled materials, where algae refused to grow, were also protected from fungal growth. At the same time, ATH had no inhibiting effect on the growth. In fact, the results indicated that ATH might even support biofilm growth. The commercial material tested showed a somewhat different behaviour. On these samples fungi started to grow before algae, indicating that some component or combination of components added to the rubber promoted the ability of the material to support fungal growth. While algae use carbon dioxide as a source of carbon through photosynthesis, fungi need a source of organic carbon to grow. It seems that some additives used in the production of the commercial material tested functioned as such a source, and hence made the resulting material more sensitive to the fungal growth. Further improvements can be obtained by addition of so-called 'biocides', i.e. active ingredients that kill or inhibit reproduction of microorganisms.[52] This method has, for example, been suggested in reference 55.

No evidence was found that the biofilms examined in this study were able to degrade the SIR matrix. When samples from the microenvironment chambers were cleaned of biological growth and analysed with infrared spectroscopy, no difference in spectra could be observed between samples treated with microorganisms and virgin ones. Similar results were obtained after additional exposure to lignin-degrading fungus, *Phanerochaete chrysosporium*. Neither significant differences in surface composition of the tested rubbers nor LMW degradation products were found in the GC-MS analysis performed after the treatment. This indicates that the probability that the SIR matrix of a high-voltage insulator could degrade under the direct influence of microorganisms is very small.

Even though SIRs seem to be very stable against direct microbial attack, composite insulators may still function as support for a biofilm. The unwanted deposition of biological matter is referred to as fouling. Microorganisms can use pollutants from the surroundings to gain nutrition, or, where algae are concerned, grow by means of photosynthesis. This ability of a biofilm to use an external carbon source makes laboratory testing difficult. Materials that seem to be inert in degradation studies may nevertheless encourage growth when exposed to outdoor conditions. The hydrophobicity of the silicone materials incubated in the microenvironment chambers was therefore tested. The results clearly showed that samples covered with microorganisms were more hydrophilic than virgin materials. However, after cleaning the surface of the infected materials, the hydrophobicity could be somewhat restored. The materials containing larger amounts of filler were not as easy to restore as were less-filled materials. Filler particles cause surface roughening that made the cleaning technique not effective enough to remove the biofilm completely from the rough surface. However, if the surface of an SIR insulator is efficiently cleaned, there is a good possibility that the surface properties of the insulator can be restored completely.

16.5 Ageing of insulator cores

Degradation processes developing in the core of composite insulators are not common[20] but may lead to mechanical failures. These failures can be divided into three categories: normal breaking of the GFR core under overload; slipping of end-fittings; and brittle fracture. The GFR rod is considered to be completely elastic up to the damage limit, where the glass fibres start to break irreversibly. It has been found that, as long as the load is below this damage limit, the insulator lifetime remains unaffected.[2] To ensure reliable operation in service, the ultimate strength is set equal to the damage limit, which can be accurately detected in the laboratory using acoustic methods. Experiments showed[56] that composite insulators have very stable time–load characteristics and that there is no significant reduction of their ultimate strength during service time.

The fact that brittle fracture can occur, even at low service loads, is still a cause for concern because a large number of the earlier produced composite insulators still remain in service. The fracture surface has two distinctly different regions: one characterised by a sharp and smooth cut perpendicular to the core axis and one where the core breaks at its damage limit. The smooth region usually occupies at least half of the core cross-section.[57] Brittle fracture is believed to be caused by stress corrosion and can be initiated by simultaneous application of mechanical stress and diluted acids.[58–60] It predominantly occurs close to the high-voltage end where both electrical and mechanical stresses are high. Defective end sealing, improper choice of bonding resin, weak interface between fibres and resin, and finally, use of glass fibres containing boron, are intimately connected with this type of failure.[61] Insulators with boron-free glass fibres were introduced in 1983, and since then no brittle fracture events have been reported in insulators equipped with cores containing this type of fibre. Earlier failures of composite insulators had mainly been related to inferior quality of interfaces in their construction.

16.6 Ageing at insulator interfaces

What differentiates most composite insulators from their ceramic counterparts is the existence of different interfaces, which in the worst cases may become weak points in the construction. Interfaces in composite insulators can be of a different nature, e.g. macroscopic as well as microscopic ones. Macroscopic interfaces of the solid–solid type can be found between the GFR core and the housing (sheath), between the GFR core and the metal fittings, and between the housing and the metal fitting. These interfaces also form the triple point between the GFR core, the fitting and the housing, which is one of the most sensitive points in insulator construction. Insufficient quality of this point appeared to be responsible for many of the reported insulator failures caused by moisture ingress, including brittle fracture of the GFR core.[62] Microscopic interfaces are found in the GFR core between epoxy resin and glass fibres and in the housing material between the polymeric matrix and the filler. In addition, depending on the manufacturing technique, there may be interfaces between the sheath and sheds, i.e. between the same or similar polymeric systems. Such interfaces may also appear as a result of possible minor repairs of damaged housings, for example in the form of cut-marks, or loss of shed pieces.

The interface between the GFR core and the sheath is the largest internal macroscopic interface in a composite insulator. Lack or loss of adhesion (delamination) of this interface, which is in direct relation to the bonding strength between two different polymeric systems, may result in development of internal degradation or even internal tracking. In order to achieve a reliable, long-term stability of interfaces in composite insulators, it is essential to avoid defect formation during manufacturing processes by careful selection of adequate primers and adhesives. The adhesion bonds must be able to resist a variety of stresses. Possible scenarios that may lead to loss of adhesion and to formation of interfacial voids include ageing processes due to hydrolysis, partial discharges and thermal decomposition.

Investigations on the quality of interfaces formed during shed repairs on large hollow-core insulators were reported in reference 63. The influence of harsh environmental and electric stresses on mechanical properties (mechanical strength and elongation at break) of the joints obtained either by cured HTV SIR or by RTV glue was tested. The tensile testing showed that the cured interfaces maintained their mechanical strength after ageing better than the RTV-glued interfaces. The reduction of elongation at break was also large after storage in a highly humid environment, indicating that moisture alone can affect the mechanical properties of the joints.

Deterioration mechanisms inside interfacial defects between unbounded epoxy core and SIR housings were also studied when exposed to partial discharges.[64] Optical inspections of the tested areas revealed that erosion was localised on the epoxy substrate as small cavities surrounded by acidic droplets of decomposition by-products in the vicinity of the discharges. Weakening of adhesion was also detected locally. At the same time, cracking patterns could be observed on the SIR unbounded surface, most probably caused by joint action of electrical discharges and ozone, indicating oxidative cross-linking on the surface of this material. Infrared analyses suggested a build-up of a thin degraded layer on the epoxy substrate as well as degradation of the primer.

It is important to stress that the presence of voids and delamination sites at the interfaces may not necessarily be harmful to an insulator. Everything depends on the defect size and on the type of stresses that are involved. An insulator containing minor defects can remain in service for years before the faults start to endanger its functionality. However, if the growth rate of voids at the interface is high, the risk of loss of desired properties becomes higher. The duration of such a process can vary significantly and it is therefore essential to attain knowledge on the rate of defect growth in composite insulation systems in a long-term perspective.

16.7 Future trends

The long-term experiences with composite insulators of different designs and made of different housing materials have led to new and better manufacturing techniques. New materials improved the performance, but still the impact of realistic service conditions brings new challenges. SIR, the most widely used type of housing material today, will maintain its domination in future composite insulator technology, albeit with various modifications. Therefore, further development of knowledge regarding its performance is needed and the main interests are related to understanding the ageing mechanisms under diverse and special stress conditions, the long-term preservation and recovery of hydrophobic surface properties, as well as the resistance to biological fouling. These three areas cannot be treated separately – the problems are interrelated and require a multidisciplinary approach from scientists and engineers.

Regarding ageing mechanisms, research interests are concentrated on applications of composite insulators in DC voltage transmission, which is gradually being introduced to the world's electric power systems. The DC stress imposes new requirements on material properties. Another area of interest is in materials for the control of electric field distribution to enable optimal design of high-voltage apparatus. In particular, obtaining the required level of electric conductivity and dielectric response is of importance, and this can be achieved by incorporating new types of fillers and additives into the rubber. It is also foreseen that nano-fillers may become advantageous. It is not yet known how all these new types of fillers, additives and necessary wetting agents will influence the parameters important for securing good and long-term performance of new SIR compositions. This implies that multi-parameter ageing studies, involving variations in material compositions and in applied stresses, will continue worldwide.

The new fillers may also influence the hydrophobic properties of SIR. In order to understand the phenomenon better, we should look to explanations provided by botanists.[65] Analogous to the self-cleaning properties of plant leaves, bird feathers and the chitin bodies and wings of insects, it is an interaction of the fine microstructure and roughness of the polluted surface, together with the hydrophobic properties of the LMW PDMS species, that contribute to the self-cleaning ability of SIR composite insulators. One should therefore try to make use of this phenomenon for producing insulators characterised by even better hydrophobic properties. A properly patterned insulator surface could result in the so-called 'super-hydrophobic property'; this can be obtained in different ways, for example, by addition of special fillers, by micro-machining of surfaces in casting moulds or by laser treatment. Preliminary attempts have already shown encouraging results.[66] It is also important, in parallel, to gain a better understanding of the conditions for release and production of LMW siloxane oligomers, which are the primary source for encapsulation of contamination particles and therefore for hydrophobic recovery on SIR surfaces. Among other types of base polymers potentially able to secure better hydrophobicity than PDMS, fluorosilicone compounds should be named; they are, however, far more expensive. The problem of material costs is also important when considering the new types of fillers.

Forcing the super-hydrophobic property on insulator surfaces may also help in increasing the resistance to biological fouling, which otherwise can be achieved by modifying the material with anti-fouling additives. The latter, again may change the material's ageing stability. In this way, the circle of interactions gets closed, indicating that the future developments of composite insulator technology will be very strongly dependent on obtaining reliable material modification through the use of optimised quantities of quality fillers and additives.

16.8 Acknowledgements

The content of this chapter is based on information available in literature published worldwide. For illustrating specific issues, more detailed considerations are presented, which are the outcome of a long lasting cooperation between academic and industrial organisations in Sweden. I would therefore like to express sincere thanks to my coworkers and colleagues involved in this cooperation. In particular, Prof. Sigbritt Karlsson, Prof. Ulf Gedde, Dr Henrik Hillborg and all the postgraduate students are acknowledged for their dedicated work that contributed greatly to the process of building the knowledge presented above.

16.9 References

1 LOOMS J S T, *Insulators for high voltages*, IEE Power Engineering Series 7, Peter Peregrinus Ltd, London, 1988.
2 GORUR R S, CHERNEY E A and BURNHAM J T, *Outdoor insulators,* Ravi S. Gorur Inc., Phoenix, AZ, 1999.
3 MACKEVICH J and SHAH M, 'Polymer outdoor insulating materials. Part I: Comparison of porcelain and polymer electrical insulation', *IEEE Electrical Insulation Magazine*, 1997, **13** (3), 5–12.
4 CARLSON S J and SEMLYEN J A, *Siloxane polymers*, Polymer Science and Technology Series, PTR Prentice Hall Inc., Englewood Cliffs, NJ, 1993.
5 HILLBORG H and GEDDE U W, 'Hydrophobicity changes in silicone rubbers', *IEEE Trans. on Dielectrics and Electrical Insulation*, 1999, **6** (5), 703–717.
6 SÖRQVIST T, *Polymeric outdoor insulators – a long term study*, PhD Thesis, Chalmers University of Technology, Göteborg, Sweden, 1997.

7 KARADY G G, 'Outdoor insulations', in *6th International Symposium on High Voltage Engineering*, New Orleans, USA, vol. 2, paper 30.01, 1995.

8 KUNDE K, HENNINGS R, KÜHL M, SCHÜTZ A, JANSSEN H and STIETZL U, 'New experience with composite insulators', Cigré session, Paris, France, 15-206, CIGRE (International Council on Large Electric Systems), Paris, France, 1998.

9 HACKAM R, 'Outdoor HV composite polymeric insulators', *IEEE Trans. on Dielectrics and Electrical Insulation*, 1999, **6** (5), 557–585.

10 WOSLOO W L, MACEY R E and DE TOURREIL C, *The practical guide to outdoor high voltage insulators*, Johannesburg, Crown Publications for Eskom, 2004.

11 GUBANSKI S M and WANKOWICZ J G, 'Distribution of natural pollution surface layers on SIR insulators and their UV absorption', *IEEE Trans. on Electrical Insulation*, 1989, **24** (4), 689–697.

12 KIM S H, CHERNEY E A and HACKAM R, 'The loss and recovery of hydrophobicity of RTV SIR insulator coatings', *IEEE Trans. on Power Delivery*, 1990, **5** (3), 1491–1500.

13 GUBANSKI S M and VLASTÓS A E, 'Wettability of naturally aged silicone and EDPM composite insulators', *IEEE Trans. on Power Delivery*, 1990, **5** (3), 1527–1535.

14 CHANG J W and GORUR R S, 'Surface recovery of SIR used for HV outdoor insulation', *IEEE Trans. on Dielectrics and Electrical Insulation*, 1994, **1** (6), 1039–1046.

15 KIM S H, CHERNEY E A and HACKAM R, 'Suppression mechanism of leakage current on RTV coated porcelain and SIR insulators', *IEEE Trans. on Power Delivery*, 1991, **6** (4), 1549–1556.

16 GUSTAVSSON T G, GUBANSKI S M and LAMBRECHT J, 'Hydratization of the PDMS backbone during water immersion test', in *Proceedings of the IEEE Conference on Electrical Insulation and Dielectric Phenomena*, Atlanta, USA, vol. 1, pp. 269–272, 1998.

17 SIRAIT K T, SALAMA, SUWARNO and KAERNER H C, 'The effect of natural tropical climate on the surface properties of silicone rubber', in *Proceedings of the International Symposium on Electrical Insulating Materials*, Toyohashi, Japan, pp. 453–456, 1998.

18 KIKUCHI T, NISHIMURA S, NAGAO M, IZUMI K, KUBOTA Y and SAKATA M, 'Survey on the use of non-ceramic composite insulators', *IEEE Trans. on Dielectrics and Electrical Insulation*, 1999, **6** (5), 548–556.

19 CIGRÉ WORKING GROUP 22.03, 'Worldwide service experience with HV composite insulators', *Electra*, 1990, **130**, 67–77.

20 CIGRÉ WORKING GROUP 22.03, 'Worldwide service experience with HV composite insulators', *Electra*, 2000, **191**, 27–43.

21 SU F and JIA Y, 'Faults of composite insulators in service and research of on-line detection method', in *Proceedings of the 11th International Symposium on High Voltage Engineering*, London, UK, vol. 4, pp. 208–211, 1999.

22 ZHICHENG G, FUZENG Z, GUOLI W, ZHIDONG J and LIMING W, 'Challenge of ultra high voltage transmission technology in China', in *Proceedings of the 14th International Symposium on High Voltage Engineering*, Beijing, China, Paper E-03, 2005.

23 SHAH M, KARADY G G and BROWN R L, 'Flashover mechanism of SIR insulators used for outdoor insulation – II', *IEEE Trans. on Power Delivery*, 1995, **10** (4), 1972–1978.

24 REYNDERS J P, JANDRELL I R and REYNDERS S M, 'Review of aging and recovery of SIR insulation for outdoor use', *IEEE Trans. on Dielectrics and Electrical Insulation*, 1999, **6** (5), 620–631.

25 STARR WT, 'polymeric outdoor insulation', *IEEE Trans. on Electrical Insulation*, 1990, **25** (1), 125–136.

26 PHILLIPS A J, CHILDS D J and SCHNEIDER H M, 'Water drop corona effects on full-scale 500 kV non-ceramic insulators', *IEEE Trans. on Power Delivery*, 1999, **14** (1), 258–265.

27 PHILLIPS A J, CHILDS D J and SCHNEIDER H M, 'Ageing of non-ceramic insulators due to corona from water drops', *IEEE Trans. on Power Delivery*, 1999, **14** (3), 1081–1089.

28 MORENO V M and GORUR R S, 'Effect of long-term corona on non-ceramic outdoor insulator housing materials', *IEEE Trans. on Dielectrics and Electrical Insulation*, 2001, **8** (1), 117–128.

29 SPELLMAN C A, YOUNG H M, HADDAD, ROWLANDS A R and WATERS R T, 'Survey of polymeric insulator ageing factors', in *Proceedings of the 11th International Symposium on High Voltage Engineering*, London, UK, vol. 4, pp. 160–163, 1999.

30 MORENO V M, GORUR R S and KROESE A, 'Impact of corona on the long-term performance of nonceramic insulators', *IEEE Trans. on Dielectrics and Electrical Insulation*, 2003, **10** (1), 80–95.

31 HILLBORG H, *Loss and recovery of hydrophobicity of polydimethylsiloxane after exposure to electrical discharges*, PhD Thesis, Department of Polymer Technology, Royal Institute of Technology, Stockholm, Sweden, 2001.

32 FATEH-ALAVI K, *Stabilizers in crosslinked polydimethylsiloxane*, PhD Thesis, Department of Polymer Technology, Royal Institute of Technology, Stockholm, Sweden, 2003.

33 WU S, *Polymer interface and adhesion*, Marcel Dekker Inc., New York and Basel, 1982.

34 LIU H, CASH G, BIRTWHISTLE D and GEORGE G, 'Characterization of a severely degraded silicone elastomer HV insulator – an aid to development of lifetime assessment techniques', *IEEE Trans. on Dielectrics and Electrical Insulation*, 2005, **12** (3), 478–486.

35 HILLBORG H and GEDDE U W, 'Hydrophobicity recovery of polydimethylsiloxane after exposure to corona discharges', *Polymer*, 1998, **39** (10), 1991–1998.

36 HILLBORG H, ANKNER J F, GEDDE U W, SMITH G D, YASUDA H K and WIKSTRÖM K, 'Crosslinked polydimethylsiloxane exposed to oxygen plasma studied by neutron reflectometry and other surface specific techniques', *Polymer*, 2000, **41**, 6851–6863.

37 VLASTÓS A E and GUBANSKI S M, 'Surface structural changes of naturally aged silicone and EPDM insulators', *IEEE Trans. on Power Delivery*, 1991, **5** (2), 888–900.

38 YOSHIMURA N, KUMAGAI S and NISHIMURA S, 'Electrical and environmental aging of SIR used in outdoor insulation', *IEEE Trans. on Dielectrics and Electrical Insulation*, 1999, **6** (5), 632–650.

39 GUSTAVSSON T G, HILLBORG H, GUBANSKI S M, GEDDE U W and KARLSSON S, 'Ageing of SIR materials under ac and dc voltages in a coastal environment', *IEEE Trans. on Dielectrics and Electrical Insulation*, 2001, **8** (6), 1029–1039.

40 GUSTAVSSON T G, *SIR insulators – impact of material formulations in coastal environment*, PhD Thesis, Department of Electric Power Engineering, Chalmers University of Technology, Gothenburg Sweden, 2002.

41 KUMAGAI S and YOSHIMURA N, 'Polydimethylsiloxane and alumina trihydrate system subjected to dry-band discharges or high temperature part I: chemical structure', *IEEE Trans. on Dielectrics and Electrical Insulation*, 2004, **11** (4), 691–700.

42 KUMAGAI S and YOSHIMURA N, 'Polydimethylsiloxane and alumina trihydrate system subjected to dry-band discharges or high temperature part II: electrical insulation', *IEEE Trans. on Dielectrics and Electrical Insulation*, 2004, **11** (4), 701–707.

43 MAYER L, JAYARAM S and CHERNEY E A, 'Thermal conductivity of filled SIR and its relationship to erosion resistance in the inclined plane test', *IEEE Trans. on Dielectrics and Electrical Insulation*, 2004, **11** (4), 620–630.

44 JANSSEN H, HERDEN A and KÄRNER H C, 'The loss and recovery of hydrophobicity on SIR surfaces', in *Proceedings of the 10th International Symposium on High Voltage Engineering*, Montreal, Canada, vol. 3, pp. 145–148, 1997.

45 KIM S H and HACKAM R, 'Formation of silicone fluid at the surface of RTV SIR coating due to heat', in *Proceedings of the IEEE Conference on Electrical Insulation and Dielectric Phenomena*, Pocono Manor, Pennsylvania, USA, pp. 605–611, 1993.

46 MCAFEE R D, HEATON R D, KING J M and FALSTER A U, 'A study of biological contaminants on high voltage porcelain insulators', *Electric Power Systems Research*, 1997, **42** (1), 35–39.

47 FERNANDO M A R M, *Performance of non-ceramic insulators in tropical environments*, PhD Thesis, Department of Electric Power Engineering, Chalmers University of Technology, Göteborg, Sweden, 1999.

48 FERNANDO M A R M and GUBANSKI S M, 'Performance of nonceramic insulators under tropical field conditions', *IEEE Trans. on Power Delivery*, 2000, **15** (1), 355–360.

49 GORUR R S, MONTESINOS ROBERSON J R, BURNHAM J and HILL R, 'Mold growth on nonceramic insulators and its impact on electrical performance', *IEEE Trans. on Power Delivery*, 2003, **18** (2), 559–563.

50 DERNFALK A, *Diagnostic methods for composite insulators with biological growth*, PhD Thesis, Chalmers University of Technology, Göteborg, Sweden, 2004.

51 RACKLIFFE G B, LEE R E, FRITZ D E and HARMON R W, 'Performance evaluation of 15-kV polymeric insulators for dead-end type applications on distribution systems', *IEEE Trans. on Power Delivery*, 1989, **4** (2), 1223–1231.

52 WOLF A, 'Mould fungus growth on sanitary sealants', *Construction and Building Materials*, 1998, **3**, 145–151.

53 WALLSTRÖM S, *Biofilms on SIR materials for outdoor high voltage insulation*, PhD Thesis, Department of Fibre and Polymer Technology, Royal Institute of Technology, Stockholm, Sweden, 2005.

54 WALLSTRÖM S and KARLSSON K, 'Biofilms on SIR insulators; microbial composition and diagnostics of removal by use of ESEM/EDS', *Polymer Degradation and Stability*, 2004, **85** (2), 841–846.

55 GUBANSKI S M, FERNANDO M A R M, PIETR S J, MATULA J and KYARUZI A, 'Effects of biological contamination on insulator performance', in *Proceedings of the 6th*

International Conference on Properties and Applications of Dielectric Materials, Xi'an, China, vol. 2, pp. 797–801, 2000.

56 BERNSTORF R A, 'Time-load testing of nonceramic insulators with fibreglass core rod – a 20 year summary', in *Proceedings of the 1999 IEEE Transmission and Distribution Conference*, New Orleans, USA, vol. 2, pp. 823–826, 1999.

57 XIDONG L and JIANJU D, 'Analysis of the acid sources of a field brittle fractured composite insulator', *IEEE Trans. on Dielectrics and Electrical Insulation*, 2006, **13** (4), 870–876.

58 DE TOURREIL C, THÉVENET G, BROCARD E, SIAMPIRINGUE N and PICHON N, 'Determination of the brittle fracture process of field failed HV insulators', in *Proceedings of the 14th International Symposium on High Voltage Engineering*, Beijing, China, D-28, 2005.

59 KUMOSA M S, KUMOSA L S and ARMENTROUT D L, 'Failure analyses of nonceramic insulators: Part 1: Brittle fracture characteristics', *IEEE Electrical Insulation Magazine*, 2005, **21** (3), 14–27.

60 KUMOSA M S, KUMOSA L S and ARMENTROUT D L, 'Failure analyses of nonceramic insulators: Part II – The brittle fracture model and failure prevention', *IEEE Electrical Insulation Magazine*, 2005, **21** (4), 28–41.

61 KUHL M, 'FRP rods for brittle fracture resistant composite insulators', *IEEE Trans. on Dielectrics and Electrical Insulation*, 2001, **8** (4), 181–190.

62 BURNHAM J T, BAKER T, BERNSTORF A, DE TOURREIL C, GEORGE J M, GORUR R S, HARTINGS R, HILL B, JAGTIANI A, MCQUARRIE T, MITCHELL D, RUFF D, SCHNEIDER H, SHAFFNER D, YU J and VARNER J, 'IEEE task force report: Brittle fracture in nonceramic insulators', *IEEE Trans. on Power Delivery*, 2002, **17** (3), 848–856.

63 ANDERSSON J, GUBANSKI S M and HILLBORG H, 'Properties of interfaces in silicone rubber', *IEEE Trans. on Dielectrics and Electrical Insulation*, 2007, **14** (1), 137–145.

64 ANDERSSON J, HILLBORG H and GUBANSKI S M, 'Deterioration of internal interfaces between silicone and epoxy resin', in *Conference Record of the 2006 IEEE International Symposium on Electrical Insulation*, Toronto, Canada, pp. 527–530, 2006.

65 BARTHLOTT W and NEINHUIS C, 'Purity of the sacred lotus, or escape from contamination in biological surfaces', *Planta*, 1997, **202**, 1–8.

66 HOLGERSSON P, *Preparation and evaluation of PDMS polymer surfaces with increased hydrophobicity by pulsed laser irradiation*, Technical Licentiate Thesis, Department of Applied Physics, Chalmers University of Technology, Gothenburg, Sweden, 2005.

17

Ageing of composites in the chemical processing industry

R. MARTIN, Materials Engineering
Research Laboratory Ltd, UK

17.1 Introduction

In the chemical processing industry, the operating environments for equipment such as reactor vessels, storage tanks, scrubbers (Fig. 17.1), stacks, piping, valves, etc. may be extremely harsh. In many instances corrosion-resistant alloys – including highly alloyed stainless steel-, titanium- and nickel-based alloys – have to be used and even these can corrode in these environments. One solution adopted in this industry is the use of fibre reinforced plastics (FRPs) in manufacturing this equipment.

FRP materials are used to a large extent in plants that manufacture chlorine, chlorate and concentrated acids (e.g. sulphuric, hydrochloric, hydrofluoric, nitric) as well as metal chloride solutions (e.g. NaCl, $FeCl_3$, $AlCl_3$, $MgCl_2$, $NiCl_2$). FRPs are now also used in desulphurization plants (flue gas ducting, scrubbers, etc.) which also have applications in oil and gas production. In many instances a thermoplastic liner is used for metallic and FRP pipe to act as a corrosion or permeation barrier, or both. The thermoplastic lining may be nylon, polypropylene or, for more aggressive service, poly vinylidine fluoride (PVDF).

Because of the environmentally hostile nature of some of the chemicals being transported or stored, failure is unacceptable. However, failures do occur (Fig. 17.2) (Bergman 2004). These failures may not only be very costly, but also present a health and safety risk to the workers at the plant, the local residents and to the environment. The consequences and liability of equipment failure, even minor leaks, are becoming increasingly severe and have resulted in a very strict regulatory climate. Inevitably, cost-effectiveness requirements on inspection and operation departments exert a need to increase time between and for the duration of inspections. This puts more emphasis on the need to have robust methods to determine how much life remains in the equipment or the setting of the inspection schedule. Customary methods are necessarily very cautious and doubtless result in safe equipment being taken out of service.

17.1 FRP precipitator and tower; courtesy of Dow Chemicals.

Although FRP is often used to solve problems of corrosion on various metallic materials, FRP materials can also be affected by these fluids. Prediction methods for FRPs are less well established than methods for corrosion prediction in metals. The term 'corrosion' is often used to discuss such degradation of FRPs, although strictly speaking they do not corrode but chemically age, which causes physical property changes. These changes may lead to increased diffusion, hydrolysis and ultimately cracking, blisters and other damage, discussed further below. Eventually this degradation may continue and lead to actual material loss from the inside of the vessels (Fig. 17.3) (Bergman 1995). This material loss can be likened to corrosion in metals. Nevertheless, FRP has been used in this industry with great success apart from a small number of failures. These may arise when the relationship between the degradation mechanisms and time is not fully understood.

A continuing shortcoming for FRPs in this industry, when competing with metals during materials selection processes, is that the corrosion science of metals is much more developed than that of FRP. While FRP materials have been used for about 40–50 years in structural applications, and although their use has increased tremendously in recent years, the problem of environmental attack has seldom been systematically evaluated

17.2 Failure in alkaline aqueous solution at 70 °C causing a fatality; courtesy of G. Bergman.

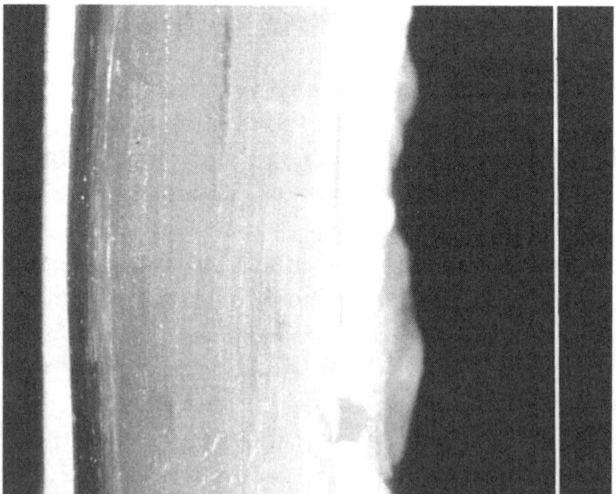

17.3 Uniform corrosion in an FRP pipe after 14 years of service. The white line shows the original thickness; courtesy of G. Bergman.

from a combined corrosion and ageing approach. This chapter describes the current techniques used in the chemical processing industry.

17.2 Examples of use of fibre reinforced plastics in the chemical processing industry

FRP tanks, including process tanks, vats and underground storage tanks, have several advantages besides corrosion resistance, they are comparatively light and have the potential to be monolithic in construction. Often FRPs are used in unison with a steel tank, as a protective liner, hence benefiting from the strength of the steel and the corrosion properties of the FRP (Roberts 1980). Other applications include reaction vessels, exhaust ducting, scrubbers and towers, chimney stacks, valves and pumps, agitators, gas accumulators, electrostatic precipitators and cryogenic pressure vessels (Pritchard 1989). Pipework in a chemical plant is commonly manufactured from FRP. In a plant, the pipes may run overhead, underground or even under the sea for several kilometres.

Vinyl ester-based FRPs are used in many applications and have replaced much of the rubber-lined steel construction. One of the first applications by BASF was in production equipment for chlorine in its chemical plant in Ludwigshafen (Anon. 1994). Conditions are very corrosive with both chlorine and water present. By fabricating the ducts and cell covers from FRP, the lifetime was increased from 6 to 8 years. Scrubbers are also made from FRP and are 35 m high and 9.5 m in diameter and have been in operation since 1988. In the incineration plant, stacks and scrubbers operate at 100 °C. All pre-cleaning tanks at Ludwigshafen are fabricated from FRP. The water being treated produces sulphur dioxide that condenses inside the covers forming acid. This acid destroyed the concrete lids and these have been replaced by FRP lids that have been in use for over 20 years and their strength has reduced by only 20%.

FRP has become the material of choice for handling saturated salt brine and depleted brine recycle systems in chlor-alkali production. Brine feeding pipes are usually fabricated from glass/vinyl esters whereas depleted chlorinated brine requires a modified glass/vinyl ester for the higher temperatures. However, if the brine contains high quantities of sodium hypochlorite solution, the FRP must be lined with a fluorocarbon (Anon. 1991). The production of chlor-alkali requires high-purity-saturated brine solutions free of any metal-ion contamination. Hence, FRP components (piping, tanks, pumps, valves, etc.) are the preferred non-metallic materials for construction for these installations (Nelson 1988,Talbot 1990). FRP has been used extensively for hot–wet chlorine gas collection headers since the 1950s. The resin systems are 'chlorendic' unsaturated polyester resins. The headers are hand laid-up with a 'corrosion allowance'.

17.3 Types of fibre reinforced plastic

The types of resin and fibre used depend very much on the application being considered. Thermosetting resins are suitable for large mouldings and for filament wound pipes. They are often partially hydrophilic, absorbing up to 3% moisture, more than semi-crystalline thermoplastics. The most important resin systems used for corrosion protection include polyesters, epoxies, bismaleimides, phenolics and vinyl esters (Pritchard 1989). Vinyl esters and epoxies are perhaps the most common materials used in the chemical processing industry and are used in similar applications (Marshall *et al.* 1982). Epoxies have certain advantages in that they have lower cure shrinkage, adhere better to steel substrates and offer better chemical resistance than polyesters towards hot alkalis. For higher temperature applications, modified epoxies such as bismaleimides may be used (Scola 1984). Vinyl esters are favoured because they achieve a compromise between epoxies and polyesters offering good corrosion resistance with moderate relative cost (Zweben *et al.* 1985). Phenolics are commonly used resin systems for liners with their high-temperature capability; however, they are less resistant to alkalis than other resin systems.

In FRP structures used in the chemical processing industry an E-glass fibre is used as the reinforcement for the main structural part. A more chemical-resistant fibre such as ECR-glass is then used in the layer in contact with the corrosive fluid. Clearly, the ECR layer must remain intact for the life of the structure. A protective resin-rich layer is often used on the surface adjacent to the acid. The outer layer of the protective layer typically consists of approximately 95% resin and is reinforced by surfacing veils of an ECR-glass. While carbon or Kevlar fibres may be used in FRPs in other industries, they are seldom the fibre of choice for corrosion-resistant applications, largely because of their higher cost compared with glass. Table 17.1 shows typical applications for different types of glass fibre (Bergman 2007).

17.4 Types of degradation in fibre reinforced plastic

The mechanisms of corrosion of metals are mostly determined by electrochemical processes. An analogous description of the mechanisms of corrosion of FRP, or other plastics, is that they are determined by organochemical processes. Plastics are semicrystalline or amorphous, therefore physical processes like diffusion, osmosis, embrittlement, microcracking and swelling play an important role in the corrosion processes of plastics.

Degradation in composites, which may be likened to corrosion, may be defined as one of any of the following in the exposed surface:

Table 17.1 Examples of the performance of different glass fibres

Glass type	Application	One-day weight loss in 10% H_2SO_4 (%)
E-glass	General-purpose fibres	39
ECR-glass®	Used where acid corrosion resistance is desired	6.2
S-2 glass®	Used for reinforcement in composite structural applications that require stability under extreme, corrosive environments	4.1
C-glass	Used for chemical stability in corrosive acid environments	2.2
A-glass	General-purpose fibres	0.4

- cracks;
- pitting;
- thickness change;
- charring;
- resin loss;
- loss of fibre;
- softening;
- blistering;
- leaching;
- etching;
- delamination;
- discoloration;
- fibre blossoming.

The research activities at universities and other research institutions have focused on 'ageing', and most have been restricted to studies of the degradation mechanisms in water rather than in organic solvents (Hogg and Hull 1983). The published work related to chemical resistance of FRP materials in different environments has focused on immersion testing, rather than on more realistic single-sided exposure. Much of the published data quote only single-point data, e.g. 'the mass of material *X* increases by 10% in methanol after 10 days at 60 °C'. This information may be misleading because the overall trend may involve a mixture of competing mechanisms involving mass loss and mass gain. These competing mechanisms make the prediction long-term performance at lower temperatures from short-term tests at higher temperatures non-trivial. Small temperature increases above the service temperature, but below the resin's glass transition temperature, T_g, can offer useful indications of longer term behaviour. For hybrid materials, such as corrosion-resistant veils, the prediction method is still more involved.

Hydrolysis is a common physical degradation mechanism and the rate is dependent on absorption kinetics. Absorption is activated by diffusion which is temperature dependent but largely pressure independent in FRPs. The rate of diffusion is also direction dependent in an anisotropic composite with the rate being higher parallel to the fibres than normal to them. A decrease in apparent absorption during an exposure test indicates that soluble constituents or degradation products from a chemical reaction are being leached out. This can be confirmed by drying the sample after exposure to identify weight loss. Cracking can also occur with chemical attack and more so with cyclic exposure to load or temperature or both.

Furthermore, FRPs may be subject to step changes in properties leading to failure such as resin embrittlement. This failure is often termed 'environmental stress cracking (ESC)' and also applies to glass fibres when undergoing strains in excess of 2% and exposed to even dilute acids (Hogg and Hull 1980, French and Pritchard 1993). In any long-term evaluation, it is necessary to use diagnostic equipment to ensure that the cause of the change in properties of the FRP is understood.

17.5 Current methods for assessing long-term ageing of fibre reinforced plastics

Estimates of the remaining life of FRP equipment in-service most often rely on periodic visual inspection, and the intuition and experience of the inspector. The inspector searches out defects such as blistering and delamination, or signs of leaking. Often, a bright light is shone on the opposite side of the laminate to help reveal flaws. The inspector often also checks the condition of the laminate surface with a Barcol indentation tester. A drastic reduction in Barcol readings since the previous inspection indicates degradation. Based on these findings the inspector will judge whether the equipment is still safe to operate and when the next inspection should occur.

In the data sheets of different resin manufacturers, the chemical or ageing resistance in different environments is presented in the form of a recommendation of use up to a certain maximum temperature or is assessed in terms of 'resistant', 'limited resistant' and 'not resistant' (e.g. Derakane data sheet: *Derakane Epoxy Vinyl Ester Resins Chemical Resistance Guide*, Ashland Chemical (http://www.derakane.com/)). However, this information does not give the user sufficient information about the possible types of attack, how fast any attack will proceed into the wall and what damage it may cause. It is therefore not possible to make any life prediction assessment from these data when designing a product for a certain lifetime. Once again, this puts more emphasis on the experience of the designers based on current and previous operational issues and on the inspections of the equipment during use.

Suitable accelerated testing methodologies using short-term experiments to obtain long-term data are required. The aim is to obtain relationships that describe the influence of exposure time, temperature, concentration, stress, etc. on the different ageing processes through the thickness of the FRP. The approaches for diffusion and ageing are based on identifying the time, temperature and concentration dependence of the life-determining deterioration mechanism. The results can then be extrapolated to longer times for the different temperatures and predictions made for the behaviour at different concentrations.

Such models could be used to predict the reduction in baseline properties of the materials as a result of ageing and can be complemented by models that allow for changes in toughness, stiffness, notch sensitivity, etc. The predicted properties may also be used as inputs into numerical models and finite element codes to identify knock-downs in strength or time to failure. The following sections describe some approaches that are available for assessing FRPs.

17.5.1 ASTM Standard for long-term usage

Materials are generally approved for long-term usage in the chemical processing industry using ASTM C581-03. This standard requires immersion of a material in a fluid at a single temperature. Various properties (Barcol hardness, flexural modulus and strength, and T_g) are determined at intervals, normally within 1 year. If the properties do not decrease by a certain amount, the material may be approved for long-term usage. However, this method does not offer an approach to allow an extrapolation to longer-term usage at a variety of temperatures and is therefore little more than a screening test.

17.5.2 The Arrhenius relationship

Another approach for long-term ageing of polymers is to conduct accelerated ageing using temperature as an accelerator and utilize the Arrhenius relationship that relates time and temperature to a change in properties. FRPs have a polymer matrix that is usually thermoset or thermoplastic. The class of polymer depends on the chemical structure of its long molecules and the way in which they pack together within the normal kinetic nature of materials. Free space exists between molecular chains to a greater or lesser degree for the three classes, to lead to a balance between stiffness and flexibility. However, this same free space can mean that polymers can absorb fluids to which they are exposed, especially those with similar solubility parameters. Such absorption can physically weaken the polymer to provide one effect of ageing. In addition, the fluid might chemically attack the polymer to provide an additional effect. The kinetics of these two ageing effects is governed by (a) diffusion and (b) chemical kinetics, both of which are governed by Arrhenius relationships with regard to the influence of temperature. Hence, accelerated testing can be performed at elevated temperatures, with the results being extrapolated back to service temperature for life-prediction purposes. Diffusion characteristics can be measured by liquid mass uptake or gas permeation experiments. Chemical kinetics, classically involving concentrations of reactants and products, can employ the fact that for cross-linked polymers, the concentration of cross-links is approximately proportional to the modulus or stiffness. Hence, measurements of changes in modulus from ageing can be plotted logarithmically against linear time (for first-order reactions) at each temperature. From a series of such ageing plots at different temperatures, times to attain the same degree of modulus change can be used to develop the Arrhenius plot:

$$\ln\frac{1}{t_{95}} = A\,\frac{-E_a}{R}\,\frac{1}{T} \qquad [17.1]$$

where t_{95} is the time for a property to reach 95% of its original value (although other values can be used), T is the absolute temperature, R is the gas constant (8.314 J/mole/K), A is constant and E_a is the activation energy. This expression holds well where there is only one degradation mechanism taking place; however, this is seldom the situation in chemical processing environments.

17.5.3 Using a semi-empirical corrosion approach

Another approach to the long-term ageing of FRPs is to relate the effects of ageing to those of corrosion in metals because on the macro-level the results are very similar (Bergman 2001). Many types of corrosion found in metals can also be found in FRPs, such as uniform corrosion (material loss), localized corrosion (pitting), selective corrosion, stress corrosion, corrosion fatigue, erosion corrosion and layer corrosion (delamination). Carbon fibre reinforced plastics are conductive and galvanic corrosion effects may arise when being coupled to a metal component. Some types of corrosion in metals, such as crevice corrosion, may not be found in FRPs but some ageing effects can cause damage in FRPs that is not observed in metals – such as swelling, osmosis blistering, microcracking and physical property changes.

For metals, there are well-developed testing methods and evaluation techniques for each type of corrosion and the data for specific materials can easily be compared and be used in the selection of materials and in the corrosion design. Analogous methods and techniques for testing and evaluation of the corrosion resistance of FRP materials must be worked out, and key procedures for obtaining the relevant ageing data and relationships for design have to be developed.

Based on the results of corrosion analyses of samples taken from FRP equipment used for different chlorine dioxide environments in various pulp mill applications over a period of 20 years, a semi-empirical relationship for uniform corrosion behaviour of FRP in chlorine dioxide environments has been established (Bergman 2004):

$$\mathbf{\Phi} = Bt^{\alpha}cAe^{-E_a/(RT)} \qquad [17.2]$$

where: Φ is the depth of corrosion (mm); B is a special factor for the case of protective deposits on the surface (usually $B = 0$ or 1); t is the time in service (years), α is a factor that depends on the thickness and the degree of degradation of the corroded surface layer (usually α is between 0.5 and 1); c is the concentration of chlorine dioxide (g/l); A is a material constant that depends on the type of resin, the degree of curing and the laminate structure; E_a is the activation energy of the rate-controlling step of the corrosion process (J/mole); R is the general gas constant (8.3 J/mole/K); T is the temperature (K).

This expression is not proven for other applications and environments, although the premise should hold.

17.6 Case studies of ageing assessment approaches

17.6.1 ASTM C581

To demonstrate the use of ASTM C581, an FRP laminate that may be used in the linings of tanks and vessels used to transport acids was characterized. Typically, plaques of materials, nominally measuring 100 × 125 mm are immersed in concentrated acids such as HCl or H_2SO_4 for periods up to 12 months. Periodically, Barcol hardness, mass change and flexural properties are measured on specimens cut from these plaques. Schematic data for a typical test programme are shown in Figs 17.4 to 17.6. The mass change is shown in Fig. 17.4, illustrating that the material shows weight loss early in the exposures and that the rate of weight loss begins to reduce as the test progresses.

The change in Barcol hardness, shown in Fig. 17.5, demonstrates that there is an initial increase, indicating that some form of local hardening has occurred. After 4 months' exposure the hardness begins to decrease, leading to a 10–15% decrease from the starting values after 12 months. This indicates that there is more than one ageing mechanism occurring and the

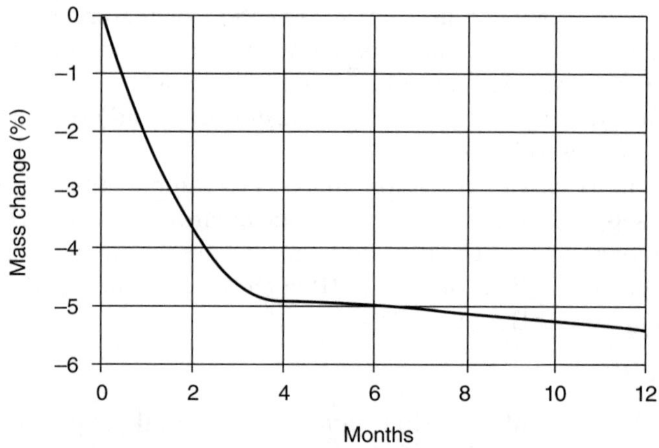

17.4 Mass change of an FRP in concentrated acid.

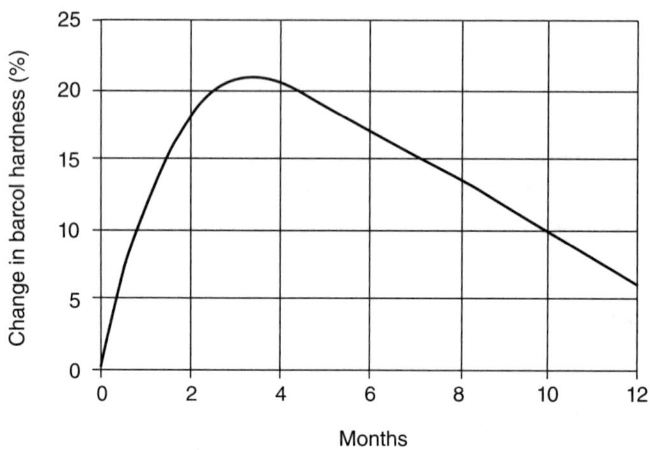

17.5 Percentage change in Barcol hardness of an FRP in concentrated acid.

resulting change in properties cannot be taken from only the end data point.

The flexural modulus and strength losses are shown in Fig. 17.6. The overall trend is that there is minimal degradation in properties for 6 months and then the strength and modulus decrease, perhaps from materials degradation. The flexure test is a through-thickness test measuring surface and bulk properties. Hence, surface diffusion and ageing, such as observed in Fig. 17.5, may not be revealed by flexural data. Therefore, it is necessary to understand competing ageing mechanisms and to be able to differentiate the different effects.

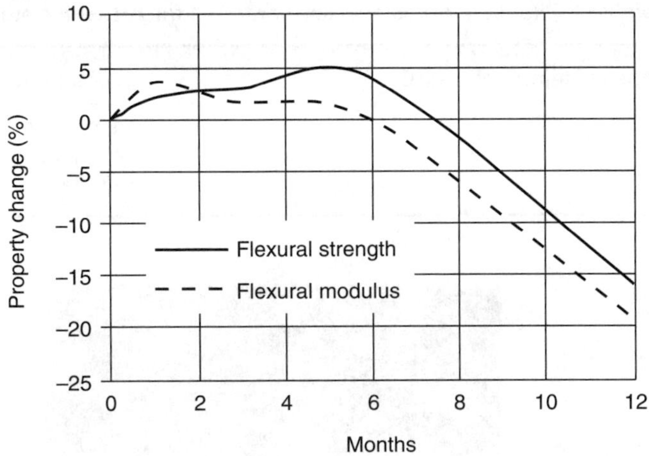

17.6 Change in flexural properties (strength and modulus) of an FRP in concentrated acid.

17.6.2 Arrhenius relationship

This section presents a case study on the use of Arrhenius plots as a means to accelerate the ageing in FRP materials, using the different test parameters used in ASTM C581, thus identifying if such an approach is valid (Martin 2008). The shorter-term exposures would be able to predict the longer-term exposures. If so, then it would be possible to replace the long exposure specified in ASTM C581 with several shorter exposures. The key to this approach being successful is that the elevated temperature exposures do not cause a change in the ageing mechanisms observed at lower temperatures. Therefore, as a first step, all exposure temperatures must be below the T_g.

For this case study, a Derakane 411 material supplied Free Issue by Dow was used. This is a vinyl ester resin and the laminates were fabricated with two plies of 1½ oz (42 g) E-glass chopped strand mat, with a 10 mil (0.25 mm) C-glass veil. This material is reported to have a maximum working temperature when exposed to concentrated (37%) hydrochloric acid (HCl) of 65 °C and 80 °C in HCl at a concentration of less than 20%. To ensure property changes would occur, exposure temperatures in excess of these temperatures were chosen for exposure to 37% HCl. The temperatures were 40 °C, 60 °C and 80 °C. The exposure work was funded by MTI Inc., USA.

The specimens were provided as 120 mm × 125 mm plaques. The edges of the plaques were coated with Derakane resin to prevent any adverse effects of the exposure fluid attacking through the edges. All exposures were conducted in sealed, acid-resistant glass vessels in an oven. The

Table 17.2 Exposure times and temperatures for Arrhenius study

Exposure temperature (°C)	Exposure time (days)
40	28, 84, 180
60	14, 28, 42
80	3, 7, 14

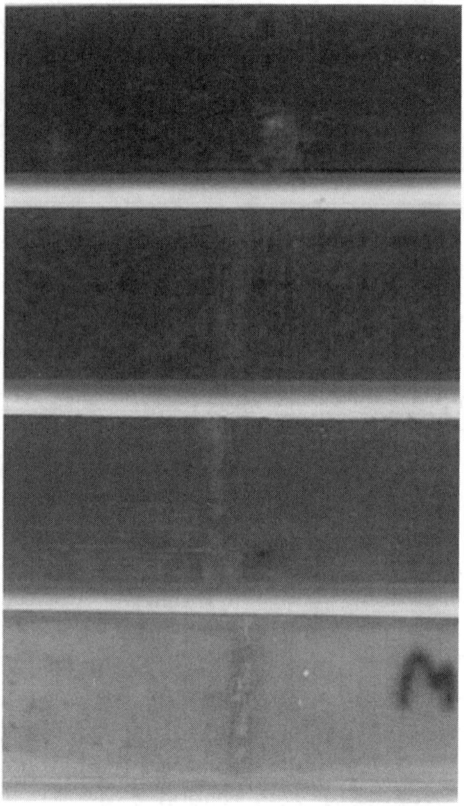

17.7 Appearance of Derakane 411 FRP flexure specimens after exposures in concentrated HCl.

specimens were placed in the vessels using 7 mm thick spacers so that the acid could reach all surfaces. The test temperatures and durations are given in Table 17.2. Before and after the exposures, thickness and weight in air and water were measured to determine any changes. Following the ageing and the weighing, suitable specimens were cut from the plaques. A summary of the appearances of the specimens is given in Fig. 17.7.

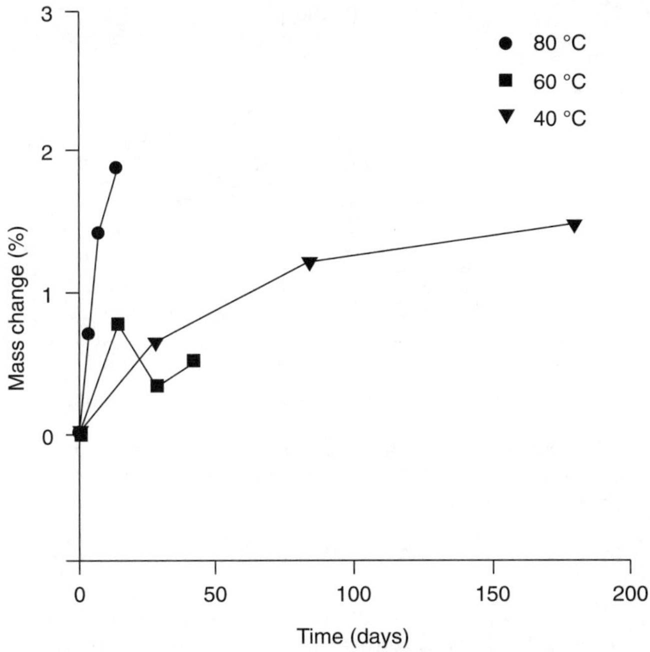

17.8 Percentage mass change of Derakane 411 FRP after exposures in concentrated HCl.

The mass changes are presented in Fig. 17.8, showing an increase in mass. Because of the C-glass surface veil, there may well be two diffusion rates complicating the conventional Fickian-type diffusion coefficient generation and time to saturation. Barcol hardness measurements were also taken, very small changes were measured for all exposures and therefore are not shown. The results of the flexure tests are given in Fig. 17.9 along with the approximate time to give a 5% and 10% decrease in flexural strength, which can be used for plotting in the Arrhenius approach.

For the 40 °C exposure, the modulus initially increased before decreasing. This is not atypical for fluid exposure where the elevated temperature exposure may result in further cross-linking. At the higher temperature exposures of 60 °C and 80 °C, the modulus and strength both decrease with time. This is consistent with the swelling of the material with HCl solution absorption. The significantly lower strength and modulus after 2 weeks at 80 °C is a result of the blisters that developed.

One method of establishing whether the physical changes from exposure noted above are a result of chemical changes is to measure the T_g. A differential scanning calorimeter (DSC) was used to identify T_g. Fragments were taken from the post-exposed specimens and placed in a container. The

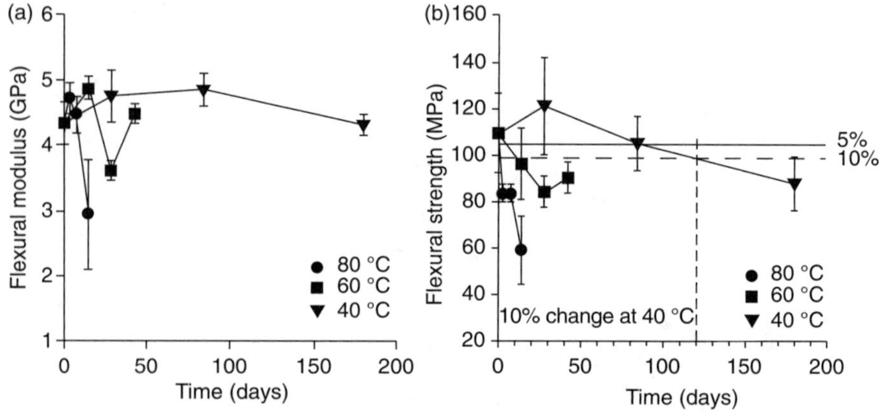

17.9 (a) Flexural strength and (b) modulus changes with exposure time of Derakane 411 FRP after exposures in concentrated HCl (error bars show standard deviation).

T_g values are given in Fig. 17.10. The T_g also decreased with time, and the rate of this decrease increased with temperature, although some scatter is seen in the data.

There is an overall trend from test to test of a property change with exposure time, the rate of which increases with time. It is this general concept that lays the foundation for developing accelerated ageing approaches using the Arrhenius equation. However, as for the work in Section 17.6.1, the data show that more than one mechanism may be occurring, thus potentially invalidating the Arrhenius approach, as discussed below. Nevertheless, these data were plotted using the Arrhenius approach and are shown for a 5% and 10% change of flexural strength and T_g in Fig. 17.11, plotted against the reciprocal of the absolute exposure temperature.

A linear fit to the T_g data points is shown but the fit of the curve is less than ideal because of the scatter in the data. The values at 80 °C act to lower the activation energy to about 10 kcal/mole, which is what would be expected for physical changes (e.g. single swelling). Because such a small sample is used for T_g analysis, this can affect interpretation where physical and/or chemical changes occur simultaneously.

In order to improve confidence in the use of this curve for extrapolation to longer durations at lower temperatures, several further exposures should be conducted. With an increased number of exposure temperatures, and a larger number of replicate samples, a statistical band could be placed on these curves. There is a better straight-line fit to the flexural strength data because of the greater number of replicate tests conducted. Therefore, the

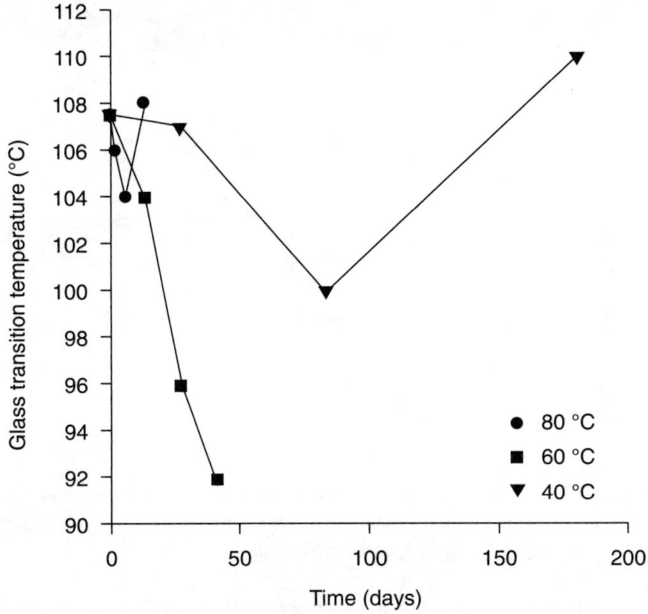

17.10 Changes in T_g with time of Derakane 411 FRP after exposures in concentrated HCl.

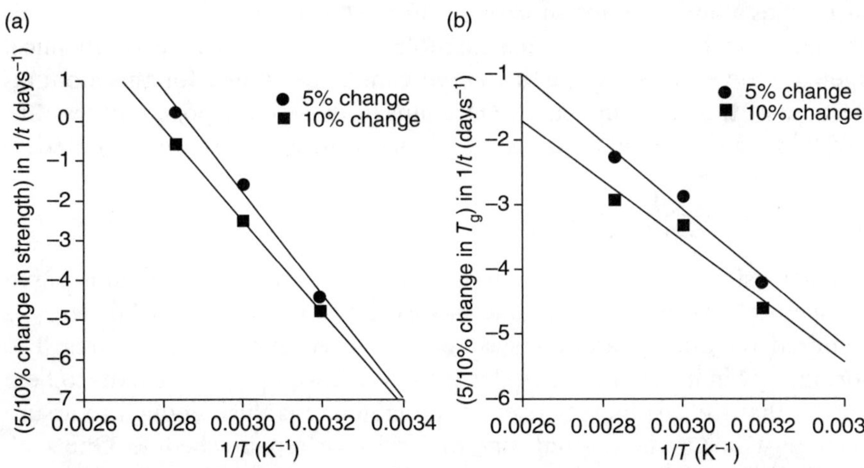

17.11 Arrhenius plot of the changes flexural in strength (a) and T_g (b).

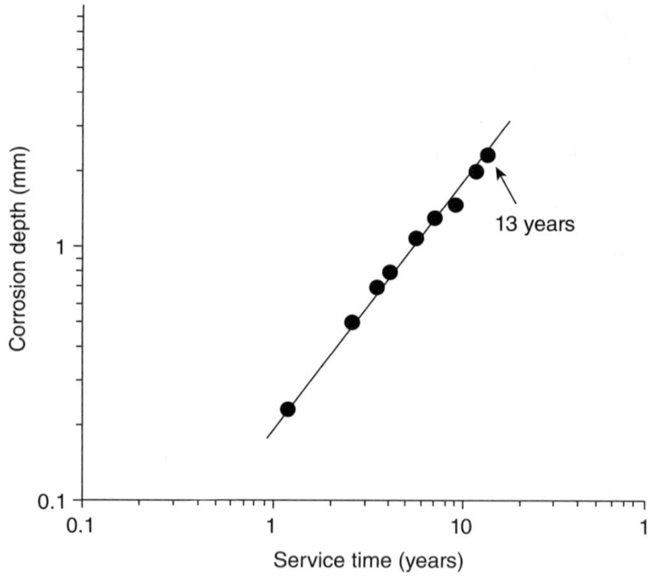

17.12 Corrosion depth versus service time in FRP pipes used for cold, concentrated ClO_2, solution (Bergman 2001).

shorter term, higher temperature exposure tests are able to predict the properties at lower temperature exposures. The energy of activation for this change is in the region of 20 kcal/mole, which is indicative of chemical ageing. However, an Arrhenius curve cannot be drawn for the modulus because of the initial increase in modulus in the early exposure times, thus invalidating this method unless the different mechanisms are separated.

17.6.3 Corrosion approach

The corrosion approach illustrated in the references by Bergman is based on semi-empirical data from one in-service exposure and further work is required on other materials systems and environmental exposures. The formula given in Equation [17.2] was used to develop design curves to help ensure that the chemical barrier of the laminate is thick enough to assure reliability during the desired time of service or repair schedule. One such curve (Bergman 2001) is shown in Fig. 17.12.

17.7 Concluding remarks

While the use of FRPs in the chemical processing industry is perhaps one of the oldest applications of these materials in a hostile environment, the method for assessing long-term ageing is still largely based on field experience and inspections rather than on a deeper understanding of the mechanisms of degradation. Despite this, these materials have been used for the most part with great success and they will continue to be so used. However, with tighter legislation for human and environmental protection, along with a desire to reduce inspection costs and extend the usable life of existing equipment and infrastructure, the industry would benefit from the development of a more rigorous long-term exposure methodology.

17.8 References

ANON. The strong advance of corrosion resistant GRP applications, *Reinforced Plastics,* 1991, **35**(1), 26–34.

ANON. GRP solves BASF's corrosion problems, *Reinforced Plastics*, 1994, **38**(9), 24–26.

ASTM C581-94 Standard Practice for Determining Chemical Resistance of Thermosetting Resins Used in Glass-Fiber-Reinforced Structures Intended for Liquid Service, American Society for Testing of Materials, West Conshocken, PA.

BERGMAN, G. *Corrosion of Plastics and Rubber in Process Equipment – Experiences from the Pulp and Paper Industry*, TAPPI Press, Atlanta, GA, 1995.

BERGMAN, G. Take the guesswork out of FRP corrosion, *Chemical Engineering Progress*, 2001, **97**(12), 54–59.

BERGMAN, G. Unexpected stress corrosion failures of high quality FRP process equipment pipes. In Corrosion 2004, Houston, TX, NACE Paper 04611.

BERGMAN, G. Environmental stress-corrosion cracking of fibreglass: Lessons learned from failures in the chemical industry, *Journal of Hazardous Materials*, 2007, **142**(3), 695–704.

FRENCH, M.A. and PRITCHARD, G. The fracture surfaces of hybrid fibre composites, *Composites Science and Technology*, 1993, **47**, 257–263.

HOGG, P.J. and HULL, D. Micromechanisms of crack growth in composite materials under corrosive environments, *Metal Science*, 1980, **14**, 120.

HOGG, P.J. and HULL, D. Corrosion and environmental deterioration of GRP. In: B. Harris, Editor, *Developments in GRP Technology*, vol. 1, Applied Science Publishers, London, UK, 1983, pp. 37–90.

MARSHALL, G.P. KISBENYI, M. HARRISON, D. and PINZELLI, R. Environmental stress corrosion of chemically resistant polyester resins and glass reinforced laminates in acid. In *Proceedings of the 37th Annual Conference, Reinforced Plastics/Composites Institute*, The Society of the Plastics Industry, Inc., January 11–15 1982, Session 9-D, pp. 1–8.

MARTIN, R.H. MORGAN, G. LEWAN, M. Accelerating the Environmental Conditioning of FRP in the Chemical Processing Industry. Materials Technology Institute, Inc., St Louis, 1998.

NELSON, J.K. FRP applications in the chlorine industry. Western SPI Conference, April 1988.

PRITCHARD, G. Reinforced plastics in anti-corrosion applications. In *Advanced Composites*, Elsevier Applied Science Publishers London, 1989, pp. 163–196.

ROBERTS, R.C. Service and performance of GRP and thermoplastic lined GRP pipes and vessels in chemical process plants. In *Proceedings of the Institution of Chemical Engineers Symposium on Reinforced Plastics Constructed Equipment in the Chemical Process Industry*, Manchester, UK, March, 1980, Paper 10.

SCOLA, D.A. *Developments in Reinforced Plastics* – 4, Ed. G. Pritchard, Elsevier Applied Science Publishers, London, 1984, Chapter 5.

TALBOT, R.C. A review of FRP performance in the chlor/alkali, pulp and paper, and mining industries. Western SPI Conference, April 1990.

ZWEBEN, C. RODINI, B. and THAW, C. Advanced composite materials for process industries and corrosion resistant applications. In *Proceedings of the Advances in Materials Technology for the Process Industries*, National Association of Corrosion Engineers, Texas, 1985.

18
Ageing of composites in underwater applications

D. CHOQUEUSE and P. DAVIES,
IFREMER Brest Centre, France

18.1 Introduction

This chapter addresses the ageing of composites in water. The mechanisms and modelling of hydrothermal ageing of composites have already been developed in Chapter 8 of this volume, and one of the main application sectors concerned by the ageing of composites in water has already been described in Chapter 12 'Ageing of composites in marine vessels'. This chapter will therefore focus on the underwater applications of composites, including the deep sea applications of these materials. The main sectors concerned with this topic are the offshore industry, military sub-sea applications and oceanography for deep sea exploration. A new area, which will expand in the near future, is renewable ocean energy.

Over the last 10 years, demand for the use of composites in deep offshore applications has been increasing rapidly. Light weight is critical for submarine structures, in order to facilitate their underwater deployment, and various specific properties (buoyancy, thermal insulation, non-corrosion, etc.) strongly favour the use of composites. In the military sectors (submarine applications), the use of composites has focused on external parts, the pressure hulls of submarines are still metallic. However, some very large components (sonar domes, external decks) are made from composite materials. Finally, there is also currently considerable interest in the development of devices to convert marine energy (waves, tidal motion, currents) to electricity. Although marine energy has been used for many years, the increase in oil prices and the European directives on renewable energy have fostered a number of innovative designs for ocean energy conversion devices in recent years.[1] This is potentially an activity where composite materials can play a significant role, as corrosion resistance and light weight are of critical importance. Difficult access for maintenance makes long-term reliability a major design consideration.

In Section 18.2 of this chapter, the environmental parameters of the deep sea environment will be recalled, as these are the parameters affecting the ageing of composites. In Section 18.3 the mechanisms involved in the long-term behaviour of composite materials will be discussed. To conclude the chapter, three case studies are presented; the first focuses on a long (10-year) study of glass/epoxy composites, the second illustrates the influence of hydrostatic pressure on ageing of a range of materials and the third case study examines syntactic foams, essential for the offshore industry and a key element in many underwater structures.

18.2 Deep sea environmental parameters

18.2.1 Chemical composition and properties of sea water

The principal components of sea water

The composition of sea water is governed by Marcet's principle;[2] Marcet remarked that, while for specimens sampled in different areas of the oceans the global salt content was different, all of them held the same component in the same proportion, as shown in Table 18.1. This is now known as 'Marcet's principle' or 'Dittmar's law', and justifies the concept of salinity of sea water. The ionic composition of continental water differs quantitatively and qualitatively from sea water. Lakes and rivers have varying mineralization depending on their origin and hydrolytic regimes (Table 18.2).

The equation of state for sea water was published by UNESCO[3] and allows the principal properties of seawater to be determined if the temperature, the salinity and the pressure are known. For sea water, a multi-electrolytic solution, the ionic force is equivalent to a 0.7 M NaCl solution,

Table 18.1 concentrations of species present in sea water with salinity of 35 g/l

Species	Concentration (g/kg)
H_2O	964.85
Na^+	10.77
Mg^{2+}	1.29
Ca^{2+}	0.4121
K^+	0.399
Sr^{2+}	0.0079
B_{total}	0.0045
Cl^-	19.354
SO_4^{2-}	2.712
$HCO_3^- + CO_3^{2-}$	0.118–0.146
Br^-	0.0673
F^-	0.0013

Table 18.2 Continental water salinity

	Salinity (g/l)
River	0.1
Lake Superior (USA–Canada)	0.06
Caspian Sea	12.8
Global ocean	34.85
Dead Sea	276.0

Table 18.3 Pressure versus depth

Pressure (MPa)	Depth (m)
5	496
10	991
20	1977
50	4908
100	9712

but due to its complexity no relationship has been established between conductivity and ionic composition. Therefore, the activity and ionic forces of seawater are quite complex to establish.

18.2.2 Pressure and depth effects

The hydrostatic equilibrium of a fluid in a gravity field can be simply formulated as $dp/dz = g\rho$ where p is pressure, z is depth, g is gravity and ρ is specific gravity. For oceanographic applications, considering the constancy of the specific gravity, the following expression can be written:

$$p \approx g\rho z + p_0$$

This can be expressed following the AFNOR standard XP P 10-812 by:

$$p = 0.0101z + 0.05 \times 10^{-6} z^2$$

where p is pressure in MPa and z is depth in metres.

Table 18.3 gives some examples of the correspondence between pressure and depth for a water of salinity 35 g/l, $t = 0\,°C$ at 30° latitude. The parameters that are greatly influenced by depth are the temperature and the dissolved oxygen content. The temperature difference between shallow and deep water can reach 20 °C. The dissolved oxygen content is variable from one ocean to another and depends on the depth. It is less than 1 ppm in deep sea water and equal to 8–9 ppm in shallow water, because of the algal photosynthesis activity. Some parameters (conductivity, viscosity, etc.) are closely linked to the water structure modifications and these are less affected by increasing pressure. These parameters vary by less than 5%. Sea water is modelled as a solvent (water molecules) in which ions (inorganic salts), organic compounds, organometallic complexes and micro-organisms are dissolved.

The properties of shallow water are strongly influenced by energetic exchanges between the atmosphere and the ocean, but the stable stratification of the ocean confers time-dependent properties to deep sea water that are only influenced by deep sea water circulation.

Temperature

Temperature decreases with depth. This decrease is not linear except for low depths. Shallow water (0–100 m) presents a large temperature gradient in summer time. Temperature then decreases down to 500 m. Deep sea water temperature decreases because of the high pressure effect.

Salinity

Shallow water salinity depends on evaporation and precipitation. Deep sea water salinity depends on ocean circulation, and there is not a clear variation of salinity with depth.

Dissolved oxygen contents

The oxygen in sea water comes from the atmosphere and photosynthesis. It is consumed by organic substance oxidation and the breathing of living organisms. The dissolved oxygen content of shallow water can vary very quickly. It is close to saturation and oversaturation can occur due to photosynthesis. The dissolved oxygen content of deep sea water depends on ocean circulation. As cold deep water comes from polar areas (shallow water), the dissolved oxygen content can be high. The minimum oxygen contents measured in middle-depth sea are probably due to high organic substance oxidation.

pH

Sea water pH varies from 7.5 to 8.5. It is regulated by the equilibrium of the carbonic gases system, bicarbonates and carbonates; pH is connected to the dissolved oxygen content (photosynthesis increases pH by decreasing CO_2 content), low oxygen content water has a lower pH.

Oceanic currents

Marine water movements are identified by their nature, geographic location and duration, by the boundary conditions set by sea floors and coasts, and by their volume of moved water. Shallow water currents are more important than deep ones because of energetic exchanges with the atmosphere and stable stratification.

Fouling

Any materials immersed in sea water will potentially be subject to biological attack. The organisms involved depend on the substrates. These can be categorized in three groups: slime (involved in microbiological damage), hard fouling (animals with calcium carbonate shells) and soft fouling (algae and animals with soft structures). These animals may be involved in damage development due to increasing load, surface alteration and reduced exchanges between the surfaces and the medium. The majority of shallow water organisms are also present in deep sea. Nevertheless, their density is then very low, slime development duration is much less and fouling is less hard.

Other parameters

The specific volume, the compressibility and the thermal expansion coefficient depend on the temperature, salinity and pressure. Changes in these values with pressure increase are very small. In order to give an idea of the variation of the environmental parameters of the ocean, an analysis of data collected in different data bases[4] is given (mean annual values) in Tables 18.4 to 18.7), and the surface temperature of the ocean is shown in Fig. 18.1. Six locations of potential interest for offshore exploitation have been chosen.

Table 18.4 Locations selected for date on physical, properties, reported in Tables 18.5 to 18.7

	Maximum depth (m)	Latitude range	Longitude range
Gulf of Guinea	5000	N 13°–S 30°	W 18°–E 18°
Gulf of Mexico (GOM)	4000	N 31°–N 20°	W 98°–W 79°
East of Brazil	5000	S 19°–S 30°	W 43°–W 29°
West of Scotland	3000	N 63°–N 53°	W 15°–E 1°
North west of Norway	3500	N 70°–N 55°	W 9°–E 1°
North west of Australia	6000	S 5°–S 25°	E105°–E 130°

Table 18.5 Temperature: maximum and minimum values (1° resolution)

Depth	Guinea	GOM	Brazil	Scotland	Norway	Australia
100 m	14/22	20/26	18/23	5/11	1/10	19/24
500 m	6/10	8/16	8/10	1/10	1/9	7/10
1000 m	3/5	4/6	3/4	0/7	0/7	4/5
2000 m	3	4	3/4	−1/4	−1/1	2/3
3000 m	2/3	4	3		−1	2/3

Table 18.6 Salinity (Practical Salinity Unit, PSU): maximum and minimum values (1 g/l resolution)

Depth	Guinea	GOM	Brazil	Scotland	Norway	Australia
100 m	35/37	36	36/37	35	35	34/35
500 m	35	35/36	35	35	35	35
1000 m	34/35	35	34/35	35	35	35
2000 m	35	35	35	35	35	35
3000 m	35	35	35		35	35

Table 18.7 Dissolved oxygen: maximum and minimum values (1 ml/l resolution)

Depth	Guinea	GOM	Brazil	Scotland	Norway	Australia
100 m	1/5	3/5	5	6/7	6/7	3/5
500 m	1/4	2/4	4/5	5/7	6/7	2/5
1000 m	3/4	4	4/5	5/6	5/7	2/3
2000 m	5/6	5/6	5/6	6/7	7	2/3
3000 m	5/6	4/5	6		7	2/4

18.3 Ageing of composites in water

In this section a brief review will first be given of the phenomena involved during the ageing of polymer matrix resins and their reinforcement (glass). The kinetics of water absorption will then be described and illustrated by various examples. Finally, the influence of ageing on mechanical properties will be discussed.

18.3.1 Hydrothermal ageing mechanisms of polymers

Ageing of polymers consists of a slow and irreversible change in the material properties,[5,6] due either to their own inherent instability or to environ-

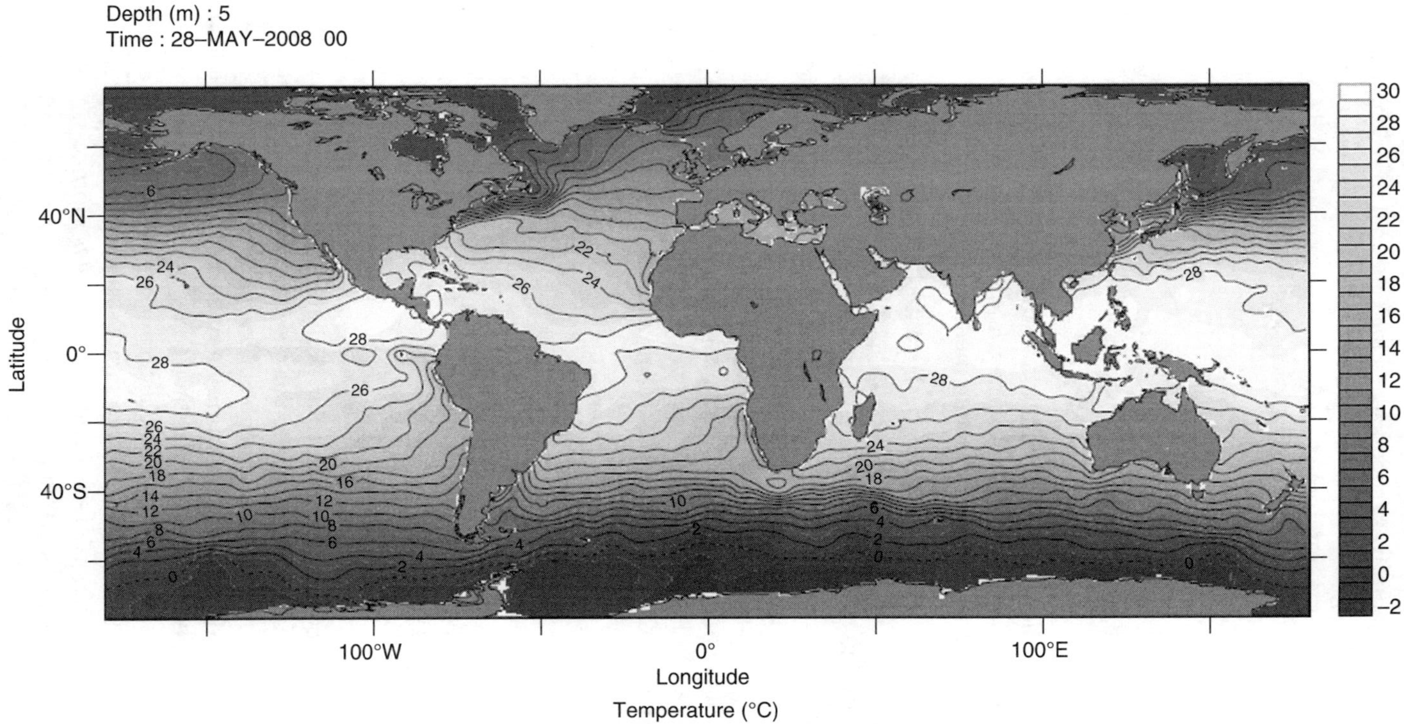

18.1 Sub-surface temperatures of the ocean (5 m depth). (Source: Coriolis).

mental effects. This degradation may concern the chemical structure of the macromolecule or of additives (chemical ageing), the material composition (introduction or release of small molecules) or its physical state (degree of crystallinity, free volume, internal stresses). A polymer exposed to water can evolve due to both chemical and physical ageing.

Physical ageing

Physical ageing corresponds to a modification of the characteristics of the material without any change of its chemical structure. This can result in:

- modification of the spatial configuration of the macromolecules;
- transport phenomena (liquid ingress, migration of additives);
- surface phenomena (crazing in certain media).

Physical ageing is due to the instability of the material below its glass transition temperature (T_g). Loss of additives (evaporation, weeping, extraction, etc.) can modify the physical properties of the material. If plasticizers are lost, a decrease in elongation to break, T_g and Young's modulus can be observed.

Swelling may occur due to liquid (or vapour) uptake, which induces an increase of volume and may cause internal stresses. Plasticization is a consequence of the diffusion of water, which modifies the structure and composition of the material generating a parallel decrease of the T_g and of the modulus in the glassy state.

Chemical ageing

Ageing is related to structural changes: polymeric chain breakage by hydrolysis and evolution of small molecules produced by the degradation process. The hydrolysis reaction can be written as

$$\sim X—Y\sim \rightarrow \sim X—OH + HY\sim$$

Hydrolysis induces cuts in the molecular skeleton and a decrease of the cross-link density or of the mean molecular mass between links. Hydrolysis phenomena are generally limited at ambient temperatures and the process is controlled by the diffusion of water in the material; their effects add to the physical effect. Hydrolysis can be catalyzed by specific chemical agents: H^+ and OH^- ions, etc. This is why the chemical composition of the ageing environment affects the kinetics. Esters, amides and imides are the main hydrolyzable groups. In order to identify a hydrolysis process, material samples must be reconditioned (in general by drying) in order to eliminate the reversible plasticization effect.

Absorption kinetics

The model most frequently used to describe the water absorption of polymeric materials is the diffusion model described by Fick. An initial linear weight gain is observed, proportional to the square root of exposure time, followed by a plateau. Theoretical analyses have been performed by Crank.[7] More complex models have been proposed by Langmuir and others,[8] which take into account two types of water inside the material: free water and linked water. These models describe a diffusion process coupled to a chemical reaction.

Solubility parameters

The affinity between a liquid and a material is stronger when the difference in solubility parameters between liquid and material is small. The hydrophilic nature of a material is defined as its affinity with water, the diffusivity is the speed of penetration of the water into the material. This affinity can be defined as the saturation level reached at equilibrium at a relative humidity close to 100%. The hydrophilic response is a function of the polarity of the chemical groups of the material and of their concentration. Examples for different polymeric materials are reported in Table 18.8. The hydrophilic nature of a polymer is linked with the partial solubility parameter of the hydrogen link. As this parameter is difficult to determine, an empirical rule is used which states that the water absorption can be estimated by summing molar fractions.

Table 18.8 Hydrophilic affinity of polymers

Hydrophilic affinity	Example	$M_{saturation}$
Very low	Polydimethylsiloxane Polyethylene	<0.5%
Low	Styrene cross-linked UP Styrene cross-linked vinylesters Anydride cross-linked epoxy	0.5–1.5%
Moderate	Vinylesters Amine low cross-link density epoxy Polyimide	1.5–3%
High	Amine high cross-link density epoxy Polyimide	<3%

$M_{saturation}$, % weight gain at saturation; UP, Un saturated polyester.

Parameters acting on ageing mechanisms

- *Temperature.* Chemical reactions are thermally activated and the increase in diffusivity with temperature can be described by an Arrhenius law.
- *pH.* This is linked to hydrolysis and the associated chemical reactions.
- *Nature of water.* In a liquid environment, the water content at equilibrium is linked with the chemical potential of water. This means that the saturation level decreases when the concentration of dissolved matter in the water (in particular, salt) increases.

18.3.2 Ageing of glass in contact with water

Dissolution of glass is a well-known phenomenon in nuclear power plant engineering.[9] The mechanisms involved are the following:

- hydration which corresponds to the ingress of water molecules into the glass;
- ionic exchange, where cations included in the glass (for example, Na^+) are replaced by the proton of water;
- hydrolysis of the cation–oxygen–cation links of the glass to create hydroxyl links (reversible phenomena).

The kinetics and nature of glass degradation are strongly governed by the composition of the glass. For improved chemical resistance, the glass industry uses borosilicate glass, but in the composite glass fibre industry E-glass is generally used. Some fibres, for example, Advantex® glass, have been specially developed to improve long-term resistance in contact with water. Typical glass compositions are shown in Table 18.9 below. The effects of

Table 18.9 Typical glass compositions (weight %)

	E-glass fibre	Glass microsphere	Borosilicate	Bottle glass	Advantex®
SiO_2	54	77.6	81	73.3	59–62
Al_2O_3	15	1.15	2	1.5	12–15
B_2O_3	8	1.15	11.4		<0.2
As_2O_3			0.3		
Fe_2O_3			0.15	0.04	
SO_3		0.7		0.4	
TiO_2			0.05	0.02	
MgO	5	0.25	0.2	0.1	1–4
CaO	17	6.25	0.3	9.8	20–24
Na_2O_3		9.65	4.5	14.2	
ZnO		2.55			
K_2O			0.1	0.6	

humidity and water on sub-critical crack growth in glass fibres have been studied,[10,11] and a long-term behaviour model has been proposed.[12]

18.3.3 Ageing of composite material in deep sea

Fibre reinforced plastics (FRPs) are now commonly used in the fabrication of the structure of deep sea devices: frames of manned or unmanned vehicles, containers, propellers, etc. These FRPs are selected because of their good specific mechanical properties and their resistance to sea water. Some other properties – such as damping of acoustic waves, non-magnetic behaviour, transparency to radiation and electrical resistance – are sometimes also required. In addition, for specific low-density applications and low thermal conductivity, materials such as syntactic foams have been developed for buoyancy and passive thermal insulation, particularly in deep offshore field applications. These will be described in more detail in Section 18.6.

In spite of their attractive properties, these materials can be affected by degradation mechanisms, for example:

- ageing due to water absorption;
- mechanical damage due to static loading under hydrostatic pressure, sometimes combined with creep, and dynamic and impact loads;
- biological attack, in certain cases.

With time, the physical and mechanical properties can be affected by the attacks on the resin and fibres, and degradation of the interface between matrix and fibre or reinforcement. Different mechanisms can be expected depending on the nature of the resin, the reinforcement and the properties of the environment. These FRP materials are sensitive to the humid environment. The water molecule, by nature, has a strong tendency to create intermolecular hydrogen bonds in polymers with negative atoms. Water can go through the material in different ways: by diffusion through the resin, by capillary action (wicking) along the matrix–reinforcement interface and by filling of internal defects (pores, voids, cracks, etc.).

Water absorption kinetics in polymers have been widely studied. Tests are generally performed in tap water, and usually show Fickian diffusion for the material where the water uptake does not generate irreversible degradation. The presence of non-reversible damage results in more complex kinetics. It should be noted that, in order to accelerate water uptake and mechanical degradation, the acceleration parameter that is most commonly used is temperature. It is generally assumed that the maximum temperature for accelerated tests should be inferior to 30 °C below T_g. This can pose a problem when the matrix material is not completely cured after the manufacturing process, particularly where the industrial process does not include post-curing. This is classically encountered for large naval structures, where curing at high temperatures is too difficult (cost, availability of equipment, etc.). The material will then tend to post-cure naturally in service, due to sun and heat exposure. This under-curing is not the case for aeronautical structures, and this must be taken into account, because the accelerating approach used in the aeronautical industry cannot be extrapolated directly to marine structures. When the latter are manufactured with polyester and vinylester resins, they are generally not post-cured. For epoxy systems, the trend is to use prepreg systems which will be cured, but repairing, and secondary bonding are often performed with room-temperature cure systems with no post-curing.

For submarine structures, it may be considered that natural post-curing will not occur, as service temperatures are low, so post-curing of the material is strongly recommended in order to avoid unexpected material behaviour in service.

Regarding the testing of ageing, another point that should be considered is the chemical attack of the glass mentioned above. For temperatures above 70 °C,[13] detrimental attack of the glass has to be considered; this sets an upper limit to the temperature of accelerated ageing tests. This was confirmed recently by tests on syntactic foam, which will be described in Section 18.6. Carbon fibre, on the other hand, is generally very stable in contact with water. However, carbon fibre composites may pose problems when used underwater and particularly in sea water when the composite material is directly in contact with metallic parts. Corrosion problems may be induced due to electrochemical coupling between the materials generating galvanic corrosion.[14]

A final point to consider is the use of composite materials in contact with metallic structures that are subjected to cathodic protection. Disbonding of the composite structure may occur due to local chemical modification of the medium, as a result of electrochemical reactions. Cathodic protection can significantly increase the local pH of the media (pH 13–14), which will cause problems during long-term contact with composite materials. This phenomenon is called 'cathodic disbonding' and is well known in pipe protection engineering.[15]

18.4 Case study 1: composite tubes

In the following case studies, some examples of both water absorption kinetics and the evolution of the mechanical properties of typical marine materials will be described, based on data from previous studies performed at IFREMER, France.

Case study 1 involves a major study performed to evaluate glass/epoxy composite tubes for cooling water applications. This included long-term ageing of tubes at different temperatures and internal pressures, and studies

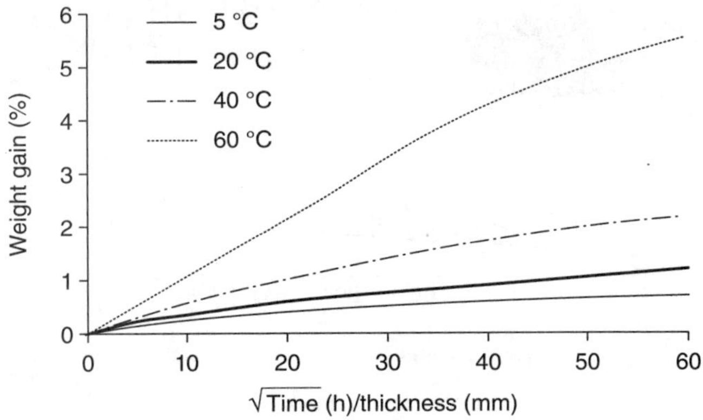

18.2 Typical absorption curve: 2.2 mm thick glass/epoxy contact moulded plates.

of different resin hardeners and the influence of liners. Some of the results have been published previously.[16,17] The results shown hereafter are from tests on plates used within that study to examine water absorption kinetics and property degradation. In a complementary study, part of another large programme on the use of composites for naval applications,[18] filament wound glass and carbon fibre reinforced epoxy were immersed in water at different temperatures for periods up to 1 year. These materials are intended for underwater applications.

The absorption curve of a 2.2 mm thick glass/epoxy composite plate made by contact moulding is presented in Fig. 18.2. After 2 years of immersion in deionized water, the absorption curves exhibit a Fickian behaviour with a saturation level of around 5–6% by weight. Using classical analysis, the diffusion coefficients can be calculated for each temperature and then the activation temperature factor for Arrhenius's law can be determined, which allows long-term behaviour of the material to be predicted. This approach is traditionally used, and the relation between water uptake and mechanical degradation may then be established using different approaches (see below). However, when long-term *in situ* tests have been performed, the correlation with extrapolations from this type of accelerated test is usually poor, and when the degradation kinetics induced by water absorption in a system are not well established, long-term testing of the material is strongly recommended, in order to guarantee long-term behaviour.

In order to illustrate this, an example is presented (Fig. 18.3), in which the weight of an FRP plate (glass/epoxy) immerged in water at different temperatures (20, 40 and 60 °C) has been followed for more than 10 years. This material was representative of a material used for fluid transport tubes, in which cooling water was to flow at temperatures up to 60 °C. The plates

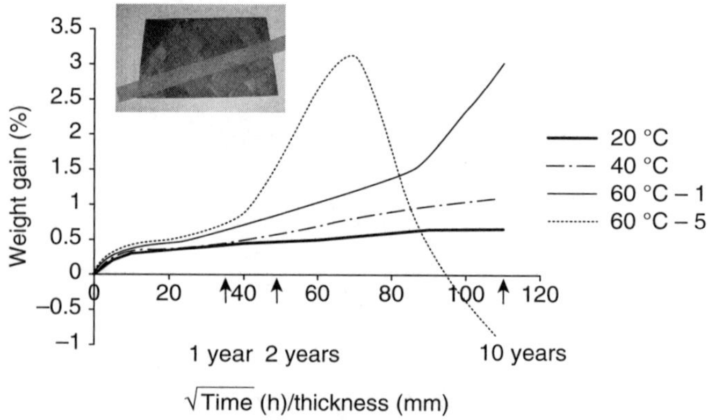

18.3 Long-term absorption curve: glass/epoxy filament wound plate; 60 °C – 1 and 60 °C – 5 refer to plates 1 and 5, respectively.

were manufactured by the filament winding technology in order to use the same process as the final product. During the first year, the water uptake kinetics seem to follow a Fickian behaviour. Over this period, the initial diffusion coefficient can be determined and, by using an Arrhenius law, the activation coefficient of the diffusion with temperature can be established. This approach is based on the equivalence principle and is often used to predict long-term behaviour. However, for longer periods (from 2 years), degradation phenomena can be initiated and the water uptake kinetics differ from the previously established Fickian model. In addition, two nominally identical samples (plates 1 and 5) from the same manufacturing batch, immersed at the same temperature (60 °C) showed different behaviour. The activation of the irreversible degradation phenomena (matrix hydrolysis here) started at quite different times (1 year for one plate and about 3 years for another one). This difference in behaviour has also been observed for samples immersed at 40 °C. No significant differences in terms of physico-chemical analysis (differential scanning calorimetry (DSC), infrared spectroscopy (IR), etc.) were detected when the material was in its initial state.

Different techniques may be used to evaluate the degradation of the material. Classically, quasi-static mechanical tests are used, and specimens are sampled at different stages of ageing. This requires a very large number of specimens. It must be assumed that, for an anisotropic material, water ingress may generate anisotropic damage, so it is very important to measure the mechanical parameters in such a way as to access this damage anisotropy. Ideally, for composite material, the recommended quasi-static testing would be: tensile at 0, 90 and 45°; flexure at 0° (to examine compression); and a through-thickness test such as interlaminar shear stress (ILSS) or,

Table 18.10 Typical loss in strength after 3 years' ageing

	Flexural strength (MPa)		ILSS (MPa)	
	35°	55°	35°	55°
Before ageing	410	127	44	19.5
Panel 1	231	87	30	13
Panel 5	302	120	43	18.5

preferably, an interlaminar fracture test (mode I or mode II). Unfortunately this requires hundreds of samples if degradation kinetics are to be determined, and very few published studies for the marine industry have been able to afford such a complete characterization. For the two studies described above (glass/epoxy plates and influence of hydrostatic pressure) the mechanical test programme was more limited.

For the glass/epoxy plates aged at 60 °C (which exhibit hydrolysis phenomena identified after 10 years' exposure), quasi-static tests on samples show the level of possible degradation after 3 years' exposure in water (Table 18.10). From these results it may be noted that:

- hydrolysis clearly affects the mechanical properties of the material;
- in flexure there is a clear drop in strength, but it may be assumed that this property is governed by the surface degradation, which is not really representative of the global degradation of the material through its total thickness;
- the loss of strength in interlaminar shear is significant for the material with high weight gain followed (after more than 10 years' exposure) by loss of weight.

The results obtained during a second series of tests allows the differences in behaviour to be examined for material made with glass fibre (E-glass) and material made of carbon fibre (T700S), in contact with water. The materials are made with the same epoxy resin and 175 mm internal diameter 5 mm thick tubes were manufactured by filament winding at ±55° to the tube axis. Samples were exposed for up to 1 year in deionized water. Absorption curves and a graph showing the evolution of apparent ILSS (measured on curved specimens cut in the circumferential direction) are reported in Figs 18.4 and 18.5.

Water uptake curves show small differences in the kinetics of absorption between the two materials even if the carbon/epoxy tube seems to absorb more water than the glass/epoxy material. On the other hand, the loss of ILSS properties appears to be more severe for the glass/epoxy material, particularly for exposure at temperatures up to 60 °C. As the same resin is

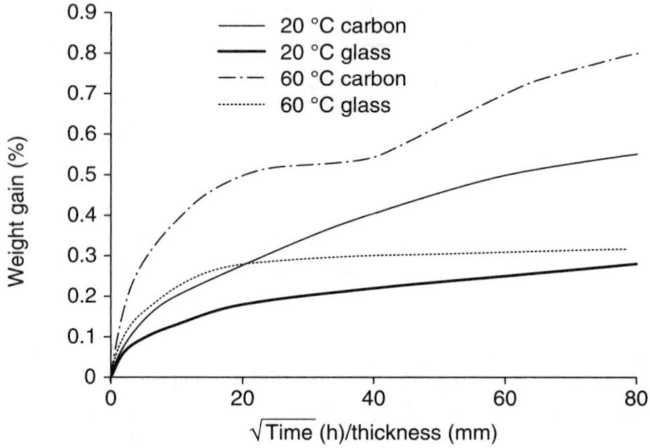

18.4 Absorption curve (1 year): comparison between glass/epoxy and carbon/epoxy ILSS samples cut from filament wound tubes.

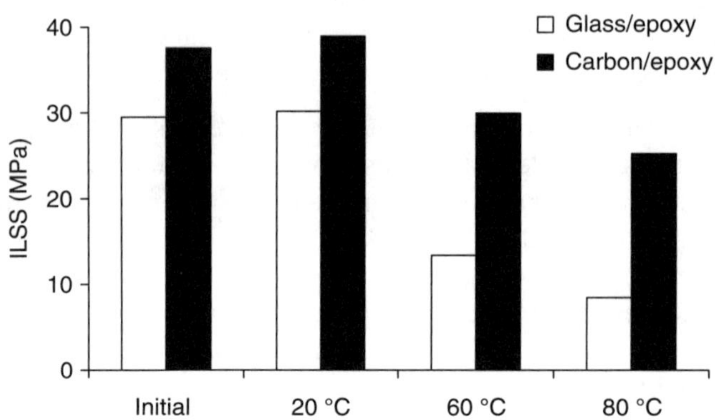

18.5 Evolution of ILSS after 1 year of immersion at different temperatures: comparison between glass/epoxy and carbon/epoxy tubes.

used, this indicates either enhanced matrix–resin debonding or degradation of the glass fibre when exposed to water at 60 °C. These examples underline the difficulty in predicting long-term behaviour in contact with water in terms of water uptake or associated damage. Laboratory studies, using the time–temperature equivalence principle are generally established based on a time period that is very short in comparison with the expected lifetime. This is quite inadequate for the prediction of the kinetics of degradation for the irreversible phenomena. In addition, numerous studies have shown that there is no correlation between water uptake and mechanical property evolution.[19,20] It must be recalled that, for deep underwater applications, particularly for the offshore industry, the lifetime requirement is very long, a minimum lifetime of 20–25 years is usually specified.

It should also be noted that most studies on water absorption kinetics of immersed materials are performed by placing the material in distilled or tap water at atmospheric pressure. In order to obtain a better estimation of the real behaviour of composite underwater structures, a study was carried out to evaluate the difference in material behaviour between laboratory and real underwater environments. For this purpose, tests were first performed to compare the water absorption kinetics in tap water and sea water (see Chapter 12). For most of the polymers studied, the difference in behaviour is small, but a decrease (about 10%) of the diffusion coefficient and in the saturation level noted in tap water is generally observed in sea water. For some polymers, in particular the polyurethanes, more significant differences have been noted. These could be explained in chemical terms by the equilibrium of the activity or chemical potential (ionic force) of the solvent (water or sea water) with the liquid (water) diffusing in the material. The diffusion kinetics are strongly dependent on the nature of the material and in particular on the nature of the matrix.

18.5 Case study 2: composite material for deep sea applications

Another study performed at IFREMER examined the influence of pressure on ageing.[20] Five candidate materials for marine applications were aged for 2 years. The composition of the materials is reported in Table 18.11. These materials are all representative of materials for marine structures. For a given class of material, it may be assumed that differences will be observed for materials provided by different suppliers; however, the results obtained give information of typical behaviour for the different classes of material and particularly for different types of resin. Orthotropic reinforcement (90% of fibres in the longitudinal (L) direction and 10% in the transverse (T) direction) has been chosen in order to examine anisotropy of the induced damage.

Material 1 is widely used in the marine environment, particularly for pleasure boats. It is generally completed with a gel-coat in order to limit water diffusion through the material (see Chapter 12, Section 12.2). Materials 2 and 3 are used for structures where better performance in terms of long-term behaviour and mechanical performance is required. Material 2 is popular for large naval structures produced using the infusion technique. Epoxy is the resin generally used for high-performance naval structures and prepreg material is now widely used for manufacturing sandwich structures in the marine industry. Epoxy is also widely used in the filament winding

Table 18.11 Composition of materials tested

	1	2	3	4	5
Resin	Isophthalic polyester	Vinylester	Epoxy	Epoxy	PEEK
Fibres	E-glass woven 90/10	E-glass woven 90/10	E-glass woven 90/10	E-glass woven 90/10	Carbon XAS UD 50/50
Fibre content (% weight)	59	61	52	60	50
Thickness (mm)	2.9	2.8	3.3	2.3	3.3
T_g (°C)	110	114	82	123	
Fabrication technique	Contact moulding*	Contact moulding*	Contact moulding*	Prepreg, 0.9 bar	Prepreg, autoclave

*These materials were post-cured at 60 °C for 24 hours to stabilize them. PEEK, poly ether ether ketone; UD, unidirectional.

process. The PEEK thermoplastic material reinforced by carbon fibre was selected as showing good potential for high-performance underwater structures in the future.

The study aimed to determine the kinetics of absorption of the materials and their degradation due to long-term contact with water. The main originality in this study was the ageing of samples under hydrostatic pressure, in order to evaluate the potential of the materials for deep sea applications. The importance of pressure on ageing was recognized over 40 years ago by Fried,[21] but since then published results on pressure effects have often been rather contradictory. The immersion conditions are reported in Table 18.12. The details of the experimental procedures are described in reference 22. The evolution of the weight of specimens for all the materials tested at the four temperatures at atmospheric pressure are presented in Figs 18.6 to 18.10 (mean results from all the test plates in the same condition). Classically, as in the previous figures, the curves are normalized on the horizontal axis by using the parameter square root of time (in hours) divided by the thickness of the sample (in mm). For the vertical axis, the evolution of the weight of samples is reported. It may be noted that, in some publications, in order to allow a comparison between behaviour of different materials with different matrix resins, the weight evolution related to the weight of resin is used. This accounts for different fibre contents between materials and the assumption is made that the water evolution of the composite is limited to the matrix alone.

Table 18.12 Ageing conditions

Immersion fluid	De-ionized water (renewed every 15 days)
Hydrostatic pressure	Atmospheric (simulation of immersed surface condition)
	100 bar (simulation of 1000 m of immersion)
Temperature	5 °C (mean water temperature of deep sea)
	20 °C reference
	40 °C (accelerating test)
	60 °C (accelerating test)

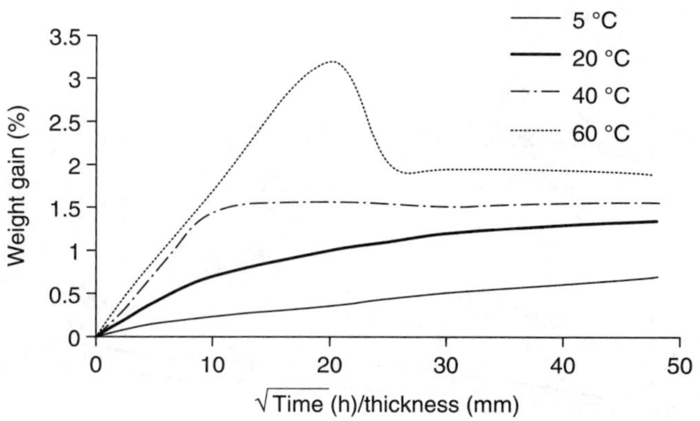

18.6 Typical absorption curve (2 years) for glass/polyester composite.

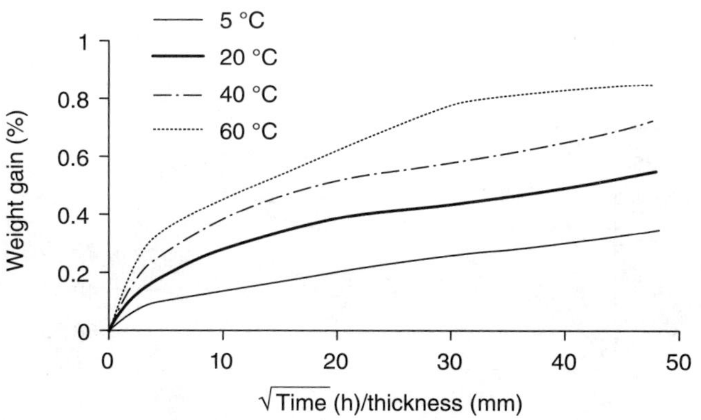

18.7 Typical absorption curve (2 years) for glass/vinylester composite.

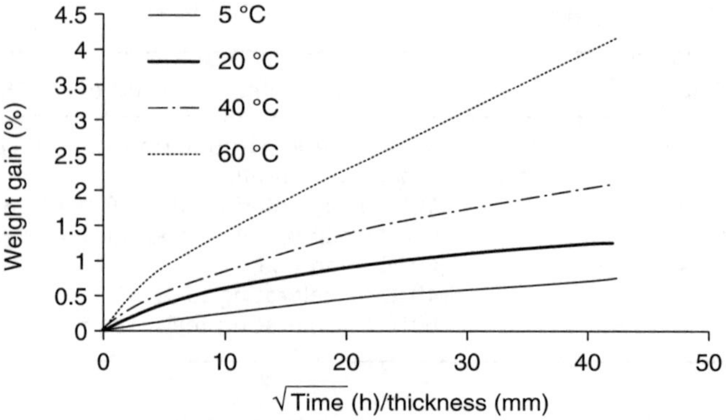

18.8 Typical absorption curve (2 years) for glass/epoxy composite (contact moulding).

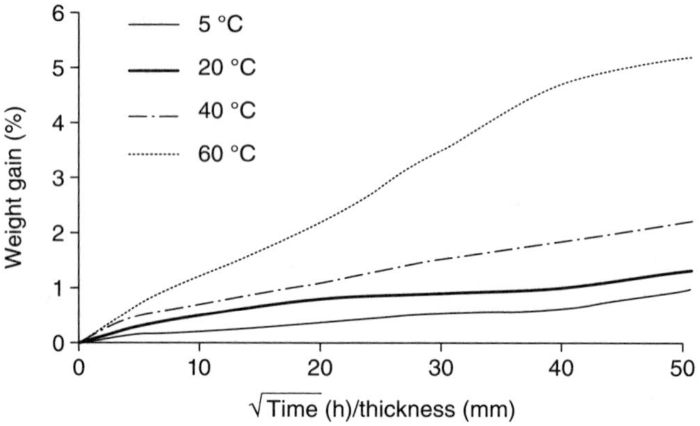

18.9 Typical absorption curve (2 years) for glass/epoxy composite (prepreg).

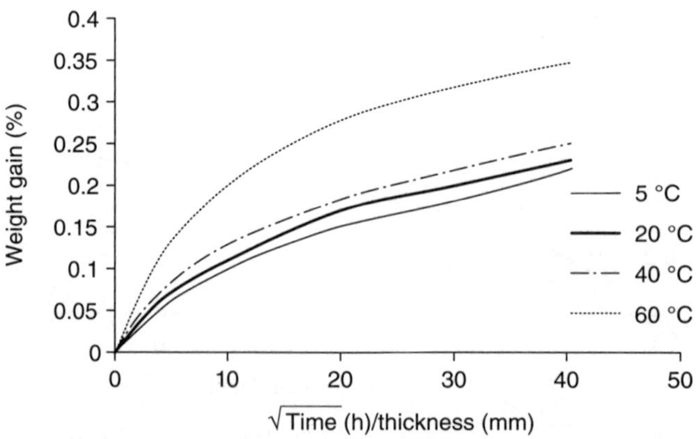

18.10 Typical absorption curve (2 years) for carbon/PEEK composite.

The description of the absorption phenomena by a Fickian model, which requires the moisture level to reach a maximum value asymptotically, is not appropriate as a saturation level is not generally reached even after 2 years' immersion of quite thin material (around 3 mm). The curves obtained are not as smooth as the usual classical description of absorption phenomena. This may be attributed to operational conditions (repeated weighing, wiping procedure, etc.), but also to the complexity of the weight evolution process of the materials. A saturation plateau is never obtained and there is no stabilization of the material weight gain. Thus the information on the stabilized maximum stage of degradation by water, which is generally provided by the saturation values, is not available here. The maximum moisture content cannot be determined, so the determination of the mass diffusion coefficient is not possible. Even if for some materials the fitting of a Fickian model works quite well, no single model could be applied to all the samples.

In order to estimate the state of the material after ageing, drying of samples was performed to determine the residual weight. The evolution of the T_g was also determined (Table 18.13). With respect to resin type, the epoxy material absorbed more water than the others (up to 5% after 2 years at 60 °C) and a residual weight gain after drying was noted. The T_g values for the epoxies (in particular, for prepreg) are significantly lowered after immersion in hot water, and residual weight after drying is noted. This suggests that degradation of the resin has occurred. For vinylester material and PEEK material, the water uptake is very low, as is the variation of the parameters highlighting residual physico-chemical degradation. This indicates that these materials are more stable in contact with water. Isophthalic polyester material exhibits a particular behaviour during immersion at 60 °C. After 2 months' immersion, degradation of the material has clearly occurred, as confirmed by the evolution of T_g and the loss of weight after

Table 18.13 Evolution of physico-chemical properties after 2 years of immersion

Temperature of ageing (°C)	Polyester		Vinylester		Epoxy (contact)		Epoxy (prepreg)	
	ΔT_g (°C)	Residual weight (%)	ΔT_g (°C)	Residual weight (%)	ΔT_g (°C)	Residual weight (%)	ΔT_g (°C)	Residual weight (%)
5 °C	+1	+0.1	+3	+0.1	0	+0.2	0	+0.2
20 °C	0	+0.1	+3	+0.1	−1	+0.6	−1	+0.5
40 °C	−1	−0.1	0	+0.1	0	+1	−10	+1.4
60 °C	−5	−0.5	−1	+0.1	0	+3.2	−9	+4.2

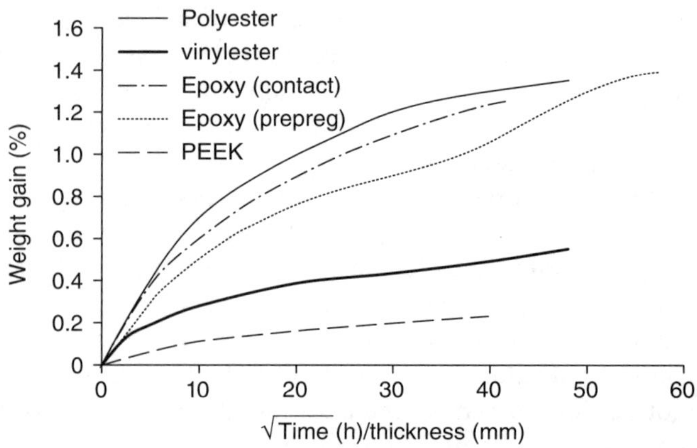

18.11 Absorption curve (2 years) comparison of material behaviour for materials immersed at 20 °C.

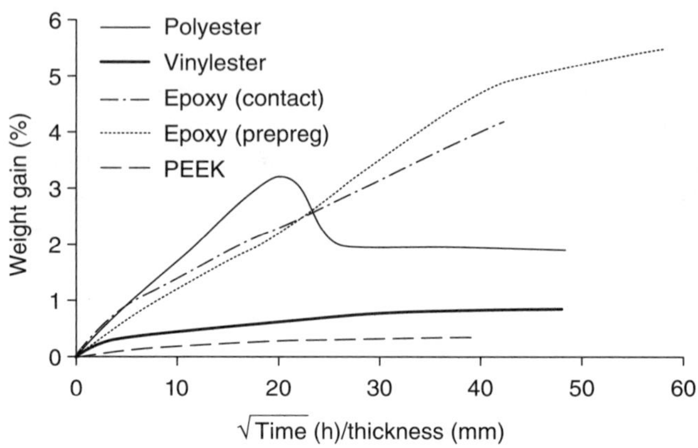

18.12 Absorption curve (2 years) comparison of material behaviour for materials immersed at 60 °C.

drying. The long-term behaviour of isophthalic polyester materials in contact with water can be very poor and additional precautions are needed in order to guarantee stability of the material. This confirms the necessity of using a gel-coat to protect the material against water ingress (see Chapter 12).

Figures 18.11 and 18.12 give a comparison of weight evolution for materials exposed at 20 °C and 60 °C. It may be noted that the materials can be divided into two groups, epoxy/polyester and vinylester/PEEK, where the latter appear to show good long-term behaviour in the presence of water.

Concerning the influence of pressure, Table 18.14 shows the diffusion coefficients at atmospheric pressure and at 10 MPa. It is interesting to note that increasing pressure results in faster diffusion rates for polyester- and epoxy-based materials, a slower rate for vinylester and no clear trend for PEEK. However, it must be emphasized that the amount of water uptake is not a useful indication of the loss of mechanical properties, as will be shown below.

In order to limit the number of specimens used to identify the degradation of the materials, non-destructive mechanical methods can be used. For example, in one study at IFREMER a vibration method was developed and applied. This technique is based on the modal analysis of the vibration of plates that had already been developed to evaluate in-plane anisotropic properties of composite materials.[23,24] The sample was freely suspended and excited by tapping. Using accelerometers bonded to the plate, its natural frequencies and mode-shapes were identified, which allowed determination of the stiffness matrix parameters, D_{11} and D_{22}, values from bending modes, and D_{66}, from torsion modes. Plates with dimensions of 5×20 cm^2 were used for this study. Figures 18.13 and 18.14 show examples of results obtained on materials immersed for 1 year at 20 °C and 60 °C.

Another non-destructive technique, based on ultrasonic wave speed measurements, can also be used and appears very promising, in particular, to evaluate through-thickness degradation of the material.[25] Examples of results obtained on 2.5 mm thick samples of composite and resin in their initial state and after 6 months of ageing at 60 °C are reported in Table 18.15. Although no significant evolution of the properties of the neat resin was measured, the elastic properties of their composites evolved during immersion.

These results highlight the anisotropic nature of damage induced by immersion of composites in water, and confirm the difficulty in extrapolating the properties of composites after ageing from results obtained on the neat resin. Finally, tests were also performed on the samples aged under hydrostatic pressure, these indicated no influence of a 10 MPa hydrostatic pressure on residual properties. This confirms the results shown in Chapter 12.

18.6 Case study 3: syntactic foam for deep sea and offshore applications

For deep offshore exploitation, there is a need for materials with specific functional properties and this provides a wide field of use for composites. In this context, syntactic foams, which can be used alone or as the core of composite sandwich structures, are particularly attractive owing to their excellent specific properties, i.e. buoyancy and thermal insulation.[26,27]

Table 18.14 Influence of pressure on initial slope of weight gain plots (10^{-2} mm $h^{-0.5}$)

	Polyester		Vinylester		Epoxy		Epoxy prepreg		C/PEEK	
Temperature	0.1 MPa	10 MPa	0.1 MPa	10 MPa	0.1 MPa	10 MPa	0.1 MPa	10 MPa	0.1 MPa	10 MPa
5 °C	1.5	3.2	0.7	0.9	15.0	18.7	13.4	16.4	0.6	0.6
20 °C	4.6	4.5	1.2	1.0	27.5	26.9	20.9	23.8	0.7	1.1
40 °C	12.8	14.3	3.4	2.3	68.1	80.0	40.3	74.6	1.0	1.5
60 °C	16.6	17.7	4.6	3.5	109.0	152.5	120.9	185.1	2.1	1.5

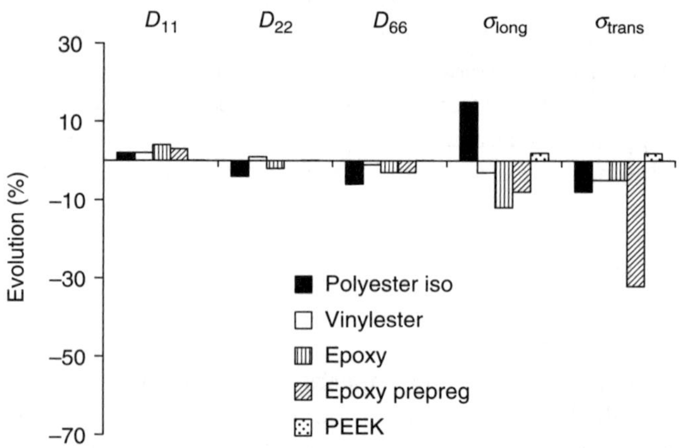

18.13 Evolution of the mechanical properties measured by vibration (D_{11}, D_{22}, D_{66}) and flexural tests (σ_{long}, σ_{trans}) for material immersed for 2 years at 20 °C.

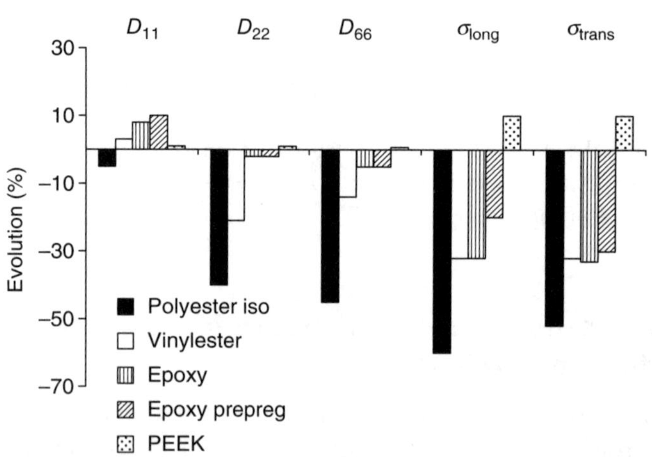

18.14 Evolution of the mechanical properties measured by vibration (D_{11}, D_{22}, D_{66}) and flexural tests (σ_{long}, σ_{trans}) for material immersed for 2 years at 60 °C.

Table 18.15 Evolution of mechanical properties determined by ultrasonic measurement, after 6 months of immersion

	Epoxy resin	Polyester resin	Vinylester resin	Glass/ epoxy	Glass/ polyester	Glass/ vinylester
E_1 (GPa)						
Initial	4.9	4.5	4.2	26.7	28.8	29.1
Aged	4.9	4.5	4.1	25.1	29.3	28.9
E_2 (GPa)						
Initial				12.7	14.1	13.9
Aged				11.4	12	13.2
E_3 (GPa)						
Initial				8.8	9.0	9.6
Aged				7.6	7.0	9.2
G_{12} (GPa)						
Initial	1.8	1.7	1.6	5.2	4.9	4.9
Aged	1.8	1.7	1.5	5.2	4.5	4.7
G_{13} (GPa)						
Initial				2.7	2.7	3.1
Aged				2.2	2.5	3.0

Initially this material was used for buoyancy for ultra-deep underwater exploration in the marine science domain in the 1970s. Over the last 30 years the material has primarily been used as a buoyancy material for supporting marine risers or sub-sea equipment. Now its use is growing rapidly in the domain of flow assurance, where thermal insulation is needed due to potential risk of blockage by paraffin or hydrates that can be formed when the internal fluid temperature falls below a critical level. In some applications (riser towers) both properties, buoyancy and thermal insulation, are required and in this case the quantity of material needed can be very large (up to 10 000 m^3 for one field). When weight is critical, a syntactic foam core in a sandwich with fibre reinforced composite facings is one of the most efficient structures available, and these are currently being considered for deep water pressure vessels.

Syntactic foam is a composite material made of thin, hollow glass microspheres embedded in a polymeric matrix (Fig. 18.15). Glass microspheres (typically 100–200 μm in diameter, with 1–2 μm of glass wall thickness, density around 20–40 kg/m^3) confer to the material the specific properties and a high compressive strength. Typical foam properties are reported in Table 18.16. Different grades of microspheres can be used. These differ in their apparent specific gravity, which is directly correlated to hydrostatic compression strength. Taking into account the requirements in terms of rigidity, and the manufacturing and installation processes, different matrix polymers can be used such as epoxy, polyurethane and polypropylene.

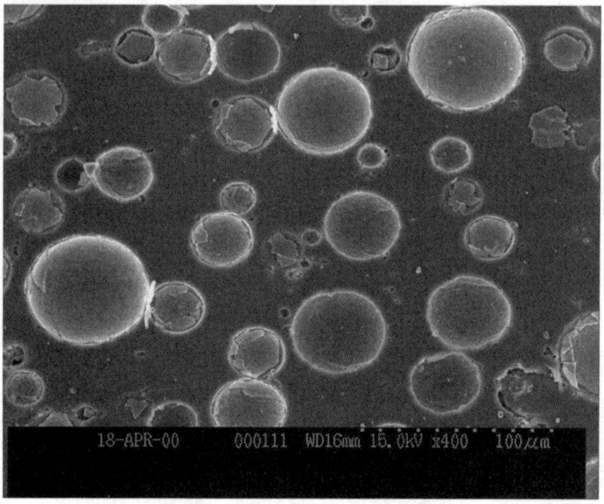

18.15 Scanning electron microscope view of syntactic foam.

Table 18.16 Typical syntactic foam properties for two matrix resins

	Units	Epoxy	Polyurethane
Specific gravity	kg/m^3	720	860
Thermal conductivity	W/m K	0.13	0.17
Heat capacity	J/kg °C	1.2	1.4
Glass transition temperature	°C	143	–37
Uniaxial compression modulus	GPa at 20 °C	2.8	0.21
Ultimate strength	MPa at 20 °C	75	12
Crush pressure	MPa at 50 °C	90	38

For thermal insulation, the material requirements can be very severe. For different applications such as pipe coating, jacketing, etc., it must be considered that the material could be in continuous contact with hot water under high hydrostatic pressure (this is the case for casings with defects or disbonding in the pipe coating). For offshore exploitation under about 3000 metres of water depth, the material requirements can be expressed as shown in Table 18.17.

Thermal insulation is the key parameter, and improvement of the insulation capability can bring significant economic gains. However, there are three environmental parameters – water, temperature and pressure – and, depending on the insulation system design, these may combine to generate detrimental degradation of the material. Material selection in terms of long-

Table 18.17 Insulation system requirements

Temperature range	Minimum: 4°C Maximum: 130–150°C
Operating condition	Contact with water (cold and hot)
Maximum pressure	30 MPa (3000 m)
Expected lifetime	20–30 years without maintenance
Thermal conductivity	As low as possible, <0.18 W/m K

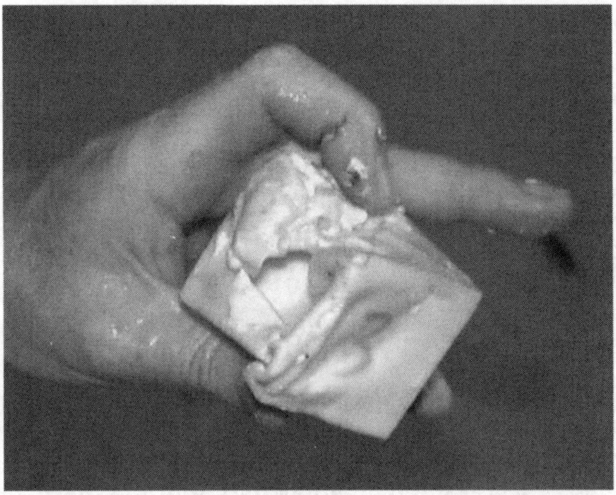

18.16 Catastrophic degradation (hydrolysis) of syntactic foam exposed to high temperature and pressure.

term durability in contact with water is then critical, and this has been studied extensively in recent projects.[28–31] Within these programmes, a large number of ageing tests have being performed, under various combinations of environmental parameters. These tests aimed to follow the water uptake of the material placed in water and subjected to pressure and temperature during a long-term exposure period (up to 14 months). The results obtained showed that, depending on the test conditions (temperature and pressure), detrimental water uptake had occurred (sometimes up to 40% weight gain) and sometimes total degradation of the material resulted (Fig. 18.16).

It is generally assumed that, for composite material, the water uptake is governed by the diffusion through the polymeric matrix and that some interfacial water diffusion may increase the water uptake potential of the material. However, taking into account the levels of water uptake attained here, additional phenomena must be involved. Chemical attack of the glass can be considered and, taking into account the water levels and the coupling

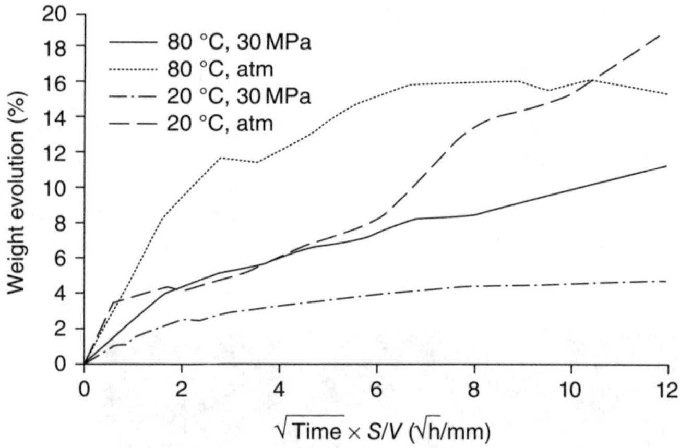

18.17 Absorption curve for polyurethane syntactic foam: effect of temperature–pressure coupling (10 mm cube, 14 months in sea water). *S*, surface area; *V*, volume.

effect of pressure and temperature, mechanical damage of the material seems to be initiated (Fig. 18.17). A good correlation has been found between percentage of water uptake and evolution of the thermal conductivity of the materials (Fig. 18.18). This could be simply expressed as follows:

$$\Delta\lambda = vf_w \lambda_w$$

where vf_w is the water volume fraction and λ_w the thermal conductivity of the water.

Glass microspheres can also suffer to some extent from hydrothermal effects. If the glass used to manufacture the microsphere contains too much free sodium or some other water-soluble ion, it will be exposed to degradation processes: gradual loss of properties, breakage and dissolution. Taking into account the absorption kinetics, which cannot be simply described by a Fickian process, the use of samples of different size ratios (surface (*S*)/volume (*V*)) is essential to describe and to model the absorption phenomena. In the above-mentioned programme, cubes of 10, 20, 50 and 100 mm edge length have been used to provide data for modelling (Fig. 18.19). During these programmes, a strong influence of the immersion medium was observed, and particularly a significant difference between behaviour in immersion in sea water and in deionized water for the polyurethane-based material (material 4) (Fig. 18.20). This highlights the importance of careful selection of ageing test conditions, if relevant results are to be obtained.

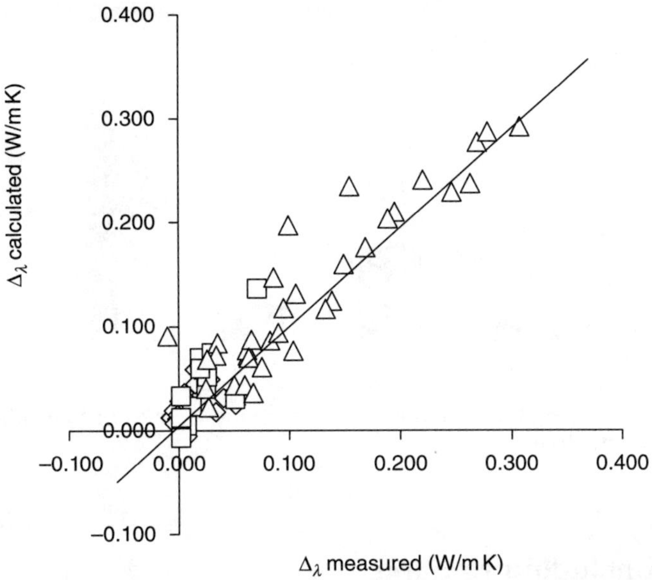

18.18 Correlation between calculated and measured evolution, due to water uptake, of the conductivity coefficient of syntactic foam.

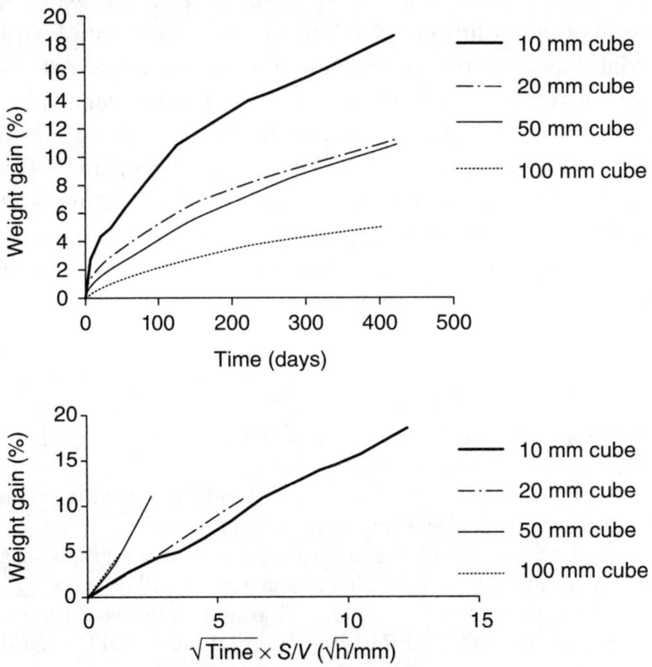

18.19 Typical absorption curves for epoxy syntactic foams for samples with different *S*/*V* ratios (300 bar, 80 °C, sea water, 14 months). Sample lengths refer to edge length of cubes. *S*, surface area; *V*, volume.

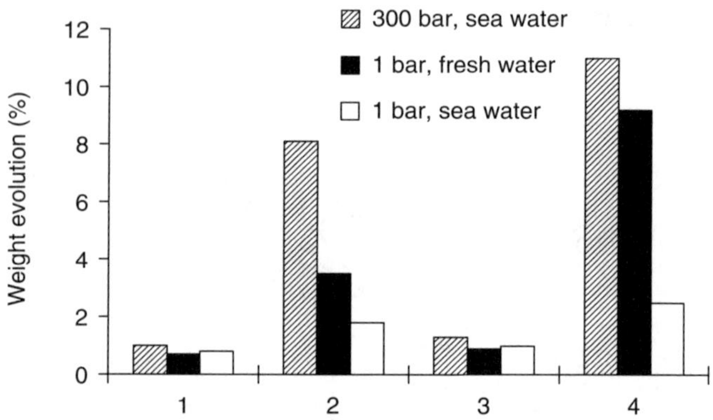

18.20 Effect of environment on absorption kinetics of four different syntactic foams.

18.7 Concluding remarks

This chapter has presented results from tests performed at IFREMER over the last 15 years to study the long-term behaviour of composite materials in underwater applications. The use of composites underwater has been limited so far, metallic solutions exist and the increased complexity of composite materials has slowed their introduction. However, recent discoveries of deepwater oil reserves and the pressure to develop renewable marine energy will encourage composite use, and for the success of these applications long-term durability is critical. The experience gained from these previous studies underlines the care required in designing appropriate ageing tests, a concerted effort is necessary to develop standard test procedures for underwater use of composites based on realistic loading conditions if costly failures are to be avoided.

18.8 References

1 OCEAN ENERGY CONVERSION IN EUROPE, 2006, published by the European 'Coordinated Action on Ocean Energy' group.
2 COPIN-MONTÉGUT G., 'Chimie de l'eau de mer', collection Synthèses, publication Institut Océanographique, 1996 (in French).
3 UNESCO 81, 10th Report of the Joint Panel on Oceanographic Tables and Standards, Technical Papers on Marine Science no. 36, 1981.
4 LEMOINE L., BELLOUIS M., SALAUN C., 'Oceanographic data collect for the selection of a test site for Wadjip DC', IFREMER Internal report DITI/GO/MM/99-08, 1999.
5 VERDU J., 'Vieillissement chimique des plastiques: aspects généraux', Techniques de l'Ingénieur, AM3 151, 2002 (in French).

6 FAYOLLE B., VERDU J., 'Vieillissement physique des matériaux polymères', Techniques de l'Ingénieur, COR 108, 2005 (in French).
7 CRANK J., '*The Mathematics of Diffusion*', First Edition, Clarendon Press, Oxford, 1956.
8 CARTER H. G., KIBLER K. G., 'Langmuir type model for anomalous moisture diffusion in composite resins', *Journal of Composite Materials*, **12**, 118–131, 1978.
9 GIMENEZ N., 'Vieillissement hydrolitique de mousses syntactiques époxyde-amide/verre pour l'isolation thermique sous hautes pressions: mécanismes de degradation et simulation de la prise d'eau', PhD Thesis, INSA de Lyon, France, 2006 (in French).
10 CHARLES R., 'Static fatigue of glass', *Journal of Applied Physics*, **29**(11), 1549–1560, 1958.
11 PRICE J., HULL D., 'Propagation of stress corrosion cracks in aligned glass fibre composite materials', *Journal of Materials Science*, **18**, 2798–2810, 1983.
12 PAUCHARD V., CHATEAUMINOIS A., GROSJEAN F., ODRU P., 'In situ analysis of delayed fibre failure within water-aged GFRP under static fatigue conditions', *International Journal of Fatigue*, **24**, 447–454, 2002.
13 CHATEAUMINOIS A., CHABERT B., SOULIER J. P., VINCENT L., 'Hygrothermal ageing effects on the static fatigue of glass/epoxy composites', *Composites*, **24**(7), 547–555, 1993.
14 ALIAS M. N., BROWN R., 'Corrosion behavior of carbon fibre composites in the marine environment', *Corrosion Science*, **35**(1–4), 395–402, 1993.
15 SHARMAN J. D. B., SYKES J. M., HANDYSIDE T., 'Cathodic disbonding of chlorinated rubber coatings from steel', *Corrosion Science*, **35**(5–8), 1375–1383, 1993.
16 DAVIES P., BAIZEAU R., CHOQUEUSE D., SALMON L., NAGOT F., 'Ageing and long term behavior of composite tubes', in *Recent Developments in Durability Analysis of Composite Systems*, Proceedings of DURACOSYS 1999, Brussels, Belgium, pp. 143–152.
17 PERREUX D., CHOQUEUSE D., DAVIES P., 'Anomalies in moisture absorption of glass fibre reinforced epoxy tubes', *Composites Part A: Applied Science and Manufacturing*, **33**(2), 147–154, 2002.
18 DAVIES P., MAZEAS F., CHOQUEUSE D., 'Summary of environmental ageing tests performed at Ifremer', IFREMER Internal report TD-11341-9701, 1997.
19 REIFSNIDER K., DILLARD D., CARDON A., '*Progress in Durability Analysis of Composites systems*', Proceedings of DURACOSYS 1997, Blacksburg, Virginia.
20 WEITSMAN Y. J., 'Fluids effect in polymeric composite – An overview', in *Proceedings of DURACOSYS 1997*, Blacksburg, Virginia, pp. 25–30.
21 FRIED N., 'Degradation of composite material', in *Proceedings of the 5th Symposium on Naval Structure Mechanics*, May 1967, pp. 813–837.
22 CHOQUEUSE D., DAVIES P., MAZEAS F., BAIZEAU R., 'Ageing of composites in water', in *High Temperature and Environmental Effects on Polymeric Composites*, ASTM STP 1302, editors T. Gates, A. Zureick, 1997.
23 SOL H., 'Identification of anisotropic plate rigidity using free vibration data', PhD Thesis, Vrije Universiteit Brussel, Belgium, October 1968.
24 TSOUVALIS N., 'A non destructive method for evaluation of anisotropic plate elastic properties', IFREMER Internal report DITI 91-17 GO/MM, 1991.
25 HOSTEN B., BASTE S., 'Ultrasonic evaluation of anisotropic damage induced in glass/epoxy compositers by water immersion', *QNDE*, **11B**, 1539–1546, 1992.

26 WATKINS L., HERSHEY E., 'Syntactic foam thermal insulation for ultra-deep oil and gas pipe', Offshore Technology Conference, Houston, 2002, OTC 13134.
27 RUCKEBUSCH J. M., 'Microspheres creuses de verre pour mousses syntactiques', Les Techniques de l'Ingénieur, Traité Matériaux Non Métalliques, A 2130 (in French).
28 CHOQUEUSE D., CHOMARD A., BUCHERIE C., 'Insulation material for ultradeepsea flow assurance: evaluation of material properties', in Offshore Technology Conference, Houston, 2002, OTC 14115.
29 CHOQUEUSE D., CHOMARD A., CHAUCHOT P., 'How to provide relevant data for the prediction of long term behavior of insulation materials under hot/wet conditions?', in Offshore Technology Conference, 2004, OTC 16503.
30 CHOQUEUSE D., CHOMARD A., BUCHERIE C., 'Thermal insulation for ultra deep pipelines. A research and evaluation program', Deep Offshore Technology Conference, New Orleans, 2000.
31 SAUVANT-MOYNOT V., GIMENEZ N,, SAUTEREAU M., 'Hydrolytic ageing of syntactic foams for thermal insulation in deep water: degradation mechanisms and water uptake model', *Journal of Materials Science*, **41**, 4047–4054, 2006.

Index